THE OXFORD AUTHORS

General Editor: Frank Kermode

JOHN KEATS was born in London in 1795. Orphaned early, he studied medicine at Guy's Hospital and, in 1816, became one of a new generation of qualified apothecaries. Medicine remained for him a standard of effective action against suffering, but he soon abandoned the profession in order to live for and by his poetry. He published three volumes during his lifetime and, while many of his contemporaries were prompt to recognize his greatness, snobbery and political hostility led the Tory press to vilify and patronize him as a 'Cockney poet'. He died of TB at the age of twenty-five. Financial anxieties and the loss of those he loved most had tried him persistently, yet he dismissed the concept of life as a vale of tears and substituted the concept of a 'vale of Soul-making'. His poetry and his remarkable letters reveal a spirit of questing vitality and profound understanding and his final volume, which contains the great odes and the unfinished *Hyperion*, attests to an astonishing maturity of power.

ELIZABETH COOK studied at the Warburg Institute and has taught at the universities of Essex and Leeds. She now works freelance. Her study of late Renaissance poetry *Seeing Through Words* was published by Yale University Press in 1986, and she has recently edited Ben Jonson's *The Alchemist*.

THE OXFORD AUTHORS

JOHN KEATS

EDITED BY

ELIZABETH COOK

Oxford New York

OXFORD UNIVERSITY PRESS

1990

Oxford University Press, Walton Street, Oxford OX2 6DP
Oxford New York Toronto
Delhi Bombay Calcutta Madras Karachi
Petaling Jaya Singapore Hong Kong Tokyo
Nairobi Dar es Salaam Cape Town
Melbourne Auckland
and associated companies in
Berlin Ibadan

Oxford is a trade mark of Oxford University Press

British Library Cataloguing in Publication Data
Keats, John, 1795–1821
John Keats. – (The Oxford authors)
I. Title II. Cook, Elizabeth, 1952–
821.7
ISBN 0–19–254194–3
ISBN 0–19–281931–3 (Pbk.)

Library of Congress Cataloging in Publication Data
Keats, John, 1795–1821.
[Selections. 1990]
John Keats / edited by Elizabeth Cook.
p. cm.—(The Oxford authors)
Includes bibliographical references.
I. Cook, Elizabeth, 1952– . II. Title. III. Series.
PR4832.C66 1990 821'.7—dc20 89–49034
ISBN 0–19–254194–3
ISBN 0–19–281931–3 (Pbk.)

Typeset by Wyvern Typesetting Ltd, Bristol
Printed in Great Britain by
Richard Clay Ltd.
Bungay, Suffolk

For Margaret Cook

CONTENTS

PROSE

ABBREVIATIONS

The three volumes of poetry published by Keats during his lifetime are referred to as *1817*, *Endymion*, and *1820*.

1848	R. M. Milnes (ed.), *The Poetical Works of John Keats* (London, 1848).
W1–3	Richard Woodhouse's transcripts of Keats's poems. *W1* and *W2* are in Harvard University Library; *W3* is in the Morgan Library.
Baldwin, *Pantheon*	Edward Baldwin (pseudonym of William Godwin), *The Pantheon: or Ancient History of the Gods of Greece and Rome* (London, 1806).
Barnard	John Barnard (ed.), *John Keats: The Complete Poems* (Harmondsworth, 1973, 2nd edn. 1977).
Bate	Walter Jackson Bate, *John Keats* (London, 1979).
Blackwood's	*Blackwood's Edinburgh Magazine.*
Finney	C. L. Finney, *The Evolution of Keats's Poetry* (New York, 1936).
Forman, *Letters*	H. B. Forman (ed.), *The Letters of John Keats* (London, 1895).
Forman (1883)	H. B. Forman (ed.), *The Poetical Works and Other Writings of John Keats*, 4 vols. (London, 1883).
FQ	Edmund Spenser, *The Faerie Queene.*
Gittings, *John Keats*	Robert Gittings, *John Keats* (London, 1968).
Gittings, *Letters*	Robert Gittings (ed.), *The Letters of John Keats* (Oxford, 1970, repr. 1975).
Hazlitt, *Works*	P. P. Howe (ed.), *The Complete Works of William Hazlitt*, 21 vols. (London, 1930–4).
Jack	Ian Jack, *Keats and the Mirror of Art* (Oxford, 1967).
Jeffrey	John Jeffrey, Georgiana Keats's second

husband, who supplied R. M. Milnes with some material for *1848*.

KSJ — *Keats–Shelley Journal*.

KC — H. E. Rollins (ed.), *The Keats Circle: Letters and Papers 1816–1879* (Cambridge, Mass., 1965).

L — H. E. Rollins (ed.), *The Letters of John Keats*, 2 vols. (Cambridge, Mass., 1958).

Lowell — A. Lowell, *John Keats* (London, 1925).

MLN — *Modern Language Notes*.

N&Q — *Notes and Queries*.

Partridge — E. B. Partridge, *A Dictionary of Slang and Unconventional English* (rev. edn., London, 1949).

PDWJ — *The Plymouth and Devenport Weekly Journal*.

PL — John Milton, *Paradise Lost*.

QR — *The Quarterly Review*.

Recollections — Charles and Mary Cowden-Clarke, *Recollections of Writers* (London, 1878).

TLS — *The Times Literary Supplement*.

The references to Shakespeare included on pp. 333–6 are to Keats's own facsimile of the First Folio. All Shakespeare references in the notes are to G. Blakemore Evans (ed.), *The Riverside Shakespeare* (Boston, Mass., 1974).

INTRODUCTION

KEATS conceived of history as a process of actualizing the world's sum total of what is knowable and thinkable. In Stoic fashion he postulates a finite quantity of world-stuff of which Milton has used up an unfairly large portion, thereby depleting not only his contemporaries, but posterity as well:

as a certain bulk of Water was instituted at the Creation—so very likely a certain portion of intellect was spun forth into the thin Air for the Brains of Man to prey upon it . . . That which is contained in the Pacific can't lie in the hollow of the Caspian—that which was in Miltons head could not find Room in Charles the seconds—he like a Moon attracted Intellect to its flow—it has not ebbd yet—but has left the shore pebble all bare—I mean all Bucks Authors of Hengist and Castelreaghs of the present day—who without Miltons gormandizing might have been all wise Men.[1]

Six weeks later, in a letter to J. H. Reynolds, Keats observes that the individual mind moves from the apparent biases of early life to a realization that 'Every department of knowedge [is] excellent and calculated towards a great whole.' He goes on in this letter to elaborate on his conception of human life as 'a large Mansion of Many Apartments'. In his programme for a life the whole building, 'dark Passages' included, must be explored and known.[2]

He writes with the assumption that a certain quota of qualities, capacities, and experiences is allotted to each individual: Shelley 'has his Quota of good qualities';[3] 'This is the second black eye I have had since leaving school—during all my school days I never had one at all—we must eat a peck before we die.'[4] In a letter to Fanny Brawne he expresses the wish that their quota of delight might be compressed within three days:

I almost wish we were butterflies and liv'd but three summer days—three such days with you I could fill with more delight than fifty common years could ever contain.[5]

Within twenty months of writing this Keats was dead. He had known

[1] Letter of 24 Mar. 1818 to James Rice.
[2] 3 May 1818 (to J. H. Reynolds).
[3] Letter of 27 Dec. 1817 to George and Tom Keats.
[4] Letter of 14 Feb.–3 May 1819 to George and Georgiana Keats.
[5] 1 July 1819.

Fanny Brawne for little more than two years and their love had never been consummated. But when one contemplates Keats's life one is struck not only by its sad brevity but by the extraordinary and triumphant fullness of its achievement. It is as if he had completed his allotted quota of experience and knowledge. His letters evince the gusto and persistence with which he entered into the self-appointed task of 'Soul-making',[6] relishing or testing each turn that life offered. The modest programme which he set himself in 'Sleep and Poetry' (composed in 1816) was to prove optimistic in the matter of years:

> Oh, for ten years, that I may overwhelm
> Myself in poesy; so I may do the deed
> That my own soul has to itself decreed.

But to read Keats's poetry through in chronological sequence (the principle of this volume) is to be impressed with the astonishing speed with which it matures. Keats effectively produced his life's work in two years; the greater part of it in one.[7] It is as if his cells had intimation of the tuberculosis that was to kill him and his whole organism accelerated its work in response.

With many writers the written work—the *œuvre*—is all that the public should decently contemplate; all else is gossip and distraction. But this is not so with Keats for whom the work of living and the work of writing are continuous, but not identical, endeavours. The letters occupy the median point: they are the registers, and often the occasions, of the work of Keats's life of which his life's work—his poetry—is only (a very important) part. Keats's letters are his best biography. But what we learn from them is not only about Keats's life; it is also *about life*.

This is not so true of the final letters, written by a man cornered between fervent love and the certainty of imminent death in a manner with which he was all too familiar. It is with some misgiving that several of these letters—and the last poems to Fanny—are included in the present volume since they represent not so much Keats's work as his un-work, an unmaking not for a public to witness. Practically, of course, it is too late for such reservations since these poems and letters have long been public property. And perhaps rightly so, for by now every known circumstance of Keats's life has come to be part of his meaning. As Charles Brown wrote to Fanny Brawne in 1829 when

[6] Letter of 14 Feb.–3 May 1819 to George and Georgiana Keats.

[7] For an account of 21 Sept. 1818 to 21 Sept. 1819 see Robert Gittings, *John Keats: the Living Year* (London, 1954).

requesting her permission to include letters alluding to her in the *Life* of Keats that he was preparing: 'As his love for you formed so great a part of him, we may be doing him an injustice in being silent on it.'[8]

In June 1818, when one brother, Tom, was dying of tuberculosis and the other, George, planning to sail with his new bride for America, Keats wrote to his friend Bailey: 'My Love for my Brothers from the early loss of our parents and even for earlier Misfortunes has grown into a affection "passing the Love of Women".'[9] The poem 'To my Brothers', composed on Tom's birthday two years earlier, evokes a mood of tender and secure intimacy. Personal affections, necessary to all, were for Keats refuges of security and warmth in a world which had whirred him from many friends. Friendship played a large part in his life: his letters tacitly celebrate it and it is the occasion of several poems. He was clearly a nice friend to have, and his friends loved him jealously and protectively.[10] These men—Clarke, Reynolds, Rice, Haslam, Dilke, Bailey, Woodhouse, Haydon, Hunt, Severn—are now known to us principally because they were Keats's friends. Haydon and Severn attempted (and usually failed) to live by their painting; Hunt divided his time between political journalism and literature; the others were middle-class professional men—lawyers, business men, amateurs of literature in the sense that it was their love but not their livelihood.

Richard Abbey—the guardian whom, with misguided forethought, their grandmother had provided for the Keats children—was anxious to see his wards respectably and gainfully employed. Among the careers which he at various times proposed to Keats were tea-brokering (Abbey's own business), hatting, and bookselling. For his ward John to attempt to live as a poet he thought preposterous.[11]

But before the moment of 'self will'[12] when Keats, legally of age and so free to choose, declared his intention to be a poet and live by it, he

[8] Quoted in Maurice Buxton Forman (ed.), *The Letters of John Keats* (Oxford, 1935), p. lxi.

[9] 10 June 1818.

[10] The jealousy and possessiveness became particularly acute after Keats's death. On 14 Aug. 1821 Brown wrote to Severn that Reynolds wished 'he should shine as the dear friend of poor Keats . . . when the fact is, he was no dear friend to Keats, nor did Keats think him so' (Jack Stillinger (ed.), *The Letters of Charles Armitage Brown* (Cambridge, Mass., 1966), p. 86). Both Brown and Reynolds had their friendship with Keats recorded on their tombs.

[11] See *KC* i. 307–8.

[12] 'In no period of my life have I acted with any self will, but in throwing up the apothecary profession' (letter of 22 Sept. 1819 to Charles Brown).

had chosen the medical profession. Apprenticed first to the surgeon Joseph Hammond he became a student at Guy's Hospital, attending the lectures of the brilliant and tender-hearted surgeon Sir Astley Cooper. He recovered some of his fees by working as a dresser, cleaning and dressing infected wounds, attending the heart-rending operations that took place in the days before anaesthetic. He worked for William Lucas, an undertrained surgeon whom Cooper described as 'neat-handed, but rash in the extreme.'[13]

But 'honey | Can't be got without hard money'. Financial anxieties dogged Keats all his life. If the money due to the Keats children from their grandmother's estate had materialized he might have lived modestly without needing to earn. But the idea of living by poetry—then a highly marketable commodity—was not as far-fetched as it might be now. In 'throwing up the apothecary profession' in 1817 Keats was reasonable in hoping he might support himself through poetry.

In the summer of 1819, pressed for money by George whose business in America had failed, wishing to marry Fanny Brawne but unable to see how he could support her, he considered resuming his medical career by becoming a surgeon on an Indiaman or a South Sea whaler. But he never acted on this and, as Dilke's comment makes clear, the idea was as much a metaphor for reckless despair as a practical plan.[14] Yet medicine was to remain for Keats a standard of effective action. Apollo, god of healing as well as of poetry, was his tutelary deity. In the self-castigating encounter with Moneta in 'The Fall of Hyperion' the poet—'a sage, | A humanist, *physician to all men*'—is distinguished from the dreamer-narrator. Keats conceived of an art which could in some way be 'friend to man',[15] which could enter the suffering world and mitigate its pain: 'were it my choice I would reject a petrarchal coronation—on account of my dying day, and because women have Cancers.'[16]

John Gibson Lockhart, reviewing *Endymion* for *Blackwood's Edinburgh Magazine*, seized upon the fact of Keats's apothecary past

[13] C. L. Feltoe, *Memorials of John Flint South*, p. 52; quoted in Gittings, *John Keats*, p. 64.

[14] Dilke recalled that while Keats '*wrote* about surgeon of an Indiaman, [he] *talked* about a South sea Whaler, and, as if to bid defiance to fortune, would have fixed on something more hateful, could his imagination have helped him to it' (*KC* ii. 223).

[15] The phrase occurs both in the 'Ode on a Grecian Urn' (l. 48) and, describing Milton, in a letter to James Rice of 24 Mar. 1818.

[16] Letter of 10 June 1818 to Benjamin Bailey.

(revealed to him by an unwitting Bailey)[17] as if such a revelation destroyed all dignity:

It is a better and a wiser thing to be a starved apothecary than a starved poet; so back to the shop, Mr. John, back to 'plasters, pills, and ointment boxes,' &c. But, for Heaven's sake, young Sangrado, be a little more sparing of extenuatives and soporifics in your practice than you have been in your poetry.[18]

Lockhart's social condescension is not only contemptible; it is misinformed. An Act had been passed in 1815 empowering the Society of Apothecaries to examine all apothecaries in England and Wales. Until this time apothecaries had been viewed rather as drug salesmen than as medical experts: they were paid for their wares, not their time or expertise.[19] When Keats passed the exams he sat in the Apothecaries' Hall in 1816 he became one of a new generation of qualified apothecaries, the forerunners of the modern general practitioner.

But that of course is not the point. For the social snobbery in which Lockhart couched his attack on Keats, and on others in what he called the 'Cockney School of Poetry', thinly veiled the real source of his opposition, which was political. Charles Cowden Clarke, Keats's schoolmaster and friend, recalled that 'with regard to Keats's political opinions I have little doubt that his whole civil creed was comprised in the master principle of "universal liberty"—viz. "Equal and stern justice to all, from the duke to the dustman."'[20] Clarke probably shared those views. While teaching at his father's school at Enfield he had introduced Keats to Leigh Hunt's liberal weekly, the *Examiner*, and it was to Clarke that Keats showed the first poem he had shown to another, 'Written on the day that Mr. Leigh Hunt Left Prison'.

Hunt and his brother John were released from prison in February 1815 having served two-year sentences on account of an article in the *Examiner* in which the Prince Regent was described as

a violator of his word, a libertine over head and ears in debt and disgrace, a despiser of domestic ties, the companion of gamblers and demireps, a man who has just closed half a century without a single claim on the gratitude of his country or the respect of posterity![21]

This was no more than the truth, but it was enough to imprison the Hunts, brought to trial by the Attorney General Lord Ellenborough. In the years that followed the French Revolution the monarchy and the

[17] See *KC* i. 34–5, 245–7, ii. 286–8. [18] *Blackwood's*, Aug. 1818.
[19] Llewellyn Woodward, *The Age of Reform* (Oxford, 1962), p. 18.
[20] *Recollections*, p. 156. [21] *Examiner*, 22 Mar. 1812, p. 179.

ministers of the Tory government were paranoiacally fearful of a repetition in Britain. Any questioning of the status quo, any request for parliamentary reform, any attempt to give the unpropertied labouring man a voice, was apt to be labelled 'sedition'. It is a word that is heard again and again in the second decade of the century—particularly in the peace that followed Waterloo when massive unemployment and poverty unmitigated by adequate Parish Relief brought the country into an acute state of polarization. The poem that Keats wrote on the day Hunt left prison was as much a declaration of political as poetic allegiance.

Charles Brown, publishing what is probably the last poem Keats wrote, 'In after time a sage of mickle lore', in a Devonshire paper eighteen years after Keats's death, wrote that 'he died with his pen wielded in the cause of Reform' though he 'never [before] wrote a line of a political tendency'.[22] Yet Brown also recalled that in 1819 the tragedy *Otho the Great*, on which he had collaborated with Keats, 'was sent to Drury Lane Theatre, not with his name . . . so utterly had it become a by-word of reproach in literature'.[23]

The reason for this reproach was that the Tory press had identified Keats with the 'seditious' enemy. In dedicating his first volume of poems (*1817*) to Hunt—by then a personal friend—Keats advertized and made explicit what would only have been evident through thoughtful reading. When his second volume, *Endymion*, came out the reviewer for the *British Critic* was able to go straight for the 'jacobinical apostrophe' at the beginning of Book iii. Not surprisingly, Lockhart singled out this passage too. What was most insulting about Lockhart's attacks was the implication that Keats's political views were second hand:

> We had almost forgot to mention, that Keats belongs to the Cockney School of Politics, as well as the Cockney School of Poetry.
>
> It is fit that he who holds *Rimini* to be the first poem, should believe the *Examiner* to be the first politician of the day. . . . Hear how their bantling has already learned to lisp sedition . . .

—and he quotes *Endymion* iii. 1–22. Keats's views were shared, but they were none the less his own. Several remarks in his letters show that he had hopes of making some direct contribution: 'I am ambitious of doing the world some good: if I should be spared that may be the work of maturer years' (letter of 27 October 1818); 'I hope sincerely I

[22] *PDWJ*, 4 July 1839. [23] *KC* ii. 66.

shall be able to put a Mite of help to the Liberal side of the Question before I die' (letter of 22 September 1819). After his death an anonymous correspondent in the *Morning Chronicle* wrote that 'His love of freedom was ardent and grand. He once said, that if he should live a few years, he would go over to South America, and write a Poem on Liberty.'[24]

Some form of action in the furtherance of 'universal liberty' was, like medicine, a standard of effectiveness against which Keats measured his poetry. Writing to Haydon in October 1819, six weeks after the Peterloo massacre and two weeks after Henry Hunt had arrived in London for trial and been greeted by huge and cheering crowds of supporters, Keats says that 'I have no doubt that if I had written Othello I should have been cheered by as good a mob as Hunt.' Milton, in his combination of active republicanism with great poetic achievement, was an exemplary figure to Keats. It was this combination that made Milton 'an active friend to Man all his Life and . . . since his death'.[25]

Keats shared with many of similar views a notion of patriotism that embraced popular freedom. Several icons of patriotism emerge repeatedly, not only in Keats's writing but in that of his contemporaries and in liberal and radical journalism of the time. They include King Alfred, William Tell, Robert Burns, Robin Hood, the Polish patriot Tadeusz Kosciusko, and, amongst English republicans, Algernon Sidney, Francis Vane, and, of course, John Milton.[26] Keats's exchange of Robin Hood sonnets with Reynolds must be seen in this context of shared political assumptions in which Robin Hood, in resisting the Norman yoke, is a champion of popular freedom whom it was entirely apposite to invoke at a time when 'Gagging Acts', provocateurs, and vindictive prosecutions were used in an attempt to stifle all opposition to the landed minority.

In his journal-letter to his brother and sister-in-law of October 1818 Keats looks at the politicians of the day and finds them puny besides those of the Cromwellian era:

[24] *Morning Chronicle*, 27 July 1821. The correspondent may have been Clarke.

[25] See n. 15 above.

[26] *The London Alfred, or People's Recorder* was an ultra-radical weekly published at this time; Schiller's play *Wilhelm Tell* (1804) had been translated into English and James Sheridan Knowles was to write another play about Tell in 1825; Samuel Bamford compares the 'gentlemen patriots' of his day unfavourably with 'the gallant Sydney—he who thumbed the cold iron that was to behead him' in *Passages from the Life of a Radical* (1884: repr. Oxford, 1984, p. 177); see also notes on Keats's 'Robin Hood' (below, p. 580) and Marilyn Butler, *Romantics, Rebels and Reactionaries* (Oxford, 1981), p. 149.

There is in truth nothing manly or sterling in any part of the Government. . . . there are none prepared to suffer in obscurity for their Country—the motives of our worst Men are interest and of our best Vanity—We have no Milton, no Algernon Sidney—Governors in these days loose the title of Man in exchange for that of Diplomat and Minister . . . all the departments of Government have strayed far from Spimpicity [*sic*] which is the greatest of Strength.

So too in America. Of Franklin and Washington he writes:

how are they to be compared to those our countrey men Milton and the two Sidneys—The one is a philosophical Quaker full of mean and thrifty maxims the other sold the very Charger who had taken him through all his Battles—Those Americans are great but they are not sublime Man.

Keats expressed both his patriotism and his antipathy towards his host county in a letter from Devonshire earlier in 1818:

Had England been a large devonshire we should not have won the Battle of Waterloo . . . a Devonshirer standing on his native hills is not a distinct object—he does not show against the light. . . . I like, I love England, I like its strong Men—Give me a 'long brown plain' for my Morning so I may meet with some of Edmund Iron side's descendants—Give me a barren mould so I may meet with some shadowing of Alfred in the shape of a Gipsey, a Huntsman or as Shepherd. Scenery is fine—but human nature is finer—The Sward is richer for the tread of a real, nervous english foot. . . . I shall never be able to relish entirely any devonshire scenery—Homer is very fine, Achilles is fine, Diomed is fine, Shakespeare is fine, Hamlet is fine, Lear is fine, but dwindled englishmen are not fine.[27]

Keats's criterion here for 'fineness' is the kind of distinctness which allows a figure to 'show against the light': an energy which lends reality to those who possess it. This vocabulary of clarity and energy (the Greek word ἐνάϱγεια brings the two together) is entirely characteristic of Keats, who apprehended reality as in a continual process of manifestation. He writes of a face 'swelling into reality';[28] of his purpose to 'clamber through the Clouds and exist'.[29] The companions, rivals, and mentors of Keats's mental universe are all possessed of that energy which allows them to show against the light. They are all 'distinct objects'; but they are as likely to be dead or legendary as alive:

I feel more and more every day, as my imagination strengthens, that I do not live in this world alone but in a thousand worlds . . . According to my state of mind I

27 Letter of 13 Mar. 1818 to Benjamin Bailey.
28 Letter of 21, ?27 Dec. 1817 to George and Tom Keats.
29 Letter of 8 Apr. 1818 to Benjamin Haydon.

am with Achilles shouting in the Trenches or with Theocritus in the Vales of Sicily.[30]

Any account of the context in which Keats worked must extend to the context he made for himself. There is nothing antiquarian about his sense of the past. Joseph Severn reports him as saying 'there is no *now* or *then* for the Holy Ghost.'[31] He writes to his brother and sister-in-law that to him 'manners and customs long since passed whether among the Babylonians or the Bactrians are as real, or eveven more real than those among which I now live.'[32]

While still at school he translated a large portion, if not all, of the *Aeneid*.[33] The act of translation (etymologically a 'carrying over') is typical of Keats's relation to the past. The Elgin Marbles, Chapman's Homer, a nightingale's song, the little town on a Grecian urn provide direct access to its other countries. Like many of his contemporaries Keats thought Chatterton's medieval pastiche to be the purest English,[34] but his love for Chatterton (to whom he dedicated *Endymion*) must have taken in the recognition of another for whom the past was not dead.

Keats's first and last known poems were written in imitation of Spenser—as much a Middle English pasticheur as was Chatterton. The first of these poems is derivative— not only of Spenser but of eighteenth-century poetry; it is the work of a young poet trying to learn what he has to say and how to say it by saying little and in the manner of someone else. The last poem ('In after time a sage of mickle lore') uses the Spenserian idiom with a relaxed confidence to engage not only with Spenser but with contemporary politics.

But Keats's most joyful relationship with the past was with Shakespeare. In May 1817 he writes that he is 'very near Agreeing with Hazlit that Shakespeare is enough for us',[35] a year later, after a period of deep and delighted discovery, that he has 'great reason to be content, for thank God I can read and perhaps understand Shakespeare to his depths'.[36] While Keats was eventually, for all his admiration, to feel crowded out by Milton—that they could not breathe the same air and

[30] Letter of 14–31 Oct. 1818 to George and Georgiana Keats.

[31] William Sharp, *The Life and Letters of Joseph Severn* (London, 1892), p. 208.

[32] Letter of 16 Dec. 1818–4 Jan. 1819 to George and Georgiana Keats.

[33] *KC* ii, 147.

[34] Letter of 21 Sept. 1819 to J. H. Reynolds and 17–27 Sept. 1819 to George and Georgiana Keats.

[35] Letter of 11 May 1817 to Benjamin Haydon.

[36] Letter of 27 Feb. 1818 to John Taylor.

live[37]—the abundance of Shakespeare's achievement was a continual source of wonder and delight. Even when Keats remarks that Shakespeare 'has left nothing to say about nothing or any thing'[38] the tone is gleeful, not oppressed. Shakespeare's phrases enter Keats's letters (and occasionally the poems) in a way that is too intimate for them to be called quotations. The many references to 'poor Tom' during the months that Keats was nursing his youngest brother, or his telling Fanny Brawne how his 'sense ached at' a vision of her which had haunted him, have none of the archness of quotation. Shakespeare's presence deepens and amplifies what are in no way literary sentiments.

Keats would like to have known 'in what position Shakespeare sat when he began "To be or not to be"'.[39] For him bodily gesture is continuous with spiritual attitude. It is clear that Keats experienced poetry in a way which was intensely physical. Clarke wrote that he 'ramped' through the *Faerie Queene* 'like a young horse turned into a Spring meadow'.[40] Friends remarked on the way in which he would look up with a 'delighted stare' when some passage of his reading particularly pleased him. One such stare was occasioned by the lines in Chapman's *Odyssey* describing the shipwrecked Odysseus, 'The sea had soak'd his heart through'. Other lines that thrilled him are the 'sea-shouldering whale' of the *Faerie Queene* (II. xii. 23) and 'See how the surly Warwick mans the wall' from *3 Henry VI* (V. i. 17). What these passages share is a sense of the animate defining itself by resistance to the inanimate, giving a strong impression of substantial physical presence.

It is almost certainly Spenser's same sea-shouldering whale that surfaces again in Keats's comment on a particularly telling sentence of Hazlitt's which 'appears to me like a Whale's back in a Sea of Prose'.[41] Just as a face might swell into reality, or he 'clamber through the Clouds' into existence, or Achilles be 'distinct' in a way that 'dwindled englishmen' are not, so language may sometimes rise out of the sea into distinction.

This intensely physical sense of language extends to the idea that a poem is possessed of an almost muscular strength. Keats writes of the

[37] 'I have but lately stood on my guard against Milton. Life to him would be death to me' (24 Sept. 1819 (in journal-letter of 17–27 Sept. 1819 to George and Georgiana Keats)).

[38] Letter of 22 Nov. 1817 to J. H. Reynolds.

[39] 12 Mar. 1819 (in journal-letter of 14 Feb.–3 May 1819 to George and Georgiana Keats).

[40] *KC* ii. 148–9.

[41] Letter of 10 May 1817 to Leigh Hunt.

ode 'growing like Atlas stronger from its load' and of the 'Atlas lines' of *Endymion*.[42] And he writes of himself as an athlete of the imagination, 'trying [himself] with mental weights'; of his 'imagination strengthening'.[43]

His programme for the composition of *Endymion*, whose length he foresaw at the outset as approximately 4,000 lines, suggests that he envisaged this 'trial of [his] Powers of Imagination'[44] as a kind of distance running. And it is curious to see that during the months of its composition Keats was physically always on the move—from the Isle of Wight to Margate, to Canterbury, to Hampstead, to Oxford, to Burford Bridge. It is as if the plan to cover a great many pages (and to compose at the rate of about fifty lines a day) was matched by a desire for topographical mileage.

Images of journeying, of taking steps with metrical feet, came naturally to Keats's very physical imagination: two or three narrative poems 'would be a famous gradus ad Parnassum altissimum'; *Endymion* i. 777–81 are 'a regular steeping of the Imagination towards a Truth'; *Endymion* itself may serve as a 'Pioneer'; in 'Hyperion' 'the march of passion and endeavour will be undeviating.'[45]

Those last images of military endeavour recur when he writes that he will undertake the journey to Rome which, it was hoped, would save him from the death-dealing of another English winter, 'with the sensation of marching against a Battery'.[46] In the 'flint-worded' letter to Fanny Brawne, sent from Winchester at a time of high energy and wilful single-mindedness, he writes: 'I can no more use soothing words to you than if I were at this moment engaged in a charge of Cavalry.'[47]

Such images would have come easily to one who had passed the greater part of his life during the Napoleonic wars and betray no militaristic leanings. (In fact Keats's sentiments were anti-militaristic.)[48] But Keats, like Reynolds and Hazlitt, was an enthusiast

[42] 'To Charles Cowden Clarke' 63; *Endymion* iii. 685.

[43] Letter of 21 Mar. 1819 to Benjamin Haydon and 17–27 Sept. 1819 to George and Georgiana Keats.

[44] Letter of 8 Oct. 1817 to Benjamin Bailey.

[45] Letter of 17 Nov. 1819, 30 Jan. 1818, and 27 Feb. 1818 to John Taylor; of 23 Jan. 1818 to Benjamin Haydon.

[46] Letter of 13 Aug. 1820 to John Taylor; see also his letter to Shelley of 16 Aug. 1820: 'I must either voyage or journey to Italy as a soldier marches up to a battery.'

[47] 16 Aug. 1819.

[48] See, for example, his remarks to Reynolds about the barracks on the road from Cowes to Newport (letter of 17, 18 Apr. 1817); 'To My Brother George' (epistle)

of boxing and his schoolfellow Edward Holmes was later to recall that while at school Keats's '*penchant* was for fighting. . . . He was a boy whom any one from his extraordinary vivacity & personal beauty might easily have fancied would become great—but rather in some military capacity than in literature.'[49] It is a mark of Keats's wholeness that he made such an impression. When Thomas Carlyle (born in the same year as Keats though he lived to become a 'Victorian') came to investigate the nature of heroism he was to find its possessors inscribed with it in every aspect of their lives: 'All that a man does is physiognomical of him. You may see how a man would fight, by the way in which he sings'.[50] Keats's writing—his letters and his poetry—is physiognomical of a spirit so alive that we continue to apprehend it as a real presence.

130; when he was at Naples he was upset by the presence of soldiers on the stage (Bate, p. 670).

[49] *KC* ii. 163–4.

[50] From 'The Hero as Poet', in *On Heroes, Hero Worship, and the Heroic in History* (1841: repr. London, 1908, p. 338).

ACKNOWLEDGEMENTS

To John Barnard, Stephen Gill, and to my General Editor Frank Kermode I owe particular thanks for their help and encouragement during the preparation of this volume. Each has been most generous in sharing his knowledge and experience and they have helped me greatly. I would also like to thank Mrs Christina Gee, the Curator of Keats House, Hampstead, and Roberta Davies, Librarian of Keats House, for their kind advice and assistance during the many days I enjoyed working in the Keats Memorial Library. That library is an invaluable resource.

I am also grateful to the curators and staff of the following libraries and learned societies who have kindly furnished me with copies of manuscripts and other materials: the Berg Collection, New York Public Library; Bristol Central Library; Buffalo and Erie County Public Library; Devon County Library; the Historical Society of Pennsylvania; the Houghton Library, Harvard University Library; Maine Historical Society; Trinity College Library, Cambridge; University of Bristol Library; the Wisbech and Fenland Museum; Yale University Library.

CHRONOLOGY

1795 Keats born at Swan and Hoop Livery Stables, Moorfields Pavement, London (31 October); Thomas Carlyle born.

1797 George Keats born (28 February).

1799 Tom Keats born (18 November). The Keats family move to Craven Street, off City Road.

1801 Edward Keats born (28 April).

1802 Death of Edward Keats. *Edinburgh Review* founded.

1802 Peace of Amiens. Fanny Keats born (3 June). Keats attends Clarke's school at Enfield (George and Tom attend later).

1804 Keats's father Thomas Keats is killed in a riding accident (16 April). Napoleon is crowned Emperor. Keats's mother marries William Rawlings (June). The Keats children move in with their maternal grandparents, John and Alice Jennings, at Ponders End, Enfield.

1805 Death of John Jennings (8 March). The family move to Lower Edmonton. Battle of Trafalgar.

1809 *Quarterly Review* founded.

1810 Keats's mother dies (10 March).

1811 Prince of Wales made Regent. Keats is apprenticed to the surgeon Joseph Hammond of Edmonton.

1812 Lord Liverpool becomes Prime Minister with Castlereagh as Leader of the Commons and Foreign Secretary.

1813 Leigh Hunt and his brother John imprisoned for libelling the Regent.

1814 Keats composes his earliest known poems. Napoleon resigns as Emperor. Death of Alice Jennings (19 December). Richard Abbey becomes official guardian of the Keats children. Wordsworth's *Excursion* published.

1815 Keats enters Guy's Hospital as a student. Leigh Hunt released from prison. Napoleon Emperor again for a 'hundred days'. Battle of Waterloo. Keats becomes a dresser.

1816 'To Solitude' ('O Solitude! if I must with thee dwell') published in the *Examiner*. Keats passes qualifying examinations and, when of age in October, is licensed to practise as an apothecary. Holiday in Margate. By mid-November is living at 76 Cheapside with Tom and George. Meets Leigh Hunt, Haydon, and Reynolds. Writes 'On First Looking into Chapman's Homer' and 'Sleep and Poetry'. Spa Fields meetings for parliamentary reform (December).

1817 *1817* published by C. and J. Ollier. Suspension of Habeas Corpus. Trial and acquittal of the publisher William Hone. Keats and brothers move to 1 Well Walk, Hampstead (March). Keats composes *Endymion* while at Carisbrooke on the Isle of Wight (15–24 April), Margate (till mid-May), Canterbury (mid-May), Hastings (late May), Hampstead (from 10 June), Oxford (early September until early October)—where he was Bailey's guest. He completes Book iv at Burford Bridge, Surrey (late November). Reads Shakespeare deeply all year. Reviews theatre for the *Indicator* in Reynolds's absence.

1818 Habeas Corpus restored. Revises *Endymion*. Attends Hazlitt's lectures on the English Poets at the Surrey Institution (13 January–3 March). Joins Tom at Teignmouth (March–April). Composes 'Isabella'. George marries Georgiana Wylie (May). *Endymion* published by Taylor and Hessey: it is harshly reviewed by *Blackwood's* in May. The *British Critic* in June, and the *Quarterly* in October. Accompanies George and Georgiana to Liverpool from where they depart for America (23 June). Keats sets off on a walking tour of the North—the Lakes, Scotland, and Northern Ireland—with Brown. Returns to London from Inverness because of ill health (8 August). Finds Tom very ill. Meets Fanny Brawne during summer. Begins *Hyperion* (November). Tom Keats dies (1 December). Keats moves into Wentworth Place as Brown's tenant. Spends Christmas day with the Brawnes.

1819 Visits Chichester and Bedhampton (January). Writes 'St Agnes Eve'. The publisher Richard Carlile is arrested in February (tried in October). Wordsworth publishes *Peter Bell*. Byron publishes Cantos i–ii of *Don Juan*. Keats composes 'La belle dame sans merci' (April). Acute financial worries lead him to consider work as a ship's surgeon. Composes odes on Grecian Urn, Melancholy, Nightingale, and Indolence. Accompanies Rice to Isle of Wight (late June). Begins work on verse tragedy *Otho the Great* and on 'Lamia'. Accompanies Brown to Winchester (mid-August). Works on the 'Fall of Hyperion'. Peterloo Massacre (16 August). Keats returns briefly to London in mid-September and witnesses Henry Hunt's arrival for trial (13 September). Composes 'Ode to Autumn'. Returns to Hampstead and visits Fanny Brawne. Plans to work as a journalist and briefly (mid-October) lodges at 25 College Street, Westminster. Back in Wentworth Place by 22 October. Parliament passes the 'Six Acts'. Keats is engaged to Fanny Brawne by end of the year.

1820 Thomas Wooler (editor of the *Black Dwarf*) imprisoned. George Keats returns to London to raise funds (leaves 28 January). Death of George III. Keats returns to Hampstead from London by stage-coach and has a severe haemorrhage (3 February). Arrest of 'Cato Street Conspirators' (set up by government *provocateur* Edwards). Keats declared 'out of danger' (March). Revises 'Lamia'. Brown lets Wentworth Place for the summer and Keats moves to 2 Wesleyan Place, Kentish Town. Queen Caroline (absent since 1814) returns to England (6 June). Keats moves into Leigh Hunt's house in Mortimer Terrace, Kentish Town

(23 June). *1820* published by Taylor and Hessey (1 or 2 July). It is thought another English winter would kill Keats and his doctor orders him to go to Italy. Coronation of George IV (19 July). Bill of Pains and Penalties (to deprive Queen Caroline of title of Consort) debated between July and November. Keats leaves Hunt's house (12 August) and stays with the Brawnes at Wentworth Place. *1820* favourably reviewed though public interest in the Queen's affairs leads to neglect of all other publications. Keats and Severn sail from Gravesend in the *Maria Crowther* (18 September), reach Naples harbour (21 October), and are kept in quarantine on board for ten days because of a typhus epidemic in London. 31 October (Keats's birthday) Keats and Severn take rooms above a trattoria in Vico S. Giuseppe, Naples. Keats and Severn travel (6 or 7 November) by carriage to Rome, arriving 15 November. They take rooms at No. 26 Piazza di Spagna. Keats writes his last known letter (10 December).

1821 23 February, Keats dies at 11 p.m. News of his death reaches London on 17 March.

NOTE ON THE TEXT

ONLY about a third of the 150 or so poems in the Keats canon were seen through the press by Keats himself. The other two-thirds—which include much of his comic and of his politically explicit verse—Keats decided against offering for publication. In the case of those poems that were published during his lifetime, my choice of text has been the latest published form in which Keats saw the poem. I have extended this rule to journal publication as well as to the three volumes *1817*, *Endymion*, and *1820*. The only instance in which this rule has not been followed is 'La belle dame sans merci'. Keats's draft of this poem, and not the version printed in the *Indicator*, is (justly, in my opinion) the better known. This is the text I give here; the *Indicator* version is printed in an Appendix.

Those poems which Keats did not see published have come down to us from a variety of sources: Keats's own holograph drafts and fair copies, transcripts by friends and relatives (whose reliability varies enormously), and various printed sources. In deciding which source text to use I am deeply indebted to Jack Stillinger who in *The Text of Keats's Poems* (1974) and in his subsequent edition of Keats's *Poems* (1978) presents his informed and considered arguments for and against each transcript and state of text. Prior to his work editors had frequently created Keats's poems from a patchwork of different source texts. The copy texts I have used in the cases of poems not published during Keats's lifetime are identified in the notes to the poems.

I have not attempted to make those poems which Keats decided against publishing appear as editorially polished as those which he saw through the press. It misrepresents a poem which may as yet exist in a raw or provisional state to tidy it up with editorial punctuation so that it appears homogeneous with other, published, poems. Keats's tendency was to be sparing of punctuation and I have added as little as is consonant with intelligibility, since the openness of his lightly punctuated poems is part of his style. Where Keats has ended a poem without any stop I have inserted a dash—a punctuation mark Keats often favoured. Some of Keats's contemporary transcribers, however, were as officious as his subsequent editors in adding punctuation marks, and when these contemporary transcribers have provided my copy texts I have felt obliged to follow them.

Keats's spelling and his use of capitals are often expressive in their idiosyncrasy. His use of capitals in the initials of various words can suggest a particular gusto for the word in question while his renowned misspellings are often, as Christopher Ricks observes, 'indications of how his imagination is working'. The capitalization of substantives was still a common practice in the early nineteenth century and, in editing the texts of Keats's poems, I have retained his capitals where they are so used but not when they are the initial letters of other parts of speech. I have retained idiosyncratic spelling in the poems only where a deliberate joke may have been intended ('Mrs. Reynoldse's Cat', 'Anthropopagi').

In the case of Keats's prose the punctuation, spelling, and capitalization of the copy texts have on the whole been retained. But whereas Keats frequently opens a quotation with single quotes and closes with double (and vice versa), or fails to close quotation marks that he has opened, I have silently regularized to single quotes, opening and closing (but I have not supplied any quotation marks to those quotations which Keats has himself left unmarked.) I have retained most of Keats's misspellings in his marginal notes and his letters. They often indicate an aural imagination—a word misspelt as its homophone ('*two* and from')—and of a quick mind, racing ahead of a hand which swallows letters and compresses words in the attempt to keep up. It is nearly always clear what word was intended ('enthusiam' for 'enthusiasm', 'anhilated' for 'annihilated'). Only in a few cases where it seemed to me that the original orthography stood in the way of comprehension have I silently emended or else explained in a note.

Robert Gittings based his *Letters of John Keats* (Oxford, 1970, rev. ed. 1975) on a re-examination of many of the manuscripts. I am grateful to him and to Oxford University Press for permission to base the greater part of my letter texts on his. In all other instances I have established a text through examination of the copy text. As with the poems, the nature of the copy text (whether holograph, scribal copy, or prior publication) for each prose piece is identified in the notes.

The degree sign (°) indicates a note at the end of the book. More general headnotes are not cued.

Imitation of Spenser

Now Morning from her orient chamber came,
And her first footsteps touch'd a verdant hill;
Crowning its lawny crest with amber flame,
Silv'ring the untainted gushes of its rill;
Which, pure from mossy beds, did down distill,
And after parting beds of simple flowers,
By many streams a little lake did fill,
Which round its marge reflected woven bowers,
And, in its middle space, a sky that never lowers.

There the king-fisher saw his plumage bright 10
Vieing with fish of brilliant dye below;
Whose silken fins, and golden scales light
Cast upward, through the waves, a ruby glow:
There saw the swan his neck of arched snow,
And oar'd himself along with majesty;
Sparkled his jetty eyes; his feet did show
Beneath the waves like Afric's ebony,
And on his back a fay reclined voluptuously.

Ah! could I tell the wonders of an isle
That in that fairest lake had placed been, 20
I could e'en Dido of her grief beguile;
Or rob from aged Lear his bitter teen:
For sure so fair a place was never seen,
Of all that ever charm'd romantic eye:
It seem'd an emerald in the silver sheen
Of the bright waters; or as when on high,
Through clouds of fleecy white, laughs the cœrulean sky.

And all around it dipp'd luxuriously
Slopings of verdure through the glossy tide,
Which, as it were in gentle amity, 30
Rippled delighted up the flowery side;
As if to glean the ruddy tears, it tried,
Which fell profusely from the rose-tree stem!
Haply it was the workings of its pride,
In strife to throw upon the shore a gem
Outvieing all the buds in Flora's diadem.

Song: 'Stay, ruby breasted warbler, stay'

TUNE—'Julia to the Wood Robin'

1

Stay, ruby breasted warbler, stay,
And let me see thy sparkling Eye,
Oh brush not yet the pearl strung spray,
Nor bow thy pretty head to fly.

2

Stay while I tell thee fluttering thing,
That thou of Love an emblem art,
Yes! patient plume thy little wing,
Whilst I my thoughts to thee impart.

3

When Summer Nights the dews bestow,
And Summer Suns enrich the day, 10
Thy Notes, the blossoms charm to blow,
Each opes delighted at thy lay.

4

So when in youth the Eye's dark glance
Speaks pleasure from its circle bright,
The tones of Love our joys enhance,
And make superiour each delight.

5

And when bleak storms resistless rove,
And ev'ry rural bliss destroy,
Nought comforts then the leafless grove,
But thy soft note—its only joy— 20

6

E'en so the words of Love beguile,
When Pleasure's Tree no longer bears,
And draw a soft endearing smile,
Amid the gloom of grief and tears.

'Fill for me a brimming Bowl'

'What wondrous beauty! From this moment I efface from my mind all women.'—TERENCE's *Eunuch*, Act 2, Sc. 4.

Fill for me a brimming Bowl,
And let me in it drown my Soul:
But put therein some drug design'd,
To banish Woman from my Mind.
For I want not the Stream inspiring,
That heats the Sense with lewd desiring;
But, I want as deep a draught
As e'er from Lethe's waves was quaft,
From my despairing Breast to charm
The Image of the fairest form 10
That e'er my rev'ling Eyes beheld
That e'er my wand'ring Fancy spell'd!

'Tis vain—away I cannot chace
The melting softness of that face—
The beaminess of those bright Eyes—
That breast Earth's only Paradise!

My sight will never more be blest
For all I see has lost its Zest;
Nor with delight can I explore
The classic Page—the Muse's lore. 20

Had she but known how beat my heart
And with one Smile reliev'd its smart,
I should have felt a sweet relief
I should have felt 'the Joy of Grief'!
Yet as a Tuscan 'mid the Snow
Of Lapland thinks on sweet Arno;°
So for ever shall she be
The Halo of my Memory.

To Lord Byron

Byron, how sweetly sad thy melody
  Attuning still the soul to tenderness
  As if soft Pity with unusual stress
Had touch'd her plaintive Lute; and thou, being by
Hadst caught the tones, nor suffered them to die.
  O'ershading sorrow doth not make thee less
  Delightful: thou thy griefs dost dress
With a bright Halo, shining beamily.
As when a cloud a golden moon doth veil
  Its sides are tinged with a resplendent glow 10
Through the dark robe oft amber rays prevail,
  And like fair veins in sable marble flow.
Still warble dying swan,—still tell the tale
  The enchanting tale—the tale of pleasing woe.

Written on the Day That Mr. Leigh Hunt Left Prison

What though, for showing truth to flatter'd state,
  Kind Hunt was shut in prison, yet has he,
  In his immortal spirit, been as free
As the sky-searching lark, and as elate.
Minion of grandeur! think you he did wait?
  Think you he nought but prison walls did see,
  Till, so unwilling, thou unturn'dst the key?
Ah, no! far happier, nobler was his fate!
In Spenser's halls he strayed, and bowers fair,
  Culling enchanted flowers; and he flew 10
With daring Milton through the fields of air:
  To regions of his own his genius true
Took happy flights. Who shall his fame impair
  When thou art dead, and all thy wretched crew?

To Hope

When by my solitary hearth I sit,
  And hateful thoughts enwrap my soul in gloom;
When no fair dreams before my 'mind's eye' flit,
  And the bare heath of life presents no bloom;
    Sweet Hope, ethereal balm upon me shed,
    And wave thy silver pinions o'er my head.

Whene'er I wander, at the fall of night,
  Where woven boughs shut out the moon's bright ray,
Should sad Despondency my musings fright,
  And frown, to drive fair Cheerfulness away, 10
    Peep with the moon-beams through the leafy roof,
    And keep that fiend Despondence far aloof.

Should Disappointment, parent of Despair,
  Strive for her son to seize my careless heart;
When, like a cloud, he sits upon the air,
  Preparing on his spell-bound prey to dart:
    Chace him away, sweet Hope, with visage bright,
    And fright him as the morning frightens night!

Whene'er the fate of those I hold most dear°
  Tells to my fearful breast a tale of sorrow, 20
O bright-eyed Hope, my morbid fancy cheer;
  Let me awhile thy sweetest comforts borrow:
    Thy heaven-born radiance around me shed,
    And wave thy silver pinions o'er my head!

Should e'er unhappy love my bosom pain,
  From cruel parents, or relentless fair;
O let me think it is not quite in vain
  To sigh out sonnets to the midnight air!
    Sweet Hope, ethereal balm upon me shed,
    And wave thy silver pinions o'er my head! 30

In the long vista of the years to roll,
  Let me not see our country's honour fade:

O let me see our land retain her soul,
Her pride, her freedom; and not freedom's shade.
From thy bright eyes unusual brightness shed—
Beneath thy pinions canopy my head!

Let me not see the patriot's high bequest,
Great Liberty! how great in plain attire!
With the base purple of a court oppress'd,
Bowing her head, and ready to expire: 40
But let me see thee stoop from heaven on wings
That fill the skies with silver glitterings!

And as, in sparkling majesty, a star
Gilds the bright summit of some gloomy cloud;
Brightening the half veil'd face of heaven afar:
So, when dark thoughts my boding spirit shroud,
Sweet Hope, celestial influence round me shed,
Waving thy silver pinions o'er my head.

Ode to Apollo

1

In thy Western Halls of gold
When thou sittest in thy state,
Bards, that erst sublimely told
Heroic deeds, and sung of Fate,
With fervour seize their adamantine lyres,
Whose cords are solid rays, and twinkle radiant fires.

2

There Homer with his nervous arms
Strikes the twanging harp of war,
And even the Western splendour warms
While the trumpets sound afar; 10
But, what creates the most intense surprize,
His soul looks out through renovated eyes.°

3

Then, through thy Temple wide, melodious swells
The sweet majestic tone of Maro's lyre;°
The soul delighted on each accent dwells,—
Enraptured dwells,—not daring to respire,
The while he tells of grief, around a funeral pyre.°

4

'Tis awful silence then again:
Expectant stand the spheres;
Breathless the laurel'd peers; 20
Nor move, till ends the lofty strain,
Nor move till Milton's tuneful thunders cease,
And leave once more the ravish'd heavens in peace.

5

Thou biddest Shakspeare wave his hand,
And quickly forward spring
The Passions—a terrific band—
And each vibrates the string
That with its tyrant temper best accords,
While from their Master's lips pour forth the inspiring words.

6

A silver trumpet Spenser blows, 30
And as its martial notes to silence flee,
From a virgin Chorus flows
A hymn in praise of spotless Chastity.°
'Tis still!—Wild warblings from the Æolian lyre°
Enchantment softly breathe, and tremblingly expire.

7

Next, thy Tasso's ardent Numbers°
Float along the pleased air,
Calling Youth from idle Slumbers,
Rousing them from pleasure's lair:—
Then o'er the strings his fingers gently move, 40
And melt the soul to pity and to love.

8

But when *Thou* joinest with the Nine,
And all the Powers of Song combine,
We listen here on earth:
The dying tones that fill the air,
And charm the ear of Evening fair,
From thee, great God of Bards, receive their heavenly birth.

'Oh Chatterton! how very sad thy fate!'

Oh Chatterton! how very sad thy fate!
Dear Child of Sorrow! Son of Misery!
How soon the film of death obscur'd that Eye,
Whence Genius wildly flash'd, and high debate!
How soon that voice, majestic, and elate,
Melted in dying murmurs! O how nigh
Was night to thy fair Morning! Thou didst die
A half-blown flower, which cold blasts amate.°
But this is past—Thou art among the Stars
Of highest Heaven; to the rolling spheres 10
Thou sweetly singest—nought thy hymning mars
Above the ingrate world and human fears.
On Earth the good Man base detraction bars
From thy fair Name, and waters it with Tears!

Lines Written on 29 May, the Anniversary of Charles's Restoration, on Hearing the Bells Ringing

Infatuate Britons, will you still proclaim
His memory, your direst, foulest shame?
Nor Patriots revere?
Ah! when I hear each traitorous lying bell,
'Tis gallant Sydney's, Russell's, Vane's sad knell,°
That pains my wounded ear—

On receiving a curious Shell, and a Copy of Verses, from [some] *Ladies*

Hast thou from the caves of Golconda, a gem°
  Pure as the ice-drop that froze on the mountain?
Bright as the humming-bird's green diadem,
  When it flutters in sun-beams that shine through a fountain?

Hast thou a goblet for dark sparkling wine?
  That goblet right heavy, and massy, and gold?
And splendidly mark'd with the story divine
  Of Armida the fair, and Rinaldo the bold?°

Hast thou a steed with a mane richly flowing?
  Hast thou a sword that thine enemy's smart is? 10
Hast thou a trumpet rich melodies blowing?
  And wear'st thou the shield of the fam'd Britomartis?°

What is it that hangs from thy shoulder, so brave,
  Embroidered with many a spring peering flower?
Is it a scarf that thy fair lady gave?
  And hastest thou now to that fair lady's bower?

Ah! courteous Sir Knight, with large joy thou art crown'd;
  Full many the glories that brighten thy youth!
I will tell thee my blisses, which richly abound
  In magical powers to bless, and to sooth. 20

On this scroll thou seest written in characters fair
  A sun-beamy tale of a wreath, and a chain;
And, warrior, it nurtures the property rare
  Of charming my mind from the trammels of pain.

This canopy mark: 'tis the work of a fay;°
  Beneath its rich shade did King Oberon languish,
When lovely Titania was far, far away,
  And cruelly left him to sorrow, and anguish.

There, oft would he bring from his soft sighing lute
Wild strains to which, spell-bound, the nightingales listened; 30
The wondering spirits of heaven were mute,
And tears 'mong the dewdrops of morning oft glistened.

In this little dome, all those melodies strange,
Soft, plaintive, and melting, for ever will sigh;
Nor e'er will the notes from their tenderness change;
Nor e'er will the music of Oberon die.

So, when I am in a voluptuous vein,
I pillow my head on the sweets of the rose,
And list to the tale of the wreath, and the chain,
Till its echoes depart; then I sink to repose. 40

Adieu, valiant Eric! with joy thou art crown'd;°
Full many the glories that brighten thy youth,
I too have my blisses, which richly abound
In magical powers, to bless and to sooth.

To George Felton Mathew

Sweet are the pleasures that to verse belong,
And doubly sweet a brotherhood in song;
Nor can remembrance, Mathew! bring to view
A fate more pleasing, a delight more true
Than that in which the brother Poets joy'd,°
Who with combined powers, their wit employ'd
To raise a trophy to the drama's muses.
The thought of this great partnership diffuses
Over the genius loving heart, a feeling
Of all that's high, and great, and good, and healing. 10

Too partial friend! fain would I follow thee
Past each horizon of fine poesy;
Fain would I echo back each pleasant note
As o'er Sicilian seas, clear anthems float
'Mong the light skimming gondolas far parted,
Just when the sun his farewell beam has darted:

But 'tis impossible; far different cares
Beckon me sternly from soft 'Lydian airs,'°
And hold my faculties so long in thrall,
That I am oft in doubt whether at all 20
I shall again see Phœbus in the morning:
Or flush'd Aurora in the roseate dawning!
Or a white Naiad in a rippling stream;
Or a rapt seraph in a moonlight beam;
Or again witness what with thee I've seen,
The dew by fairy feet swept from the green,
After a night of some quaint jubilee
Which every elf and fay had come to see:
When bright processions took their airy march
Beneath the curved moon's triumphal arch. 30

But might I now each passing moment give
To the coy muse, with me she would not live
In this dark city, nor would condescend
'Mid contradictions her delights to lend.
Should e'er the fine-eyed maid to me be kind,
Ah! surely it must be whene'er I find
Some flowery spot, sequester'd, wild, romantic,
That often must have seen a poet frantic;
Where oaks, that erst the Druid knew, are growing,
And flowers, the glory of one day, are blowing; 40
Where the dark-leav'd laburnum's drooping clusters
Reflect athwart the stream their yellow lustres,
And intertwined the cassia's arms unite,°
With its own drooping buds, but very white.
Where on one side are covert branches hung,
'Mong which the nightingales have always sung
In leafy quiet: where to pry, aloof,
Atween the pillars of the sylvan roof,
Would be to find where violet beds were nestling,
And where the bee with cowslip bells was wrestling. 50
There must be too a ruin dark, and gloomy,
To say 'joy not too much in all that's bloomy.'

Yet this is vain—O Mathew lend thy aid
To find a place where I may greet the maid—
Where we may soft humanity put on,
And sit, and rhyme and think on Chatterton;

And that warm-hearted Shakspeare sent to meet him
Four laurell'd spirits, heaven-ward to intreat him.
With reverence would we speak of all the sages
Who have left streaks of light athwart their ages: 60
And thou shouldst moralize on Milton's blindness,
And mourn the fearful dearth of human kindness
To those who strove with the bright golden wing
Of genius, to flap away each sting
Thrown by the pitiless world. We next could tell
Of those who in the cause of freedom fell;
Of our own Alfred, of Helvetian Tell;°
Of him whose name to ev'ry heart's a solace,
High-minded and unbending William Wallace.
While to the rugged north our musing turns 70
We well might drop a tear for him, and Burns.

Felton! without incitements such as these,
How vain for me the niggard Muse to tease:
For thee, she will thy every dwelling grace,
And make 'a sun-shine in a shady place'.°
For thou wast once a flowret blooming wild,
Close to the source, bright, pure, and undefil'd,°
Whence gush the streams of song: in happy hour
Came chaste Diana from her shady bower,
Just as the sun was from the east uprising; 80
And, as for him some gift she was devising,
Beheld thee, pluck'd thee, cast thee in the stream
To meet her glorious brother's greeting beam.
I marvel much that thou hast never told°
How, from a flower, into a fish of gold
Apollo chang'd thee; how thou next didst seem
A black-eyed swan upon the widening stream;
And when thou first didst in that mirror trace
The placid features of a human face:
That thou hast never told thy travels strange, 90
And all the wonders of the mazy range
O'er pebbly crystal, and o'er golden sands;
Kissing thy daily food from Naiad's pearly hands.°

'O Solitude! if I must with thee dwell'

O Solitude! if I must with thee dwell,
  Let it not be among the jumbled heap
  Of murky buildings; climb with me the steep,—
Nature's observatory—whence the dell,
Its flowery slopes, its river's crystal swell,
  May seem a span; let me thy vigils keep
  'Mongst boughs pavillion'd, where the deer's swift leap
Startles the wild bee from the fox-glove bell.
But though I'll gladly trace these scenes with thee,
  Yet the sweet converse of an innocent mind, 10
  Whose words are images of thoughts refin'd,
Is my soul's pleasure; and it sure must be
  Almost the highest bliss of human-kind,
When to thy haunts two kindred spirits flee.

'Woman! when I behold thee flippant, vain'

Woman! when I behold thee flippant, vain,
  Inconstant, childish, proud, and full of fancies;
  Without that modest softening that enhances
The downcast eye, repentant of the pain
That its mild light creates to heal again:
  E'en then, elate, my spirit leaps, and prances,
  E'en then my soul with exultation dances
For that to love, so long, I've dormant lain:
But when I see thee meek, and kind, and tender,
  Heavens! how desperately do I adore 10
Thy winning graces;—to be thy defender
  I hotly burn—to be a Calidore—°
A very Red Cross Knight—a stout Leander—
  Might I be loved by thee like these of yore.

Light feet, dark violet eyes, and parted hair;
  Soft dimpled hands, white neck, and creamy breast,
  Are things on which the dazzled senses rest
Till the fond, fixed eyes, forget they stare.

From such fine pictures, heavens! I cannot dare
To turn my admiration, though unpossess'd 20
They be of what is worthy,—though not drest
In lovely modesty, and virtues rare.
Yet these I leave as thoughtless as a lark;
These lures I straight forget,—e'en ere I dine,
Or thrice my palate moisten: but when I mark
Such charms with mild intelligences shine,
My ear is open like a greedy shark,
To catch the tunings of a voice divine.

Ah! who can e'er forget so fair a being?
Who can forget her half retiring sweets? 30
God! she is like a milk-white lamb that bleats
For man's protection. Surely the All-seeing,
Who joys to see us with his gifts agreeing,
Will never give him pinions, who intreats
Such innocence to ruin,—who vilely cheats
A dove-like bosom. In truth there is no freeing
One's thoughts from such a beauty; when I hear
A lay that once I saw her hand awake,
Her form seems floating palpable, and near;
Had I e'er seen her from an arbour take 40
A dewy flower, oft would that hand appear,
And o'er my eyes the trembling moisture shake.

*To ******: 'Had I a man's fair form, then might my sighs'*

Had I a man's fair form, then might my sighs
Be echoed swiftly through that ivory shell
Thine ear, and find thy gentle heart; so well
Would passion arm me for the enterprize:
But ah! I am no knight whose foeman dies;
No cuirass glistens on my bosom's swell:
I am no happy shepherd of the dell
Whose lips have trembled with a maiden's eyes.

Yet must I dote upon thee,—call thee sweet,
  Sweeter by far than Hybla's honied roses 10
    When steep'd in dew rich to intoxication.
Ah! I will taste that dew, for me 'tis meet,
  And when the moon her pallid face discloses,
    I'll gather some by spells, and incantation.

'Give me women wine and snuff'

Give me women wine and snuff
Untill I cry out hold enough!
You may do so sans objection
Till y[e] day of resurrection°
For bless my beard they aye shall be
My beloved Trinity—

'I am as brisk'

I am as brisk
As a bottle of Wisk-
Ey and as nimble
As a Milliner's thimble—

'O grant that like to Peter I'

O grant that like to Peter I
May like to Peter B.
And tell me lovely Jesus Y
Old Jonah went to C.

*To ****: 'Hadst thou liv'd in days of old'*

Hadst thou liv'd in days of old,
O what wonders had been told
Of thy lively countenance,°
And thy humid eyes that dance
In the midst of their own brightness;
In the very fane of lightness.°
Over which thine eyebrows, leaning,
Picture out each lovely meaning:
In a dainty bend they lie,
Like to streaks across the sky, 10
Or the feathers from a crow,
Fallen on a bed of snow.
Of thy dark hair that extends
Into many graceful bends:
As the leaves of Hellebore
Turn to whence they sprung before.
And behind each ample curl
Peeps the richness of a pearl.
Downward too flows many a tress
With a glossy waviness; 20
Full, and round like globes that rise
From the censer to the skies
Through sunny air. Add too, the sweetness
Of thy honied voice; the neatness
Of thine ankle lightly turn'd:
With those beauties, scarce discern'd,
Kept with such sweet privacy,
That they seldom meet the eye
Of the little loves that fly°
Round about with eager pry. 30
Saving when, with freshening lave,
Thou dipp'st them in the taintless wave;
Like twin water lillies, born
In the coolness of the morn.
O, if thou hadst breathed then,
Now the Muses had been ten.°
Couldst thou wish for lineage higher
Than twin sister of Thalia?

At least for ever, evermore,
Will I call the Graces four.° 40

Hadst thou liv'd when chivalry
Lifted up her lance on high,
Tell me what thou wouldst have been?
Ah! I see the silver sheen°
Of thy broidered, floating vest
Cov'ring half thine ivory breast;°
Which, O heavens! I should see,
But that cruel destiny
Has placed a golden cuirass there;
Keeping secret what is fair. 50
Like sunbeams in a cloudlet nested
Thy locks in knightly casque are rested:
O'er which bend four milky plumes
Like the gentle lilly's blooms
Springing from a costly vase.
See with what a stately pace
Comes thine alabaster steed;
Servant of heroic deed!
O'er his loins, his trappings glow
Like the northern lights on snow.° 60
Mount his back! thy sword unsheath!
Sign of the enchanter's death;
Bane of every wicked spell;
Silencer of dragon's yell.
Alas! thou this wilt never do:
Thou art an enchantress too,
And wilt surely never spill
Blood of those whose eyes can kill.

Specimen of an Induction to a Poem

Lo! I must tell a tale of chivalry;
For large white plumes are dancing in mine eye.
Not like the formal crest of latter days:
But bending in a thousand graceful ways;

So graceful, that it seems no mortal hand,
Or e'en the touch of Archimago's wand,°
Could charm them into such an attitude.
We must think rather, that in playful mood,
Some mountain breeze had turned its chief delight,
To show this wonder of its gentle might. 10
Lo! I must tell a tale of chivalry;
For while I muse, the lance points slantingly
Athwart the morning air: some lady sweet,
Who cannot feel for cold her tender feet,
From the worn top of some old battlement
Hails it with tears, her stout defender sent:
And from her own pure self no joy dissembling,
Wraps round her ample robe with happy trembling.
Sometimes, when the good Knight his rest would take,
It is reflected, clearly, in a lake, 20
With the young ashen boughs, 'gainst which it rests,
And th' half seen mossiness of linnets' nests.
Ah! shall I ever tell its cruelty,
When the fire flashes from a warrior's eye,
And his tremendous hand is grasping it,
And his dark brow for very wrath is knit?
Or when his spirit, with more calm intent,
Leaps to the honors of a tournament,
And makes the gazers round about the ring
Stare at the grandeur of the ballancing? 30
No, no! this is far off:—then how shall I
Revive the dying tones of minstrelsy,
Which linger yet about lone gothic arches,
In dark green ivy, and among wild larches?
How sing the splendour of the revelries,
When buts of wine are drunk off to the lees?
And that bright lance, against the fretted wall,
Beneath the shade of stately banneral,°
Is slung with shining cuirass, sword, and shield,
Where ye may see a spur in bloody field?° 40
Light-footed damsels move with gentle paces
Round the wide hall, and show their happy faces;
Or stand in courtly talk by fives and sevens:
Like those fair stars that twinkle in the heavens.

Yet must I tell a tale of chivalry:
Or wherefore comes that steed so proudly by?
Wherefore more proudly does the gentle knight
Rein in the swelling of his ample might?

Spenser! thy brows are arched, open, kind,
And come like a clear sun-rise to my mind; 50
And always does my heart with pleasure dance,
When I think on thy noble countenance:
Where never yet was ought more earthly seen
Than the pure freshness of thy laurels green.
Therefore, great bard, I not so fearfully
Call on thy gentle spirit to hover nigh
My daring steps: or if thy tender care,
Thus startled unaware,
Be jealous that the foot of other wight
Should madly follow that bright path of light 60
Trac'd by thy lov'd Libertas; he will speak,°
And tell thee that my prayer is very meek;
That I will follow with due reverence,
And start with awe at mine own strange pretence.
Him thou wilt hear; so I will rest in hope
To see wide plains, fair trees and lawny slope:
The morn, the eve, the light, the shade, the flowers;
Clear streams, smooth lakes, and overlooking towers.

Calidore: A Fragment

Young Calidore is paddling o'er the lake;
His healthful spirit eager and awake
To feel the beauty of a silent eve,
Which seem'd full loath this happy world to leave;
The light dwelt o'er the scene so lingeringly.
He bares his forehead to the cool blue sky,
And smiles at the far clearness all around,
Until his heart is well nigh over wound,
And turns for calmness to the pleasant green
Of easy slopes, and shadowy trees that lean 10

So elegantly o'er the waters' brim
And show their blossoms trim.
Scarce can his clear and nimble eye-sight follow
The freaks, and dartings of the black-wing'd swallow,
Delighting much, to see it half at rest,
Dip so refreshingly its wings, and breast
'Gainst the smooth surface, and to mark anon,
The widening circles into nothing gone.

And now the sharp keel of his little boat°
Comes up with ripple, and with easy float, 20
And glides into a bed of water lillies:
Broad leav'd are they and their white canopies
Are upward turn'd to catch the heavens' dew.
Near to a little island's point they grew;
Whence Calidore might have the goodliest view
Of this sweet spot of earth. The bowery shore°
Went off in gentle windings to the hoar
And light blue mountains: but no breathing man°
With a warm heart, and eye prepared to scan
Nature's clear beauty, could pass lightly by 30
Objects that look'd out so invitingly
On either side. These, gentle Calidore
Greeted, as he had known them long before.

The sidelong view of swelling leafiness,
Which the glad setting sun in gold doth dress;
Whence ever and anon the jay outsprings,
And scales upon the beauty of its wings.

The lonely turret, shatter'd, and outworn,
Stands venerably proud; too proud to mourn
Its long lost grandeur: fir trees grow around, 40
Aye dropping their hard fruit upon the ground.

The little chapel with the cross above
Upholding wreaths of ivy; the white dove,
That on the windows spreads his feathers light,
And seems from purple clouds to wing its flight.

Green tufted islands casting their soft shades
Across the lake; sequester'd leafy glades,
That through the dimness of their twilight show
Large dock leaves, spiral foxgloves, or the glow°
Of the wild cat's eyes, or the silvery stems 50
Of delicate birch trees, or long grass which hems
A little brook. The youth had long been viewing
These pleasant things, and heaven was bedewing
The mountain flowers, when his glad senses caught
A trumpet's silver voice. Ah! it was fraught
With many joys for him: the warder's ken
Had found white coursers prancing in the glen:
Friends very dear to him he soon will see;
So pushes off his boat most eagerly,
And soon upon the lake he skims along, 60
Deaf to the nightingale's first under-song;
Nor minds he the white swans that dream so sweetly:
His spirit flies before him so completely.

And now he turns a jutting point of land,
Whence may be seen the castle gloomy, and grand:
Nor will a bee buzz round two swelling peaches,
Before the point of his light shallop reaches°
Those marble steps that through the water dip:
Now over them he goes with hasty trip,
And scarcely stays to ope the folding doors: 70
Anon he leaps along the oaken floors
Of halls and corridors.

Delicious sounds! those little bright-eyed things
That float about the air on azure wings,
Had been less heartfelt by him than the clang
Of clattering hoofs; into the court he sprang,
Just as two noble steeds, and palfreys twain,
Were slanting out their necks with loosened rein;
While from beneath the threat'ning portcullis
They brought their happy burthens. What a kiss, 80
What gentle squeeze he gave each lady's hand!
How tremblingly their delicate ancles spann'd!
Into how sweet a trance his soul was gone,
While whisperings of affection

Made him delay to let their tender feet
Come to the earth; with an incline so sweet
From their low palfreys o'er his neck they bent:
And whether there were tears of languishment,
Or that the evening dew had pearl'd their tresses,
He feels a moisture on his cheek, and blesses 90
With lips that tremble, and with glistening eye
All the soft luxury
That nestled in his arms. A dimpled hand,
Fair as some wonder out of fairy land,
Hung from his shoulder like the drooping flowers
Of whitest Cassia, fresh from summer showers:
And this he fondled with his happy cheek
As if for joy he would not further seek;
When the kind voice of good Sir Clerimond
Came to his ear, like something from beyond 100
His present being: so he gently drew
His warm arms, thrilling now with pulses new,
From their sweet thrall, and forward gently bending,
Thank'd heaven that his joy was never ending;
While 'gainst his forehead he devoutly press'd
A hand heaven made to succour the distress'd;
A hand that from the world's bleak promontory°
Had lifted Calidore for deeds of glory.

Amid the pages, and the torches' glare,
There stood a knight, patting the flowing hair 110
Of his proud horse's mane: he was withal
A man of elegance, and stature tall:
So that the waving of his plumes would be
High as the berries of a wild ash tree,
Or as the winged cap of Mercury.
His armour was so dexterously wrought
In shape, that sure no living man had thought
It hard, and heavy steel: but that indeed
It was some glorious form, some splendid weed,°
In which a spirit new come from the skies 120
Might live, and show itself to human eyes.
'Tis the far-fam'd, the brave Sir Gondibert,°
Said the good man to Calidore alert;

While the young warrior with a step of grace
Came up,—a courtly smile upon his face,
And mailed hand held out, ready to greet
The large-eyed wonder, and ambitious heat
Of the aspiring boy; who as he led
Those smiling ladies, often turned his head
To admire the visor arched so gracefully 130
Over a knightly brow; while they went by
The lamps that from the high-roof'd hall were pendent,
And gave the steel a shining quite transcendent.

Soon in a pleasant chamber they are seated;
The sweet-lipp'd ladies have already greeted
All the green leaves that round the window clamber,
To show their purple stars, and bells of amber,
Sir Gondibert has doff'd his shining steel,
Gladdening in the free, and airy feel
Of a light mantle; and while Clerimond 140
Is looking round about him with a fond,
And placid eye, young Calidore is burning
To hear of knightly deeds, and gallant spurning
Of all unworthiness; and how the strong of arm
Kept off dismay, and terror, and alarm
From lovely woman: while brimful of this,
He gave each damsel's hand so warm a kiss,
And had such manly ardour in his eye,
That each at other look'd half staringly;
And then their features started into smiles 150
Sweet as blue heavens o'er enchanted isles.

Softly the breezes from the forest came,
Softly they blew aside the taper's flame;
Clear was the song from Philomel's far bower;
Grateful the incense from the lime-tree flower;°
Mysterious, wild, the far heard trumpet's tone;
Lovely the moon in ether, all alone:
Sweet too the converse of these happy mortals,
As that of busy spirits when the portals
Are closing in the west; or that soft humming 160
We hear around when Hesperus is coming.
Sweet be their sleep. * * * * * * * * * * * *

'To one who has been long in city pent'

To one who has been long in city pent,
'Tis very sweet to look into the fair
And open face of heaven,—to breathe a prayer
Full in the smile of the blue firmament.
Who is more happy, when, with heart's content,
Fatigued he sinks into some pleasant lair
Of wavy grass, and reads a debonair
And gentle tale of love and languishment?
Returning home at evening, with an ear
Catching the notes of Philomel,—an eye 10
Watching the sailing cloudlet's bright career,
He mourns that day so soon has glided by:
E'en like the passage of an angel's tear
That falls through the clear ether silently.

'Oh! how I love, on a fair summer's eve'

Oh! how I love, on a fair summer's eve,
When streams of light pour down the golden West,
And on the balmy Zephyrs tranquil rest
The silver clouds, far—far away to leave
All meaner thoughts, and take a sweet reprieve
From little cares:—to find, with easy quest,
A fragrant wild, with Nature's beauty drest,
And there into delight my soul deceive.—
There warm my breast with patriotic lore,
Musing on Milton's fate—on Sydney's bier—° 10
Till their stern forms before my mind arise:
Perhaps on the Wing of poesy upsoar,—
Full often dropping a delicious tear,
When some melodious sorrow spells mine eyes.—°

To My Brother George

Many the wonders I this day have seen:°
The sun, when first he kist away the tears
That fill'd the eyes of morn;—the laurel'd peers
Who from the feathery gold of evening lean;—
The ocean with its vastness, its blue green,
Its ships, its rocks, its caves, its hopes, its fears,—
Its voice mysterious, which whoso hears
Must think on what will be, and what has been.
E'en now, dear George, while this for you I write,
Cynthia is from her silken curtains peeping° 10
So scantly, that it seems her bridal night,
And she her half-discover'd revels keeping.
But what, without the social thought of thee,°
Would be the wonders of the sky and sea?

To My Brother George

Full many a dreary hour have I past,°
My brain bewilder'd, and my mind o'ercast
With heaviness; in seasons when I've thought
No spherey strains by me could e'er be caught°
From the blue dome, though I to dimness gaze
On the far depth where sheeted lightning plays;
Or, on the wavy grass outstretch'd supinely,
Pry 'mong the stars, to strive to think divinely:
That I should never hear Apollo's song.
Though feathery clouds were floating all along 10
The purple west, and, two bright streaks between,
The golden lyre itself were dimly seen:
That the still murmur of the honey bee
Would never teach a rural song to me:
That the bright glance from beauty's eyelids slanting
Would never make a lay of mine enchanting,
Or warm my breast with ardour to unfold
Some tale of love and arms in time of old.

But there are times, when those that love the bay,°
Fly from all sorrowing far, far away; 20
A sudden glow comes on them, nought they see
In water, earth, or air, but poesy.
It has been said, dear George, and true I hold it,
(For knightly Spenser to Libertas told it,)
That when a Poet is in such a trance,
In air he sees white coursers paw, and prance,
Bestridden of gay knights, in gay apparel,
Who at each other tilt in playful quarrel,
And what we, ignorantly, sheet-lightning call,
Is the swift opening of their wide portal, 30
When the bright warder blows his trumpet clear,
Whose tones reach nought on earth but Poet's ear.
When these enchanted portals open wide,
And through the light the horsemen swiftly glide,
The Poet's eye can reach those golden halls,
And view the glory of their festivals:
Their ladies fair, that in the distance seem
Fit for the silv'ring of a seraph's dream;
Their rich brimm'd goblets, that incessant run
Like the bright spots that move about the sun; 40
And, when upheld, the wine from each bright jar
Pours with the lustre of a falling star.
Yet further off, are dimly seen their bowers,
Of which no mortal eye can reach the flowers;
And 'tis right just, for well Apollo knows
'Twould make the Poet quarrel with the rose.
All that's reveal'd from that far seat of blisses,
Is, the clear fountains' interchanging kisses,
As gracefully descending, light and thin,
Like silver streaks across a dolphin's fin, 50
When he upswimmeth from the coral caves,
And sports with half his tail above the waves.

These wonders strange he sees, and many more,
Whose head is pregnant with poetic lore.
Should he upon an evening ramble fare
With forehead to the soothing breezes bare,
Would he naught see but the dark, silent blue
With all its diamonds trembling through and through?

Or the coy moon, when in the waviness
Of whitest clouds she does her beauty dress, 60
And staidly paces higher up, and higher,
Like a sweet nun in holy-day attire?
Ah yes! much more would start into his sight—
The revelries, and mysteries of night:
And should I ever see them, I will tell you
Such tales as needs must with amazement spell you.°

These are the living pleasures of the bard:
But richer far posterity's award.
What does he murmur with his latest breath,
While his proud eye looks through the film of death? 70
'What though I leave this dull, and earthly mould,
Yet shall my spirit lofty converse hold
With after times.—The patriot shall feel
My stern alarum, and unsheath his steel;
Or, in the senate thunder out my numbers
To startle princes from their easy slumbers.
The sage will mingle with each moral theme
My happy thoughts sententious; he will teem°
With lofty periods when my verses fire him,
And then I'll stoop from heaven to inspire him. 80
Lays have I left of such a dear delight
That maids will sing them on their bridal night.
Gay villagers, upon a morn of May,
When they have tired their gentle limbs with play,
And form'd a snowy circle on the grass,
And plac'd in midst of all that lovely lass
Who chosen is their queen,—with her fine head
Crowned with flowers purple, white, and red:
For there the lily, and the musk-rose, sighing,
Are emblems true of hapless lovers dying: 90
Between her breasts, that never yet felt trouble,
A bunch of violets full blown, and double,
Serenely sleep:—she from a casket takes
A little book,—and then a joy awakes
About each youthful heart,—with stifled cries,
And rubbing of white hands, and sparkling eyes:
For she's to read a tale of hopes, and fears;
One that I foster'd in my youthful years:

The pearls, that on each glist'ning circlet sleep,
Gush ever and anon with silent creep, 100
Lured by the innocent dimples. To sweet rest
Shall the dear babe, upon its mother's breast,
Be lull'd with songs of mine. Fair world, adieu!
Thy dales, and hills, are fading from my view:
Swiftly I mount, upon wide spreading pinions,
Far from the narrow bounds of thy dominions.
Full joy I feel, while thus I cleave the air,
That my soft verse will charm thy daughters fair,
And warm thy sons!' Ah, my dear friend and brother,
Could I, at once, my mad ambition smother, 110
For tasting joys like these, sure I should be
Happier, and dearer to society.
At times, 'tis true, I've felt relief from pain
When some bright thought has darted through my brain:
Through all that day I've felt a greater pleasure
Than if I'd brought to light a hidden treasure.
As to my sonnets, though none else should heed them,
I feel delighted, still, that you should read them.
Of late, too, I have had much calm enjoyment,
Stretch'd on the grass at my best lov'd employment 120
Of scribbling lines for you. These things I thought
While, in my face, the freshest breeze I caught.
E'en now I'm pillow'd on a bed of flowers
That crowns a lofty clift, which proudly towers
Above the ocean-waves. The stalks, and blades,
Chequer my tablet with their quivering shades.
On one side is a field of drooping oats,
Through which the poppies show their scarlet coats;
So pert and useless, that they bring to mind
The scarlet coats that pester human-kind.° 130
And on the other side, outspread, is seen
Ocean's blue mantle streak'd with purple, and green.
Now 'tis I see a canvass'd ship, and now
Mark the bright silver curling round her prow.
I see the lark down-dropping to his nest,
And the broad winged sea-gull never at rest;
For when no more he spreads his feathers free,
His breast is dancing on the restless sea.

Now I direct my eyes into the west,
Which at this moment is in sunbeams drest: 140
Why westward turn? 'Twas but to say adieu!°
'Twas but to kiss my hand, dear George, to you!

To Charles Cowden Clarke

Oft have you seen a swan superbly frowning,
And with proud breast his own white shadow crowning;
He slants his neck beneath the waters bright
So silently, it seems a beam of light
Come from the galaxy: anon he sports,—
With outspread wings the Naiad Zephyr courts,°
Or ruffles all the surface of the lake
In striving from its crystal face to take
Some diamond water drops, and them to treasure
In milky nest, and sip them off at leisure. 10
But not a moment can he there insure them,
Nor to such downy rest can he allure them;
For down they rush as though they would be free,
And drop like hours into eternity.
Just like that bird am I in loss of time,
Whene'er I venture on the stream of rhyme;
With shatter'd boat, oar snapt, and canvass rent,
I slowly sail, scarce knowing my intent;
Still scooping up the water with my fingers,
In which a trembling diamond never lingers. 20

By this, friend Charles, you may full plainly see
Why I have never penn'd a line to thee:
Because my thoughts were never free, and clear,
And little fit to please a classic ear;
Because my wine was of too poor a savour
For one whose palate gladdens in the flavour
Of sparkling Helicon:—small good it were
To take him to a desert rude, and bare,
Who had on Baiæ's shore reclin'd at ease,°
While Tasso's page was floating in a breeze 30

That gave soft music from Armida's bowers,
Mingled with fragrance from her rarest flowers:
Small good to one who had by Mulla's stream°
Fondled the maidens with the breasts of cream;
Who had beheld Belphœbe in a brook,
And lovely Una in a leafy nook,
And Archimago leaning o'er his book:
Who had of all that's sweet tasted, and seen,
From silv'ry ripple, up to beauty's queen;
From the sequester'd haunts of gay Titania, 40
To the blue dwelling of divine Urania:
One who, of late, had ta'en sweet forest walks
With him who elegantly chats, and talks—
The wrong'd Libertas,—who has told you stories°
Of laurel chaplets, and Apollo's glories;
Of troops chivalrous prancing through a city,
And tearful ladies made for love, and pity:
With many else which I have never known.
Thus have I thought; and days on days have flown
Slowly, or rapidly—unwilling still 50
For you to try my dull, unlearned quill.
Nor should I now, but that I've known you long;
That you first taught me all the sweets of song:
The grand, the sweet, the terse, the free, the fine;
What swell'd with pathos, and what right divine:
Spenserian vowels that elope with ease,°
And float along like birds o'er summer seas;
Miltonian storms, and more, Miltonian tenderness;
Michael in arms, and more, meek Eve's fair slenderness.
Who read for me the sonnet swelling loudly 60
Up to its climax and then dying proudly?
Who found for me the grandeur of the ode,
Growing, like Atlas, stronger from its load?
Who let me taste that more than cordial dram,
The sharp, the rapier-pointed epigram?
Shew'd me that epic was of all the king,
Round, vast, and spanning all like Saturn's ring?
You too upheld the veil from Clio's beauty,
And pointed out the patriot's stern duty;°
The might of Alfred, and the shaft of Tell; 70
The hand of Brutus, that so grandly fell

Upon a tyrant's head. Ah! had I never seen,
Or known your kindness, what might I have been?
What my enjoyments in my youthful years,
Bereft of all that now my life endears?
And can I e'er these benefits forget?
And can I e'er repay the friendly debt?
No, doubly no;—yet should these rhymings please,
I shall roll on the grass with two-fold ease:
For I have long time been my fancy feeding 80
With hopes that you would one day think the reading
Of my rough verses not an hour misspent;
Should it e'er be so, what a rich content!
Some weeks have pass'd since last I saw the spires
In lucent Thames reflected:—warm desires
To see the sun o'er peep the eastern dimness,
And morning shadows streaking into slimness
Across the lawny fields, and pebbly water;
To mark the time as they grow broad, and shorter;
To feel the air that plays about the hills, 90
And sips its freshness from the little rills;
To see high, golden corn wave in the light
When Cynthia smiles upon a summer's night,
And peers among the cloudlet's jet and white,
As though she were reclining in a bed
Of bean blossoms, in heaven freshly shed.
No sooner had I stepp'd into these pleasures
Than I began to think of rhymes and measures:
The air that floated by me seem'd to say
'Write! thou wilt never have a better day.' 100
And so I did. When many lines I'd written,
Though with their grace I was not oversmitten,
Yet, as my hand was warm, I thought I'd better
Trust to my feelings, and write you a letter.
Such an attempt required an inspiration
Of a peculiar sort,—a consummation;—
Which, had I felt, these scribblings might have been
Verses from which the soul would never wean:
But many days have past since last my heart
Was warm'd luxuriously by divine Mozart;° 110
By Arne delighted, or by Handel madden'd;
Or by the song of Erin pierc'd and sadden'd:

What time you were before the music sitting,
And the rich notes to each sensation fitting;
Since I have walk'd with you through shady lanes
That freshly terminate in open plains,
And revel'd in a chat that ceased not
When at night-fall among your books we got:
No, nor when supper came, nor after that,—
Nor when reluctantly I took my hat; 120
No, nor till cordially you shook my hand
Mid-way between our homes:—your accents bland°
Still sounded in my ears, when I no more
Could hear your footsteps touch the grav'ly floor.
Sometimes I lost them, and then found again;
You chang'd the footpath for the grassy plain.
In those still moments I have wish'd you joys
That well you know to honour:—'Life's very toys
With him,' said I, 'will take a pleasant charm;
It cannot be that ought will work him harm.' 130
These thoughts now come o'er me with all their might:—
Again I shake your hand,—friend Charles, good night.

On First Looking into Chapman's Homer

Much have I travell'd in the realms of gold,°
  And many goodly states and kingdoms seen;
  Round many western islands have I been
Which bards in fealty to Apollo hold.
Oft of one wide expanse had I been told
  That deep-brow'd Homer ruled as his demesne;
  Yet did I never breathe its pure serene°
Till I heard Chapman speak out loud and bold:
Then felt I like some watcher of the skies
  When a new planet swims into his ken; 10
Or like stout Cortez when the eagle eyes°
  He star'd at the Pacific—and all his men
Look'd at each other with a wild surmise—
  Silent, upon a peak in Darien.

'Keen, fitful gusts are whisp'ring here and there'

Keen, fitful gusts are whisp'ring here and there
  Among the bushes half leafless, and dry;
  The stars look very cold about the sky,
And I have many miles on foot to fare.
Yet feel I little of the cool bleak air,
  Or of the dead leaves rustling drearily,
  Or of those silver lamps that burn on high,
Or of the distance from home's pleasant lair:
For I am brimfull of the friendliness
  That in a little cottage I have found; 10
Of fair-hair'd Milton's eloquent distress,
  And all his love for gentle Lycid drown'd;°
Of lovely Laura in her light green dress,°
  And faithful Petrarch gloriously crown'd.

Sleep and Poetry

As I lay in my bed slepe full unmete°
Was unto me, but why that I ne might
Rest I ne wist, for there n'as erthly wight
(As I suppose) had more of hertis ese
Than I, for I n'ad sicknesse nor disese.

CHAUCER

What is more gentle than a wind in summer?
What is more soothing that the pretty hummer
That stays one moment in an open flower,
And buzzes cheerily from bower to bower?
What is more tranquil than a musk-rose blowing
In a green island, far from all men's knowing?
More healthful than the leafiness of dales?
More secret than a nest of nightingales?
More serene than Cordelia's countenance?
More full of visions than a high romance? 10
What, but thee Sleep? Soft closer of our eyes!
Low murmurer of tender lullabies!

Light hoverer around our happy pillows!
Wreather of poppy buds, and weeping willows!
Silent entangler of a beauty's tresses!
Most happy listener! when the morning blesses
Thee for enlivening all the cheerful eyes
That glance so brightly at the new sun-rise.

But what is higher beyond thought than thee?
Fresher than berries of a mountain tree? 20
More strange, more beautiful, more smooth, more regal,
Than wings of swans, than doves, than dim-seen eagle?
What is it? And to what shall I compare it?
It has a glory, and nought else can share it:
The thought thereof is awful, sweet, and holy,
Chacing away all worldliness and folly;
Coming sometimes like fearful claps of thunder,
Or the low rumblings earth's regions under;
And sometimes like a gentle whispering
Of all the secrets of some wond'rous thing 30
That breathes about us in the vacant air;
So that we look around with prying stare,
Perhaps to see shapes of light, aerial lymning,°
And catch soft floatings from a faint-heard hymning;
To see the laurel wreath, on high suspended,
That is to crown our name when life is ended.
Sometimes it gives a glory to the voice,
And from the heart up-springs, rejoice! rejoice!
Sounds which will reach the Framer of all things,
And die away in ardent mutterings. 40

No one who once the glorious sun has seen,
And all the clouds, and felt his bosom clean
For his great Maker's presence, but must know
What 'tis I mean, and feel his being glow:
Therefore no insult will I give his spirit,
By telling what he sees from native merit.

O Poesy! for thee I hold my pen
That am not yet a glorious denizen
Of thy wide heaven—Should I rather kneel
Upon some mountain-top until I feel 50

A glowing splendour round about me hung,
And echo back the voice of thine own tongue?
O Poesy! for thee I grasp my pen
That am not yet a glorious denizen
Of thy wide heaven; yet, to my ardent prayer,
Yield from thy sanctuary some clear air,
Smoothed for intoxication by the breath
Of flowering bays, that I may die a death
Of luxury, and my young spirit follow
The morning sun-beams to the great Apollo 60
Like a fresh sacrifice; or, if I can bear
The o'erwhelming sweets, 'twill bring to me the fair
Visions of all places: a bowery nook
Will be elysium—an eternal book
Whence I may copy many a lovely saying
About the leaves, and flowers—about the playing
Of nymphs in woods, and fountains; and the shade
Keeping a silence round a sleeping maid;
And many a verse from so strange influence
That we must ever wonder how, and whence 70
It came. Also imaginings will hover
Round my fire-side, and haply there discover
Vistas of solemn beauty, where I'd wander
In happy silence, like the clear Meander°
Through its lone vales; and where I found a spot
Of awfuller shade, or an enchanted grot,
Or a green hill o'erspread with chequered dress
Of flowers, and fearful from its loveliness,
Write on my tablets all that was permitted,
All that was for our human senses fitted. 80
Then the events of this wide world I'd seize
Like a strong giant, and my spirit teaze
Till at its shoulders it should proudly see
Wings to find out an immortality.

Stop and consider! life is but a day;°
A fragile dew-drop on its perilous way
From a tree's summit; a poor Indian's sleep
While his boat hastens to the monstrous steep
Of Montmorenci. Why so sad a moan?°
Life is the rose's hope while yet unblown; 90

The reading of an ever-changing tale;
The light uplifting of a maiden's veil;
A pigeon tumbling in clear summer air;
A laughing school-boy, without grief or care,
Riding the springy branches of an elm.

O for ten years, that I may overwhelm°
Myself in poesy; so I may do the deed
That my own soul has to itself decreed.
Then will I pass the countries that I see
In long perspective, and continually 100
Taste their pure fountains. First the realm I'll pass
Of Flora, and old Pan: sleep in the grass,°
Feed upon apples red, and strawberries,
And choose each pleasure that my fancy sees;
Catch the white-handed nymphs in shady places,
To woo sweet kisses from averted faces,—
Play with their fingers, touch their shoulders white
Into a pretty shrinking with a bite
As hard as lips can make it: till agreed,
A lovely tale of human life we'll read. 110
And one will teach a tame dove how it best
May fan the cool air gently o'er my rest;
Another, bending o'er her nimble tread,
Will set a green robe floating round her head,
And still will dance with ever varied ease,
Smiling upon the flowers and the trees:
Another will entice me on, and on
Through almond blossoms and rich cinnamon;
Till in the bosom of a leafy world
We rest in silence, like two gems upcurl'd
In the recesses of a pearly shell. 120

And can I ever bid these joys farewell?
Yes, I must pass them for a nobler life,
Where I may find the agonies, the strife
Of human hearts: for lo! I see afar,
O'er sailing the blue cragginess, a car°
And steeds with streamy manes—the charioteer°
Looks out upon the winds with glorious fear:

And now the numerous tramplings quiver lightly
Along a huge cloud's ridge; and now with sprightly 130
Wheel downward come they into fresher skies,
Tipt round with silver from the sun's bright eyes.
Still downward with capacious whirl they glide;
And now I see them on a green-hill's side
In breezy rest among the nodding stalks.
The charioteer with wond'rous gesture talks
To the trees and mountains; and there soon appear
Shapes of delight, of mystery, and fear,
Passing along before a dusky space
Made by some mighty oaks: as they would chase 140
Some ever-fleeting music on they sweep.
Lo! how they murmur, laugh, and smile, and weep:
Some with upholden hand and mouth severe;
Some with their faces muffled to the ear
Between their arms; some, clear in youthful bloom,
Go glad and smilingly athwart the gloom;
Some looking back, and some with upward gaze;
Yes, thousands in a thousand different ways
Flit onward—now a lovely wreath of girls
Dancing their sleek hair into tangled curls; 150
And now broad wings. Most awfully intent
The driver of those steeds is forward bent,
And seems to listen: O that I might know
All that he writes with such a hurrying glow.

The visions all are fled—the car is fled
Into the light of heaven, and in their stead
A sense of real things comes doubly strong,°
And, like a muddy stream, would bear along°
My soul to nothingness: but I will strive
Against all doubtings, and will keep alive 160
The thought of that same chariot, and the strange
Journey it went.

Is there so small a range
In the present strength of manhood, that the high
Imagination cannot freely fly
As she was wont of old? prepare her steeds,
Paw up against the light, and do strange deeds

Upon the clouds? Has she not shewn us all?
From the clear space of ether, to the small°
Breath of new buds unfolding? From the meaning
Of Jove's large eye-brow, to the tender greening 170
Of April meadows? Here her altar shone,
E'en in this isle; and who could paragon°
The fervid choir that lifted up a noise
Of harmony, to where it aye will poise
Its mighty self of convoluting sound,
Huge as a planet, and like that roll round,
Eternally around a dizzy void?
Ay, in those days the Muses were nigh cloy'd
With honors; nor had any other care
Than to sing out and sooth their wavy hair. 180

Could all this be forgotten? Yes, a scism
Nurtured by foppery and barbarism,
Made great Apollo blush for this his land.
Men were thought wise who could not understand
His glories: with a puling infant's force
They sway'd about upon a rocking horse,°
And thought it Pegasus. Ah dismal soul'd!
The winds of heaven blew, the ocean roll'd
Its gathering waves—ye felt it not. The blue
Bared its eternal bosom, and the dew 190
Of summer nights collected still to make
The morning precious: beauty was awake!
Why were ye not awake? But ye were dead
To things ye knew not of,—were closely wed
To musty laws lined out with wretched rule
And compass vile: so that ye taught a school
Of dolts to smooth, inlay, and clip, and fit,
Till, like the certain wands of Jacob's wit,°
Their verses tallied. Easy was the task:
A thousand handicraftsmen wore the mask 200
Of Poesy. Ill-fated, impious race!
That blasphemed the bright Lyrist to his face,°
And did not know it,—no, they went about,
Holding a poor, decrepid standard out
Mark'd with most flimsy mottos, and in large
The name of one Boileau!°

O ye whose charge
It is to hover round our pleasant hills!
Whose congregated majesty so fills
My boundly reverence, that I cannot trace°
Your hallowed names, in this unholy place, 210
So near those common folk; did not their shames
Affright you? Did our old lamenting Thames
Delight you? Did ye never cluster round
Delicious Avon, with a mournful sound,
And weep? Or did ye wholly bid adieu
To regions where no more the laurel grew?
Or did ye stay to give a welcoming
To some lone spirits who could proudly sing°
Their youth away, and die? 'Twas even so:
But let me think away those times of woe: 220
Now 'tis a fairer season; ye have breathed°
Rich benedictions o'er us; ye have wreathed
Fresh garlands: for sweet music has been heard
In many places;—some has been upstirr'd
From out its crystal dwelling in a lake,
By a swan's ebon bill; from a thick brake,°
Nested and quiet in a valley mild,
Bubbles a pipe; fine sounds are floating wild
About the earth: happy are ye and glad.

These things are doubtless: yet in truth we've had° 230
Strange thunders from the potency of song;
Mingled indeed with what is sweet and strong,
From majesty: but in clear truth the themes
Are ugly clubs, the Poets Polyphemes
Disturbing the grand sea. A drainless shower
Of light is poesy; 'tis the supreme of power;
'Tis might half slumb'ring on its own right arm.
The very archings of her eye-lids charm
A thousand willing agents to obey,
And still she governs with the mildest sway: 240
But strength alone though of the Muses born°
Is like a fallen angel: trees uptorn,°
Darkness, and worms, and shrouds, and sepulchres
Delight it; for it feeds upon the burrs,

And thorns of life; forgetting the great end°
Of poesy, that it should be a friend
To sooth the cares, and lift the thoughts of man.

Yet I rejoice: a myrtle fairer than°
E'er grew in Paphos, from the bitter weeds
Lifts its sweet head into the air, and feeds 250
A silent space with ever sprouting green.
All tenderest birds there find a pleasant screen,
Creep through the shade with jaunty fluttering,
Nibble the little cupped flowers and sing.
Then let us clear away the choaking thorns
From round its gentle stem; let the young fawns,
Yeaned in after times, when we are flown,°
Find a fresh sward beneath it, overgrown
With simple flowers: let there nothing be
More boisterous than a lover's bended knee; 260
Nought more ungentle than the placid look
Of one who leans upon a closed book;
Nought more untranquil than the grassy slopes
Between two hills. All hail delightful hopes!
As she was wont, th' imagination
Into most lovely labyrinths will be gone,
And they shall be accounted poet kings
Who simply tell the most heart-easing things.
O may these joys be ripe before I die.

Will not some say that I presumptuously 270
Have spoken? that from hastening disgrace
'Twere better far to hide my foolish face?
That whining boyhood should with reverence bow
Ere the dread thunderbolt could reach? How!
If I do hide myself, it sure shall be
In the very fane, the light of Poesy:
If I do fall, at least I will be laid
Beneath the silence of a poplar shade;
And over me the grass shall be smooth shaven;
And there shall be a kind memorial graven. 280
But off Despondence! miserable bane!
They should not know thee, who athirst to gain

A noble end, are thirsty every hour.
What though I am not wealthy in the dower
Of spanning wisdom; though I do not know
The shiftings of the mighty winds that blow
Hither and thither all the changing thoughts
Of man: though no great minist'ring reason sorts
Out the dark mysteries of human souls
To clear conceiving: yet there ever rolls 290
A vast idea before me, and I glean
Therefrom my liberty; thence too I've seen
The end and aim of Poesy. 'Tis clear
As any thing most true; as that the year
Is made of the four seasons—manifest
As a large cross, some old cathedral's crest,
Lifted to the white clouds. Therefore should I
Be but the essence of deformity,
A coward, did my very eye-lids wink
At speaking out what I have dared to think. 300
Ah! rather let me like a madman run
Over some precipice; let the hot sun
Melt my Dedalian wings, and drive me down°
Convuls'd and headlong! Stay! an inward frown
Of conscience bids me be more calm awhile.
An ocean dim, sprinkled with many an isle,
Spreads awfully before me. How much toil!
How many days! what desperate turmoil!
Ere I can have explored its widenesses.
Ah, what a task! upon my bended knees, 310
I could unsay those—no, impossible!
Impossible!

For sweet relief I'll dwell
On humbler thoughts, and let this strange assay
Begun in gentleness die so away.
E'en now all tumult from my bosom fades:
I turn full hearted to the friendly aids
That smooth the path of honour; brotherhood,
And friendliness the nurse of mutual good;
The hearty grasp that sends a pleasant sonnet
Into the brain ere one can think upon it; 320

The silence when some rhymes are coming out;
And when they're come, the very pleasant rout:°
The message certain to be done to-morrow—
'Tis perhaps as well that it should be to borrow
Some precious book from out its snug retreat,
To cluster round it when we next shall meet.
Scarce can I scribble on; for lovely airs
Are fluttering round the room like doves in pairs;
Many delights of that glad day recalling,
When first my senses caught their tender falling. 330
And with these airs come forms of elegance
Stooping their shoulders o'er a horse's prance,°
Careless, and grand—fingers soft and round
Parting luxuriant curls;—and the swift bound
Of Bacchus from his chariot, when his eye
Made Ariadne's cheek look blushingly.
Thus I remember all the pleasant flow
Of words at opening a portfolio.°

Things such as these are ever harbingers
To trains of peaceful images: the stirs 340
Of a swan's neck unseen among the rushes:
A linnet starting all about the bushes:
A butterfly, with golden wings broad parted,
Nestling a rose, convuls'd as though it smarted
With over pleasure—many, many more,
Might I indulge at large in all my store
Of luxuries: yet I must not forget
Sleep, quiet with his poppy coronet:
For what there may be worthy in these rhymes
I partly owe to him: and thus, the chimes 350
Of friendly voices had just given place
To as sweet a silence, when I 'gan retrace
The pleasant day, upon a couch at ease.
It was a poet's house who keeps the keys
Of pleasure's temple. Round about were hung
The glorious features of the bards who sung
In other ages—cold and sacred busts
Smiled at each other. Happy he who trusts
To clear Futurity his darling fame!
Then there were fauns and satyrs taking aim 360

At swelling apples with a frisky leap
And reaching fingers, 'mid a luscious heap
Of vine leaves. Then there rose to view a fane
Of liny marble, and thereto a train
Of nymphs approaching fairly o'er the sward:
One, loveliest, holding her white hand toward
The dazzling sun-rise: two sisters sweet
Bending their graceful figures till they meet
Over the trippings of a little child:
And some are hearing, eagerly, the wild 370
Thrilling liquidity of dewy piping.
See, in another picture, nymphs are wiping
Cherishingly Diana's timorous limbs;—
A fold of lawny mantle dabbling swims
At the bath's edge, and keeps a gentle motion
With the subsiding crystal: as when ocean
Heaves calmly its broad swelling smoothiness o'er
Its rocky marge, and balances once more
The patient weeds; that now unshent by foam°
Feel all about their undulating home. 380

Sappho's meek head was there half smiling down°
At nothing; just as though the earnest frown
Of over thinking had that moment gone
From off her brow, and left her all alone.

Great Alfred's too, with anxious, pitying eyes,
As if he always listened to the sighs
Of the goaded world; and Kosciusko's worn°
By horrid suffrance—mightily forlorn.

Petrarch, outstepping from the shady green,
Starts at the sight of Laura; nor can wean 390
His eyes from her sweet face. Most happy they!
For over them was seen a free display
Of out-spread wings, and from between them shone
The face of Poesy: from off her throne
She overlook'd things that I scarce could tell.
The very sense of where I was might well

Keep Sleep aloof: but more than that there came
Thought after thought to nourish up the flame
Within my breast; so that the morning light
Surprised me even from a sleepless night; 400
And up I rose refresh'd, and glad, and gay,
Resolving to begin that very day
These lines; and howsoever they be done,
I leave them as a father does his son.

To My Brothers

Small, busy flames play through the fresh laid coals,
  And their faint cracklings o'er our silence creep
  Like whispers of the household gods that keep
A gentle empire o'er fraternal souls.
And while, for rhymes, I search around the poles,
  Your eyes are fix'd, as in poetic sleep,
  Upon the lore so voluble and deep,
That aye at fall of night our care condoles.°
This is your birth-day Tom, and I rejoice
  That thus it passes smoothly, quietly. 10
Many such eves of gently whisp'ring noise
  May we together pass, and calmly try
What are this world's true joys,—ere the great voice,
  From its fair face, shall bid our spirits fly.

Addressed to [*Haydon*]

Great spirits now on earth are sojourning;
  He of the cloud, the cataract, the lake,
  Who on Helvellyn's summit, wide awake,°
Catches his freshness from Archangel's wing:
He of the rose, the violet, the spring,
  The social smile, the chain for Freedom's sake:
  And lo!—whose stedfastness would never take

A meaner sound than Raphael's whispering.
And other spirits there are standing apart
Upon the forehead of the age to come; 10
These, these will give the world another heart,
And other pulses. Hear ye not the hum
Of mighty workings?—°
Listen awhile ye nations, and be dumb.

To G.A.W.

Nymph of the downward smile, and sidelong glance,
In what diviner moments of the day
Art thou most lovely? When gone far astray
Into the labyrinths of sweet utterance?
Or when serenely wand'ring in a trance
Of sober thought? Or when starting away,
With careless robe, to meet the morning ray,
Thou spar'st the flowers in thy mazy dance?
Haply 'tis when thy ruby lips part sweetly,
And so remain, because thou listenest: 10
But thou to please wert nurtured so completely
That I can never tell what mood is best.
I shall as soon pronounce which Grace more neatly
Trips it before Apollo than the rest.

To Kosciusko

Good Kosciusko, thy great name alone
Is a full harvest whence to reap high feeling;
It comes upon us like the glorious pealing
Of the wide spheres—an everlasting tone.
And now it tells me, that in worlds unknown,
The names of heroes, burst from clouds concealing,
And change to harmonies, for ever stealing°
Through cloudless blue, and round each silver throne.

It tells me too, that on a happy day,
  When some good spirit walks upon the earth, 10
    Thy name with Alfred's and the great of yore
  Gently commingling, gives tremendous birth
To a loud hymn, that sounds far, far away
    To where the great God lives for evermore.

'I stood tip-toe upon a little hill'

Places of nestling green for Poets made.

Story of Rimini°

I stood tip-toe upon a little hill,
The air was cooling, and so very still,
That the sweet buds which with a modest pride
Pull droopingly, in slanting curve aside,
Their scantly leaved, and finely tapering stems,
Had not yet lost those starry diadems
Caught from the early sobbing of the morn.
The clouds were pure and white as flocks new shorn,
And fresh from the clear brook; sweetly they slept
On the blue fields of heaven, and then there crept 10
A little noiseless noise among the leaves,
Born of the very sigh that silence heaves:
For not the faintest motion could be seen
Of all the shades that slanted o'er the green.
There was wide wand'ring for the greediest eye,
To peer about upon variety;
Far round the horizon's crystal air to skim,
And trace the dwindled edgings of its brim;
To picture out the quaint, and curious bending
Of a fresh woodland alley, never ending; 20
Or by the bowery clefts, and leafy shelves,
Guess where the jaunty streams refresh themselves.
I gazed awhile, and felt as light, and free
As though the fanning wings of Mercury
Had played upon my heels: I was light-hearted,
And many pleasures to my vision started;

So I straightway began to pluck a posey
Of luxuries bright, milky, soft and rosy.

A bush of May flowers with the bees about them;
Ah, sure no tasteful nook would be without them; 30
And let a lush laburnum oversweep them,
And let long grass grow round the roots to keep them
Moist, cool and green; and shade the violets,
That they may bind the moss in leafy nets.

A filbert hedge with wild briar overtwined,
And clumps of woodbine taking the soft wind
Upon their summer thrones; there too should be
The frequent chequer of a youngling tree,
That with a score of light green brethren shoots
From the quaint mossiness of aged roots: 40
Round which is heard a spring-head of clear waters
Babbling so wildly of its lovely daughters
The spreading blue bells: it may haply mourn
That such fair clusters should be rudely torn
From their fresh beds, and scattered thoughtlessly
By infant hands, left on the path to die.

Open afresh your round of starry folds,°
Ye ardent marigolds!
Dry up the moisture from your golden lids,
For great Apollo bids 50
That in these days your praises should be sung
On many harps, which he has lately strung;
And when again your dewiness he kisses,
Tell him, I have you in my world of blisses:
So haply when I rove in some far vale,
His mighty voice may come upon the gale.

Here are sweet peas, on tip-toe for a flight:
With wings of gentle flush o'er delicate white,
And taper fingers catching at all things,
To bind them all about with tiny rings. 60

Linger awhile upon some bending planks°
That lean against a streamlet's rushy banks,

And watch intently Nature's gentle doings:
They will be found softer than ring-dove's cooings.
How silent comes the water round that bend;
Not the minutest whisper does it send
To the o'erhanging sallows: blades of grass
Slowly across the chequer'd shadows pass.
Why, you might read two sonnets, ere they reach
To where the hurrying freshnesses aye preach 70
A natural sermon o'er their pebbly beds;
Where swarms of minnows show their little heads,
Staying their wavy bodies 'gainst the streams,
To taste the luxury of sunny beams
Temper'd with coolness. How they ever wrestle
With their own sweet delight, and never nestle
Their silver bellies on the pebbly sand.
If you but scantily hold out the hand,
That very instant not one will remain;
But turn your eye, and they are there again. 80
The ripples seem right glad to reach those cresses,
And cool themselves among the em'rald tresses;
The while they cool themselves, they freshness give,
And moisture, that the bowery green may live:
So keeping up an interchange of favours,
Like good men in the truth of their behaviours,
Sometimes goldfinches one by one will drop
From low hung branches; little space they stop;
But sip, and twitter, and their feathers sleek;°
Then off at once, as in a wanton freak: 90
Or perhaps, to show their black, and golden wings,
Pausing upon their yellow flutterings.
Were I in such a place, I sure should pray
That nought less sweet might call my thoughts away,
Than the soft rustle of a maiden's gown
Fanning away the dandelion's down;
Than the light music of her nimble toes
Patting against the sorrel as she goes.
How she would start, and blush, thus to be caught
Playing in all her innocence of thought. 100
O let me lead her gently o'er the brook,
Watch her half-smiling lips, and downward look;

O let me for one moment touch her wrist;
Let me one moment to her breathing list;
And as she leaves me may she often turn
Her fair eyes looking through her locks auburne.

What next? A tuft of evening primroses,
O'er which the mind may hover till it dozes;
O'er which it well might take a pleasant sleep,
But that 'tis ever startled by the leap 110
Of buds into ripe flowers; or by the flitting
Of diverse moths, that aye their rest are quitting;
Or by the moon lifting her silver rim
Above a cloud, and with a gradual swim
Coming into the blue with all her light.
O Maker of sweet poets, dear delight
Of this fair world, and all its gentle livers;
Spangler of clouds, halo of crystal rivers,
Mingler with leaves, and dew and tumbling streams,
Closer of lovely eyes to lovely dreams, 120
Lover of loneliness, and wandering,
Of upcast eye, and tender pondering!
Thee must I praise above all other glories
That smile us on to tell delightful stories.
For what has made the sage or poet write°
But the fair paradise of Nature's light?
In the calm grandeur of a sober line,
We see the waving of the mountain pine;
And when a tale is beautifully staid,
We feel the safety of a hawthorn glade: 130
When it is moving on luxurious wings,
The soul is lost in pleasant smotherings:
Fair dewy roses brush against our faces,
And flowering laurels spring from diamond vases;°
O'er head we see the jasmine and sweet briar,
And bloomy grapes laughing from green attire;
While at our feet, the voice of crystal bubbles
Charms us at once away from all our troubles:
So that we feel uplifted from the world,
Walking upon the white clouds wreath'd and curl'd. 140
So felt he, who first told, how Psyche went
On the smooth wind to realms of wonderment;

What Psyche felt, and Love, when their full lips
First touch'd; what amorous, and fondling nips
They gave each other's cheeks; with all their sighs,
And how they kist each other's tremulous eyes:
The silver lamp,—the ravishment,—the wonder—°
The darkness,—loneliness,—the fearful thunder;
Their woes gone by, and both to heaven upflown,
To bow for gratitude before Jove's throne. 150
So did he feel, who pull'd the boughs aside,
That we might look into a forest wide,
To catch a glimpse of Fawns, and Dryades
Coming with softest rustle through the trees;
And garlands woven of flowers wild, and sweet,
Upheld on ivory wrists, or sporting feet:
Telling us how fair, trembling Syrinx fled
Arcadian Pan, with such a fearful dread.
Poor nymph,—poor Pan,—how he did weep to find,
Nought but a lovely sighing of the wind 160
Along the reedy stream; a half heard strain,
Full of sweet desolation—balmy pain.

What first inspired a bard of old to sing
Narcissus pining o'er the untainted spring?
In some delicious ramble, he had found
A little space, with boughs all woven round;
And in the midst of all, a clearer pool
Than e'er reflected in its pleasant cool,
The blue sky here, and there, serenely peeping
Through tendril wreaths fantastically creeping. 170
And on the bank a lonely flower he spied,
A meek and forlorn flower, with naught of pride,
Drooping its beauty o'er the watery clearness,
To woo its own sad image into nearness:
Deaf to light Zephyrus it would not move;
But still would seem to droop, to pine, to love.
So while the Poet stood in this sweet spot,
Some fainter gleamings o'er his fancy shot;
Nor was it long ere he had told the tale
Of young Narcissus, and sad Echo's bale. 180

Where had he been, from whose warm head out-flew°
That sweetest of all songs, that ever new,

That aye refreshing, pure deliciousness,
Coming ever to bless
The wanderer by moonlight? to him bringing
Shapes from the invisible world, unearthly singing
From out the middle air, from flowery nests,
And from the pillowy silkiness that rests
Full in the speculation of the stars.
Ah! surely he had burst our mortal bars; 190
Into some wond'rous region he had gone,
To search for thee, divine Endymion!

He was a Poet, sure a lover too,
Who stood on Latmus' top, what time there blew
Soft breezes from the myrtle vale below;
And brought in faintness solemn, sweet, and slow
A hymn from Dian's temple; while upswelling,
The incense went to her own starry dwelling.
But though her face was clear as infant's eyes,
Though she stood smiling o'er the sacrifice, 200
The Poet wept at her so piteous fate,
Wept that such beauty should be desolate:
So in fine wrath some golden sounds he won,
And gave meek Cynthia her Endymion.

Queen of the wide air; thou most lovely queen
Of all the brightness that mine eyes have seen!
As thou exceedest all things in thy shine,
So every tale, does this sweet tale of thine.
O for three words of honey, that I might
Tell but one wonder of thy bridal night! 210

Where distant ships do seem to show their keels,
Phœbus awhile delayed his mighty wheels,
And turned to smile upon thy bashful eyes,
Ere he his unseen pomp would solemnize.
The evening weather was so bright, and clear,
That men of health were of unusual cheer;
Stepping like Homer at the trumpet's call,
Or young Apollo on the pedestal:°
And lovely women were as fair and warm,
As Venus looking sideways in alarm. 220

The breezes were ethereal, and pure,
And crept through half closed lattices to cure
The languid sick; it cool'd their fever'd sleep,
And soothed them into slumbers full and deep.
Soon they awoke clear eyed: nor burnt with thirsting,
Nor with hot fingers, nor with temples bursting:
And springing up, they met the wond'ring sight
Of their dear friends, nigh foolish with delight;
Who feel their arms, and breasts, and kiss and stare,
And on their placid foreheads part the hair. 230
Young men, and maidens at each other gaz'd
With hands held back, and motionless, amaz'd
To see the brightness in each other's eyes;
And so they stood, fill'd with a sweet surprise,
Until their tongues were loos'd in poesy.
Therefore no lover did of anguish die:
But the soft numbers, in that moment spoken,
Made silken ties, that never may be broken.
Cynthia! I cannot tell the greater blisses,
That follow'd thine, and thy dear shepherd's kisses: 240
Was there a Poet born?—but now no more,
My wand'ring spirit must no further soar.—

Written in Disgust of Vulgar Superstition

The Church bells toll a melancholy round,
  Calling the people to some other prayers,
  Some other gloominess, more dreadfull cares,
More heark'ning to the Sermon's horrid sound—
Surely the mind of Man is closely bound
  In some black spell; seeing that each one tears
  Himself from fireside joys and Lydian airs,°
And converse high of those with glory crown'd—
Still, still they toll, and I should feel a damp,—
  A chill as from a tomb, did I not know 10
That they are dying like an outburnt lamp;
  That 'tis their sighing, wailing ere they go°
  Into oblivion;—that fresh flowers will grow,
And many glories of immortal stamp—

On the Grasshopper and Cricket°

The poetry of earth is never dead:
  When all the birds are faint with the hot sun,
  And hide in cooling trees, a voice will run
From hedge to hedge about the new-mown mead;
That is the Grasshopper's—he takes the lead
  In summer luxury,—he has never done
  With his delights; for when tired out with fun
He rests at ease beneath some pleasant weed.
The poetry of earth is ceasing never:
  On a lone winter evening, when the frost 10
    Has wrought a silence, from the stove there shrills
The Cricket's song, in warmth increasing ever,
  And seems to one in drowsiness half lost,
    The Grasshopper's among some grassy hills.

'God of the golden bow'

God of the golden bow,
  And of the golden Lyre,
And of the golden hair
  And of the golden fire,
    Charioteer
    Round the patient year;
  Where? Where slept thine ire,
When like a blank ideot I put on thy Wreath
    Thy Laurel—thy glory—
    The Light of thy story— 10
Or was I a worm too low-creeping for death—
    O Delphic Apollo?

The Thunderer grasp'd and grasp'd
  The Thunderer frown'd and frown'd
The Eagle's feathery mane
  For Wrath became stiffened; the sound
    Of breeding thunder
    Went drowsily under
  Muttering to be unbound.

O why didst thou pity and beg for a worm? 20
Why touch thy soft Lute
Till the thunder was mute?
Why was I not crush'd—such a pitiful germ?°
O delphic Apollo!

The Pleiades were up
Watching the silent air,
The seeds and roots in Earth
Were swelling for summer fare;
The ocean its neighbour
Was at his old Labor 30
When—who—who did dare
To tie for a moment thy plant round his brow
And grin and look proudly
And blaspheme so loudly
And live for that honor to stoop for thee now
O delphic Apollo?

'After dark vapors have oppress'd our plains'

After dark vapors have oppress'd our plains
For a long dreary season, comes a day
Born of the gentle SOUTH, and clears away
From the sick heavens all unseemly stains,
The anxious Month, relieving of its pains,
Takes as a long lost right the feel of May:
The eyelids with the passing coolness play
Like Rose leaves with the drip of Summer rains.
And calmest thoughts come round us; as of leaves
Budding—fruit ripening in stillness—Autumn Suns 10
Smiling at Eve upon the quiet sheaves—
Sweet Sappho's Cheek—a sleeping infant's breath—
The gradual Sand that through an hour-glass runs—
A woodland Rivulet—a Poet's death.°

'This pleasant Tale is like a little Copse'

This pleasant Tale is like a little Copse:°
The honied Lines do freshly interlace
To keep the Reader in so sweet a place,
So that he here and there full-hearted stops;
And oftentimes he feels the dewy drops
Come cool and suddenly against his face,
And by the wand'ring Melody may trace
Which way the tender-legged Linnet hops.
O what a power has white Simplicity!
What mighty power has this gentle story! 10
I, that do ever feel a thirst for glory,
Could at this moment be content to lie
Meekly upon the grass, as those whose sobbings°
Were heard of none beside the mournful Robins.

To Leigh Hunt, Esq.

Glory and loveliness have passed away;°
For if we wander out in early morn,
No wreathed incense do we see upborne
Into the east, to meet the smiling day:
No crowd of nymphs soft voic'd and young, and gay,°
In woven baskets bringing ears of corn,
Roses, and pinks, and violets, to adorn
The shrine of Flora in her early May.
But there are left delights as high as these,
And I shall ever bless my destiny, 10
That in a time, when under pleasant trees
Pan is no longer sought, I feel a free
A leafy luxury, seeing I could please
With these poor offerings, a man like thee.

On seeing the Elgin Marbles

My spirit is too weak—mortality
Weighs heavily on me like unwilling sleep,
And each imagined pinnacle and steep
Of godlike hardship, tells me I must die
Like a sick eagle looking at the sky.°
Yet 'tis a gentle luxury to weep
That I have not the cloudy winds to keep,
Fresh for the opening of the morning's eye.
Such dim-conceived glories of the brain
Bring round the heart an undescribable feud;° 10
So do these wonders a most dizzy pain,
That mingles Grecian grandeur with the rude
Wasting of old time—with a billowy main—
A sun—a shadow of a magnitude.

On a Leander which Miss Reynolds my kind friend gave me

Come hither all sweet Maidens soberly
Downlooking—aye and with a chastened Light
Hid in the fringes of your eyelids white—
And meekly let your fair hands joined be.
So gentle are ye that ye could not see
Untouch'd a Victim of your beauty bright—
Sinking away to his young spirit's Night,
Sinking bewilder'd mid the dreary Sea—
'Tis young Leander toiling to his Death.
Nigh swooning he doth purse his weary Lips 10
For Hero's cheek and smiles against her smile.
O horrid dream—see how his body dips
Dead heavy—Arms and shoulders gleam awhile—
He's gone—upbubbles all his amourous breath—

On the Sea

It keeps eternal whisperings around
Desolate shores,—and with its mighty swell
Gluts twice ten thousand caverns,—till the spell°
Of Hecate leaves them their old shadowy sound.
Often 'tis in such gentle temper found,
That scarcely will the very smallest shell
Be lightly moved, from where it sometime fell,
When last the winds of heaven were unbound.
Ye, that have your eye-balls vex'd and tired,°
Feast them upon the wideness of the sea;— 10
Or are your hearts disturb'd with uproar rude,
Or fed too much with cloying melody,—
Sit ye near some old cavern's mouth and brood
Until ye start, as if the sea nymphs quired.°

'Unfelt unheard unseen'

Unfelt unheard unseen
I've left my little Queen
Her languid arms in silver slumber dying—
Ah! through their nestling touch
Who, who could tell how much
There is for Madness—cruel or complying?

Those faery lids how sleek
Those lips how moist—they speak
In ripest quiet shadows of sweet sounds;
Into my fancy's ear 10
Melting a Burden dear
How 'Love doth know no fulness nor no bounds'

True tender Monitors
I bend unto your laws
This sweetest day for dalliance was born;
So without more ado
I'll feel my heaven anew
For all the blushing of the hasty morn—

'You say you love; but with a voice'

You say you love; but with a voice
  Chaster than a nun's, who singeth
The soft vespers to herself
  While the chime-bell ringeth—
    O love me truly!

You say you love; but with a smile
  Cold as sunrise in September;
As you were Saint Cupid's Nun,
  And kept his weeks of Ember.
    O love me truly! 10

You say you love; but then your lips
  Coral tinted teach no blisses,
More than coral in the sea—
  They never pout for kisses—
    O love me truly!

You say you love; but then your hand
  No soft squeeze for squeeze returneth,
It is like a Statue's, dead,—
  While mine for passion burneth—
    O love me truly! 20

O breathe a word or two of fire!
  Smile, as if those words should burn me,
Squeeze as lovers should—O kiss
  And in thy heart inurn me—
    O love me truly!

'Hither hither Love'

Hither hither Love
'Tis a shady Mead.
Hither, hither Love
Let us feed and feed.

Hither hither sweet
'Tis a cowslip bed,
Hither hither sweet
'Tis with dew bespread.

Hither hither dear
By the breath of Life, 10
Hither hither dear
Be the summer's wife.

Though one moment's pleasure
In one moment flies,
Though the passion's treasure
In one moment dies;

Yet it has not pass'd
Think how near, how near,
And while it doth last
Think how dear how dear— 20

Hither hither hither
Love this boon has sent,
If I die and wither
I shall die content—

Endymion: A Poetic Romance

'The stretched metre of an antique song'°

INSCRIBED TO THE MEMORY OF THOMAS CHATTERTON

Preface

Knowing within myself the manner in which this Poem has been produced, it is not without a feeling of regret that I make it public.

What manner I mean, will be quite clear to the reader, who must soon perceive great inexperience, immaturity, and every error denoting a feverish attempt, rather than a deed accomplished. The two first books, and indeed the two last, I feel sensible are not of such completion as to warrant their passing the press; nor should they if I thought a year's castigation would do them any good;—it will not: the foundations are too sandy. It is just that this youngster should die away: a sad thought for me, if I had not some hope that while it is dwindling I may be plotting, and fitting myself for verses fit to live.

This may be speaking too presumptuously, and may deserve a punishment: but no feeling man will be forward to inflict it: he will leave me alone, with the conviction that there is not a fiercer hell than the failure in a great object. This is not written with the least atom of purpose to forestall criticisms of course, but from the desire I have to conciliate men who are competent to look, and who do look with a zealous eye, to the honour of English literature.

The imagination of a boy is healthy, and the mature imagination of a man is healthy; but there is a space of life between, in which the soul is in a ferment, the character undecided, the way of life uncertain, the ambition thick-sighted: thence proceeds mawkishness, and all the thousand bitters which those men I speak of must necessarily taste in going over the following pages.

I hope I have not in too late a day touched the beautiful mythology of Greece, and dulled its brightness: for I wish to try once more, before I bid it farewel.

Teignmouth, April 10, 1818

BOOK I

A thing of beauty is a joy for ever:
Its loveliness increases; it will never
Pass into nothingness; but still will keep
A bower quiet for us, and a sleep
Full of sweet dreams, and health, and quiet breathing.
Therefore, on every morrow, are we wreathing
A flowery band to bind us to the earth,
Spite of despondence, of the inhuman dearth
Of noble natures, of the gloomy days,
Of all the unhealthy and o'er-darkened ways 10
Made for our searching: yes, in spite of all,
Some shape of beauty moves away the pall
From our dark spirits. Such the sun, the moon,
Trees old, and young sprouting a shady boon
For simple sheep; and such are daffodils
With the green world they live in; and clear rills
That for themselves a cooling covert make
'Gainst the hot season; the mid forest brake,
Rich with a sprinkling of fair musk-rose blooms:
And such too is the grandeur of the dooms 20
We have imagined for the mighty dead;
All lovely tales that we have heard or read:
An endless fountain of immortal drink,
Pouring unto us from the heaven's brink.

Nor do we merely feel these essences
For one short hour; no, even as the trees
That whisper round a temple become soon
Dear as the temple's self, so does the moon,
The passion poesy, glories infinite,
Haunt us till they become a cheering light 30
Unto our souls, and bound to us so fast,
That, whether there be shine, or gloom o'ercast,
They alway must be with us, or we die.

Therefore, 'tis with full happiness that I
Will trace the story of Endymion.
The very music of the name has gone

Into my being, and each pleasant scene
Is growing fresh before me as the green
Of our own vallies: so I will begin
Now while I cannot hear the city's din; 40
Now while the early budders are just new,
And run in mazes of the youngest hue
About old forests; while the willow trails
Its delicate amber; and the dairy pails
Bring home increase of milk. And, as the year
Grows lush in juicy stalks, I'll smoothly steer
My little boat, for many quiet hours,°
With streams that deepen freshly into bowers.
Many and many a verse I hope to write,
Before the daisies, vermeil rimm'd and white, 50
Hide in deep herbage; and ere yet the bees
Hum about globes of clover and sweet peas,
I must be near the middle of my story.
O may no wintry season, bare and hoary,
See it half finished: but let Autumn bold,
With universal tinge of sober gold,
Be all about me when I make an end.
And now at once, adventuresome, I send
My herald thought into a wilderness:
There let its trumpet blow, and quickly dress 60
My uncertain path with green, that I may speed
Easily onward, thorough flowers and weed.

Upon the sides of Latmos was outspread
A mighty forest; for the moist earth fed
So plenteously all weed-hidden roots
Into o'er-hanging boughs, and precious fruits.
And it had gloomy shades, sequestered deep,
Where no man went; and if from shepherd's keep
A lamb strayed far a-down those inmost glens,
Never again saw he the happy pens 70
Whither his brethren, bleating with content,
Over the hills at every nightfall went.
Among the shepherds, 'twas believed ever,
That not one fleecy lamb which thus did sever
From the white flock, but pass'd unworried
By angry wolf, or pard with prying head,

Until it came to some unfooted plains
Where fed the herds of Pan: ay great his gains
Who thus one lamb did lose. Paths there were many,
Winding through palmy fern, and rushes fenny, 80
And ivy banks; all leading pleasantly
To a wide lawn, whence one could only see
Stems thronging all around between the swell
Of turf and slanting branches: who could tell
The freshness of the space of heaven above,
Edg'd round with dark tree tops? through which a dove
Would often beat its wings, and often too
A little cloud would move across the blue.

Full in the middle of this pleasantness
There stood a marble altar, with a tress 90
Of flowers budded newly; and the dew
Had taken fairy phantasies to strew
Daisies upon the sacred sward last eve,
And so the dawned light in pomp receive.
For 'twas the morn: Apollo's upward fire
Made every eastern cloud a silvery pyre
Of brightness so unsullied, that therein
A melancholy spirit well might win
Oblivion, and melt out his essence fine
Into the winds: rain-scented eglantine 100
Gave temperate sweets to that well-wooing sun;
The lark was lost in him; cold springs had run
To warm their chilliest bubbles in the grass;
Man's voice was on the mountains; and the mass
Of nature's lives and wonders puls'd tenfold,
To feel this sun-rise and its glories old.

Now while the silent workings of the dawn
Were busiest, into that self-same lawn
All suddenly, with joyful cries, there sped
A troop of little children garlanded; 110
Who gathering round the altar, seemed to pry
Earnestly round as wishing to espy
Some folk of holiday: nor had they waited
For many moments, ere their ears were sated

With a faint breath of music, which ev'n then
Fill'd out its voice, and died away again.
Within a little space again it gave
Its airy swellings, with a gentle wave,
To light-hung leaves, in smoothest echoes breaking
Through copse-clad vallies,—ere their death, o'ertaking 120
The surgy murmurs of the lonely sea.

And now, as deep into the wood as we
Might mark a lynx's eye, there glimmered light
Fair faces and a rush of garments white,
Plainer and plainer shewing, till at last
Into the widest alley they all past,
Making directly for the woodland altar.
O kindly muse! let not my weak tongue faulter
In telling of this goodly company,°
Of their old piety, and of their glee: 130
But let a portion of ethereal dew
Fall on my head, and presently unmew
My soul; that I may dare, in wayfaring,
To stammer where old Chaucer used to sing.

Leading the way, young damsels danced along,
Bearing the burden of a shepherd song;°
Each having a white wicker over brimm'd
With April's tender younglings: next, well trimm'd,
A crowd of shepherds with as sunburnt looks
As may be read of in Arcadian books; 140
Such as sat listening round Apollo's pipe,°
When the great deity, for earth too ripe,
Let his divinity o'er-flowing die
In music, through the vales of Thessaly:
Some idly trailed their sheep-hooks on the ground,
And some kept up a shrilly mellow sound
With ebon-tipped flutes: close after these,
Now coming from beneath the forest trees,
A venerable priest full soberly,
Begirt with ministring looks: alway his eye 150
Stedfast upon the matted turf he kept,
And after him his sacred vestments swept.

From his right hand there swung a vase, milk-white,
Of mingled wine, out-sparkling generous light;
And in his left he held a basket full
Of all sweet herbs that searching eye could cull:
Wild thyme, and valley-lilies whiter still
Than Leda's love, and cresses from the rill.°
His aged head, crowned with beechen wreath,
Seem'd like a poll of ivy in the teeth 160
Of winter hoar. Then came another crowd
Of shepherds, lifting in due time aloud
Their share of the ditty. After then appear'd,
Up-followed by a multitude that rear'd
Their voices to the clouds, a fair wrought car,
Easily rolling so as scarce to mar
The freedom of three steeds of dapple brown:
Who stood therein did seem of great renown
Among the throng. His youth was fully blown,
Shewing like Ganymede to manhood grown; 170
And, for those simple times, his garments were
A chieftain king's: beneath his breast, half bare,
Was hung a silver bugle, and between
His nervy knees there lay a boar-spear keen.
A smile was on his countenance; he seem'd,
To common lookers on, like one who dream'd
Of idleness in groves Elysian:
But there were some who feelingly could scan
A lurking trouble in his nether lip,
And see that oftentimes the reins would slip 180
Through his forgotten hands: then would they sigh,
And think of yellow leaves, of owlet's cry,
Of logs piled solemnly.—Ah, well-a-day,
Why should our young Endymion pine away!

Soon the assembly, in a circle rang'd
Stood silent round the shrine: each look was chang'd
To sudden veneration: women meek
Beckon'd their sons to silence; while each cheek
Of virgin bloom paled gently for slight fear.
Endymion too, without a forest peer, 190
Stood, wan, and pale, and with an awed face,
Among his brothers of the mountain chase.

In midst of all, the venerable priest
Eyed them with joy from greatest to the least,
And, after lifting up his aged hands,
Thus spake he: 'Men of Latmos! shepherd bands!
Whose care it is to guard a thousand flocks:
Whether descended from beneath the rocks
That overtop your mountains; whether come
From vallies where the pipe is never dumb; 200
Or from your swelling downs, where sweet air stirs
Blue hare-bells lightly, and where prickly furze
Buds lavish gold; or ye, whose precious charge
Nibble their fill at ocean's very marge,
Whose mellow reeds are touch'd with sounds forlorn
By the dim echoes of old Triton's horn:
Mothers and wives! who day by day prepare
The scrip, with needments, for the mountain air;°
And all ye gentle girls who foster up
Udderless lambs, and in a little cup 210
Will put choice honey for a favoured youth:
Yea, every one attend! for in good truth
Our vows are wanting to our great god Pan.
Are not our lowing heifers sleeker than
Night-swollen mushrooms? Are not our wide plains
Speckled with countless fleeces? Have not rains
Green'd over April's lap? No howling sad
Sickens our fearful ewes; and we have had
Great bounty from Endymion our lord.
The earth is glad: the merry lark has pour'd 220
His early song against yon breezy sky,
That spreads so clear o'er our solemnity.'

Thus ending, on the shrine he heap'd a spire
Of teeming sweets, enkindling sacred fire;
Anon he stain'd the thick and spongy sod
With wine, in honour of the shepherd-god.
Now while the earth was drinking it, and while
Bay leaves were crackling in the fragrant pile,
And gummy frankincense was sparkling bright
'Neath smothering parsley, and a hazy light 230
Spread greyly eastward, thus a chorus sang:

'O thou, whose mighty palace roof doth hang°

From jagged trunks, and overshadoweth
Eternal whispers, glooms, the birth, life, death
Of unseen flowers in heavy peacefulness;
Who lov'st to see the hamadryads dress
Their ruffled locks where meeting hazels darken;
And through whole solemn hours dost sit, and hearken
The dreary melody of bedded reeds—
In desolate places, where dank moisture breeds 240
The pipy hemlock to strange overgrowth;
Bethinking thee, how melancholy loth
Thou wast to lose fair Syrinx—do thou now,
By thy love's milky brow!
By all the trembling mazes that she ran,
Hear us, great Pan!

'O thou, for whose soul-soothing quiet, turtles
Passion their voices cooingly 'mong myrtles,°
What time thou wanderest at eventide
Through sunny meadows, that outskirt the side 250
Of thine enmossed realms: O thou, to whom
Broad leaved fig trees even now foredoom
Their ripen'd fruitage; yellow girted bees
Their golden honeycombs; our village leas
Their fairest blossom'd beans and poppied corn;
The chuckling linnet its five young unborn,
To sing for thee; low creeping strawberries
Their summer coolness; pent up butterflies
Their freckled wings; yea, the fresh budding year
All its completions—be quickly near, 260
By every wind that nods the mountain pine,
O forester divine!

'Thou, to whom every fawn and satyr flies
For willing service; whether to surprise
The squatted hare while in half sleeping fit;
Or upward ragged precipices flit
To save poor lambkins from the eagle's maw;
Or by mysterious enticement draw
Bewildered shepherds to their path again;
Or to tread breathless round the frothy main, 270
And gather up all fancifullest shells
For thee to tumble into Naiads' cells,

And, being hidden, laugh at their out-peeping;
Or to delight thee with fantastic leaping,
The while they pelt each other on the crown
With silvery oak apples, and fir cones brown—
By all the echoes that about thee ring,
Hear us, O satyr king!

'O Hearkener to the loud clapping shears,
While ever and anon to his shorn peers 280
A ram goes bleating: Winder of the horn,
When snouted wild-boars routing tender corn
Anger our huntsmen: Breather round our farms,°
To keep off mildews, and all weather harms:
Strange ministrant of undescribed sounds,°
That come a swooning over hollow grounds,
And wither drearily on barren moors:
Dread opener of the mysterious doors
Leading to universal knowledge—see,
Great son of Dryope, 290
The many that are come to pay their vows
With leaves about their brows!

'Be still the unimaginable lodge
For solitary thinkings; such as dodge
Conception to the very bourne of heaven,
Then leave the naked brain: be still the leaven,
That spreading in this dull and clodded earth
Gives it a touch ethereal—a new birth:
Be still a symbol of immensity;
A firmament reflected in a sea; 300
An element filling the space between;
An unknown—but no more: we humbly screen
With uplift hands our foreheads, lowly bending,
And giving out a shout most heaven rending,
Conjure thee to receive our humble pæan,
Upon thy Mount Lycean!'

Even while they brought the burden to a close,
A shout from the whole multitude arose,
That lingered in the air like dying rolls
Of abrupt thunder, when Ionian shoals 310

Of dolphins bob their noses through the brine.
Meantime, on shady levels, mossy fine,
Young companies nimbly began dancing
To the swift treble pipe, and humming string.
Aye, those fair living forms swam heavenly
To tunes forgotten—out of memory:
Fair creatures! whose young children's children bred
Thermopylæ its heroes—not yet dead,
But in old marbles ever beautiful.
High genitors, unconscious did they cull° 320
Time's sweet first-fruits—they danc'd to weariness,
And then in quiet circles did they press
The hillock turf, and caught the latter end
Of some strange history, potent to send
A young mind from its bodily tenement.
Or they might watch the quoit-pitchers, intent
On either side; pitying the sad death°
Of Hyacinthus, when the cruel breath
Of Zephyr slew him,—Zephyr penitent,
Who now, ere Phœbus mounts the firmament, 330
Fondles the flower amid the sobbing rain.
The archers too, upon a wider plain,
Beside the feathery whizzing of the shaft,
And the dull twanging bowstring, and the raft°
Branch down sweeping from a tall ash top,
Call'd up a thousand thoughts to envelope
Those who would watch. Perhaps, the trembling knee
And frantic gape of lonely Niobe,
Poor, lonely Niobe! when her lovely young
Were dead and gone, and her caressing tongue 340
Lay a lost thing upon her paly lip,
And very, very deadliness did nip
Her motherly cheeks. Arous'd from this sad mood
By one, who at a distance loud halloo'd,
Uplifting his strong bow into the air,
Many might after brighter visions stare:
After the Argonauts, in blind amaze°
Tossing about on Neptune's restless ways,
Until, from the horizon's vaulted side,
There shot a golden splendour far and wide, 350
Spangling those million poutings of the brine

With quivering ore: 'twas even an awful shine
From the exaltation of Apollo's bow;
A heavenly beacon in their dreary woe.
Who thus were ripe for high contemplating,
Might turn their steps towards the sober ring
Where sat Endymion and the aged priest
'Mong shepherds gone in eld, whose looks increas'd
The silvery setting of their mortal star.
There they discours'd upon the fragile bar 360
That keeps us from our homes ethereal;
And what our duties there: to nightly call
Vesper, the beauty-crest of summer weather;
To summon all the downiest clouds together
For the sun's purple couch; to emulate
In ministring the potent rule of fate
With speed of fire-tailed exhalations;
To tint her pallid cheek with bloom, who cons
Sweet poesy by moonlight: besides these,
A world of other unguess'd offices. 370
Anon they wander'd, by divine converse,
Into Elysium; vieing to rehearse
Each one his own anticipated bliss.
One felt heart-certain that he could not miss
His quick gone love, among fair blossom'd boughs,
Where every zephyr-sigh pouts, and endows
Her lips with music for the welcoming.
Another wish'd, mid that eternal spring,
To meet his rosy child, with feathery sails,
Sweeping, eye-earnestly, through almond vales: 380
Who, suddenly, should stoop through the smooth wind,
And with the balmiest leaves his temples bind;
And, ever after, through those regions be
His messenger, his little Mercury.
Some were athirst in soul to see again
Their fellow huntsmen o'er the wide champaign
In times long past; to sit with them, and talk
Of all the chances in their earthly walk;
Comparing, joyfully, their plenteous stores
Of happiness, to when upon the moors, 390
Benighted, close they huddled from the cold,
And shar'd their famish'd scrips. Thus all out-told°

Their fond imaginations,—saving him
Whose eyelids curtain'd up their jewels dim,
Endymion: yet hourly had he striven
To hide the cankering venom, that had riven
His fainting recollections. Now indeed
His senses had swoon'd off: he did not heed
The sudden silence, or the whispers low,
Or the old eyes dissolving at this woe, 400
Or anxious calls, or close of trembling palms,
Or maiden's sigh, that grief itself embalms:
But in the self-same fixed trance he kept,
Like one who on the earth had never stept,
Aye, even as dead-still as a marble man,°
Frozen in that old tale Arabian.

Who whispers him so pantingly and close?
Peona, his sweet sister: of all those,°
His friends, the dearest. Hushing signs she made,
And breath'd a sister's sorrow to persuade 410
A yielding up, a cradling on her care.
Her eloquence did breathe away the curse:
She led him, like some midnight spirit nurse
Of happy changes in emphatic dreams,
Along a path between two little streams,—
Guarding his forehead, with her round elbow,
From low-grown branches, and his footsteps slow
From stumbling over stumps and hillocks small;
Until they came to where these streamlets fall,
With mingled bubblings and a gentle rush, 420
Into a river, clear, brimful, and flush
With crystal mocking of the trees and sky.
A little shallop, floating there hard by,
Pointed its beak over the fringed bank;
And soon it lightly dipt, and rose, and sank,
And dipt again, with the young couple's weight,—
Peona guiding, through the water straight,
Towards a bowery island opposite;
Which gaining presently, she steered light
Into a shady, fresh, and ripply cove, 430
Where nested was an arbour, overwove
By many a summer's silent fingering;

To whose cool bosom she was used to bring
Her playmates, with their needle broidery,
And minstrel memories of times gone by.

So she was gently glad to see him laid
Under her favourite bower's quiet shade,
On her own couch, new made of flower leaves,
Dried carefully on the cooler side of sheaves
When last the sun his autumn tresses shook, 440
And the tann'd harvesters rich armfuls took.
Soon was he quieted to slumbrous rest:
But, ere it crept upon him, he had prest
Peona's busy hand against his lips,
And still, a sleeping, held her finger-tips
In tender pressure. And as a willow keeps
A patient watch over the stream that creeps
Windingly by it, so the quiet maid
Held her in peace: so that a whispering blade
Of grass, a wailful gnat, a bee bustling 450
Down in the blue-bells, or a wren light rustling
Among sere leaves and twigs, might all be heard.

O magic sleep! O comfortable bird,
That broodest o'er the troubled sea of the mind
Till it is hush'd and smooth! O unconfin'd
Restraint! imprisoned liberty! great key
To golden palaces, strange minstrelsy,
Fountains grotesque, new trees, bespangled caves,
Echoing grottos, full of tumbling waves
And moonlight; aye, to all the mazy world° 460
Of silvery enchantment!—who, upfurl'd
Beneath thy drowsy wing a triple hour,
But renovates and lives?—Thus, in the bower,
Endymion was calm'd to life again.
Opening his eyelids with a healthier brain,
He said: 'I feel this thine endearing love°
All through my bosom: thou art as a dove
Trembling its closed eyes and sleeked wings
About me; and the pearliest dew not brings
Such morning incense from the fields of May, 470
As do those brighter drops that twinkling stray

From those kind eyes,—the very home and haunt
Of sisterly affection. Can I want
Aught else, aught nearer heaven, than such tears?
Yet dry them up, in bidding hence all fears
That, any longer, I will pass my days
Alone and sad. No, I will once more raise
My voice upon the mountain-heights; once more
Make my horn parley from their foreheads hoar:
Again my trooping hounds their tongues shall loll 480
Around the breathed boar: again I'll poll
The fair-grown yew tree, for a chosen bow:
And, when the pleasant sun is getting low,
Again I'll linger in a sloping mead
To hear the speckled thrushes, and see feed
Our idle sheep. So be thou cheered sweet,
And, if thy lute is here, softly intreat
My soul to keep in its resolved course.'

Hereat Peona, in their silver source,
Shut her pure sorrow drops with glad exclaim, 490
And took a lute, from which there pulsing came
A lively prelude, fashioning the way
In which her voice should wander. 'Twas a lay
More subtle cadenced, more forest wild
Than Dryope's lone lulling of her child;
And nothing since has floated in the air
So mournful strange. Surely some influence rare
Went, spiritual, through the damsel's hand;
For still, with Delphic emphasis, she spann'd
The quick invisible strings, even though she saw 500
Endymion's spirit melt away and thaw
Before the deep intoxication.
But soon she came, with sudden burst, upon
Her self-possession—swung the lute aside,
And earnestly said: 'Brother, 'tis vain to hide
That thou dost know of things mysterious,
Immortal, starry; such alone could thus
Weigh down thy nature. Hast thou sinn'd in aught
Offensive to the heavenly powers? Caught
A Paphian dove upon a message sent?° 510
Thy deathful bow against some deer-herd bent,

Sacred to Dian? Haply, thou hast seen°
Her naked limbs among the alders green;
And that, alas! is death. No, I can trace
Something more high perplexing in thy face!'

Endymion look'd at her, and press'd her hand,
And said, 'Art thou so pale, who wast so bland°
And merry in our meadows? How is this?
Tell me thine ailment: tell me all amiss!—
Ah! thou hast been unhappy at the change 520
Wrought suddenly in me. What indeed more strange?
Or more complete to overwhelm surmise?
Ambition is no sluggard: 'tis no prize,
That toiling years would put within my grasp,
That I have sigh'd for: with so deadly gasp
No man e'er panted for a mortal love.
So all have set my heavier grief above
These things which happen. Rightly have they done:
I, who still saw the horizontal sun
Heave his broad shoulder o'er the edge of the world, 530
Out-facing Lucifer, and then had hurl'd
My spear aloft, as signal for the chace—
I, who, for very sport of heart, would race
With my own steed from Araby; pluck down
A vulture from his towery perching; frown
A lion into growling, loth retire—
To lose, at once, all my toil breeding fire,
And sink thus low! but I will ease my breast
Of secret grief, here in this bowery nest.

'This river does not see the naked sky, 540
Till it begins to progress silverly
Around the western border of the wood,
Whence, from a certain spot, its winding flood
Seems at the distance like a crescent moon:
And in that nook, the very pride of June,
Had I been used to pass my weary eves;
The rather for the sun unwilling leaves
So dear a picture of his sovereign power,
And I could witness his most kingly hour,
When he doth tighten up the golden reins, 550

And paces leisurely down amber plains
His snorting four. Now when his chariot last
Its beams against the zodiac-lion cast,
There blossom'd suddenly a magic bed
Of sacred ditamy, and poppies red:°
At which I wondered greatly, knowing well
That but one night had wrought this flowery spell;
And, sitting down close by, began to muse
What it might mean. Perhaps, thought I, Morpheus,
In passing here, his owlet pinions shook; 560
Or, it may be, ere matron Night uptook
Her ebon urn, young Mercury, by stealth,
Had dipt his rod in it: such garland wealth
Came not by common growth. Thus on I thought,
Until my head was dizzy and distraught.
Moreover, through the dancing poppies stole
A breeze, most softly lulling to my soul;
And shaping visions all about my sight
Of colours, wings, and bursts of spangly light;
The which became more strange, and strange, and dim, 570
And then were gulph'd in a tumultuous swim:
And then I fell asleep. Ah, can I tell
The enchantment that afterwards befel?
Yet it was but a dream: yet such a dream°
That never tongue, although it overteem
With mellow utterance, like a cavern spring,
Could figure out and to conception bring
All I beheld and felt. Methought I lay
Watching the zenith, where the milky way
Among the stars in virgin splendour pours; 580
And travelling my eye, until the doors
Of heaven appear'd to open for my flight,
I became loth and fearful to alight
From such high soaring by a downward glance:
So kept me stedfast in that airy trance,
Spreading imaginary pinions wide.
When, presently, the stars began to glide,
And faint away, before my eager view:
At which I sigh'd that I could not pursue,
And dropt my vision to the horizon's verge; 590
And lo! from opening clouds, I saw emerge

The loveliest moon, that ever silver'd o'er
A shell for Neptune's goblet: she did soar
So passionately bright, my dazzled soul
Commingling with her argent spheres did roll
Through clear and cloudy, even when she went
At last into a dark and vapoury tent—
Whereat, methought, the lidless-eyed train
Of planets all were in the blue again.
To commune with those orbs, once more I rais'd 600
My sight right upward: but it was quite dazed
By a bright something, sailing down apace,
Making me quickly veil my eyes and face:
Again I look'd, and, O ye deities,
Who from Olympus watch our destinies!
Whence that completed form of all completeness?
Whence came that high perfection of all sweetness?
Speak stubborn earth, and tell me where, O where
Has thou a symbol of her golden hair?
Not oat-sheaves drooping in the western sun; 610
Not—thy soft hand, fair sister! let me shun
Such follying before thee—yet she had,
Indeed, locks bright enough to make me mad;
And they were simply gordian'd up and braided,°
Leaving, in naked comeliness, unshaded,
Her pearl round ears, white neck, and orbed brow;
The which were blended in, I know not how,
With such a paradise of lips and eyes,
Blush-tinted cheeks, half smiles, and faintest sighs,
That, when I think thereon, my spirit clings 620
And plays about its fancy, till the stings
Of human neighbourhood envenom all.
Unto what awful power shall I call?
To what high fane?—Ah! see her hovering feet,
More bluely vein'd, more soft, more whitely sweet
Than those of sea-born Venus, when she rose
From out her cradle shell. The wind out-blows
Her scarf into a fluttering pavilion;
'Tis blue, and over-spangled with a million
Of little eyes, as though thou wert to shed, 630
Over the darkest, lushest blue-bell bed,
Handfuls of daisies.'—'Endymion, how strange!

Dream within dream!'—'She took an airy range,
And then, towards me, like a very maid,
Came blushing, waning, willing, and afraid,
And press'd me by the hand: Ah! 'twas too much;
Methought I fainted at the charmed touch,
Yet held my recollection, even as one
Who dives three fathoms where the waters run
Gurgling in beds of coral: for anon, 640
I felt upmounted in that region
Where falling stars dart their artillery forth,
And eagles struggle with the buffeting north
That balances the heavy meteor-stone;—
Felt too, I was not fearful, nor alone,
But lapp'd and lull'd along the dangerous sky.
Soon, as it seem'd, we left our journeying high,
And straightway into frightful eddies swoop'd;°
Such as ay muster where grey time has scoop'd
Huge dens and caverns in a mountain's side: 650
There hollow sounds arous'd me, and I sigh'd
To faint once more by looking on my bliss—
I was distracted; madly did I kiss
The wooing arms which held me, and did give
My eyes at once to death: but 'twas to live,
To take in draughts of life from the gold fount
Of kind and passionate looks; to count, and count°
The moments, by some greedy help that seem'd
A second self, that each might be redeem'd
And plunder'd of its load of blessedness. 660
Ah, desperate mortal! I ev'n dar'd to press
Her very cheek against my crowned lip,
And, at that moment, felt my body dip
Into a warmer air: a moment more,
Our feet were soft in flowers. There was store
Of newest joys upon that alp. Sometimes
A scent of violets, and blossoming limes,
Loiter'd around us; then of honey cells,
Made delicate from all white-flower bells;
And once, above the edges of our nest, 670
An arch face peep'd,—an Oread as I guess'd.

'Why did I dream that sleep o'er-power'd me

In midst of all this heaven? Why not see,
Far off, the shadows of his pinions dark,
And stare them from me? But no, like a spark
That needs must die, although its little beam
Reflects upon a diamond, my sweet dream
Fell into nothing—into stupid sleep.
And so it was, until a gentle creep,
A careful moving caught my waking ears, 680
And up I started: Ah! my sighs, my tears,
My clenched hands;—for lo! the poppies hung
Dew-dabbled on their stalks, the ouzel sung°
A heavy ditty, and the sullen day
Had chidden herald Hesperus away,
With leaden looks: the solitary breeze
Bluster'd, and slept, and its wild self did teaze
With wayward melancholy; and I thought,
Mark me, Peona! that sometimes it brought
Faint fare-thee-wells, and sigh-shrilled adieus!— 690
Away I wander'd—all the pleasant hues
Of heaven and earth had faded: deepest shades
Were deepest dungeons; heaths and sunny glades
Were full of pestilent light; our taintless rills
Seem'd sooty, and o'er-spread with upturn'd gills
Of dying fish; the vermeil rose had blown
In frightful scarlet, and its thorns out-grown
Like spiked aloe. If an innocent bird
Before my heedless footsteps stirr'd, and stirr'd
In little journeys, I beheld in it 700
A disguis'd demon, missioned to knit
My soul with under darkness; to entice
My stumblings down some monstrous precipice:
Therefore I eager followed, and did curse
The disappointment. Time, that aged nurse,
Rock'd me to patience. Now, thank gentle heaven!
These things, with all their comfortings, are given
To my down-sunken hours, and with thee,
Sweet sister, help to stem the ebbing sea
Of weary life.'

Thus ended he, and both 710
Sat silent: for the maid was very loth

To answer; feeling well that breathed words
Would all be lost, unheard, and vain as swords
Against the enchased crocodile, or leaps
Of grasshoppers against the sun. She weeps,
And wonders; struggles to devise some blame;
To put on such a look as would say, *Shame*
On this poor weakness! but, for all her strife,
She could as soon have crush'd away the life
From a sick dove. At length, to break the pause, 720
She said with trembling chance: 'Is this the cause?
This all? Yet it is strange, and sad, alas!
That one who through this middle earth should pass
Most like a sojourning demi-god, and leave
His name upon the harp-string, should achieve
No higher bard than simple maidenhood,°
Singing alone, and fearfully,—how the blood
Left his young cheek; and how he used to stray
He knew not where; and how he would say, *nay,*
If any said 'twas love: and yet 'twas love; 730
What could it be but love? How a ring-dove
Let fall a sprig of yew tree in his path;
And how he died: and then, that love doth scathe
The gentle heart, as northern blasts do roses;
And then the ballad of his sad life closes
With sighs, and an alas!—Endymion!
Be rather in the trumpet's mouth,—anon
Among the winds at large—that all may hearken!
Although, before the crystal heavens darken,
I watch and dote upon the silver lakes 740
Pictur'd in western cloudiness, that takes
The semblance of gold rocks and bright gold sands,
Islands, and creeks, and amber-fretted strands
With horses prancing o'er them, palaces
And towers of amethyst,—would I so tease
My pleasant days, because I could not mount
Into those regions? The Morphean fount
Of that fine element and visions, dreams,
And fitful whims of sleep are made of, streams
Into its airy channels with so subtle, 750
So thin a breathing, not the spider's shuttle,
Circled a million times within the space

Of a swallow's nest-door, could delay a trace,
A tinting of its quality: how light
Must dreams themselves be; seeing they're more slight
Than the mere nothing that engenders them!
Then wherefore sully the entrusted gem
Of high and noble life with thoughts so sick?
Why pierce high-fronted honour to the quick
For nothing but a dream?' Hereat the youth 760
Look'd up: a conflicting of shame and ruth
Was in his plaited brow: yet, his eyelids
Widened a little, as when Zephyr bids
A little breeze to creep between the fans
Of careless butterflies: amid his pains
He seem'd to taste a drop of manna-dew,
Full palatable; and a colour grew
Upon his cheek, while thus he lifeful spake.

'Peona! ever have I long'd to slake
My thirst for the world's praises: nothing base, 770
No merely slumberous phantasm, could unlace
The stubborn canvas for my voyage prepar'd—
Though now 'tis tatter'd; leaving my bark bar'd
And sullenly drifting: yet my higher hope
Is of too wide, too rainbow-large a scope,
To fret at myriads of earthly wrecks.°
Wherein lies happiness? In that which becks
Our ready minds to fellowship divine,
A fellowship with essence; till we shine,
Full alchemiz'd, and free of space. Behold 780
The clear religion of heaven! Fold
A rose leaf round thy finger's taperness,
And soothe thy lips: hist, when the airy stress
Of music's kiss impregnates the free winds,
And with a sympathetic touch unbinds
Eolian magic from their lucid wombs:°
The old songs waken from enclouded tombs;
Old ditties sigh above their father's grave;
Ghosts of melodious prophecyings rave
Round every spot where trod Apollo's foot; 790
Bronze clarions awake, and faintly bruit,
Where long ago a giant battle was;°

And, from the turf, a lullaby doth pass
In every place where infant Orpheus slept.
Feel we these things?—that moment have we stept
Into a sort of oneness, and our state
Is like a floating spirit's. But there are
Richer entanglements, enthralments far
More self-destroying, leading, by degrees,
To the chief intensity: the crown of these 800
Is made of love and friendship, and sits high
Upon the forehead of humanity.
All its more ponderous and bulky worth
Is friendship, whence there ever issues forth
A steady splendour; but at the tip-top,
There hangs by unseen film, an orbed drop
Of light, and that is love: its influence,
Thrown in our eyes, genders a novel sense,
At which we start and fret; till in the end,
Melting into its radiance, we blend, 810
Mingle, and so become a part of it,—
Nor with aught else can our souls interknit
So wingedly: when we combine therewith,
Life's self is nourish'd by its proper pith,
And we are nurtured like a pelican brood.°
Aye, so delicious is the unsating food,
That men, who might have tower'd in the van
Of all the congregated world, to fan
And winnow from the coming step of time
All chaff of custom, wipe away all slime 820
Left by men-slugs and human serpentry,
Have been content to let occasion die,
Whilst they did sleep in love's elysium.
And, truly, I would rather be struck dumb,
Than speak against this ardent listlessness:
For I have ever thought that it might bless
The world with benefits unknowingly;
As does the nightingale, upperched high,
And cloister'd among cool and bunched leaves—
She sings but to her love, nor e'er conceives 830
How tiptoe Night holds back her dark-grey hood.
Just so may love, although 'tis understood°
The mere commingling of passionate breath,

Produce more than our searching witnesseth:
What I know not: but who, of men, can tell
That flowers would bloom, or that green fruit would swell
To melting pulp, that fish would have bright mail,
The earth its dower of river, wood, and vale,
The meadows runnels, runnels pebble-stones,
The seed its harvest, or the lute its tones, 840
Tones ravishment, or ravishment its sweet,
If human souls did never kiss and greet?

'Now, if this earthly love has power to make
Men's being mortal, immortal; to shake
Ambition from their memories, and brim
Their measure of content; what merest whim,
Seems all this poor endeavour after fame,
To one, who keeps within his stedfast aim
A love immortal, an immortal too.
Look not so wilder'd; for these things are true, 850
And never can be born of atomies
That buzz about our slumbers, like brain-flies,
Leaving us fancy-sick. No, no, I'm sure,
My restless spirit never could endure
To brood so long upon one luxury,
Unless it did, though fearfully, espy
A hope beyond the shadow of a dream.
My sayings will the less obscured seem,
When I have told thee how my waking sight
Has made me scruple whether that same night 860
Was pass'd in dreaming. Hearken, sweet Peona!
Beyond the matron-temple of Latona,
Which we should see but for these darkening boughs,
Lies a deep hollow, from whose ragged brows
Bushes and trees do lean all round athwart,
And meet so nearly, that with wings outraught,
And spreaded tail, a vulture could not glide
Past them, but he must brush on every side.
Some moulder'd steps lead into this cool cell,
Far as the slabbed margin of a well, 870
Whose patient level peeps its crystal eye
Right upward, through the bushes, to the sky.
Oft have I brought thee flowers, on their stalks set

Like vestal primroses, but dark velvet
Edges them round, and they have golden pits:
'Twas there I got them, from the gaps and slits
In a mossy stone, that sometimes was my seat,
When all above was faint with mid-day heat.
And there in strife no burning thoughts to heed,
I'd bubble up the water through a reed; 880
So reaching back to boy-hood: make me ships
Of moulted feathers, touchwood, alder chips,
With leaves stuck in them; and the Neptune be
Of their petty ocean. Oftener, heavily,
When love-lorn hours had left me less a child,
I sat contemplating the figures wild
Of o'er-head clouds melting the mirror through.
Upon a day, while thus I watch'd, by flew
A cloudy Cupid, with his bow and quiver;
So plainly character'd, no breeze would shiver 890
The happy chance: so happy, I was fain
To follow it upon the open plain,
And, therefore, was just going; when, behold!
A wonder, fair as any I have told—
The same bright face I tasted in my sleep,
Smiling in the clear well. My heart did leap
Through the cool depth.—It moved as if to flee—
I started up, when lo! refreshfully,
There came upon my face, in plenteous showers,
Dew-drops, and dewy buds, and leaves, and flowers, 900
Wrapping all objects from my smothered sight,
Bathing my spirit in a new delight.
Aye, such a breathless honey-feel of bliss
Alone preserved me from the drear abyss
Of death, for the fair form had gone again.
Pleasure is oft a visitant; but pain
Clings cruelly to us, like the gnawing sloth°
On the deer's tender haunches: late, and loth,
'Tis scar'd away by slow returning pleasure.
How sickening, how dark the dreadful leisure 910
Of weary days, made deeper exquisite,
By a fore-knowledge of unslumbrous night!
Like sorrow came upon me, heavier still,
Than when I wander'd from the poppy hill:

And a whole age of lingering moments crept
Sluggishly by, ere more contentment swept
Away at once the deadly yellow spleen.
Yes, thrice have I this fair enchantment seen;
Once more been tortured with renewed life.
When last the wintry gusts gave over strife 920
With the conquering sun of spring, and left the skies
Warm and serene, but yet with moistened eyes
In pity of the shatter'd infant buds,—
That time thou didst adorn, with amber studs,°
My hunting cap, because I laugh'd and smil'd,
Chatted with thee, and many days exil'd
All torment from my breast;—'twas even then,
Straying about, yet, coop'd up in the den
Of helpless discontent,—hurling my lance
From place to place, and following at chance, 930
At last, by hap, through some young trees it struck,
And, plashing among bedded pebbles, stuck
In the middle of a brook,—whose silver ramble
Down twenty little falls, through reeds and bramble,
Tracing along, it brought me to a cave,
Whence it ran brightly forth, and white did lave
The nether sides of mossy stones and rock,—
'Mong which it gurgled blythe adieus, to mock
Its own sweet grief at parting. Overhead,
Hung a lush screen of drooping weeds, and spread 940
Thick, as to curtain up some wood-nymph's home.
"Ah! impious mortal, whither do I roam?"
Said I, low voic'd: "Ah, whither! 'Tis the grot
Of Proserpine, when Hell, obscure and hot,
Doth her resign; and where her tender hands
She dabbles, on the cool and sluicy sands:
Or 'tis the cell of Echo, where she sits,
And babbles thorough silence, till her wits
Are gone in tender madness, and anon,
Faints into sleep, with many a dying tone 950
Of sadness. O that she would take my vows,
And breathe them sighingly among the boughs,
To sue her gentle ears for whose fair head,
Daily, I pluck sweet flowerets from their bed,
And weave them dyingly—send honey-whispers

Round every leaf, that all those gentle lispers
May sigh my love unto her pitying!
O charitable echo! hear, and sing
This ditty to her!—tell her"—so I stay'd
My foolish tongue, and listening, half afraid, 960
Stood stupefied with my own empty folly,
And blushing for the freaks of melancholy.
Salt tears were coming, when I heard my name
Most fondly lipp'd, and then these accents came:
"Endymion! the cave is secreter
Than the isle of Delos. Echo hence shall stir
No sighs but sigh-warm kisses, or light noise
Of thy combing hand, the while it travelling cloys
And trembles through my labyrinthine hair."
At that oppress'd I hurried in.—Ah! where 970
Are those swift moments? Whither are they fled?
I'll smile no more, Peona; nor will wed
Sorrow the way to death; but patiently
Bear up against it: so farewel, sad sigh;
And come instead demurest meditation,
To occupy me wholly, and to fashion
My pilgrimage for the world's dusky brink.
No more will I count over, link by link,
My chain of grief: no longer strive to find
A half-forgetfulness in mountain wind 980
Blustering about my ears: aye, thou shalt see,
Dearest of sisters, what my life shall be;
What a calm round of hours shall make my days.
There is a paly flame of hope that plays
Where'er I look: but yet, I'll say 'tis naught—
And here I bid it die. Have not I caught,
Already, a more healthy countenance?
By this the sun is setting; we may chance
Meet some of our near-dwellers with my car.'

This said, he rose, faint-smiling like a star 990
Through autumn mists, and took Peona's hand:
They stept into the boat, and launch'd from land.

BOOK II

O sovereign power of love! O grief! O balm!
All records, saving thine, come cool, and calm,
And shadowy, through the mist of passed years:
For others, good or bad, hatred and tears
Have become indolent; but touching thine,
One sigh doth echo, one poor sob doth pine,
One kiss brings honey-dew from buried days.
The woes of Troy, towers smothering o'er their blaze,
Stiff-holden shields, far-piercing spears, keen blades,
Struggling, and blood, and shrieks—all dimly fades 10
Into some backward corner of the brain;
Yet, in our very souls, we feel amain
The close of Troilus and Cressid sweet.°
Hence, pageant history! hence, gilded cheat!
Swart planet in the universe of deeds!
Wide sea, that one continuous murmur breeds
Along the pebbled shore of memory!
Many old rotten-timber'd boats there be
Upon thy vaporous bosom, magnified
To goodly vessels; many a sail of pride, 20
And golden keel'd, is left unlaunch'd and dry.
But wherefore this? What care, though owl did fly
About the great Athenian admiral's mast?°
What care, though striding Alexander past°
The Indus with his Macedonian numbers?
Though old Ulysses tortured from his slumbers
The glutted Cyclops, what care?—Juliet leaning
Amid her window-flowers,—sighing,—weaning
Tenderly her fancy from its maiden snow,
Doth more avail than these: the silver flow 30
Of Hero's tears, the swoon of Imogen,°
Fair Pastorella in the bandit's den,
Are things to brood on with more ardency
Than the death-day of empires. Fearfully
Must such conviction come upon his head,
Who, thus far, discontent, has dared to tread,
Without one muse's smile, or kind behest,
The path of love and poesy. But rest,°
In chaffing restlessness, is yet more drear

Than to be crush'd, in striving to uprear 40
Love's standard on the battlements of song.
So once more days and nights aid me along,
Like legion'd soldiers.

Brain-sick shepherd prince,
What promise hast thou faithful guarded since
The day of sacrifice? Or, have new sorrows
Come with the constant dawn upon thy morrows?
Alas! 'tis his old grief. For many days,
Has he been wandering in uncertain ways:
Through wilderness, and woods of mossed oaks:
Counting his woe-worn minutes, by the strokes 50
Of the lone woodcutter; and listening still,
Hour after hour, to each lush-leav'd rill.
Now he is sitting by a shady spring,
And elbow-deep with feverous fingering
Stems the upbursting cold: a wild rose tree
Pavilions him in bloom, and he doth see
A bud which snares his fancy: lo! but now
He plucks it, dips its stalk in the water: how!
It swells, it buds, it flowers beneath his sight;
And, in the middle, there is softly pight° 60
A golden butterfly; upon whose wings
There must be surely character'd strange things,
For with wide eye he wonders, and smiles oft.

Lightly this little herald flew aloft,
Follow'd by glad Endymion's clasped hands:
Onward it flies. From languor's sullen bands
His limbs are loos'd, and eager, on he hies
Dazzled to trace it in the sunny skies.
It seem'd he flew, the way so easy was;
And like a new-born spirit did he pass 70
Through the green evening quiet in the sun,
O'er many a heath, through many a woodland dun,
Through buried paths, where sleepy twilight dreams
The summer time away. One track unseams
A wooded cleft, and, far away, the blue
Of ocean fades upon him; then, anew,
He sinks adown a solitary glen,

Where there was never sound of mortal men,
Saving, perhaps, some snow-light cadences
Melting to silence, when upon the breeze 80
Some holy bark let forth an anthem sweet,
To cheer itself to Delphi. Still his feet
Went swift beneath the merry-winged guide,
Until it reached a splashing fountain's side
That, near a cavern's mouth, for ever pour'd
Unto the temperate air: then high it soar'd,
And, downward, suddenly began to dip,
As if, athirst with so much toil, 'twould sip
The crystal spout-head: so it did, with touch
Most delicate, as though afraid to smutch° 90
Even with mealy gold the waters clear.
But, at that very touch, to disappear
So fairy-quick, was strange! Bewildered,
Endymion sought around, and shook each bed
Of covert flowers in vain; and then he flung
Himself along the grass. What gentle tongue,
What whisperer disturb'd his gloomy rest?
It was a nymph uprisen to the breast
In the fountain's pebbly margin, and she stood
'Mong lilies, like the youngest of the brood. 100
To him her dripping hand she softly kist,
And anxiously began to plait and twist
Her ringlets round her fingers, saying: 'Youth!
Too long, alas, hast thou starv'd on the ruth,
The bitterness of love: too long indeed,
Seeing thou art so gentle. Could I weed
Thy soul of care, by heavens, I would offer
All the bright riches of my crystal coffer
To Amphitrite; all my clear-eyed fish,
Golden, or rainbow-sided, or purplish, 110
Vermilion-tail'd, or finn'd with silvery gauze;
Yea, or my veined pebble-floor, that draws
A virgin light to the deep; my grotto-sands
Tawny and gold, ooz'd slowly from far lands
By my diligent springs; my level lilies, shells,
My charming rod, my potent river spells;
Yes, every thing, even to the pearly cup
Meander gave me,—for I bubbled up°

To fainting creatures in a desert wild.
But woe is me, I am but as a child 120
To gladden thee; and all I dare to say,
Is, that I pity thee; that on this day
I've been thy guide; that thou must wander far
In other regions, past the scanty bar
To mortal steps, before thou canst be ta'en
From every wasting sigh, from every pain,
Into the gentle bosom of thy love.
Why it is thus, one knows in heaven above:
But, a poor Naiad, I guess not. Farewel!
I have a ditty for my hollow cell.' 130

Hereat, she vanished from Endymion's gaze,
Who brooded o'er the water in amaze:
The dashing fount pour'd on, and where its pool
Lay, half asleep, in grass and rushes cool,
Quick waterflies and gnats were sporting still,
And fish were dimpling, as if good nor ill
Had fallen out that hour. The wanderer,
Holding his forehead, to keep off the burr°
Of smothering fancies, patiently sat down;
And, while beneath the evening's sleepy frown 140
Glow-worms began to trim their starry lamps,
Thus breath'd he to himself: 'Whoso encamps
To take a fancied city of delight,
O what a wretch is he! and when 'tis his,
After long toil and travelling, to miss
The kernel of his hopes, how more than vile:
Yet, for him there's refreshment even in toil;
Another city doth he set about,
Free from the smallest pebble-bead of doubt
That he will seize on trickling honey-combs: 150
Alas, he finds them dry; and then he foams,
And onward to another city speeds.
But this is human life: the war, the deeds,
The disappointment, the anxiety,
Imagination's struggles, far and nigh,
All human; bearing in themselves this good,
That they are still the air, the subtle food,
To make us feel existence, and to shew

How quiet death is. Where soil is men grow,
Whether to weeds or flowers; but for me, 160
There is no depth to strike in: I can see
Nought earthly worth my compassing; so stand
Upon a misty, jutting head of land—
Alone? No, no; and by the Orphean lute,
When mad Eurydice is listening to't;
I'd rather stand upon this misty peak,
With not a thing to sigh for, or to seek,
But the soft shadow of my thrice-seen love,
Than be—I care not what. O meekest dove
Of heaven! O Cynthia, ten-times bright and fair! 170
From thy blue throne, now filling all the air,
Glance but one little beam of temper'd light
Into my bosom, that the dreadful might
And tyranny of love be somewhat scar'd!
Yet do not so, sweet queen; one torment spar'd,
Would give a pang to jealous misery,
Worse than the torment's self: but rather tie
Large wings upon my shoulders, and point out
My love's far dwelling. Though the playful rout
Of Cupids shun thee, too divine art thou, 180
Too keen in beauty, for thy silver prow
Not to have dipp'd in love's most gentle stream.
O be propitious, nor severely deem
My madness impious; for, by all the stars
That tend thy bidding, I do think the bars
That kept my spirit in are burst—that I°
Am sailing with thee through the dizzy sky!
How beautiful thou art! The world how deep!
How tremulous-dazzlingly the wheels sweep
Around their axle! Then these gleaming reins, 190
How lithe! When this thy chariot attains
Its airy goal, haply some bower veils
Those twilight eyes? Those eyes!—my spirit fails—
Dear goddess, help! or the wide-gaping air
Will gulph me—help!'—At this with madden'd stare,
And lifted hands, and trembling lips he stood;
Like old Deucalion mountain'd o'er the flood,
Or blind Orion hungry for the morn.
And, but from the deep cavern there was borne

A voice, he had been froze to senseless stone; 200
Nor sigh of his, nor plaint, nor passion'd moan
Had more been heard. Thus swell'd it forth: 'Descend,
Young mountaineer! descend where alleys bend
Into the sparry hollows of the world!
Oft hast thou seen bolts of the thunder hurl'd
As from thy threshold; day by day hast been
A little lower than the chilly sheen
Of icy pinnacles, and dipp'dst thine arms
Into the deadening ether that still charms
Their marble being: now, as deep profound 210
As those are high, descend! He ne'er is crown'd°
With immortality, who fears to follow
Where airy voices lead: so through the hollow,
The silent mysteries of earth, descend!'

He heard but the last words, nor could contend
One moment in reflection: for he fled
Into the fearful deep, to hide his head
From the clear moon, the trees, and coming madness.

'Twas far too strange, and wonderful for sadness;
Sharpening, by degrees, his appetite 220
To dive into the deepest. Dark, nor light,
The region; nor bright, nor sombre wholly,
But mingled up; a gleaming melancholy;
A dusky empire and its diadems;
One faint eternal eventide of gems.
Aye, millions sparkled on a vein of gold,
Along whose track the prince quick footsteps told,
With all its lines abrupt and angular:
Out-shooting sometimes, like a meteor-star,
Through a vast antre; then the metal woof,° 230
Like Vulcan's rainbow, with some monstrous roof
Curves hugely: now, far in the deep abyss,
It seems an angry lightning, and doth hiss
Fancy into belief: anon it leads
Through winding passages, where sameness breeds
Vexing conceptions of some sudden change;
Whether to silver grots, or giant range
Of sapphire columns, or fantastic bridge

Athwart a flood of crystal. On a ridge
Now fareth he, that o'er the vast beneath 240
Towers like an ocean-cliff, and whence he seeth
A hundred waterfalls, whose voices come
But as the murmuring surge. Chilly and numb
His bosom grew, when first he, far away,
Descried an orbed diamond, set to fray
Old darkness from his throne: 'twas like the sun
Uprisen o'er chaos: and with such a stun
Came the amazement, that, absorb'd in it,
He saw not fiercer wonders—past the wit
Of any spirit to tell, but one of those 250
Who, when this planet's sphering time doth close,°
Will be its high remembrancers: who they?
The mighty ones who have made eternal day
For Greece and England. While astonishment
With deep-drawn sighs was quieting, he went
Into a marble gallery, passing through
A mimic temple, so complete and true
In sacred custom, that he well nigh fear'd
To search it inwards; whence far off appear'd,
Through a long pillar'd vista, a fair shrine, 260
And, just beyond, on light tiptoe divine,
A quiver'd Dian. Stepping awfully,
The youth approach'd; oft turning his veil'd eye
Down sidelong aisles, and into niches old.
And when, more near against the marble cold
He had touch'd his forehead, he began to thread
All courts and passages, where silence dead
Rous'd by his whispering footsteps murmured faint:
And long he travers'd to and fro, to acquaint
Himself with every mystery, and awe; 270
Till, weary, he sat down before the maw
Of a wide outlet, fathomless and dim,
To wild uncertainty and shadows grim.
There, when new wonders ceas'd to float before,
And thoughts of self came on, how crude and sore
The journey homeward to habitual self!
A mad-pursuing of the fog-born elf,°
Whose flitting lantern, through rude nettle-briar,
Cheats us into a swamp, into a fire,

Into the bosom of a hated thing. 280

What misery most drowningly doth sing
In lone Endymion's ear, now he has raught°
The goal of consciousness? Ah, 'tis the thought,
The deadly feel of solitude: for lo!
He cannot see the heavens, nor the flow
Of rivers, nor hill-flowers running wild
In pink and purple chequer, nor, up-pil'd,°
The cloudy rack slow journeying in the west.
Like herded elephants; nor felt, nor prest
Cool grass, nor tasted the fresh slumberous air; 290
But far from such companionship to wear
An unknown time, surcharg'd with grief, away,
Was now his lot. And must he patient stay,
Tracing fantastic figures with his spear?
'No!' exclaimed he, 'why should I tarry here?'
No! loudly echoed times innumerable.
At which he straightway started, and 'gan tell
His paces back into the temple's chief;
Warming and glowing strong in the belief
Of help from Dian: so that when again 300
He caught her airy form, thus did he plain,
Moving more near the while. 'O Haunter chaste
Of river sides, and woods, and heathy waste,
Where with thy silver bow and arrows keen
Art thou now forested? O woodland Queen,
What smoothest air thy smoother forehead woos?
Where dost thou listen to the wide halloos
Of thy disparted nymphs? Through what dark tree
Glimmers thy crescent? Wheresoe'er it be,
'Tis in the breath of heaven: thou dost taste 310
Freedom as none can taste it, nor dost waste
Thy loveliness in dismal elements;
But, finding in our green earth sweet contents,
There livest blissfully. Ah, if to thee
It feels Elysian, how rich to me,
An exil'd mortal, sounds its pleasant name!
Within my breast there lives a choking flame—
O let me cool it the zephyr-boughs among!
A homeward fever parches up my tongue—

O let me slake it at the running springs! 320
Upon my ear a noisy nothing rings—
O let me once more hear the linnet's note!
Before mine eyes thick films and shadows float—
O let me 'noint them with the heaven's light!
Dost thou now lave thy feet and ankles white?
O think now sweet to me the freshening sluice!
Dost thou now please thy thirst with berry-juice?
O think how this dry palate would rejoice!
If in soft slumber thou dost hear my voice,
O think how I should love a bed of flowers!— 330
Young goddess! let me see my native bowers!
Deliver me from this rapacious deep!'

Thus ending loudly, as he would o'erleap
His destiny, alert he stood: but when
Obstinate silence came heavily again,
Feeling about for its old couch of space
And airy cradle, lowly bow'd his face
Desponding, o'er the marble floor's cold thrill.
But 'twas not long; for, sweeter than the rill
To its old channel, or a swollen tide 340
To margin sallows, were the leaves he spied,
And flowers, and wreaths, and ready myrtle crowns
Up heaping through the slab: refreshment drowns
Itself, and strives its own delights to hide—
Nor in one spot alone; the floral pride
In a long whispering birth enchanted grew
Before his footsteps; as when heav'd anew
Old ocean rolls a lengthened wave to the shore,
Down whose green back the short-liv'd foam, all hoar,
Bursts gradual, with a wayward indolence. 350

Increasing still in heart, and pleasant sense,
Upon his fairy journey on he hastes;
So anxious for the end, he scarcely wastes
One moment with his hand among the sweets:
Onward he goes—he stops—his bosom beats
As plainly in his ear, as the faint charm
Of which the throbs were born. This still alarm,
This sleepy music, forc'd him walk tiptoe:

For it came more softly than the east could blow
Arion's magic to the Atlantic isles; 360
Or than the west, made jealous by the smiles
Of thron'd Apollo, could breathe back the lyre°
To seas Ionian and Tyrian.

O did he ever live, that lonely man,
Who lov'd—and music slew not? 'Tis the pest
Of love, that fairest joys give most unrest;
That things of delicate and tenderest worth
Are swallow'd all, and made a seared dearth,
By one consuming flame: it doth immerse
And suffocate true blessings in a curse. 370
Half-happy, by comparison of bliss,
Is miserable. 'Twas even so with this
Dew-dropping melody, in the Carian's ear;°
First heaven, then hell, and then forgotten clear,
Vanish'd in elemental passion.

And down some swart abysm he had gone,
Had not a heavenly guide benignant led
To where thick myrtle branches, 'gainst his head
Brushing, awakened: then the sounds again
Went noiseless as a passing noontide rain 380
Over a bower, where little space he stood;
For, as the sunset peeps into a wood,
So saw he panting light, and towards it went
Through winding alleys; and lo, wonderment!
Upon soft verdure saw, one here, one there,
Cupids a slumbering on their pinions fair.

After a thousand mazes overgone,
At last, with sudden step, he came upon
A chamber, myrtle wall'd, embowered high,°
Full of light, incense, tender minstrelsy, 390
And more of beautiful and strange beside:
For on a silken couch of rosy pride,
In midst of all, there lay a sleeping youth
Of fondest beauty; fonder, in fair sooth,
Than sighs could fathom, or contentment reach:
And coverlids gold-tinted like the peach,
Or ripe October's faded marigolds,

Fell sleek about him in a thousand folds—
Not hiding up an Apollonian curve
Of neck and shoulder, nor the tenting swerve 400
Of knee from knee, nor ankles pointing light;
But rather, giving them to the filled sight
Officiously. Sideway his face repos'd
On one white arm, and tenderly unclos'd,
By tenderest pressure, a faint damask mouth
To slumbery pout; just as the morning south
Disparts a dew-lipp'd rose. Above his head,°
Four lily stalks did their white honours wed
To make a coronal; and round him grew
All tendrils green, of every bloom and hue, 410
Together intertwin'd and trammel'd fresh:
The vine of glossy sprout; the ivy mesh,
Shading its Ethiop berries; and woodbine,
Of velvet leaves and bugle-blooms divine;
Convolvulus in streaked vases flush;
The creeper, mellowing for an autumn blush;
And virgin's bower, trailing airily;°
With others of the sisterhood. Hard by,
Stood serene Cupids watching silently.
One, kneeling to a lyre, touch'd the strings, 420
Muffling to death the pathos with his wings;
And, ever and anon, uprose to look
At the youth's slumber; while another took
A willow-bough, distilling odorous dew,
And shook it on his hair; another flew
In through the woven roof, and fluttering-wise
Rain'd violets upon his sleeping eyes.

At these enchantments, and yet many more,
The breathless Latmian wonder'd o'er and o'er; 430
Until, impatient in embarrassment,
He forthright pass'd, and lightly treading went
To that same feather'd lyrist, who straightway,
Smiling, thus whisper'd: 'Though from upper day
Thou art a wanderer, and thy presence here
Might seem unholy, be of happy cheer!
For 'tis the nicest touch of human honour,
When some ethereal and high-favouring donor

Presents immortal bowers to mortal sense;
As now 'tis done to thee, Endymion. Hence
Was I in no wise startled. So recline 440
Upon these living flowers. Here is wine,
Alive with sparkles—never, I aver,
Since Ariadne was a vintager,
So cool a purple: taste these juicy pears,
Sent me by sad Vertumnus, when his fears
Were high about Pomona: here is cream,
Deepening to richness from a snowy gleam;
Sweeter than that nurse Amalthea skimm'd
For the boy Jupiter: and here, undimm'd
By any touch, a bunch of blooming plums 450
Ready to melt between an infant's gums:
And here is manna pick'd from Syrian trees,
In starlight, by the three Hesperides.
Feast on, and meanwhile I will let thee know
Of all these things around us.' He did so,
Still brooding o'er the cadence of his lyre;
And thus: 'I need not any hearing tire
By telling how the sea-born goddess pin'd°
For a mortal youth, and how she strove to bind
Him all in all unto her doting self. 460
Who would not be so prison'd? but, fond elf,
He was content to let her amorous plea
Faint through his careless arms; content to see
An unseiz'd heaven dying at his feet;
Content, O fool! to make a cold retreat,
When on the pleasant grass such love, lovelorn,
Lay sorrowing; when every tear was born
Of diverse passion; when her lips and eyes
Were clos'd in sullen moisture, and quick sighs
Came vex'd and pettish through her nostrils small. 470
Hush! no exclaim—yet, justly mightst thou call
Curses upon his head.—I was half glad,
But my poor mistress went distract and mad,
When the boar tusk'd him: so away she flew
To Jove's high throne, and by her plainings drew
Immortal tear-drops down the thunderer's beard;
Whereon, it was decreed he should be rear'd
Each summer time to life. Lo! this is he,

That same Adonis, safe in the privacy
Of this still region all his winter-sleep. 480
Aye, sleep; for when our love-sick queen did weep
Over his waned corse, the tremulous shower
Heal'd up the wound, and, with a balmy power,
Medicined death to lengthened drowsiness:
The which she fills with visions, and doth dress
In all this quiet luxury; and hath set
Us young immortals, without any let,
To watch his slumber through. 'Tis well nigh pass'd,
Even to a moment's filling up, and fast
She scuds with summer breezes, to pant through 490
The first long kiss, warm firstling, to renew
Embower'd sports in Cytherea's isle.
Look! how those winged listeners all this while
Stand anxious: see! behold!'—This clamant word
Broke through the careful silence; for they heard
A rustling noise of leaves, and out there flutter'd
Pigeons and doves: Adonis something mutter'd,
The while one hand, that erst upon his thigh
Lay dormant, mov'd convuls'd and gradually
Up to his forehead. Then there was a hum 500
Of sudden voices, echoing, 'Come! come!
Arise! awake! Clear summer has forth walk'd
Unto the clover-sward, and she has talk'd
Full soothingly to every nested finch:
Rise, Cupids! or we'll give the blue-bell pinch
To your dimpled arms. Once more sweet life begin!'
At this, from every side they hurried in,
Rubbing their sleepy eyes with lazy wrists,
And doubling over head their little fists
In backward yawns. But all were soon alive: 510
For as delicious wine doth, sparkling, dive
In nectar'd clouds and curls through water fair,
So from the arbour roof down swell'd an air
Odorous and enlivening; making all
To laugh, and play, and sing, and loudly call
For their sweet queen: when lo! the wreathed green
Disparted, and far upward could be seen
Blue heaven, and a silver car, air-borne,
Whose silent wheels, fresh wet from clouds of morn,

Spun off a drizzling dew,—which falling chill 520
On soft Adonis' shoulders, made him still
Nestle and turn uneasily about.
Soon were the white doves plain, with necks stretch'd out,
And silken traces tighten'd in descent;
And soon, returning from love's banishment,
Queen Venus leaning downward open arm'd:
Her shadow fell upon his breast, and charm'd
A tumult to his heart, and a new life
Into his eyes. Ah, miserable strife,
But for her comforting! unhappy sight, 530
But meeting her blue orbs! Who, who can write
Of these first minutes? The unchariest muse
To embracements warm as theirs makes coy excuse.

O it has ruffled every spirit there,
Saving Love's self, who stands superb to share°
The general gladness: awfully he stands;
A sovereign quell is in his waving hands;°
No sight can bear the lightning of his bow;
His quiver is mysterious, none can know
What themselves think of it; from forth his eyes 540
There darts strange light of varied hues and dyes:
A scowl is sometimes on his brow, but who
Look full upon it feel anon the blue
Of his fair eyes run liquid through their souls.
Endymion feels it, and no more controls
The burning prayer within him: so, bent low,
He had begun a plaining of his woe.
But Venus, bending forward, said: 'My child,
Favour this gentle youth; his days are wild
With love—he—but alas! too well I see 550
Thou know'st the deepness of his misery.
Ah, smile not so, my son: I tell thee true,
That when through heavy hours I used to rue
The endless sleep of this new-born Adon',
This stranger ay I pitied. For upon
A dreary morning once I fled away
Into the breezy clouds, to weep and pray
For this my love: for vexing Mars had teaz'd
Me even to tears: thence, when a little eas'd,

Down-looking, vacant, through a hazy wood, 560
I saw this youth as he despairing stood:
Those some dark curls blown vagrant in the wind;
Those same full fringed lids a constant blind
Over his sullen eyes: I saw him throw
Himself on wither'd leaves, even as though
Death had come sudden; for no jot he mov'd,
Yet mutter'd wildly. I could hear he lov'd
Some fair immortal, and that his embrace
Had zoned her through the night. There is no trace°
Of this in heaven: I have mark'd each cheek, 570
And find it is the vainest thing to seek;
And that of all things 'tis kept secretest.
Endymion! one day thou wilt be blest:
So still obey the guiding hand that fends
Thee safely through these wonders for sweet ends.
'Tis a concealment needful in extreme;
And if I guess'd not so, the sunny beam
Thou shouldst mount up to with me. Now adieu!
Here must we leave thee.'—At these words up flew°
The impatient doves, up rose the floating car, 580
Up went the hum celestial. High afar
The Latmian saw them minish into nought;
And, when all were clear vanish'd, still he caught
A vivid lightning from that dreadful bow.
When all was darkened, with Etnean throe
The earth clos'd—gave a solitary moan—
And left him once again in twilight lone.

He did not rave, he did not stare aghast,
For all those visions were o'ergone, and past,
And he in loneliness: he felt assur'd 590
Of happy times, when all he had endur'd
Would seem a feather to the mighty prize.
So, with unusual gladness, on he hies
Through caves, and palaces of mottled ore,
Gold dome, and crystal wall, and turquois floor,
Black polish'd porticos of awful shade,
And, at the last, a diamond balustrade,
Leading afar past wild magnificence,
Spiral through ruggedest loopholes, and thence

Stretching across a void, then guiding o'er 600
Enormous chasms, where, all foam and roar,
Streams subterranean tease their granite beds;
Then heighten'd just above the silvery heads
Of a thousand fountains, so that he could dash
The waters with his spear; but at the splash,
Done heedlessly, those spouting columns rose
Sudden a poplar's height, and 'gan to enclose
His diamond path with fretwork, streaming round
Alive, and dazzling cool, and with a sound,
Haply, like dolphin tumults, when sweet shells 610
Welcome the float of Thetis. Long he dwells
On this delight; for, every minute's space,
The streams with changed magic interlace:
Sometimes like delicatest lattices,
Cover'd with crystal vines; then weeping trees,
Moving about as in a gentle wind,
Which, in a wink, to watery gauze refin'd,
Pour'd into shapes of curtain'd canopies,
Spangled, and rich with liquid broideries
Of flowers, peacocks, swans, and naiads fair. 620
Swifter than lightning went these wonders rare;
And then the water, into stubborn streams
Collecting, mimick'd the wrought oaken beams,
Pillars, and frieze, and high fantastic roof,
Of those dusk places in times far aloof
Cathedrals call'd. He bade a loth farewel
To these founts Protean, passing gulph, and dell,
And torrent, and ten thousand jutting shapes,
Half seen through deepest gloom, and griesly gapes,
Blackening on every side, and overhead 630
A vaulted dome like Heaven's, far bespread
With starlight gems: aye, all so huge and strange,
The solitary felt a hurried change
Working within him into something dreary,—
Vex'd like a morning eagle, lost, and weary,
And purblind amid foggy, midnight wolds.
But he revives at once: for who beholds
New sudden things, nor casts his mental slough?
Forth from a rugged arch, in the dusk below,
Came mother Cybele! alone—alone— 640

In sombre chariot; dark foldings thrown
About her majesty, and front death-pale,
With turrets crown'd. Four maned lions hale
The sluggish wheels; solemn their toothed maws,
Their surly eyes brow-hidden, heavy paws
Uplifted drowsily, and nervy tails
Cowering their tawny brushes. Silent sails
This shadowy queen athwart, and faints away
In another gloomy arch.

Wherefore delay,
Young traveller, in such a mournful place? 650
Art thou wayworn, or canst not further trace
The diamond path? And does it indeed end
Abrupt in middle air? Yet earthward bend
Thy forehead, and to Jupiter cloud-borne
Call ardently! He was indeed wayworn;
Abrupt, in middle air, his way was lost;
To cloud-borne Jove he bowed, and there crost
Towards him a large eagle, 'twixt whose wings,°
Without one impious word, himself he flings,
Committed to the darkness and the gloom: 660
Down, down, uncertain to what pleasant doom,
Swift as a fathoming plummet down he fell
Through unknown things; till exhaled asphodel,
And rose, with spicy fannings interbreath'd,
Came swelling forth where little caves were wreath'd
So thick with leaves and mosses, that they seem'd
Large honey-combs of green, and freshly teem'd
With airs delicious. In the greenest nook
The eagle landed him, and farewel took.

It was a jasmine bower, all bestrown 670
With golden moss. His every sense had grown
Ethereal for pleasure; 'bove his head
Flew a delight half-graspable; his tread
Was Hesperean; to his capable ears°
Silence was music from the holy spheres;
A dewy luxury was in his eyes;
The little flowers felt his pleasant sighs
And stirr'd them faintly. Verdant cave and cell

He wander'd through, oft wondering at such swell
Of sudden exaltation: but, 'Alas!' 680
Said he, 'will all this gush of feeling pass
Away in solitude? And must they wane,
Like melodies upon a sandy plain,
Without an echo? Then shall I be left
So sad, so melancholy, so bereft!
Yet still I feel immortal! O my love,
My breath of life, where art thou? High above,
Dancing before the morning gates of heaven?
Or keeping watch among those starry seven,°
Old Atlas' children? Art a maid of the waters, 690
One of shell-winding Triton's bright-hair'd daughters?
Or art, impossible! a nymph of Dian's,
Weaving a coronal of tender scions
For very idleness? Where'er thou art,
Methinks it now is at my will to start
Into thine arms; to scare Aurora's train,
And snatch thee from the morning; o'er the main
To scud like a wild bird, and take thee off
From thy sea-foamy cradle; or to doff
Thy shepherd vest, and woo thee mid fresh leaves. 700
No, no, too eagerly my soul deceives
Its powerless self: I know this cannot be.
O let me then by some sweet dreaming flee
To her entrancements: hither sleep awhile!
Hither most gentle sleep! and soothing foil
For some few hours the coming solitude.'

Thus spake he, and that moment felt endued
With power to dream deliciously; so wound
Through a dim passage, searching till he found
The smoothest mossy bed and deepest, where 710
He threw himself, and just into the air
Stretching his indolent arms, he took, O bliss!
A naked waist: 'Fair Cupid, whence is this?'
A well-known voice sigh'd, 'Sweetest, here am I!'
At which soft ravishment, with doating cry
They trembled to each other.—Helicon!
O fountain'd hill! Old Homer's Helicon!
That thou wouldst spout a little streamlet o'er

These sorry pages; then the verse would soar
And sing above this gentle pair, like lark 720
Over his nested young: but all is dark
Around thine aged top, and thy clear fount
Exhales in mists to heaven. Aye, the count
Of mighty Poets is made up; the scroll
Is folded by the Muses; the bright roll
Is in Apollo's hand: our dazed eyes
Have seen a new tinge in the western skies:
The world has done its duty. Yet, oh yet,
Although the sun of poesy is set,
These lovers did embrace, and we must weep 730
That there is no old power left to steep
A quill immortal in their joyous tears.
Long time in silence did their anxious fears
Question that thus it was; long time they lay
Fondling and kissing every doubt away;
Long time ere soft caressing sobs began
To mellow into words, and then there ran
Two bubbling springs of talk from their sweet lips.
'O known Unknown! from whom my being sips
Such darling essence, wherefore may I not 740
Be ever in these arms? in this sweet spot
Pillow my chin for ever? ever press
These toying hands and kiss their smooth excess?
Why not for ever and for ever feel
That breath about my eyes? Ah, thou wilt steal
Away from me again, indeed, indeed—
Thou wilt be gone away, and wilt not heed
My lonely madness. Speak, delicious fair!
Is—is it to be so? No! Who will dare
To pluck thee from me? And, of thine own will, 750
Full well I feel thou wouldst not leave me. Still
Let me entwine thee surer, surer—now
How can we part? Elysium! who art thou?
Who, that thou canst not be for ever here,
Or lift me with thee to some starry sphere?
Enchantress! tell me by this soft embrace,
By the most soft completion of thy face,
Those lips, O slippery blisses, twinkling eyes,
And by these tenderest, milky sovereignties—

These tenderest, and by the nectar-wine, 760
The passion'—'O dov'd Ida the divine!°
Endymion! dearest! Ah, unhappy me!
His soul will 'scape us—O felicity!
How he does love me! His poor temples beat
To the very tune of love—how sweet, sweet, sweet.
Revive, dear youth, or I shall faint and die;
Revive, or these soft hours will hurry by
In tranced dulness; speak, and let that spell
Affright this lethargy! I cannot quell
Its heavy pressure, and will press at least 770
My lips to thine, that they may richly feast
Until we taste the life of love again.
What! dost thou move? dost kiss? O bliss! O pain!
I love thee, youth, more than I can conceive;
And so long absence from thee doth bereave
My soul of any rest: yet must I hence:
Yet, can I not to starry eminence
Uplift thee; nor for very shame can own
Myself to thee. Ah, dearest, do not groan
Or thou wilt force me from this secrecy, 780
And I must blush in heaven. O that I
Had done't already; that the dreadful smiles
At my lost brightness, my impassion'd wiles,
Had waned from Olympus' solemn height,
And from all serious Gods; that our delight
Was quite forgotten, save of us alone!
And wherefore so ashamed? 'Tis but to atone
For endless pleasure, by some coward blushes:
Yet must I be a coward!—Horror rushes
Too palpable before me—the sad look 790
Of Jove—Minerva's start—no bosom shook
With awe of purity—no Cupid pinion
In reverence vailed—my crystaline dominion
Half lost, and all old hymns made nullity!
But what is this to love? O I could fly
With thee into the ken of heavenly powers,
So thou wouldst thus, for many sequent hours,
Press me so sweetly. Now I swear at once
That I am wise, that Pallas is a dunce—
Perhaps her love like mine is but unknown— 800

O I do think that I have been alone
In chastity: yes, Pallas, has been sighing,
While every eve saw me my hair uptying
With fingers cool as aspen leaves. Sweet love,
I was as vague as solitary dove,
Nor knew that nests were built. Now a soft kiss—
Aye, by that kiss, I vow an endless bliss,
An immortality of passion's thine:
Ere long I will exalt thee to the shine
Of heaven ambrosial; and we will shade 810
Ourselves whole summers by a river glade;
And I will tell thee stories of the sky,
And breathe thee whispers of its minstrelsy.
My happy love will overwing all bounds!
O let me melt into thee; let the sounds
Of our close voices marry at their birth;
Let us entwine hoveringly—O dearth
Of human words! roughness of mortal speech!
Lispings empyrean will I sometime teach
Thine honied tongue—lute-breathings, which I gasp 820
To have thee understand, now while I clasp
Thee thus, and weep for fondness—I am pain'd,
Endymion: woe! woe! is grief contain'd°
In the very deeps of pleasure, my sole life?'—
Hereat, with many sobs, her gentle strife
Melted into a languor. He return'd
Entranced vows and tears.

Ye who have yearn'd
With too much passion, will here stay and pity,
For the mere sake of truth; as 'tis a ditty
Not of these days, but long ago 'twas told° 830
By a cavern wind unto a forest old;
And then the forest told it in a dream
To a sleeping lake, whose cool and level gleam
A poet caught as he was journeying
To Phœbus' shrine; and in it he did fling
His weary limbs, bathing an hour's space,
And after, straight in that inspired place
He sang the story up into the air,
Giving it universal freedom. There

Has it been ever sounding for those ears 840
Whose tips are glowing hot. The legend cheers
Yon centinel stars; and he who listens to it
Must surely be self-doomed or he will rue it:
For quenchless burnings come upon the heart,
Made fiercer by a fear lest any part
Should be engulphed in the eddying wind.
As much as here is penn'd doth always find
A resting place, thus much comes clear and plain;
Anon the strange voice is upon the wane—
And 'tis but echo'd from departing sound, 850
That the fair visitant at last unwound
Her gentle limbs, and left the youth asleep.—
Thus the tradition of the gusty deep.

Now turn we to our former chroniclers.—°
Endymion awoke, that grief of hers
Sweet paining on his ear: he sickly guess'd
How lone he was once more, and sadly press'd
His empty arms together, hung his head,
And most forlorn upon that widow'd bed
Sat silently. Love's madness he had known: 860
Often with more than tortured lion's groan
Moanings had burst from him; but now that rage
Had pass'd away: no longer did he wage
A rough-voic'd war against the dooming stars.
No, he had felt too much for such harsh jars:
The lyre of his soul Eolian tun'd°
Forgot all violence, and but commun'd
With melancholy thought: O he had swoon'd
Drunken from pleasure's nipple; and his love
Henceforth was dove-like.—Loth was he to move 870
From the imprinted couch, and when he did,
'Twas with slow, languid paces, and face hid
In muffling hands. So temper'd, out he stray'd
Half seeing visions that might have dismay'd
Alecto's serpents; ravishments more keen
Than Hermes' pipe, when anxious he did lean
Over eclipsing eyes: and at the last
It was a sounding grotto, vaulted, vast,
O'er studded with a thousand, thousand pearls,

And crimson mouthed shells with stubborn curls, 880
Of every shape and size, even to the bulk
In which whales arbour close, to brood and sulk
Against an endless storm. Moreover too,
Fish-semblances, of green and azure hue,
Ready to snort their streams. In this cool wonder°
Endymion sat down, and 'gan to ponder
On all his life: his youth, up to the day
When 'mid acclaim, and feasts, and garlands gay,
He stept upon his shepherd throne: the look
Of his white palace in wild forest nook, 890
And all the revels he had lorded there:
Each tender maiden whom he once thought fair,
With every friend and fellow-woodlander—
Pass'd like a dream before him. Then the spur
Of the old bards to mighty deeds: his plans
To nurse the golden age 'mong shepherd clans:
That wondrous night: the great Pan-festival:
His sister's sorrow; and his wanderings all,
Until into the earth's deep maw he rush'd:
Then all its buried magic, till it flush'd 900
High with excessive love. 'And now,' thought he,
'How long must I remain in jeopardy
Of blank amazements that amaze no more?
Now I have tasted her sweet soul to the core
All other depths are shallow: essences,
Once spiritual, are like muddy lees,
Meant but to fertilize my earthly root,
And make my branches lift a golden fruit
Into the bloom of heaven: other light,
Though it be quick and sharp enough to blight 910
The Olympian eagle's vision, is dark,
Dark as the parentage of chaos. Hark!
My silent thoughts are echoing from these shells;
Or they are but the ghosts, the dying swells
Of noises far away?—list!'—Hereupon
He kept an anxious ear. The humming tone
Came louder, and behold, there as he lay,
On either side outgush'd, with misty spray,
A copious spring; and both together dash'd
Swift, mad, fantastic round the rocks, and lash'd 920

Among the conchs and shells of the lofty grot,
Leaving a trickling dew. At last they shot
Down from the ceiling's height, pouring a noise
As of some breathless racers whose hopes poize
Upon the last few steps, and with spent force
Along the ground they took a winding course.
Endymion follow'd—for it seem'd that one
Ever pursued, the other strove to shun—
Follow'd their languid mazes, till well nigh
He had left thinking of the mystery,— 930
And was now rapt in tender hoverings
Over the vanish'd bliss. Ah! what is it sings
His dream away? What melodies are these?
They sound as through the whispering of trees,
Not native in such barren vaults. Give ear!

'O Arethusa, peerless nymph! why fear
Such tenderness as mine? Great Dian, why,
Why didst thou hear her prayer? O that I°
Were rippling round her dainty fairness now,
Circling about her waist, and striving how 940
To entice her to a dive! then stealing in
Between her luscious lips and eyelids thin.
O that her shining hair was in the sun,
And I distilling from it thence to run
In amorous rillets down her shrinking form!
To linger on her lily shoulders, warm
Between her kissing breasts, and every charm
Touch raptur'd!—See how painfully I flow:
Fair maid, be pitiful to my great woe.
Stay, stay thy weary course, and let me lead, 950
A happy wooer, to the flowery mead
Where all that beauty snar'd me.'—'Cruel god,
Desist! or my offended mistress' nod
Will stagnate all thy fountains:—tease me not
With syren words—Ah, have I really got
Such power to madden thee? And is it true—
Away, away, or I shall dearly rue
My very thoughts: in mercy then away,
Kindest Alpheus, for should I obey
My own dear will, 'twould be a deadly bane. 960

O, Oread-Queen! would that thou hadst a pain°
Like this of mine, then would I fearless turn
And be a criminal.—Alas, I burn,
I shudder—gentle river, get thee hence.
Alpheus! thou enchanter! every sense
Of mine was once made perfect in these woods.
Fresh breezes, bowery lawns, and innocent floods,
Ripe fruits, and lonely couch, contentment gave;
But ever since I heedlessly did lave
In thy deceitful stream, a panting glow 970
Grew strong within me: wherefore serve me so,
And call it love? Alas, 'twas cruelty.
Not once more did I close my happy eyes
Amid the thrush's song. Away! Avaunt!
O 'twas a cruel thing.'—'Now thou dost taunt
So softly, Arethusa, that I think
If thou wast playing on my shady brink,
Thou wouldst bathe once again. Innocent maid!
Stifle thine heart no more;—nor be afraid
Of angry powers: there are deities 980
Will shade us with their wings. Those fitful sighs
'Tis almost death to hear: O let me pour
A dewy balm upon them!—fear no more,
Sweet Arethusa! Dian's self must feel
Sometimes these very pangs. Dear maiden, steal
Blushing into my soul, and let us fly
These dreary caverns for the open sky.
I will delight thee all my winding course,
From the green sea up to my hidden source
About Arcadian forests; and will shew 990
The channels where my coolest waters flow
Through mossy rocks; where, 'mid exuberant green,
I roam in pleasant darkness, more unseen
Than Saturn in his exile; where I brim
Round flowery islands, and take thence a skim
Of mealy sweets, which myriads of bees
Buzz from their honied wings: and thou shouldst please
Thyself to choose the richest, where we might
Be incense-pillow'd every summer night.
Doff all sad fears, thou white deliciousness, 1000
And let us be thus comforted; unless

Thou couldst rejoice to see my hopeless stream
Hurry distracted from Sol's temperate beam,
And pour to death along some hungry sands.'—
'What can I do, Alpheus? Dian stands
Severe before me: persecuting fate!
Unhappy Arethusa! thou wast late
A huntress free in'—At this, sudden fell
Those two sad streams adown a fearful dell.
The Latmian listen'd, but he heard no more, 1010
Save echo, faint repeating o'er and o'er
The name of Arethusa. On the verge
Of that dark gulph he wept, and said: 'I urge
Thee, gentle Goddess of my pilgrimage,
By our eternal hopes, to soothe, to assuage,
If thou art powerful, these lovers' pains;
And make them happy in some happy plains.'

He turn'd—there was a whelming sound—he stept,
There was a cooler light; and so he kept
Towards it by a sandy path, and lo! 1020
More suddenly than doth a moment go,
The visions of the earth were gone and fled—
He saw the giant sea above his head.

BOOK III

There are who lord it o'er their fellow-men°
With most prevailing tinsel: who unpen
Their baaing vanities, to browse away
The comfortable green and juicy hay
From human pastures; or, O torturing fact!
Who, through an idiot blink, will see unpack'd
Fire-branded foxes to sear up and singe°
Our gold and ripe-ear'd hopes. With not one tinge
Of sanctuary splendour, not a sight
Able to face an owl's, they still are dight 10
By the blear-eyed nations in empurpled vests,°
And crowns, and turbans. With unladen breasts,
Save of blown self-applause, they proudly mount
To their spirit's perch, their being's high account,

Their tiptop nothings, their dull skies, their thrones—
Amid the fierce intoxicating tones°
Of trumpets, shoutings, and belabour'd drums,
And sudden cannon. Ah! how all this hums,
In wakeful ears, like uproar past and gone—
Like thunder clouds that spake to Babylon, 20
And set those old Chaldeans to their tasks.—
Are then regalities all gilded masks?
No, there are throned seats unscalable
But by a patient wing, a constant spell,
Or by ethereal things that, unconfin'd,
Can make a ladder of the eternal wind,
And poise about in cloudy thunder-tents
To watch the abysm-birth of elements.
Aye, 'bove the withering of old-lipp'd Fate
A thousand Powers keep religious state, 30
In water, fiery realm, and airy bourne;
And, silent as a consecrated urn,
Hold sphery sessions for a season due.
Yet few of these far majesties, ah, few!
Have bared their operations to this globe—
Few, who with gorgeous pageantry enrobe
Our piece of heaven—whose benevolence
Shakes hand with our own Ceres; every sense
Filling with spiritual sweets to plenitude,
As bees gorge full their cells. And, by the feud 40
'Twixt Nothing and Creation, I here swear,
Eterne Apollo! that thy Sister fair
Is of all these the gentlier-mightiest.
When thy gold breath is misting in the west,
She unobserved steals unto her throne,
And there she sits most meek and most alone;
As if she had not pomp subservient;
As if thine eye, high Poet! was not bent
Towards her with the Muses in thine heart;
As if the ministring stars kept not apart, 50
Waiting for silver-footed messages.
O Moon! the oldest shades 'mong oldest trees
Feel palpitations when thou lookest in:
O Moon! old boughs lisp forth a holier din
The while they feel thine airy fellowship.

Thou dost bless every where, with silver lip
Kissing dead things to life. The sleeping kine,
Couched in thy brightness, dream of fields divine:
Innumerable mountains rise, and rise,
Ambitious for the hallowing of thine eyes; 60
And yet thy benediction passeth not
One obscure hiding-place, one little spot
Where pleasure may be sent: the nested wren
Has thy fair face within its tranquil ken,
And from beneath a sheltering ivy leaf
Takes glimpses of thee; thou art a relief
To the poor patient oyster, where it sleeps
Within its pearly house.—The mighty deeps,
The monstrous sea is thine—the myriad sea!
O Moon! far-spooming Ocean bows to thee, 70
And Tellus feels his forehead's cumbrous load.

Cynthia! where art thou now? What far abode
Of green or silvery bower doth enshrine
Such utmost beauty? Alas, thou dost pine
For one as sorrowful: thy cheek is pale
For one whose cheek is pale: thou dost bewail
His tears, who weeps for thee. Where dost thou sigh?
Ah! surely that light peeps from Vesper's eye,
Or what a thing is love! 'Tis She, but lo!
How chang'd, how full of ache, how gone in woe! 80
She dies at the thinnest cloud; her loveliness
Is wan on Neptune's blue: yet there's a stress
Of love-spangles, just off yon cape of trees,
Dancing upon the waves, as if to please
The curly foam with amorous influence.
O, not so idle: for down-glancing thence
She fathoms eddies, and runs wild about
O'erwhelming water-courses; scaring out
The thorny sharks from hiding-holes, and fright'ning
Their savage eyes with unaccustomed lightning. 90
Where will the splendor be content to reach?
O love! how potent hast thou been to teach
Strange journeyings! Wherever beauty dwells,
In gulf or aerie, mountains or deep dells,
In light, in gloom, in star or blazing sun,

Thou pointest out the way, and straight 'tis won.
Amid his toil thou gav'st Leander breath;°
Thou leddest Orpheus through the gleams of death;
Thou madest Pluto bear thin element;
And now, O winged Chieftain! thou hast sent 100
A moon-beam to the deep, deep water-world,
To find Endymion.

On gold sand impearl'd
With lily shells, and pebbles milky white,
Poor Cynthia greeted him, and sooth'd her light
Against his pallid face: he felt the charm
To breathlessness, and suddenly a warm
Of his heart's blood: 'twas very sweet; he stay'd
His wandering steps, and half-entranced laid
His head upon a tuft of straggling weeds,
To taste the gentle moon, and freshening beads, 110
Lashed from the crystal roof by fishes' tails.
And so he kept, until the rosy veils
Mantling the east, by Aurora's peering hand
Were lifted from the water's breast, and fann'd
Into sweet air; and sober'd morning came
Meekly through billows:—when like taper-flame
Left sudden by a dallying breath of air,
He rose in silence, and once more 'gan fare
Along his fated way.

Far had he roam'd,°
With nothing save the hollow vast, that foam'd 120
Above, around, and at his feet; save things
More dead than Morpheus' imaginings:
Old rusted anchors, helmets, breast-plates large
Of gone sea-warriors; brazen beaks and targe;
Rudders that for a hundred years had lost
The sway of human hand; gold vase emboss'd
With long-forgotton story, and wherein
No reveller had ever dipp'd a chin
But those of Saturn's vintage; mouldering scrolls,
Writ in the tongue of heaven, by those souls 130
Who first were on the earth; and sculptures rude
In ponderous stone, developing the mood

Of ancient Nox;—then skeletons of man,
Of beast, behemoth, and leviathan,
And elephant, and eagle, and huge jaw
Of nameless monster. A cold leaden awe
These secrets struck into him; and unless
Dian had chaced away that heaviness,
He might have died: but now, with cheered feel,
He onward kept; wooing these thoughts to steal 140
About the labyrinth in his soul of love.

'What is there in thee, Moon! that thou shouldst move
My heart so potently? When yet a child
I oft have dried my tears when thou hast smil'd.
Thou seem'dst my sister: hand in hand we went
From eve to morn across the firmament.
No apples would I gather from the tree,
Till thou hadst cool'd their cheeks deliciously:
No tumbling water ever spake romance,
But when my eyes with thine thereon could dance: 150
No woods were green enough, no bower divine,
Until thou liftedst up thine eyelids fine:
In sowing time ne'er would I dibble take,
Or drop a seed, till thou wast wide awake;
And, in the summer tide of blossoming,
No one but thee hath heard me blithly sing
And mesh my dewy flowers all the night.
No melody was like a passing spright
If it went not to solemnize thy reign.
Yes, in my boyhood, every joy and pain 160
By thee were fashion'd to the self-same end;
And as I grew in years, still didst thou blend
With all my ardours: thou wast the deep glen;
Thou wast the mountain-top—the sage's pen—
The poet's harp—the voice of friends—the sun;
Thou wast the river—thou wast glory won;
Thou wast my clarion's blast—thou wast my steed—
My goblet full of wine—my topmost deed:—
Thou wast the charm of women, lovely Moon!
O what a wild and harmonized tune 170
My spirit struck from all the beautiful!
On some bright essence could I lean, and lull

Myself to immortality: I prest
Nature's soft pillow in a wakeful rest.
But, gentle Orb! there came a nearer bliss—
My strange love came—Felicity's abyss!
She came, and thou didst fade, and fade away—
Yet not entirely; no, thy starry sway
Has been an under-passion to this hour.
Now I begin to feel thine orby power 180
Is coming fresh upon me: O be kind,
Keep back thine influence, and do not blind
My sovereign vision.—Dearest love, forgive
That I can think away from thee and live!—
Pardon me, airy planet, that I prize
One thought beyond thine argent luxuries!
How far beyond!' At this a surpris'd start
Frosted the springing verdure of his heart;
For as he lifted up his eyes to swear
How his own goddess was past all things fair, 190
He saw far in the concave green of the sea
An old man sitting calm and peacefully.°
Upon a weeded rock this old man sat,
And his white hair was awful, and a mat
Of weeds were cold beneath his cold thin feet;
And, ample as the largest winding-sheet,
A cloak of blue wrapp'd up his aged bones,
O'erwrought with symbols by the deepest groans
Of ambitious magic: every ocean-form
Was woven in with black distinctness; storm, 200
And calm, and whispering, and hideous roar
Quicksand and whirlpool, and deserted shore
Were emblem'd in the woof; with every shape
That skims, or dives, or sleeps, 'twixt cape and cape.
The gulphing whale was like a dot in the spell,
Yet look upon it, and 'twould size and swell
To its huge self; and the minutest fish
Would pass the very hardest gazer's wish,
And shew his little eye's anatomy.
Then there was pictur'd the regality 210
Of Neptune; and the sea nymphs round his state,
In beauteous vassalage, look up and wait.
Beside this old man lay a pearly wand,

And in his lap a book, the which he conn'd
So stedfastly, that the new denizen
Had time to keep him in amazed ken,
To mark these shadowings, and stand in awe.

The old man rais'd his hoary head and saw
The wilder'd stranger—seeming not to see,
His features were so lifeless. Suddenly 220
He woke as from a trance; his snow-white brows
Went arching up, and like two magic ploughs°
Furrow'd deep wrinkles in his forehead large,
Which kept as fixedly as rocky marge,
Till round his wither'd lips had gone a smile.
Then up he rose, like one whose tedious toil
Had watch'd for years in forlorn hermitage,
Who had not from mid-life to utmost age
Eas'd in one accent his o'er-burden'd soul,
Even to the trees. He rose: he grasp'd his stole, 230
With convuls'd clenches waving it abroad,
And in a voice of solemn joy, that aw'd
Echo into oblivion, he said:—

'Thou art the man! Now shall I lay my head°
In peace upon my watery pillow: now
Sleep will come smoothly to my weary brow.
O Jove! I shall be young again, be young!
O shell-borne Neptune, I am pierc'd and stung
With new-born life! What shall I do? Where go,
When I have cast this serpent-skin of woe?— 240
I'll swim to the syrens, and one moment listen
Their melodies, and see their long hair glisten;
Anon upon that giant's arm I'll be,°
That writhes about the roots of Sicily:
To northern seas I'll in a twinkling sail,
And mount upon the snortings of a whale
To some black cloud; thence down I'll madly sweep
On forked lightning, to the deepest deep,
Where through some sucking pool I will be hurl'd
With rapture to the other side of the world! 250
O, I am full of gladness! Sisters three,
I bow full hearted to your old decree!

Yes, every god be thank'd, and power benign,
For I no more shall wither, droop, and pine.
Thou art the man!' Endymion started back
Dismay'd; and, like a wretch from whom the rack
Tortures hot breath, and speech of agony,
Mutter'd: 'What lonely death am I to die
In this cold region? Will he let me freeze,
And float my brittle limbs o'er polar seas? 260
Or will he touch me with his searing hand,
And leave a black memorial on the sand?
Or tear me piece-meal with a bony saw,
And keep me as a chosen food to draw
His magian fish through hated fire and flame?
O misery of hell! resistless, tame,
Am I to be burnt up? No, I will shout,
Until the gods through heaven's blue look out!—
O Tartarus! but some few days agone
Her soft arms were entwining me, and on 270
Her voice I hung like fruit among green leaves:
Her lips were all my own, and—ah, ripe sheaves
Of happiness! ye on the stubble droop,
But never may be garner'd. I must stoop
My head, and kiss death's foot. Love! love, farewel!
Is there no hope from thee? This horrid spell
Would melt at thy sweet breath.—By Dian's hind
Feeding from her white fingers, on the wind
I see thy streaming hair! and now, by Pan,
I care not for this old mysterious man!' 280

He spake, and walking to that aged form,
Look'd high defiance. Lo! his heart 'gan warm
With pity, for the grey-hair'd creature wept.
Had he then wrong'd a heart where sorrow kept?
Had he, though blindly contumelious, brought
Rheum to kind eyes, a sting to human thought,
Convulsion to a mouth of many years?
He had in truth; and he was ripe for tears.
The penitent shower fell, as down he knelt
Before that care-worn sage, who trembling felt 290
About his large dark locks, and faultering spake:

'Arise, good youth, for sacred Phœbus' sake!
I know thine inmost bosom, and I feel
A very brother's yearning for thee steal
Into mine own: for why? thou openest
The prison gates that have so long opprest
My weary watching. Though thou know'st it not,
Thou art commission'd to this fated spot
For great enfranchisement. O weep no more;
I am a friend to love, to loves of yore: 300
Aye, hadst thou never lov'd an unknown power,
I had been grieving at this joyous hour.
But even now most miserable old,
I saw thee, and my blood no longer cold
Gave mighty pulses: in this tottering case
Grew a new heart, which at this moment plays
As dancingly as thine. Be not afraid,
For thou shalt hear this secret all display'd,
Now as we speed towards our joyous task.'

So saying, this young soul in age's mask 310
Went forward with the Carian side by side:
Resuming quickly thus; while ocean's tide
Hung swollen at their backs, and jewel'd sands
Took silently their foot-prints.

'My soul stands
Now past the midway from mortality,
And so I can prepare without a sigh
To tell thee briefly all my joy and pain.
I was a fisher once, upon this main,°
And my boat danc'd in every creek and bay;
Rough billows were my home by night and day,— 320
The sea-gulls not more constant; for I had
No housing from the storm and tempests mad,
But hollow rocks,—and they were palaces
Of silent happiness, of slumberous ease:
Long years of misery have told me so.
Aye, thus it was one thousand years ago.
One thousand years!—Is it then possible
To look so plainly through them? to dispel
A thousand years with backward glance sublime?

To breathe away as 'twere all scummy slime 330
From off a crystal pool, to see its deep,
And one's own image from the bottom peep?
Yes: now I am no longer wretched thrall,
My long captivity and moanings all
Are but a slime, a thin-pervading scum,
The which I breathe away, and thronging come
Like things of yesterday my youthful pleasures.

'I touch'd no lute, I sang not, trod no measures:
I was a lonely youth on desert shores.
My sports were lonely, 'mid continuous roars, 340
And craggy isles, and sea-mew's plaintive cry
Plaining discrepant between sea and sky.
Dolphins were still my playmates; shapes unseen
Would let me feel their scales of gold and green,
Nor be my desolation; and, full oft,
When a dread waterspout had rear'd aloft
Its hungry hugeness, seeming ready ripe
To burst with hoarest thunderings, and wipe
My life away like a vast sponge of fate,
Some friendly monster, pitying my sad state, 350
Has dived to its foundations, gulph'd it down,
And left me tossing safely. But the crown
Of all my life was utmost quietude:
More did I love to lie in cavern rude,
Keeping in wait whole days for Neptune's voice,
And if it came at last, hark, and rejoice!
There blush'd no summer eve but I would steer
My skiff along green shelving coasts, to hear
The shepherd's pipe come clear from aery steep,
Mingled with ceaseless bleatings of his sheep: 360
And never was a day of summer shine,
But I beheld its birth upon the brine:
For I would watch all night to see unfold
Heaven's gates, and Æthon snort his morning gold
Wide o'er the swelling streams: and constantly
At brim of day-tide, on some grassy lea,
My nets would be spread out, and I at rest.
The poor folk of the sea-country I blest
With daily boon of fish most delicate:

They knew not whence this bounty, and elate 370
Would strew sweet flowers on a sterile beach.

'Why was I not contented? Wherefore reach
At things which, but for thee, O Latmian!
Had been my dreary death? Fool! I began
To feel distemper'd longings: to desire
The utmost privilege that ocean's sire
Could grant in benediction: to be free
Of all his kingdom. Long in misery
I wasted, ere in one extremest fit
I plung'd for life or death. To interknit 380
One's senses with so dense a breathing stuff
Might seem a work of pain; so not enough
Can I admire how crystal-smooth it felt,
And buoyant round my limbs. At first I dwelt
Whole days and days in sheer astonishment;
Forgetful utterly of self-intent;
Moving but with the mighty ebb and flow.
Then, like a new fledg'd bird that first doth shew
His spreaded feathers to the morrow chill,
I tried in fear the pinions of my will. 390
'Twas freedom! and at once I visited
The ceaseless wonders of this ocean-bed.
No need to tell thee of them, for I see
That thou hast been a witness—it must be—
For these I know thou canst not feel a drouth,
By the melancholy corners of that mouth.
So I will in my story straightway pass
To more immediate matter. Woe, alas!
That love should be my bane! Ah, Scylla fair!
Why did poor Glaucus ever—ever dare 400
To sue thee to his heart? Kind stranger-youth!
I lov'd her to the very white of truth,
And she would not conceive it. Timid thing!
She fled me swift as sea-bird on the wing,
Round every isle, and point, and promontory,
From where large Hercules wound up his story°
Far as Egyptian Nile. My passion grew
The more, the more I saw her dainty hue
Gleam delicately through the azure clear:

Until 'twas too fierce agony to bear; 410
And in that agony, across my grief
It flash'd, that Circe might find some relief—
Cruel enchantress! So above the water
I rear'd my head, and look'd for Phœbus' daughter°.
Ææa's isle was wondering at the moon:—
It seem'd to whirl around me, and a swoon
Left me dead-drifting to that fatal power.

'When I awoke, 'twas in a twilight bower;
Just when the light of morn, with hum of bees,
Stole through its verdurous matting of fresh trees. 420
How sweet, and sweeter! for I heard a lyre,
And over it a sighing voice expire.
It ceased—I caught light footsteps; and anon
The fairest face that morn e'er look'd upon
Push'd through a screen of roses. Starry Jove!
With tears, and smiles, and honey-words she wove
A net whose thraldom was more bliss than all
The range of flower'd Elysium. Thus did fall
The dew of her rich speech: "Ah! Art awake?
O let me hear thee speak, for Cupid's sake! 430
I am so oppress'd with joy! Why, I have shed
An urn of tears, as though thou wert cold dead;
And now I find thee living, I will pour
From these devoted eyes their silver store,
Until exhausted of the latest drop,
So it will pleasure thee, and force thee stop
Here, that I too may live: but if beyond
Such cool and sorrowful offerings, thou art fond
Of soothing warmth, of dalliance supreme;
If thou art ripe to taste a long love dream; 440
If smiles, if dimples, tongues for ardour mute,
Hang in thy vision like a tempting fruit,
O let me pluck it for thee." Thus she link'd
Her charming syllables, till indistinct
Their music came to my o'er-sweeten'd soul;
And then she hover'd over me, and stole
So near, that if no nearer it had been
This furrow'd visage thou hadst never seen.

'Young man of Latmos! thus particular°
Am I, that thou may'st plainly see how far 450
This fierce temptation went: and thou may'st not
Exclaim, How then, was Scylla quite forgot?

'Who could resist? Who in this universe?
She did so breathe ambrosia; so immerse
My fine existence in a golden clime.
She took me like a child of suckling time,
And cradled me in roses. Thus condemn'd,
The current of my former life was stemm'd,
And to this arbitrary queen of sense
I bow'd a tranced vassal: nor would thence 460
Have mov'd, even though Amphion's harp had woo'd
Me back to Scylla o'er the billows rude.
For as Apollo each eve doth devise
A new appareling for western skies;
So every eve, nay every spendthrift hour
Shed balmy consciousness within that bower.
And I was free of haunts umbrageous;
Could wander in the mazy forest-house
Of squirrels, foxes shy, and antler'd deer,
And birds from coverts innermost and drear 470
Warbling for very joy mellifluous sorrow—
To me new born delights!

'Now let me borrow,
For moments few, a temperament as stern
As Pluto's sceptre, that my words not burn
These uttering lips, while I in calm speech tell
How specious heaven was changed to real hell.

'One morn she left me sleeping: half awake
I sought for her smooth arms and lips, to slake
My greedy thirst with nectarous camel-draughts;
But she was gone. Whereat the barbed shafts
Of disappointment stuck in me so sore, 480
That out I ran and search'd the forest o'er.
Wandering about in pine and cedar gloom
Damp awe assail'd me; for there 'gan to boom
A sound of moan, an agony of sound,
Sepulchral from the distance all around.

Then came a conquering earth-thunder, and rumbled
That fierce complain to silence: while I stumbled
Down a precipitous path, as if impell'd.
I came to a dark valley.—Groanings swell'd 490
Poisonous about my ears, and louder grew,
The nearer I approach'd a flame's gaunt blue,
That glar'd before me through a thorny brake.
This fire, like the eye of gordian snake,
Bewitch'd me towards; and I soon was near
A sight too fearful for the feel of fear:
In thicket hid I curs'd the haggard scene—
The banquet of my arms, my arbour queen,
Seated upon an uptorn forest root;
And all around her shapes, wizard and brute, 500
Laughing, and wailing, groveling, serpenting,
Shewing tooth, tusk, and venom-bag, and sting!
O such deformities! Old Charon's self,
Should he give up awhile his penny pelf,°
And take a dream 'mong rushes Stygian,
It could not be so phantasied. Fierce, wan,
And tyrannizing was the lady's look,
As over them a gnarled staff she shook.
Oft-times upon the sudden she laugh'd out,
And from a basket emptied to the rout 510
Clusters of grapes, the which they raven'd quick
And roar'd for more; with many a hungry lick
About their shaggy jaws. Avenging, slow,
Anon she took a branch of mistletoe,
And emptied on't a black dull-gurgling phial:
Groan'd one and all, as if some piercing trial
Was sharpening for their pitiable bones.
She lifted up the charm: appealing groans
From their poor breasts went sueing to her ear
In vain; remorseless as an infant's bier 520
She whisk'd against their eyes the sooty oil.
Whereat was heard a noise of painful toil,
Increasing gradual to a tempest rage,
Shrieks, yells, and groans of torture-pilgrimage;
Until their grieved bodies 'gan to bloat
And puff from the tail's end to stifled throat:
Then was appalling silence: then a sight

More wildering than all that hoarse affright;
For the whole herd, as by a whirlwind writhen,
Went through the dismal air like one huge Python 530
Antagonizing Boreas,—and so vanish'd.
Yet there was not a breath of wind: she banish'd
These phantoms with a nod. Lo! from the dark
Came waggish fauns, and nymphs, and satyrs stark,
With dancing and loud revelry,—and went
Swifter than centaurs after rapine bent.—
Sighing an elephant appear'd and bow'd
Before the fierce witch, speaking thus aloud
In human accent: "Potent goddess! chief
Of pains resistless! make my being brief, 540
Or let me from this heavy prison fly:
Or give me to the air, or let me die!
I sue not for my happy crown again;
I sue not for my phalanx on the plain;
I sue not for my lone, my widow'd wife;
I sue not for my ruddy drops of life,
My children fair, my lovely girls and boys!
I will forget them; I will pass these joys;
Ask nought so heavenward, so too—too high:
Only I pray, as fairest boon, to die, 550
Or be deliver'd from this cumbrous flesh,°
From this gross, detestable, filthy mesh,
And merely given to the cold bleak air.
Have mercy, Goddess! Circe, feel my prayer!"

'That curst magician's name fell icy numb
Upon my wild conjecturing: truth had come
Naked and sabre-like against my heart.
I saw a fury whetting a death-dart;
And my slain spirit, overwrought with fright,
Fainted away in that dark lair of night. 560
Think, my deliverer, how desolate
My waking must have been! disgust, and hate,
And terrors manifold divided me
A spoil amongst them. I prepar'd to flee
Into the dungeon core of that wild wood:
I fled three days—when lo! before me stood
Glaring the angry witch. O Dis, even now,

A clammy dew is beading on my brow,
At mere remembering her pale laugh, and curse.
"Ha! ha! Sir Dainty! there must be a nurse° 570
Made of rose leaves and thistledown, express,
To cradle thee my sweet, and lull thee: yes,
I am too flinty-hard for thy nice touch:
My tenderest squeeze is but a giant's clutch.
So, fairy-thing, it shall have lullabies
Unheard of yet; and it shall still its cries
Upon some breast more lily-feminine.
Oh, no—it shall not pine, and pine, and pine
More than one pretty, trifling thousand years;
And then 'twere pity, but fate's gentle shears 580
Cut short its immortality. Sea-flirt!
Young dove of the waters! truly I'll not hurt
One hair of thine: see how I weep and sigh,
That our heart-broken parting is so nigh.
And must we part? Ah, yes, it must be so.
Yet ere thou leavest me in utter woe,
Let me sob over thee my last adieus,
And speak a blessing: Mark me! Thou hast thews
Immortal, for thou art of heavenly race:
But such a love is mine, that here I chase° 590
Eternally away from thee all bloom
Of youth, and destine thee towards a tomb.
Hence shalt thou quickly to the watery vast;
And there, ere many days be overpast,
Disabled age shall seize thee; and even then
Thou shalt not go the way of aged men;
But live and wither, cripple and still breathe
Ten hundred years: which gone, I then bequeath
Thy fragile bones to unknown burial.
Adieu, sweet love, adieu!"—As shot stars fall, 600
She fled ere I could groan for mercy. Stung
And poisoned was my spirit: despair sung
A war-song of defiance 'gainst all hell.
A hand was at my shoulder to compel
My sullen steps; another 'fore my eyes
Moved on with pointed finger. In this guise
Enforced, at the last by ocean's foam
I found me; by my fresh, my native home.

Its tempering coolness, to my life akin,
Came salutary as I waded in; 610
And, with a blind voluptuous rage, I gave
Battle to the swollen billow-ridge, and drave
Large froth before me, while there yet remain'd
Hale strength, nor from my bones all marrow drain'd.

'Young lover, I must weep—such hellish spite
With dry cheek who can tell? While thus my might
Proving upon this element, dismay'd,
Upon a dead thing's face my hand I laid;
I look'd—'twas Scylla! Cursed, cursed Circe!
O vulture-witch, hast never heard of mercy? 620
Could not thy harshest vengeance be content,
But thou must nip this tender innocent
Because I lov'd her?—Cold, O cold indeed
Were her fair limbs, and like a common weed
The sea-swell took her hair. Dead as she was
I clung about her waist, nor ceas'd to pass
Fleet as an arrow through unfathom'd brine,
Until there shone a fabric crystalline,
Ribb'd and inlaid with coral, pebble, and pearl.
Headlong I darted; at one eager swirl 630
Gain'd its bright portal, enter'd, and behold!
'Twas vast, and desolate, and icy-cold;
And all around—But wherefore this to thee
Who in few minutes more thyself shalt see?—
I left poor Scylla in a niche and fled.
My fever'd parchings up, my scathing dread
Met palsy half way: soon these limbs became
Gaunt, wither'd, sapless, feeble, cramp'd, and lame.

'Now let me pass a cruel, cruel space,
Without one hope, without one faintest trace 640
Of mitigation, or redeeming bubble
Of colour'd phantasy; for I fear 'twould trouble
Thy brain to loss of reason: and next tell
How a restoring chance came down to quell
One half of the witch in me.

'On a day,
Sitting upon a rock above the spray,

I saw grow up from the horizon's brink
A gallant vessel: soon she seem'd to sink
Away from me again, as though her course
Had been resum'd in spite of hindering force— 650
So vanish'd: and not long, before arose
Dark clouds, and muttering of winds morose.
Old Eolus would stifle his mad spleen,
But could not: therefore all the billows green
Toss'd up the silver spume against the clouds.
The tempest came: I saw that vessel's shrouds
In perilous bustle; while upon the deck
Stood trembling creatures. I beheld the wreck;
The final gulphing; the poor struggling souls:
I heard their cries amid loud thunder-rolls. 660
O they had all been sav'd but crazed eld
Annull'd my vigorous cravings: and thus quell'd
And curb'd, think on't, O Latmian! did I sit
Writhing with pity, and a cursing fit
Against that hell-born Circe. The crew had gone,
By one and one, to pale oblivion;
And I was gazing on the surges prone,
With many a scalding tear and many a groan,
When at my feet emerg'd an old man's hand,
Grasping this scroll, and this same slender wand. 670
I knelt with pain—reached out my hand—had grasp'd
These treasures—touch'd the knuckles—they unclasp'd—
I caught a finger: but the downward weight
O'erpowered me—it sank. Then 'gan abate
The storm, and through chill aguish gloom outburst
The comfortable sun. I was athirst
To search the book, and in the warming air
Parted its dripping leaves with eager care.
Strange matters did it treat of, and drew on
My soul page after page, till well-nigh won 680
Into forgetfulness; when, stupefied,
I read these words, and read again, and tried
My eyes against the heavens, and read again.
O what a load of misery and pain
Each Atlas-line bore off!—a shine of hope°
Came gold around me, cheering me to cope
Strenuous with hellish tyranny. Attend!

For thou hast brought their promise to an end.

'In the wide sea there lives a forlorn wretch,
Doom'd with enfeebled carcase to outstretch 690
His loath'd existence through ten centuries,
And then to die alone. Who can devise
A total opposition? No one. So
One million times ocean must ebb and flow,
And he oppressed. Yet he shall not die,
These things accomplish'd:—If he utterly
Scans all the depths of magic, and expounds
The meanings of all motions, shapes, and sounds;
If he explores all forms and substances
Straight homeward to their symbol-essences; 700
He shall not die. Moreover, and in chief,
He must pursue this task of joy and grief
Most piously;—all lovers tempest-tost,°
And in the savage overwhelming lost,
He shall deposit side by side, until
Time's creeping shall the dreary space fulfil:°
Which done, and all these labours ripened,
A youth, by heavenly power lov'd and led,
Shall stand before him; whom he shall direct
How to consummate all. The youth elect 710
Must do the thing, or both will be destroy'd.'—

'Then,' cried the young Endymion, overjoy'd,
'We are twin brothers in this destiny!
Say, I intreat thee, what achievement high
Is, in this restless world, for me reserv'd.
What! if from thee my wandering feet had swerv'd,
Had we both perish'd?'—'Look!' the sage replied,
'Dost thou not mark a gleaming through the tide,
Of divers brilliances? 'tis the edifice
I told thee of, where lovely Scylla lies; 720
And where I have enshrined piously
All lovers, whom fell storms have doom'd to die
Throughout my bondage.' Thus discoursing, on
They went till unobscur'd the porches shone;
Which hurryingly they gain'd, and enter'd straight.
Sure never since king Neptune held his state

Was seen such wonder underneath the stars.
Turn to some level plain where haughty Mars°
Has legion'd all his battle; and behold
How every soldier, with firm foot, doth hold 730
His even breast: see, many steeled squares,
And rigid ranks of iron—whence who dares
One step? Imagine further, line by line,
These warrior thousands on the field supine:—
So in that crystal place, in silent rows,
Poor lovers lay at rest from joys and woes.—
The stranger from the mountains, breathless, trac'd
Such thousands of shut eyes in order plac'd;
Such ranges of white feet, and patient lips
All ruddy,—for here death no blossom nips. 740
He mark'd their brows and foreheads; saw their hair
Put sleekly on one side with nicest care;
And each one's gentle wrists, with reverence,
Put cross-wise to its heart.

'Let us commence,'
Whisper'd the guide, stuttering with joy, 'even now.'
He spake, and, trembling like an aspen-bough,
Began to tear his scroll in pieces small,
Uttering the while some mumblings funeral.
He tore it into pieces small as snow
That drifts unfeather'd when bleak northerns blow; 750
And having done it, took his dark blue cloak
And bound it round Endymion: then struck
His wand against the empty air times nine.—
'What more there is to do, young man, is thine:
But first a little patience; first undo
This tangled thread, and wind it to a clue.
Ah, gentle! 'tis as weak as spider's skein;
And shouldst thou break it—What, is it done so clean?
A power overshadows thee! Oh, brave!
The spite of hell is tumbling to its grave. 760
Here is a shell; 'tis pearly blank to me,
Nor mark'd with any sign or charactery—
Canst thou read aught? O read for pity's sake!
Olympus! we are safe! Now, Carian, break
This wand against yon lyre on the pedestal.'

'Twas done: and straight with sudden swell and fall
Sweet music breath'd her soul away, and sigh'd
A lullaby to silence.—'Youth! now strew
These minced leaves on me, and passing through
Those files of dead, scatter the same around, 770
And thou wilt see the issue.'—'Mid the sound
Of flutes and viols, ravishing his heart,
Endymion from Glaucus stood apart,
And scatter'd in his face some fragments light.
How lightning-swift the change! a youthful wight
Smiling beneath a coral diadem,
Out-sparkling sudden like an upturn'd gem,
Appear'd, and, stepping to a beauteous corse,
Kneel'd down beside it, and with tenderest force
Press'd its cold hand, and wept,—and Scylla sigh'd! 780
Endymion, with quick hand, the charm applied—
The nymph arose: he left them to their joy,
And onward went upon his high employ,
Showering those powerful fragments on the dead.
And, as he pass'd, each lifted up its head,
As doth a flower at Apollo's touch.
Death felt it to his inwards: 'twas too much:
Death fell a weeping in his charnel-house.
The Latmian persever'd along, and thus
All were re-animated. There arose 790
A noise of harmony, pulses and throes
Of gladness in the air—while many, who
Had died in mutual arms devout and true,
Sprang to each other madly; and the rest
Felt a high certainty of being blest.
They gaz'd upon Endymion. Enchantment
Grew drunken, and would have its head and bent.
Delicious symphonies, like airy flowers,
Budded, and swell'd, and, full-blown, shed full showers
Of light, soft, unseen leaves of sounds divine. 800
The two deliverers tasted a pure wine
Of happiness, from fairy-press, ooz'd out.
Speechless they eyed each other, and about
The fair assembly wander'd to and fro,
Distracted with the richest overflow
Of joy that ever pour'd from heaven.

—'Away!'
Shouted the new born god; 'Follow, and pay
Our piety to Neptunus supreme!'—
Then Scylla, blushing sweetly from her dream,
They led on first, bent to her meek surprise, 810
Through portal columns of a giant size,
Into the vaulted, boundless emerald.
Joyous all follow'd, as the leader call'd,
Down marble steps; pouring as easily
As hour-glass sand,—and fast, as you might see
Swallows obeying the south summer's call,
Or swans upon a gentle waterfall.

Thus went that beautiful multitude, nor far,
Ere from among some rocks of glittering spar,
Just within ken, they saw descending thick 820
Another multitude. Whereat more quick
Moved either host. On a wide sand they met,
And of those numbers every eye was wet;
For each their old love found. A murmuring rose,
Like what was never heard in all the throes
Of wind and waters: 'tis past human wit
To tell: 'tis dizziness to think of it.

This mighty consummation made, the host
Mov'd on for many a league; and gain'd, and lost
Huge sea-marks; vanward swelling in array, 830
And from the rear diminishing away,—
Till a faint dawn surpris'd them. Glaucus cried,
'Behold! behold, the palace of his pride!
God Neptune's palaces!' With noise increas'd,
They shoulder'd on towards that brightening east.
At every onward step proud domes arose
In prospect,—diamond gleams, and golden glows
Of amber 'gainst their faces levelling.
Joyous, and many as the leaves in spring,
Still onward; still the splendour gradual swell'd. 840
Rich opal domes were seen, on high upheld
By jasper pillars, letting through their shafts
A blush of coral. Copious wonder-draughts
Each gazer drank; and deeper drank more near:

For what poor mortals fragment up, as mere°
As marble was there lavish, to the vast
Of one fair palace, that far far surpass'd,
Even for common bulk, those olden three,
Memphis, and Babylon, and Nineveh.

As large, as bright, as colour'd as the bow 850
Of Iris, when unfading it doth shew
Beyond a silvery shower, was the arch
Through which this Paphian army took its march,
Into the outer courts of Neptune's state:
Whence could be seen, direct, a golden gate,
To which the leaders sped; but not half raught
Ere it burst open swift as fairy thought,
And made those dazzled thousands veil their eyes
Like callow eagles at the first sunrise.
Soon with an eagle nativeness their gaze 860
Ripe from hue-golden swoons took all the blaze,
And then, behold! large Neptune on his throne
Of emerald deep: yet not exalt alone;
At his right hand stood winged Love, and on
His left sat smiling Beauty's paragon.

Far as the mariner on highest mast
Can see all round upon the calmed vast,
So wide was Neptune's hall: and as the blue
Doth vault the waters, so the waters drew
Their doming curtains, high, magnificent, 870
Aw'd from the throne aloof;—and when storm-rent
Disclos'd the thunder-gloomings in Jove's air;
But sooth'd as now, flash'd sudden everywhere,
Noiseless, sub-marine cloudlets, glittering
Death to a human eye: for there did spring
From natural west, and east, and south, and north,
A light as of four sunsets, blazing forth
A gold-green zenith 'bove the Sea-God's head.
Of lucid depth the floor, and far outspread
As breezeless lake, on which the slim canoe 880
Of feather'd Indian darts about, as through
The delicatest air: air verily,°
But for the portraiture of clouds and sky:

This palace floor breath-air,—but for the amaze
Of deep-seen wonders motionless,—and blaze
Of the dome pomp, reflected in extremes,
Globing a golden sphere.

They stood in dreams
Till Triton blew his horn. The palace rang;
The Nereids danc'd; the Syrens faintly sang;
And the great Sea-King bow'd his dripping head. 890
Then Love took wing, and from his pinions shed
On all the multitude a nectarous dew.
The ooze-born Goddess beckoned and drew
Fair Scylla and her guides to conference;
And when they reach'd the throned eminence
She kist the sea-nymph's cheek,—who sat her down
A toying with the doves. Then,—'Mighty crown
And sceptre of this kingdom!' Venus said,
'Thy vows were on a time to Nais paid:
Behold!'—Two copious tear-drops instant fell 900
From the God's large eyes; he smil'd delectable,
And over Glaucus held his blessing hands.—
'Endymion! Ah! still wandering in the bands
Of love? Now this is cruel. Since the hour
I met thee in earth's bosom, all my power
Have I put forth to serve thee. What, not yet
Escap'd from dull mortality's harsh net?
A little patience, youth! 'twill not be long,
Or I am skilless quite: an idle tongue,
A humid eye, and steps luxurious, 910
Where these are new and strange, are ominous.
Aye, I have seen these signs in one of heaven,
When others were all blind; and were I given
To utter secrets, haply I might say
Some pleasant words:—but Love will have his day.
So wait awhile expectant. Pr'ythee soon,
Even in the passing of thine honey-moon,
Visit my Cytherea: thou wilt find
Cupid well-natured, my Adonis kind;
And pray persuade with thee—Ah, I have done, 920
All blisses be upon thee, my sweet son!'—
Thus the fair goddess: while Endymion

Knelt to receive those accents halcyon.

Meantime a glorious revelry began
Before the Water-Monarch. Nectar ran
In courteous fountains to all cups outreach'd;
And plunder'd vines, teeming exhaustless, pleach'd
New growth about each shell and pendent lyre;
The which, in disentangling for their fire,
Pull'd down fresh foliage and coverture 930
For dainty toying. Cupid, empire-sure,
Flutter'd and laugh'd, and oft-times through the throng
Made a delighted way. Then dance, and song,
And garlanding grew wild; and pleasure reign'd.
In harmless tendril they each other chain'd,
And strove who should be smother'd deepest in
Fresh crush of leaves.

O 'tis a very sin
For one so weak to venture his poor verse
In such a place as this. O do not curse,
High Muses! let him hurry to the ending. 940

All suddenly were silent. A soft blending
Of dulcet instruments came charmingly;
And then a hymn.

'King of the stormy sea!
Brother of Jove, and co-inheritor
Of elements! Eternally before
Thee the waves awful bow. Fast, stubborn rock,
At thy fear'd trident shrinking, doth unlock
Its deep foundations, hissing into foam.
All mountain-rivers lost in the wide home
Of thy capacious bosom ever flow. 950
Thou frownest, and old Eolus thy foe
Skulks to his cavern, 'mid the gruff complaint
Of all his rebel tempests. Dark clouds faint
When, from thy diadem, a silver gleam
Slants over blue dominion. Thy bright team
Gulphs in the morning light, and scuds along
To bring thee nearer to that golden song

Apollo singeth, while his chariot
Waits at the doors of heaven. Thou art not
For scenes like this: an empire stern hast thou; 960
And it hath furrow'd that large front: yet now,
As newly come of heaven, dost thou sit
To blend and interknit
Subdued majesty with this glad time.
O shell-borne King sublime!
We lay our hearts before thee evermore—
We sing, and we adore!

'Breathe softly, flutes;
Be tender of your strings, ye soothing lutes;
Nor be the trumpet heard! O vain, O vain; 970
Not flowers budding in an April rain,
Nor breath of sleeping dove, nor river's flow,—
No, nor the Eolian twang of Love's own bow,
Can mingle music fit for the soft ear
Of goddess Cytherea!
Yet deign, white Queen of Beauty, thy fair eyes
On our souls' sacrifice.

'Bright-winged Child!
Who has another care when thou hast smil'd?
Unfortunates on earth, we see at last 980
All death-shadows, and glooms that overcast
Our spirits, fann'd away by thy light pinions.
O sweetest essence! sweetest of all minions!
God of warm pulses, and dishevell'd hair,
And panting bosoms bare!
Dear unseen light in darkness! eclipser
Of light in light! delicious poisoner!
Thy venom'd goblet will we quaff until
We fill—we fill!
And by thy Mother's lips—'

Was heard no more 990
For clamour, when the golden palace door
Opened again, and from without, in shone
A new magnificence. On oozy throne
Smooth-moving came Oceanus the old,

To take a latest glimpse at his sheep-fold,
Before he went into his quiet cave
To muse for ever—Then a lucid wave,
Scoop'd from its trembling sisters of mid-sea,
Afloat, and pillowing up the majesty
Of Doris, and the Egean seer, her spouse—° 1000
Next, on a dolphin, clad in laurel boughs,
Theban Amphion leaning on his lute:
His fingers went across it—All were mute
To gaze on Amphitrite, queen of pearls,
And Thetis pearly too.—

The palace whirls
Around giddy Endymion; seeing he
Was there far strayed from mortality.
He could not bear it—shut his eyes in vain;
Imagination gave a dizzier pain.
'O I shall die! sweet Venus, be my stay! 1010
Where is my lovely mistress? Well-away!
I die—I hear her voice—I feel my wing—'
At Neptune's feet he sank. A sudden ring
Of Nereids were about him, in kind strife
To usher back his spirit into life:
But still he slept. At last they interwove
Their cradling arms, and purpos'd to convey
Towards a crystal bower far away.

Lo! while slow carried through the pitying crowd,
To his inward senses these words spake aloud; 1020
Written in star-light on the dark above:
Dearest Endymion! my entire love!
How have I dwelt in fear of fate: 'tis done—
Immortal bliss for me too hast thou won.
Arise then! for the hen-dove shall not hatch
Her ready eggs, before I'll kissing snatch
Thee into endless heaven. Awake! awake!

The youth at once arose: a placid lake
Came quiet to his eyes; and forest green,
Cooler than all the wonders he had seen, 1030
Lull'd with its simple song his fluttering breast.
How happy once again in grassy nest!

BOOK IV

Muse of my native land! loftiest Muse!°
O first-born on the mountains! by the hues
Of heaven on the spiritual air begot:
Long didst thou sit alone in northern grot,
While yet our England was a wolfish den;
Before our forests heard the talk of men;
Before the first of Druids was a child;—
Long didst thou sit amid our regions wild
Rapt in a deep prophetic solitude.
There came an eastern voice of solemn mood:—° 10
Yet wast thou patient. Then sang forth the Nine,
Apollo's garland:—yet didst thou divine
Such home-bred glory, that they cry'd in vain,
'Come hither, Sister of the Island!' Plain
Spake fair Ausonia; and once more she spake°
A higher summons:—still didst thou betake
Thee to thy native hopes. O thou hast won
A full accomplishment! The thing is done,
Which undone, these our latter days had risen
On barren souls. Great Muse, thou know'st what prison, 20
Of flesh and bone, curbs, and confines, and frets
Our spirit's wings: despondency besets
Our pillows; and the fresh to-morrow morn
Seems to give forth its light in very scorn
Of our dull, uninspired, snail-paced lives.
Long have I said, how happy he who shrives°
To thee! But then I thought on poets gone,°
And could not pray:—nor can I now—so on
I move to the end in lowliness of heart.—

'Ah, woe is me! that I should fondly part 30
From my dear native land! Ah, foolish maid!
Glad was the hour, when, with thee, myriads bade
Adieu to Ganges and their pleasant fields!
To one so friendless the clear freshet yields
A bitter coolness; the ripe grape is sour:
Yet I would have, great gods! but one short hour
Of native air—let me but die at home.'

Endymion to heaven's airy dome
Was offering up a hecatomb of vows,
When these words reach'd him. Whereupon he bows 40
His head through thorny-green entanglement
Of underwood, and to the sound is bent,
Anxious as hind towards her hidden fawn.

'Is no one near to help me? No fair dawn
Of life from charitable voice? No sweet saying
To set my dull and sadden'd spirit playing?
No hand to toy with mine? No lips so sweet
That I may worship them? No eyelids meet
To twinkle on my bosom? No one dies
Before me, till from these enslaving eyes 50
Redemption sparkles!—I am sad and lost.'

Thou, Carian lord, hadst better have been tost
Into a whirlpool. Vanish into air,
Warm mountaineer! for canst thou only bear
A woman's sigh alone and in distress?
See not her charms! Is Phœbe passionless?
Phœbe is fairer far—O gaze no more:—
Yet if thou wilt behold all beauty's store,
Behold her panting in the forest grass!
Do not those curls of glossy jet surpass 60
For tenderness the arms so idly lain
Amongst them? Feelest not a kindred pain,
To see such lovely eyes in swimming search
After some warm delight, that seems to perch
Dovelike in the dim cell lying beyond
Their upper lids?—Hist!

'O for Hermes' wand,
To touch this flower into human shape!
That woodland Hyacinthus could escape
From his green prison, and here kneeling down
Call me his queen, his second life's fair crown! 70
Ah me, how I could love!—My soul doth melt
For the unhappy youth—Love! I have felt
So faint a kindness, such a meek surrender
To what my own full thoughts had made too tender,

That but for tears my life had fled away!—
Ye deaf and senseless minutes of the day,
And thou, old forest, hold ye this for true,
There is no lightning, no authentic dew
But in the eye of love: there's not a sound,
Melodious howsoever, can confound 80
The heavens and earth in one to such a death
As doth the voice of love: there's not a breath
Will mingle kindly with the meadow air,
Till it has panted round, and stolen a share
Of passion from the heart!'—

Upon a bough
He leant, wretched. He surely cannot now
Thirst for another love: O impious,
That he can even dream upon it thus!—
Thought he, 'Why am I not as are the dead,
Since to a woe like this I have been led 90
Through the dark earth, and through the wondrous sea?
Goddess! I love thee not the less: from thee
By Juno's smile I turn not—no, no, no—
While the great waters are at ebb and flow.—
I have a triple soul! O fond pretence—
For both, for both my love is so immense,
I feel my heart is cut for them in twain.'

And so he groan'd, as one by beauty slain.
The lady's heart beat quick, and he could see
Her gentle bosom heave tumultuously. 100
He sprang from his green covert: there she lay,
Sweet as a muskrose upon new-made hay;
With all her limbs on tremble, and her eyes
Shut softly up alive. To speak he tries.
'Fair damsel, pity me! forgive that I
Thus violate thy bower's sanctity!
O pardon me, for I am full of grief—
Grief born of thee, young angel! fairest thief!
Who stolen hast away the wings wherewith
I was to top the heavens. Dear maid, sith 110
Thou art my executioner, and I feel

Loving and hatred, misery and weal,
Will in a few short hours be nothing to me,
And all my story that much passion slew me;
Do smile upon the evening of my days:
And, for my tortur'd brain begins to craze,
Be thou my nurse; and let me understand
How dying I shall kiss that lily hand.—
Dost weep for me? Then should I be content.
Scowl on, ye fates! until the firmament 120
Outblackens Erebus, and the full-cavern'd earth
Crumbles into itself. By the cloud girth
Of Jove, those tears have given me a thirst
To meet oblivion.'—As her heart would burst
The maiden sobb'd awhile, and then replied:
'Why must such desolation betide
As that thou speakest of? Are not these green nooks
Empty of all misfortune? Do the brooks
Utter a gorgon voice? Does yonder thrush,°
Schooling its half-fledg'd little ones to brush 130
About the dewy forest, whisper tales?—
Speak not of grief, young stranger, or cold snails
Will slime the rose to night. Though if thou wilt,
Methinks 'twould be a guilt—a very guilt—
Not to companion thee, and sigh away
The light—the dusk—the dark—till break of day!'
'Dear lady,' said Endymion, ''tis past:
I love thee! and my days can never last.
That I may pass in patience still speak:
Let me have music dying, and I seek 140
No more delight—I bid adieu to all.
Didst thou not after other climates call,
And murmur about Indian streams?'—Then she,
Sitting beneath the midmost forest tree,
For pity sang this roundelay—

'O Sorrow,°
Why dost borrow
The natural hue of health, from vermeil lips?—
To give maiden blushes
To the white rose bushes? 150
Or is't thy dewy hand the daisy tips?

'O Sorrow,
Why dost borrow
The lustrous passion from a falcon-eye?—
To give the glow-worm light?
Or, on a moonless night,
To tinge, on syren shores, the salt sea-spry?°

'O Sorrow,
Why dost borrow
The mellow ditties from a mourning tongue?— 160
To give at evening pale
Unto the nightingale,
That thou mayst listen the cold dews among?

'O Sorrow,
Why dost borrow
Heart's lightness from the merriment of May?—
A lover would not tread°
A cowslip on the head,
Though he should dance from eve till peep of day—
Nor any drooping flower 170
Held sacred for thy bower,
Wherever he may sport himself and play.

'To Sorrow,
I bade good-morrow,
And thought to leave her far away behind;
But cheerly, cheerly,
She loves me dearly;
She is so constant to me, and so kind:
I would deceive her
And so leave her, 180
But ah! she is so constant and so kind.

'Beneath my palm trees, by the river side,°
I sat a weeping: in the whole world wide
There was no one to ask me why I wept,—
And so I kept
Brimming the water-lily cups with tears°
Cold as my fears.

'Beneath my palm trees, by the river side,
I sat a weeping: what enamour'd bride,
Cheated by shadowy wooer from the clouds, 190
But hides and shrouds
Beneath dark palm trees by a river side?

'And as I sat, over the light blue hills°
There came a noise of revellers: the rills
Into the wide stream came of purple hue—
'Twas Bacchus and his crew!
The earnest trumpet spake, and silver thrills
From kissing cymbals made a merry din—
'Twas Bacchus and his kin!
Like to a moving vintage down they came, 200
Crown'd with green leaves, and faces all on flame;
All madly dancing through the pleasant valley,
To scare thee, Melancholy!°
O then, O then, thou wast a simple name!
And I forgot thee, as the berried holly
By shepherds is forgotten, when, in June,
Tall chestnuts keep away the sun and moon:—
I rush'd into the folly!

'Within his car, aloft, young Bacchus stood,
Trifling his ivy-dart, in dancing mood, 210
With sidelong laughing;
And little rills of crimson wine imbrued
His plump white arms, and shoulders, enough white
For Venus' pearly bite:
And near him rode Silenus on his ass,
Pelted with flowers as he on did pass
Tipsily quaffing.

'Whence came ye, merry Damsels! whence came ye!
So many, and so many, and such glee?
Why have ye left your bowers desolate, 220
Your lutes, and gentler fate?—
"We follow Bacchus! Bacchus on the wing,
A conquering!

Bacchus, young Bacchus! good or ill betide,
We dance before him thorough kingdoms wide:—
Come hither, lady fair, and joined be
        To our wild minstrelsy!"

'Whence came ye, jolly Satyrs! whence came ye!
So many, and so many, and such glee?
Why have ye left your forest haunts, why left   230
        Your nuts in oak-tree cleft?—
"For wine, for wine we left our kernel tree;
For wine we left our heath, and yellow brooms,
        And cold mushrooms;
For wine we follow Bacchus through the earth;
Great God of breathless cups and chirping mirth!—
Come hither, lady fair, and joined be
        To our mad minstrelsy!"

'Over wide streams and mountains great we went,
And, save when Bacchus kept his ivy tent,   240
Onward the tiger and the leopard pants,
        With Asian elephants:
Onward these myriads—with song and dance,
With zebras striped, and sleek Arabians' prance,
Web-footed alligators, crocodiles,
Bearing upon their scaly backs, in files,
Plump infant laughers mimicking the coil
Of seamen, and stout galley-rowers' toil:
With toying oars and silken sails they glide,
        Nor care for wind and tide.   250

'Mounted on panthers' furs and lions' manes,°
From rear to van they scour about the plains;
A three days' journey in a moment done:
And always, at the rising of the sun,
About the wilds they hunt with spear and horn,
        On spleenful unicorn.

'I saw Osirian Egypt kneel adown
        Before the vine-wreath crown!
I saw parch'd Abyssinia rouse and sing
        To the silver cymbals' ring!   260

I saw the whelming vintage hotly pierce
        Old Tartary the fierce!
The kings of Inde their jewel-sceptres vail,
And from their treasures scatter pearled hail;
Great Brahma from his mystic heaven groans,
        And all his priesthood moans;
Before young Bacchus' eye-wink turning pale.—
Into these regions came I following him,
Sick hearted, weary—so I took a whim
To stray away into these forests drear 270
        Alone, without a peer:
And I have told thee all thou mayest hear.

        'Young stranger!
        I've been a ranger
In search of pleasure throughout every clime:
        Alas, 'tis not for me!
        Bewitch'd I sure must be,
To lose in grieving all my maiden prime.

        'Come then, Sorrow!
        Sweetest Sorrow! 280
Like an own babe I nurse thee on my breast:
        I thought to leave thee
        And deceive thee,
But now of all the world I love thee best.

        'There is not one,
        No, no, not one
But thee to comfort a poor lonely maid;
        Thou art her mother,
        And her brother,
Her playmate, and her wooer in the shade.' 290

O what a sigh she gave in finishing,
And look, quite dead to every worldly thing!
Endymion could not speak, but gazed on her;
And listened to the wind that now did stir
About the crisped oaks full drearily,
Yet with as sweet a softness as might be
Remember'd from its velvet summer song.

At last he said: 'Poor lady, how thus long
Have I been able to endure that voice?
Fair Melody! kind Syren! I've no choice; 300
I must be thy sad servant evermore:
I cannot choose but kneel here and adore.
Alas, I must not think—by Phœbe, no!
Let me not think, soft Angel! shall it be so?
Say, beautifullest, shall I never think?
O thou could'st foster me beyond the brink
Of recollection! make my watchful care
Close up its bloodshot eyes, nor see despair!
Do gently murder half my soul, and I
Shall feel the other half so utterly!— 310
I'm giddy at that cheek so fair and smooth;
O let it blush so ever! let it soothe
My madness! let it mantle rosy-warm
With the tinge of love, panting in safe alarm.—
This cannot be thy hand, and yet it is;
And this is sure thine other softling—this
Thine own fair bosom, and I am so near!
Wilt fall asleep? O let me sip that tear!
And whisper one sweet word that I may know
This is this world—sweet dewy blossom!'—*Woe!* 320
Woe! Woe to that Endymion! Where is he?—
Even these words went echoing dismally
Through the wide forest—a most fearful tone,
Like one repenting in his latest moan;
And while it died away a shade pass'd by,
As of a thunder cloud. When arrows fly
Through the thick branches, poor ring-doves sleek forth
Their timid necks and tremble; so these both
Leant to each other trembling, and sat so
Waiting for some destruction—when lo, 330
Foot-feather'd Mercury appear'd sublime
Beyond the tall tree tops; and in less time
Than shoots the slanted hail-storm, down he dropt
Towards the ground; but rested not, nor stopt
One moment from his home: only the sward
He with his wand light touch'd, and heavenward
Swifter than sight was gone—even before
The teeming earth a sudden witness bore

Of his swift magic. Diving swans appear
Above the crystal circlings white and clear; 340
And catch the cheated eye in wide surprise,
How they can dive in sight and unseen rise—
So from the turf outsprang two steeds jet-black,
Each with large dark blue wings upon his back.
The youth of Caria plac'd the lovely dame
On one, and felt himself in spleen to tame
The other's fierceness. Through the air they flew,
High as the eagles. Like two drops of dew
Exhal'd to Phœbus' lips, away they are gone,
Far from the earth away—unseen, alone, 350
Among cool clouds and winds, but that the free,
The buoyant life of song can floating be
Above their heads, and follow them untir'd.—
Muse of my native land, am I inspir'd?
This is the giddy air, and I must spread
Wide pinions to keep here; nor do I dread°
Or height, or depth, or width, or any chance
Precipitous: I have beneath my glance
Those towering horses and their mournful freight.
Could I thus sail, and see, and thus await 360
Fearless for power of thought, without thine aid?—

There is a sleepy dusk, an odorous shade
From some approaching wonder, and behold
Those winged steeds, with snorting nostrils bold
Snuff at its faint extreme, and seem to tire,
Dying to embers from their native fire!

There curl'd a purple mist around them; soon,
It seem'd as when around the pale new moon
Sad Zephyr droops the clouds like weeping willow:
'Twas Sleep slow journeying with head on pillow. 370
For the first time, since he came nigh dead born
From the old womb of night, his cave forlorn
Had he left more forlorn; for the first time,
He felt aloof the day and morning's prime—
Because into his depth Cimmerian
There came a dream, shewing how a young man,
Ere a lean bat could plump its wintery skin,

Would at high Jove's empyreal footstool win
An immortality, and how espouse
Jove's daughter, and be reckon'd of his house. 380
Now was he slumbering towards heaven's gate,
That he might at the threshold one hour wait
To hear the marriage melodies, and then
Sink downward to his dusky cave again.
His litter of smooth semilucent mist,
Diversely ting'd with rose and amethyst,
Puzzled those eyes that for the centre sought;
And scarcely for one moment could be caught
His sluggish form reposing motionless.
Those two on winged steeds, with all the stress 390
Of vision search'd for him, as one would look
Athwart the sallows of a river nook
To catch a glance at silver throated eels,—
Or from old Skiddaw's top, when fog conceals°
His rugged forehead in a mantle pale,
With an eye-guess towards some pleasant vale
Descry a favourite hamlet faint and far.

These raven horses, though they foster'd are
Of earth's splenetic fire, dully drop
Their full-veined ears, nostrils blood wipe, and stop; 400
Upon the spiritless mist have they outspread
Their ample feathers, are in slumber dead,—
And on those pinions, level in mid air,
Endymion sleepeth and the lady fair.
Slowly they sail, slowly as icy isle
Upon a calm sea drifting: and meanwhile
The mournful wanderer dreams. Behold! he walks
On heaven's pavement; brotherly he talks
To divine powers: from his hand full fain
Juno's proud birds are pecking pearly grain: 410
He tries the nerve of Phœbus' golden bow,
And asketh where the golden apples grow:
Upon his arm he braces Pallas' shield,
And strives in vain to unsettle and wield
A Jovian thunderbolt: arch Hebe brings
A full-brimm'd goblet, dances lightly, sings
And tantalizes long; at last he drinks,

And lost in pleasure at her feet he sinks,
Touching with dazzled lips her starlight hand.
He blows a bugle,—an ethereal band 420
Are visible above: the Seasons four,—
Green-kyrtled Spring, flush Summer, golden store
In Autumn's sickle, Winter frosty hoar,
Join dance with shadowy Hours; while still the blast,
In swells unmitigated, still doth last
To sway their floating morris. 'Whose is this?
Whose bugle?' he inquires: they smile—'O Dis!
Why is this mortal here? Dost thou not know
Its mistress' lips? Not thou?—'Tis Dian's: lo!
She rises crescented!' He looks, 'tis she, 430
His very goddess: good-bye earth, and sea,
And air, and pains, and care, and suffering;
Good-bye to all but love! Then doth he spring
Towards her, and awakes—and, strange, o'erhead,
Of those same fragrant exhalations bred,
Beheld awake his very dream: the gods
Stood smiling; merry Hebe laughs and nods;
And Phœbe bends towards him crescented.
O state perplexing! On the pinion bed,
Too well awake, he feels the panting side 440
Of his delicious lady. He who died°
For soaring too audacious in the sun,
Where that same treacherous wax began to run,
Felt not more tongue-tied than Endymion.
His heart leapt up as to its rightful throne,
To that fair shadow'd passion puls'd its way—
Ah, what perplexity! Ah, well a day!
So fond, so beauteous was his bed-fellow,
He could not help but kiss her: then he grew
Awhile forgetful of all beauty save 450
Young Phœbe's, golden hair'd; and so 'gan crave
Forgiveness: yet he turn'd once more to look
At the sweet sleeper,—all his soul was shook,—
She press'd his hand in slumber; so once more
He could not help but kiss her and adore.
At this the shadow wept, melting away.
The Latmian started up: 'Bright goddess, stay!
Search my most hidden breast! By truth's own tongue,

I have no dædale heart: why is it wrung°
To desperation? Is there nought for me, 460
Upon the bourne of bliss, but misery?'

These words awoke the stranger of dark tresses:
Her dawning love-look rapt Endymion blesses
With 'haviour soft. Sleep yawned from underneath.
'Thou swan of Ganges, let us no more breathe
This murky phantasm! thou contented seem'st
Pillow'd in lovely idleness, nor dream'st
What horrors may discomfort thee and me.
Ah, shouldst thou die from my heart-treachery!—
Yet did she merely weep—her gentle soul 470
Hath no revenge in it: as it is whole
In tenderness, would I were whole in love!
Can I prize thee, fair maid, all price above,
Even when I feel as true as innocence?
I do, I do.—What is this soul then? Whence
Came it? It does not seem my own, and I
Have no self-passion or identity.
Some fearful end must be: where, where is it?
By Nemesis, I see my spirit flit
Alone about the dark—Forgive me, sweet: 480
Shall we away?' He rous'd the steeds: they beat
Their wings chivalrous into the clear air,
Leaving old Sleep within his vapoury lair.

The good-night blush of eve was waning slow,
And Vesper, risen star, began to throe
In the dusk heavens silverly, when they
Thus sprang direct towards the Galaxy.
Nor did speed hinder converse soft and strange—
Eternal oaths and vows they interchange,
In such wise, in such temper, so aloof 490
Up in the winds, beneath a starry roof,
So witless of their doom, that verily
'Tis well nigh past man's search their hearts to see;
Whether they wept, or laugh'd, or griev'd, or toy'd—
Most like with joy gone mad, with sorrow cloy'd.

Full facing their swift flight, from ebon streak,
The moon put forth a little diamond peak,

No bigger than an unobserved star,
Or tiny point of fairy scymetar;
Bright signal that she only stoop'd to tie 500
Her silver sandals, ere deliciously
She bow'd into the heavens her timid head.
Slowly she rose, as though she would have fled,
While to his lady meek the Carian turn'd,
To mark if her dark eyes had yet discern'd
This beauty in its birth—Despair! despair!
He saw her body fading gaunt and spare
In the cold moonshine. Straight he seiz'd her wrist;
It melted from his grasp: her hand he kiss'd,
And, horror! kiss'd his own—he was alone. 510
Her steed a little higher soar'd, and then
Dropt hawkwise to the earth.

There lies a den,
Beyond the seeming confines of the space
Made for the soul to wander in and trace
Its own existence, of remotest glooms.
Dark regions are around it, where the tombs
Of buried griefs the spirit sees, but scarce
One hour doth linger weeping, for the pierce
Of new-born woe it feels more inly smart:
And in these regions many a venom'd dart 520
At random flies; they are the proper home
Of every ill: the man is yet to come
Who hath not journeyed in this native hell.
But few have ever felt how calm and well
Sleep may be had in that deep den of all.
There anguish does not sting; nor pleasure pall:
Woe-hurricanes beat ever at the gate,
Yet all is still within and desolate.
Beset with plainful gusts, within ye hear
No sound so loud as when on curtain'd bier 530
The death-watch tick is stifled. Enter none
Who strive therefore: on the sudden it is won.
Just when the sufferer begins to burn,
Then it is free to him; and from an urn,
Still fed by melting ice, he takes a draught—
Young Semele such richness never quaft

In her maternal longing! Happy gloom!
Dark Paradise! where pale becomes the bloom
Of health by due; where silence dreariest
Is most articulate; where hopes infest; 540
Where those eyes are the brightest far that keep
Their lids shut longest in a dreamless sleep.
O happy spirit-home! O wondrous soul!
Pregnant with such a den to save the whole
In thine own depth. Hail, gentle Carian!
For, never since thy griefs and woes began,
Hast thou felt so content: a grievous feud
Hath led thee to this Cave of Quietude.
Aye, his lull'd soul was there, although upborne
With dangerous speed: and so he did not mourn 550
Because he knew not whither he was going.
So happy was he, not the aerial blowing
Of trumpets at clear parley from the east
Could rouse from that fine relish, that high feast.
They stung the feather'd horse: with fierce alarm
He flapp'd towards the sound. Alas, no charm
Could lift Endymion's head, or he had view'd
A skyey masque, a pinion'd multitude,—
And silvery was its passing: voices sweet
Warbling the while as if to lull and greet 560
The wanderer in his path. Thus warbled they,
While past the vision went in bright array.

'Who, who from Dian's feast would be away?
For all the golden bowers of the day
Are empty left? Who, who away would be
From Cynthia's wedding and festivity?
Not Hesperus: lo! upon his silver wings
He leans away for highest heaven and sings,
Snapping his lucid fingers merrily!—
Ah, Zephyrus! art here, and Flora too! 570
Ye tender bibbers of the rain and dew,
Young playmates of the rose and daffodil,
Be careful, ere ye enter in, to fill
Your baskets high
With fennel green, and balm, and golden pines,
Savory, latter-mint, and columbines,

Cool parsley, basil sweet, and sunny thyme;
Yea, every flower and leaf of every clime,
All gather'd in the dewy morning: hie
        Away! fly, fly!— 580
Crystalline brother of the belt of heaven,°
Aquarius! to whom king Jove has given°
Two liquid pulse streams 'stead of feather'd wings,
Two fan-like fountains,—thine illuminings
        For Dian play:
Dissolve the frozen purity of air;
Let thy white shoulders silvery and bare
Shew cold through watery pinions; make more bright
The Star-Queen's crescent on her marriage night:
        Haste, haste away!— 590
Castor has tamed the planet Lion, see!
And of the Bear has Pollux mastery:
A third is in the race! who is the third,
Speeding away swift as the eagle bird?
        The ramping Centaur!
The Lion's mane's on end: the Bear how fierce!
The Centaur's arrow ready seems to pierce
Some enemy: far forth his bow is bent
Into the blue of heaven. He'll be shent,
        Pale unrelentor, 600
When he shall hear the wedding lutes a playing.—
Andromeda! sweet woman! why delaying
So timidly among the stars: come hither!
Join this bright throng, and nimbly follow whither
        They all are going.
Danae's Son, before Jove newly bow'd,
Has wept for thee, calling to Jove aloud.
Thee, gentle lady, did he disenthral:
Ye shall for ever live and love, for all
        Thy tears are flowing.— 610
By Daphne's fright, behold Apollo!—'

                                        More
Endymion heard not: down his steed him bore,
Prone to the green head of a misty hill.

  His first touch of the earth went nigh to kill.

'Alas!' said he, 'were I but always borne
Through dangerous winds, had but my footsteps worn
A path in hell, for ever would I bless
Horrors which nourish an uneasiness
For my own sullen conquering: to him
Who lives beyond earth's boundary, grief is dim, 620
Sorrow is but a shadow: now I see
The grass; I feel the solid ground—Ah, me!
It is thy voice—divinest! Where?—who? who
Left thee so quiet on this bed of dew?
Behold upon this happy earth we are;
Let us ay love each other; let us fare
On forest-fruits, and never, never go
Among the abodes of mortals here below,
Or be by phantoms duped. O destiny!
Into a labyrinth now my soul would fly, 630
But with thy beauty will I deaden it.
Where didst thou melt to? By thee will I sit
For ever: let our fate stop here—a kid
I on this spot will offer: Pan will bid
Us live in peace, in love and peace among
His forest wildernesses. I have clung
To nothing, lov'd a nothing, nothing seen
Or felt but a great dream! O I have been
Presumptuous against love, against the sky,
Against all elements, against the tie 640
Of mortals each to each, against the blooms
Of flowers, rush of rivers, and the tombs
Of heroes gone! Against his proper glory
Has my own soul conspired: so my story
Will I to children utter, and repent.
There never liv'd a mortal man, who bent
His appetite beyond his natural sphere,
But starv'd and died. My sweetest Indian, here,
Here will I kneel, for thou redeemed hast
My life from too thin breathing: gone and past 650
Are cloudy phantasms. Caverns lone, farewel!
And air of visions, and the monstrous swell
Of visionary seas! No, never more
Shall airy voices cheat me to the shore
Of tangled wonder, breathless and aghast.

Adieu, my daintiest Dream! although so vast
My love is still for thee. The hour may come
When we shall meet in pure elysium.
On earth I may not love thee; and therefore
Doves will I offer up, and sweetest store 660
All through the teeming year: so thou wilt shine
On me, and on this damsel fair of mine,
And bless our simple lives. My Indian bliss!
My river-lily bud! one human kiss!
One sigh of real breath—one gentle squeeze,
Warm as a dove's nest among summer trees,
And warm with dew at ooze from living blood!
Whither didst melt? Ah, what of that!—all good
We'll talk about—no more of dreaming.—Now,
Where shall our dwelling be? Under the brow 670
Of some steep mossy hill, where ivy dun
Would hide us up, although spring leaves were none;
And where dark yew trees, as we rustle through,
Will drop their scarlet berry cups of dew?
O thou wouldst joy to live in such a place;
Dusk for our loves, yet light enough to grace
Those gentle limbs on mossy bed reclin'd:
For by one step the blue sky shouldst thou find,
And by another, in deep dell below,
See, through the trees, a little river go 680
All in its mid-day gold and glimmering.
Honey from out the gnarled hive I'll bring,
And apples, wan with sweetness, gather thee,—
Cresses that grow where no man may them see,
And sorrel untorn by the dew-claw'd stag:°
Pipes will I fashion of the syrinx flag,°
That thou mayst always know whither I roam,
When it shall please thee in our quiet home
To listen and think of love. Still let me speak;
Still let me dive into the joy I seek,— 690
For yet the past doth prison me. The rill,
Thou haply mayst delight in, will I fill
With fairy fishes from the mountain tarn,
And thou shalt feed them from the squirrel's barn.
Its bottom will I strew with amber shells,
And pebbles blue from deep enchanted wells.

Its sides I'll plant with dew-sweet eglantine,
And honeysuckles full of clear bee-wine.
I will entice this crystal rill to trace
Love's silver name upon the meadow's face. 700
I'll kneel to Vesta, for a flame of fire;
And to god Phœbus, for a golden lyre;
To Empress Dian, for a hunting spear;
To Vesper, for a taper silver-clear,
That I may see thy beauty through the night;
To Flora, and a nightingale shall light
Tame on thy finger; to the River-gods,
And they shall bring thee taper fishing-rods
Of gold, and lines of Naiads' long bright tress.
Heaven shield thee for thine utter loveliness! 710
Thy mossy footstool shall the altar be
'Fore which I'll bend, bending, dear love, to thee:
Those lips shall be my Delphos, and shall speak°
Laws to my footsteps, colour to my cheek,
Trembling or stedfastness to this same voice,
And of three sweetest pleasurings the choice:
And that affectionate light, those diamond things,
Those eyes, those passions, those supreme pearl springs,
Shall be my grief, or twinkle me to pleasure.
Say, is not bliss within our perfect seisure? 720
O that I could not doubt!'

The mountaineer
Thus strove by fancies vain and crude to clear
His briar'd path to some tranquillity.
It gave bright gladness to his lady's eye,
And yet the tears she wept were tears of sorrow;
Answering thus, just as the golden morrow
Beam'd upward from the vallies of the east:
'O that the flutter of this heart had ceas'd,
Or the sweet name of love had pass'd away.
Young feather'd tyrant! by a swift decay 730
Wilt thou devote this body to the earth:
And I do think that at my very birth
I lisp'd thy blooming titles inwardly;
For at the first, first dawn and thought of thee,
With uplift hands I blest the stars of heaven.

Art thou not cruel? Ever have I striven
To think thee kind, but ah, it will not do!
When yet a child, I heard that kisses drew
Favour from thee, and so I kisses gave
To the void air, bidding them find out love: 740
But when I came to feel how far above
All fancy, pride, and fickle maidenhood,
All earthly pleasure, all imagin'd good,
Was the warm tremble of a devout kiss,—
Even then, that moment, at the thought of this,
Fainting I fell into a bed of flowers,
And languish'd there three days. Ye milder powers,
Am I not cruelly wrong'd? Believe, believe
Me, dear Endymion, were I to weave
With my own fancies garlands of sweet life, 750
Thou shouldst be one of all. Ah, bitter strife!
I may not be thy love: I am forbidden—
Indeed I am—thwarted, affrighted, chidden,
By things I trembled at, and gorgon wrath.
Twice hast thou ask'd whither I went: henceforth
Ask me no more! I may not utter it.
Nor may I be thy love. We might commit
Ourselves at once to vengeance; we might die;
We might embrace and die: voluptuous thought!
Enlarge not to my hunger, or I'm caught 760
In trammels of perverse deliciousness.
No, no, that shall not be: thee will I bless,
And bid a long adieu.'

The Carian
No word return'd: both lovelorn, silent, wan,
Into the vallies green together went.
Far wandering, they were perforce content
To sit beneath a fair lone beechen tree;
Nor at each other gaz'd, but heavily
Por'd on its hazle cirque of shedded leaves.

Endymion! unhappy! it nigh grieves 770
Me to behold thee thus in last extreme:
Ensky'd ere this, but truly that I deem
Truth the best music in a first-born song.

Thy lute-voic'd brother will I sing ere long,°
And thou shalt aid—hast thou not aided me?
Yes, moonlight Emperor! felicity
Has been thy meed for many thousand years;
Yet often have I, on the brink of tears,
Mourn'd as if yet thou wert a forester;—
Forgetting the old tale.

He did not stir 780
His eyes from the dead leaves, or one small pulse
Of joy he might have felt. The spirit culls
Unfaded amaranth, when wild it strays
Through the old garden-ground of boyish days.
A little onward ran the very stream
By which he took his first soft poppy dream;
And on the very bark 'gainst which he leant
A crescent he had carv'd, and round it spent
His skill in little stars. The teeming tree
Had swollen and green'd the pious charactery, 790
But not ta'en out. Why, there was not a slope
Up which he had not fear'd the antelope;
And not a tree, beneath whose rooty shade
He had not with his tamed leopards play'd:
Nor could an arrow light, or javelin,
Fly in the air where his had never been—
And yet he knew it not.

O treachery!
Why does his lady smile, pleasing her eye
With all his sorrowing? He sees her not.
But who so stares on him? His sister sure! 800
Peona of the woods!—Can she endure—
Impossible—how dearly they embrace!
His lady smiles; delight is in her face:
It is no treachery.

'Dear brother mine!
Endymion, weep not so! Why shouldst thou pine
When all great Latmos so exalt will be?
Thank the great gods, and look not bitterly;
And speak not one pale word, and sigh no more.

Sure I will not believe thou hast such store
Of grief, to last thee to my kiss again. 810
Thou surely canst not bear a mind in pain,
Come hand in hand with one so beautiful.
Be happy both of you! for I will pull
The flowers of autumn for your coronals.
Pan's holy priest for young Endymion calls;
And when he is restor'd, thou, fairest dame,
Shalt be our queen. Now, is it not a shame
To see ye thus,—not very, very sad?
Perhaps ye are too happy to be glad:
O feel as if it were a common day; 820
Free-voic'd as one who never was away.
No tongue shall ask, whence come ye? but ye shall
Be gods of your own rest imperial.
Not even I, for one whole month, will pry
Into the hours that have pass'd us by,
Since in my arbour I did sing to thee.
O Hermes! on this very night will be
A hymning up to Cynthia, queen of light;
For the soothsayers old saw yesternight
Good visions in the air,—whence will befal, 830
As say these sages, health perpetual
To shepherds and their flocks; and furthermore,
In Dian's face they read the gentle lore:
Therefore for her these vesper-carols are.
Our friends will all be there from nigh and far.
Many upon thy death have ditties made;
And many, even now, their foreheads shade
With cypress, on a day of sacrifice.
New singing for our maids shalt thou devise,
And pluck the sorrow from our huntsmen's brows. 840
Tell me, my lady-queen, how to espouse
This wayward brother to his rightful joys!
His eyes are on thee bent, as thou didst poise
His fate most goddess-like. Help me, I pray,
To lure—Endymion, dear brother, say
What ails thee?' He could bear no more, and so
Bent his soul fiercely like a spiritual bow,
And twang'd it inwardly, and calmly said:
'I would have thee my only friend, sweet maid!

My only visitor! not ignorant though, 850
That those deceptions which for pleasure go
'Mong men, are pleasures real as real may be:
But there are higher ones I may not see,
If impiously an earthly realm I take.
Since I saw thee, I have been wide awake
Night after night, and day by day, until
Of the empyrean I have drunk my fill.
Let it content thee, Sister, seeing me
More happy than betides mortality.
A hermit young, I'll live in mossy cave, 860
Where thou alone shalt come to me, and lave
Thy spirit in the wonders I shall tell.
Through me the shepherd realm shall prosper well;
For to thy tongue will I all health confide.
And, for my sake, let this young maid abide
With thee as a dear sister. Thou alone,
Peona, mayst return to me. I own
This may sound strangely: but when, dearest girl,
Thou seest it for my happiness, no pearl
Will trespass down those cheeks. Companion fair! 870
Wilt be content to dwell with her, to share
This sister's love with me?' Like one resign'd
And bent by circumstance, and thereby blind
In self-commitment, thus that meek unknown:
'Aye, but a buzzing by my ears has flown,
Of jubilee to Dian:—truth I heard?
Well then, I see there is no little bird,°
Tender soever, but is Jove's own care.
Long have I sought for rest, and, unaware,
Behold I find it! so exalted too! 880
So after my own heart! I knew, I knew
There was a place untenanted in it:
In that same void white Chastity shall sit,
And monitor me nightly to lone slumber.
With sanest lips I vow me to the number
Of Dian's sisterhood; and, kind lady,
With thy good help, this very night shall see
My future days to her fane consecrate.'

As feels a dreamer what doth most create

His own particular fright, so these three felt: 890
Or like one who, in after ages, knelt
To Lucifer or Baal, when he'd pine
After a little sleep: or when in mine
Far under-ground, a sleeper meets his friends
Who know him not. Each diligently bends
Towards common thoughts and things for very fear;
Striving their ghastly malady to cheer,
By thinking it a thing of yes and no,
That housewives talk of. But the spirit-blow
Was struck, and all were dreamers. At the last 900
Endymion said: 'Are not our fates all cast?
Why stand we here? Adieu, ye tender pair!
Adieu!' Whereat those maidens, with wild stare,
Walk'd dizzily away. Pained and hot
His eyes went after them, until they got
Near to a cypress grove, whose deadly maw,
In one swift moment, would what then he saw
Engulph for ever. 'Stay!' he cried, 'ah, stay!
Turn, damsels! hist! one word I have to say.
Sweet Indian, I would see thee once again. 910
It is a thing I dote on: so I'd fain,
Peona, ye should hand in hand repair
Into those holy groves, that silent are
Behind great Dian's temple. I'll be yon,
At vesper's earliest twinkle—they are gone—
But once, once, once again—' At this he press'd
His hands against his face, and then did rest
His head upon a mossy hillock green,
And so remain'd as he a corpse had been
All the long day; save when he scantly lifted 920
His eyes abroad, to see how shadows shifted
With the slow move of time,—sluggish and weary
Until the poplar tops, in journey dreary,
Had reach'd the river's brim. Then up he rose,
And, slowly as that very river flows,
Walk'd towards the temple grove with this lament:
'Why such a golden eve? The breeze is sent
Careful and soft, that not a leaf may fall
Before the serene father of them all
Bows down his summer head below the west. 930

Now am I of breath, speech, and speed possest,
But at the setting I must bid adieu
To her for the last time. Night will strew
On the damp grass myriads of lingering leaves,
And with them shall I die; nor much it grieves
To die, when summer dies on the cold sward.
Why, I have been a butterfly, a lord
Of flowers, garlands, love-knots, silly posies,
Groves, meadows, melodies, and arbour roses;
My kingdom's at its death, and just it is 940
That I should die with it: so in all this
We miscal grief, bale, sorrow, heartbreak, woe,
What is there to plain of? By Titan's foe°
I am but rightly serv'd.' So saying, he
Tripp'd lightly on, in sort of deathful glee;
Laughing at the clear stream and setting sun,
As though they jests had been: nor had he done
His laugh at nature's holy countenance,
Until that grove appear'd, as if perchance,
And then his tongue with sober seemlihed° 950
Gave utterance as he entered: 'Ha! I said,
King of the butterflies; but by this gloom,
And by old Rhadamanthus' tongue of doom,
This dusk religion, pomp of solitude,
And the Promethean clay by thief endued,
By old Saturnus' forelock, by his head°
Shook with eternal palsy, I did wed
Myself to things of light from infancy;
And thus to be cast out, thus lorn to die,
Is sure enough to make a mortal man 960
Grow impious.' So he inwardly began
On things for which no wording can be found;
Deeper and deeper sinking, until drown'd
Beyond the reach of music: for the choir
Of Cynthia he heard not, though rough briar
Nor muffling thicket interpos'd to dull
The vesper hymn, far swollen, soft and full,
Through the dark pillars of those sylvan aisles.
He saw not the two maidens, nor their smiles,
Wan as primroses gather'd at midnight 970
By chilly finger'd spring. 'Unhappy wight!

Endymion!' said Peona, 'we are here!
What wouldst thou ere we all are laid on bier?'
Then he embrac'd her, and his lady's hand
Press'd, saying: 'Sister, I would have command,
If it were heaven's will, on our sad fate.'
At which that dark-eyed stranger stood elate
And said, in a new voice, but sweet as love,
To Endymion's amaze: 'By Cupid's dove,
And so thou shalt! and by the lily truth 980
Of my own breast thou shalt, beloved youth!'
And as she spake, into her face there came
Light, as reflected from a silver flame:
Her long black hair swell'd ampler, in display
Full golden; in her eyes a brighter day
Dawn'd blue and full of love. Aye, he beheld
Phœbe, his passion! joyous she upheld
Her lucid bow, continuing thus: 'Drear, drear
Has our delaying been; but foolish fear
Withheld me first; and then decrees of fate; 990
And then 'twas fit that from this mortal state
Thou shouldst, my love, by some unlook'd for change
Be spiritualiz'd. Peona, we shall range
These forests, and to thee they safe shall be
As was thy cradle; hither shalt thou flee
To meet us many a time.' Next Cynthia bright
Peona kiss'd, and bless'd with fair good night:
Her brother kiss'd her too, and knelt adown
Before his goddess, in a blissful swoon.
She gave her fair hands to him, and behold, 1000
Before three swiftest kisses he had told,
They vanish'd far away!—Peona went
Home through the gloomy wood in wonderment.

On Oxford

1

The Gothic looks solemn,
The plain Doric column
Supports an old Bishop and Crosier;
The mouldering arch,
Shaded o'er by a larch
Stands next door to Wilson the Hosier.

2

Vicè—that is, by turns,—
O'er pale faces mourns
The black tassell'd trencher and common hat;°
The Chantry boy sings, 10
The Steeple-bell rings,
And as for the Chancellor—*dominat.*

3

There are plenty of trees,
And plenty of ease,
And plenty of fat deer for Parsons;
And when it is venison,
Short is the benison,—
Then each on a leg or thigh fastens.

'Think not of it, sweet one, so'

Think not of it, sweet one, so;
Give it not a tear;
Sigh thou mayest, but bid it go
Any, any where.

Do not look so sad, sweet one,
Sad and fadingly:
Shed one drop then—It is gone—
Oh! 'twas born to die—

Still so pale?—then, dearest, weep;
  Weep! I'll count the tears: 10
And each one shall be a bliss
  For thee in after years.

Brighter has it left thine eyes
  Than a sunny rill:°
And thy whispering melodies
  Are tenderer still.

Yet, as all things mourn awhile
  At fleeting blisses,
Let us too!—but be our dirge
  A dirge of kisses. 20

'In drear nighted December'

In drear nighted December
  Too happy, happy tree
Thy Branches ne'er remember
  Their green felicity—
The north cannot undo them
With a sleety whistle through them
Nor frozen thawings glew them
  From budding at the prime—

In drear nighted December
  Too happy happy Brook 10
Thy bubblings ne'er remember
  Apollo's Summer look
But with a sweet forgetting
They stay their crystal fretting
Never never petting
  About the frozen time—

Ah! would 'twere so with many
A gentle girl and boy—
But were there ever any
Writh'd not of passed joy: 20
The feel of not to feel it°
When there is none to heal it
Nor numbed sense to steel it°
Was never said in rhyme—

'Before he went to live with owls and bats'

Before he went to live with owls and bats,
Nebuchadnezzar had an ugly dream,
Worse than a Housewife's, when she thinks her cream
Made a Naumachia for mice and rats:°
So scared, he sent for that 'good king of cats,"°
Young Daniel, who did straightway pluck the beam
From out his eye, and said—'I do not deem
Your sceptre worth a straw, your cushions old door mats.'
A horrid Nightmare, similar somewhat,
Of late has haunted a most valiant crew 10
Of Loggerheads and Chapmen;—we are told
That any Daniel, though he be a sot,
Can make their lying lips turn pale of hue,
By drawling out—'Ye are that head of gold!'

To Mrs. Reynoldse's Cat

Cat! who hast past thy Grand Climacteric,°
How many mice and Rats hast in thy days
Destroyed?—how many tit bits stolen? Gaze
With those bright languid segments green and prick
Those velvet ears—but prythee do not stick
Thy latent talons in me—and upraise
Thy gentle mew—and tell me all thy frays
Of Fish and Mice and Rats and tender chick.
Nay look not down, nor lick thy dainty wrists—
For all the weezy Asthma—and for all 10

Thy tail's tip is nicked off—and though the fists
   Of many a Maid have given thee many a mawl°
Still is that fur as soft as when the lists
   In youth thou enterd'st on glass bottled wall—°

Lines on seeing a Lock of Milton's hair

Chief of organic Numbers!
   Old Scholar of the Spheres!
Thy Spirit never slumbers,
   But rolls about our ears
For ever, and for ever:
O, what a mad endeavour
      Worketh he,
Who, to thy sacred and ennobled hearse,
Would offer a burnt sacrifice of verse
      And Melody. 10

How heavenward thou soundedst,
   Live Temple of sweet noise;
And discord unconfoundedst,—
   Giving delight new joys,
And pleasure nobler pinions—
O, where are thy dominions?
      Lend thine ear,
To a young delian oath,—aye, by the soul,°
By all that from thy mortal Lips did roll;
And by the Kernel of thine earthly Love, 20
Beauty, in things on earth and things above;
   When every childish fashion
      Has vanish'd from my rhyme,
   Will I, grey-gone in passion,
      Leave to an after time
      Hymning and Harmony
Of thee, and of thy works, and of thy Life;
But vain is now the burning, and the strife,
Pangs are in vain—until I grow high-rife
      With old Philosophy; 30
And mad with glimpses of futurity!

For many years my offerings must be hush'd.
When I do speak, I'll think upon this hour,
Because I feel my forehead hot and flush'd—
Even at the simplest vassal of thy Power—°
A Lock of thy bright hair—sudden it came,
And I was startled, when I caught thy name
Coupled so unaware—
Yet at the moment, temperate was my blood—
Methought I had beheld it from the flood— 40

On Sitting Down to Read King Lear *Once Again*

O Golden-tongued Romance, with serene Lute!
Fair plumed Syren, Queen of far-away!
Leave melodizing on this wintry day
Shut up thine olden Pages, and be mute.
Adieu! for, once again, the fierce dispute,
Betwixt Damnation and impassion'd clay
Must I burn through; once more humbly assay
The bitter-sweet of this Shaksperean fruit.
Chief Poet! and ye Clouds of Albion,
Begetters of our deep eternal theme! 10
When through the old oak forest I am gone,
Let me not wander in a barren dream:
But, when I am consumed in the fire,
Give me new Phœnix Wings to fly at my desire.

'When I have fears that I may cease to be'

When I have fears that I may cease to be
Before my pen has glean'd my teeming brain,
Before high piled Books in charactery
Hold like rich garners the full ripen'd grain—
When I behold upon the night's starr'd face
Huge cloudy symbols of a high romance,
And feel that I may never live to trace
Their shadows with the magic hand of Chance:

And when I feel, fair creature of an hour,
  That I shall never look upon thee more 10
Never have relish in the fairy power
  Of unreflecting Love: then on the Shore
Of the wide world I stand alone and think
Till Love and Fame to Nothingness do sink.—

'O blush not so, O blush not so'

1

O blush not so, O blush not so
  Or I shall think you knowing;
And if you smile, the blushing while
  Then Maidenheads are going.

2

There's a blush for want, and a blush for shan't°
  And a blush for having done it,
There's a blush for thought, and a blush for naught,
  And a blush for just begun it.

3

O sigh not so, O sigh not so
  For it sounds of Eve's sweet Pipin;° 10
By those loosen'd hips, you have tasted the pips
  And fought in an amorous nipping.

4

Will ye play once more, at nice cut-core
  For it only will last our youth out;
And we have the prime of the Kissing time
  We have not one sweet tooth out.

5

There's a sigh for yes, and a sigh for no,
  And a sigh for 'I can't bear it'—
O what can be done, shall we stay or run?
  O cut the sweet apple and share it— 20

'Hence Burgundy, Claret and port'

Hence Burgundy, Claret and port
  Away with old Hock and Madeira
Too couthly ye are for my sport;°
  There's a Beverage brighter and clearer—
Instead of a pitiful rummer
My Wine overbrims a whole Summer
    My bowl is the sky
    And I drink at my eye
    Till I feel in the brain
    A delphian pain—° 10
Then follow my Caius then follow°
    On the Green of the Hill,
    We will drink our fill
    Of golden sunshine,
    Till our brains intertwine
With the glory and grace of Apollo!

'God of the Meridian'

God of the Meridian
  And of the East and West
To thee my soul is flown,
  And my body is earthward press'd—
It is an awful mission
A terrible division
And leaves a gulph austere
To be fill'd with worldly fear—
Aye, when the Soul is fled
To high above our head 10
Affrighted do we gaze
After its airy maze—
As doth a Mother wild
When her young infant child°
Is in an eagle's claws—
And is not this the cause

Of Madness? God of Song
Thou bearest me along
Through sights I scarce can bear
O let me, let me share 20
With the hot Lyre and thee
The staid Philosophy.
Temper my lonely hours
And let me see thy bowers
More unalarm'd!—°

Lines on the Mermaid Tavern

Souls of Poets dead and gone,
What Elysium have ye known,
Happy field or mossy cavern,
Choicer than the Mermaid Tavern?
Have ye tippled drink more fine
Than mine host's Canary wine?
Or are fruits of Paradise
Sweeter than those dainty pies
Of venison? O generous food!
Drest as though bold Robin Hood 10
Would, with his maid Marian,
Sup and bowse from horn and can.°

I have heard that on a day
Mine host's sign-board flew away,
Nobody knew whither, till
An astrologer's old quill
To a sheepskin gave the story,
Said he saw you in your glory,
Underneath a new old sign
Sipping beverage divine, 20
And pledging with contented smack
The Mermaid in the Zodiac.

Souls of Poets dead and gone,
What Elysium have ye known,
Happy field or mossy cavern,
Choicer than the Mermaid Tavern?

Robin Hood

TO A FRIEND

No! those days are gone away,
And their hours are old and gray,
And their minutes buried all
Under the down-trodden pall
Of the leaves of many years:
Many times have winter's shears,
Frozen North, and chilling East,
Sounded tempests to the feast
Of the forest's whispering fleeces,
Since men knew nor rent nor leases.° 10

No, the bugle sounds no more,
And the twanging bow no more;
Silent is the ivory shrill
Past the heath and up the hill;
There is no mid-forest laugh,
Where lone Echo gives the half
To some wight, amaz'd to hear
Jesting, deep in forest drear.

On the fairest time of June
You may go, with sun or moon, 20
Or the seven stars to light you,
Or the polar ray to right you;
But you never may behold
Little John, or Robin bold;
Never one, of all the clan,
Thrumming on an empty can
Some old hunting ditty, while
He doth his green way beguile
To fair hostess Merriment,
Down beside the pasture Trent;° 30
For he left the merry tale
Messenger for spicy ale.

Gone, the merry morris din;
Gone, the song of Gamelyn;°

Gone, the tough-belted outlaw
Idling in the 'grenè shawe';°
All are gone away and past!
And if Robin should be cast
Sudden from his turfed grave,
And if Marian should have 40
Once again her forest days,
She would weep, and he would craze:
He would swear, for all his oaks,
Fall'n beneath the dockyard strokes,
Have rotted on the briny seas;
She would weep that her wild bees
Sang not to her—strange! that honey
Can't be got without hard money!

So it is: yet let us sing,
Honour to the old bow-string! 50
Honour to the bugle-horn!
Honour to the woods unshorn!
Honour to the Lincoln green!
Honour to the archer keen!
Honour to tight little John,°
And the horse he rode upon!
Honour to bold Robin Hood,
Sleeping in the underwood!
Honour to maid Marian,
And to all the Sherwood-clan! 60
Though their days have hurried by
Let us two a burden try.

'Time's sea hath been five years at its slow ebb'

Time's sea hath been five years at its slow ebb;—
Long hours have to and fro let creep the sand,—
Since I was tangled in thy beauty's web,
And snared by the ungloving of thy hand:
And yet I never look on midnight sky,
But I behold thine eyes' well-memoried light;
I cannot look upon the rose's dye,
But to thy cheek my soul doth take its flight:

I cannot look on any budding flower,
  But my fond ear, in fancy at thy lips, 10
And hearkening for a love-sound, doth devour
  Its sweets in the wrong sense.—Thou dost eclipse
Every delight with sweet remembering,
And grief unto my darling joys dost bring.

To the Nile

Son of the old moon-mountains African!°
  Chief of the Pyramid and Crocodile!
  We call thee fruitful, and, that very while,
A desert fills our seeing's inward span;
Nurse of swart nations since the world began,
  Art thou so fruitful? or dost thou beguile
  Such Men to honor thee, who, worn with toil,
Rest for a space 'twixt Cairo and Decan?
O may dark fancies err! they surely do;
  'Tis ignorance that makes a barren waste° 10
Of all beyond itself, thou dost bedew
  Green rushes like our rivers, and dost taste
The pleasant sun-rise, green isles hast thou too,
  And to the Sea as happily dost haste.

'Spenser, a jealous Honorer of thine'

Spenser, a jealous Honorer of thine,
  A forester deep in thy midmost trees
Did last eve ask my promise to refine
  Some English that might strive thine ear to please—
But Elfin-Poet, 'tis impossible
  For an inhabitant of wintry Earth
To rise like Phœbus with a golden quell,°
  Fire-wing'd, and make a morning in his Mirth:

It is impossible to escape from toil
O' the sudden, and receive thy spiriting:— 10
The flower must drink the nature of the soil
Before it can put forth its blossoming—
Be with me in the Summer days, and I
Will for thine honor, and his pleasure try.

'Blue!—'Tis the life of Heaven—the domain'

Blue!—'Tis the life of Heaven—the domain
Of Cynthia:—the wide palace of the Sun;
The tent of Hesperus and all his train;
The bosomer of clouds gold, grey, and dun.
Blue!—'Tis the life of waters—Ocean,
And all its vassal streams, pools numberless,
May rage, and foam, and fret, but never can
Subside, if not to dark blue nativeness.
Blue!—gentle cousin to the forest green,
Married to green in all the sweetest flowers— 10
Forget-me-not—the blue-bell—and, that Queen
Of secrecy, the violet:—What strange powers
Hast thou, as a mere shadow?—But how great,
When in an eye thou art, alive with fate!

'O thou whose face hath felt the Winter's wind'

'O thou whose face hath felt the Winter's wind;
Whose eye has seen the Snow clouds hung in Mist
And the black-elm tops 'mong the freezing Stars
To thee the Spring will be a harvest-time—
O thou whose only book has been the light
Of supreme darkness which thou feddest on
Night after night, when Phœbus was away
To thee the Spring shall be a tripple morn—
O fret not after Knowledge—I have none
And yet my song comes native with the warmth 10
O fret not after Knowledge—I have none

And yet the Evening listens—He who saddens
At thought of Idleness cannot be idle,
And he's awake who thinks himself asleep.'

Extracts from an Opera

I

O were I one of the Olympian twelve,
Their Godships should pass this into a law;
That when a man doth set himself in toil
After some beauty veiled far-away,
Each step he took should make his Lady's hand
More soft, more white, and her fair cheek more fair;
And for each briar-berry he might eat,
A kiss should bud upon the tree of love,
And pulp, and ripen, richer every hour,
To melt away upon the traveller's lips. 10

II

DAISY'S SONG

1

The Sun, with his great eye,
Sees not so much as I;
And the Moon, all silver proud,
Might as well be in a cloud.

2

And O the Spring—the Spring!
I lead the life of a King!
Couch'd in the teeming grass,
I spy each pretty lass.

3

I look where no one dares,
And I stare where no one stares, 10
And when the night is nigh,
Lambs bleat my lullaby.

III

FOLLY'S SONG

When wedding fiddles are a playing,
Huzza for folly O!
And when Maidens go a maying,
Huzza etc.
When a milk-pail is upset,
Huzza etc.
And the clothes left in the wet,
Huzza etc.
When the barrel's set abroach,
Huzza etc. 10
When Kate Eyebrow keeps a coach,
Huzza etc.
When the Pig is overroasted,
Huzza etc.
And the Cheese is overtoasted,
Huzza etc.
When Sir Snap is with his Lawyer,
Huzza etc.
And Miss Chip has kiss'd the Sawyer,
Huzza etc. 20

IV

O, I am frighten'd with most hateful thoughts!
Perhaps her voice is not a Nightingale's.
Perhaps her teeth are not the fairest pearl,
Her eye-lashes may be, for ought I know,
Not longer than the May-fly's small fan-horns;
There may not be one dimple on her hand;
And freckles many; ah! a careless Nurse,
In haste to teach the little thing to walk,
May have crumpt up a pair of Dian's legs,
And warpt the ivory of a Juno's neck. 10

V

SONG

1

The Stranger lighted from his steed,
  And ere he spake a word,
He seiz'd my Lady's lily hand,
  And kiss'd it all unheard.

2

The Stranger walk'd into the Hall,
  And ere he spake a word,
He kiss'd my Lady's cherry lips,
  And kiss'd 'em all unheard.

3

The Stranger walk'd into the bower,— 10
  But, my Lady first did go,—
Aye hand in hand into the bower,
  Where my Lord's roses blow.

4

My Lady's Maid had a silken scarf,
  And a golden ring had she,
And a kiss from the Stranger as off he went
  Again on his fair palfrey.

VI

Asleep! O sleep a little while, white Pearl,
And let me kneel, and let me pray to thee,
And let me call Heaven's blessing on thine eyes,
And let me breathe into the happy air,
That doth enfold and touch thee all about,
Vows of my slavery, my giving up,
My sudden adoration, my great love!

The Human Seasons

Four seasons fill the measure of the year;
  There are four seasons in the mind of man:
He has his lusty Spring, when fancy clear
  Takes in all beauty with an easy span:
He has his Summer, when luxuriously
  Spring's honied cud of youthful thought he loves
To ruminate, and by such dreaming nigh
  His nearest unto heaven: quiet coves
His soul has in its Autumn, when his wings
  He furleth close; contented so to look 10
On mists in idleness—to let fair things
  Pass by unheeded as a threshold brook.
    He has his Winter too of pale misfeature,
    Or else he would forget his mortal nature.

'For there's Bishop's Teign'

1

For there's Bishop's Teign°
And King's Teign
And Coomb at the clear Teign head.
Where close by the Stream
You may have your cream
All spread upon barley bread—

2

There's Arch Brook
And there's Larch Brook,
Both turning many a Mill
And cooling the drouth 10
Of the salmon's mouth
And fattening his silver gill.

3

There is Wild Wood
A mild hood
To the sheep on the lea o' the down
Where the golden furze
With its green thin spurs
Doth catch at the Maiden's gown.

4

There is Newton Marsh
With its spear grass harsh— 20
A pleasant summer level
Where the Maidens sweet
Of the Market Street
Do meet in the dusk to revel.

5

There's the Barton rich
With dyke and ditch
And hedge for the thrush to live in
And the hollow tree
For the buzzing bee
And a bank for the Wasp to hive in— 30

6

And O, and O
The Daisies blow
And the Primroses are waken'd
And the violet white
Sits in silver plight°
And the green bud's as long as the spike end.

7

Then who would go
Into dark Soho°
And chatter with dack'd hair'd critics°
When he can stay 40
For the new mown hay
And startle the dappled Prickets?°

'Where be ye going you Devon maid'

I

Where be ye going you Devon Maid
  And what have ye there i' the Basket?
Ye tight little fairy—just fresh from the dairy
  Will ye give me some cream if I ask it—

2

I love your Meads and I love your flowers
  And I love your junkets mainly°
But 'hind the door, I love kissing more
  O look not so disdainly!

3

I love your Hills and I love your dales
  And I love your flocks a bleating— 10
But O on the hether to lie together
  With both our hearts a beating—

4

I'll put your Basket all safe in a *nook*°
  And your shawl I hang up *on this willow*
And we will sigh in the daisy's eye
  And kiss on a grass green pillow.

'Over the hill and over the dale'

Over the hill and over the dale,
  And over the bourn to Dawlish—°
Where Gingerbread Wives have a scanty sale
  And gingerbread nuts are smallish—

Rantipole Betty she ran down a hill°
  And kick'd up her petticoats fairly
Says I I'll be Jack if you will be Gill—
  So she sat on the Grass debonnairly—

Here's somebody coming, here's somebody coming!
  Says I 'tis the Wind at a parley 10
So without any fuss any hawing and humming
  She lay on the grass debonnairly—

Here's somebody here and here's somebody *there*!
  Says I hold your tongue you young Gipsey.
So she held her tongue and lay plump and fair
  And dead as a venus tipsy—

O who wouldn't hie to Dawlish fair
  O who wouldn't stop in a Meadow
O who would not rumple the daisies there
  And make the wild fern for a bed do— 20

To J. H. Reynolds Esq.

Dear Reynolds, as last night I lay in bed,
There came before my eyes that wonted thread
Of shapes, and Shadows and Remembrances,
That every other minute vex and please:
Things all disjointed come from North and south,
Two witch's eyes above a cherub's mouth,
Voltaire with casque and shield and Habergeon,°
And Alexander with his night-cap on—
Old Socrates a tying his cravat;
And Hazlitt playing with Miss Edgworth's cat; 10
And Junius Brutus pretty well so so,°
Making the best of 's way towards Soho.

  Few are there who escape these visitings—
P'rhaps one or two, whose lives have patent wings;
And through whose curtains peeps no hellish nose,
No wild boar tushes, and no Mermaid's toes:°
But flowers bursting out with lusty pride;
And young Æolian harps personified,°

Some, Titian colours touch'd into real life.—
The sacrifice goes on; the pontif knife° 20
Gloams in the sun, the milk-white heifer lows,°
The pipes go shrilly, the libation flows:
A white sail shews above the green-head cliff
Moves round the point, and throws her anchor stiff.
The Mariners join hymn with those on land.—
You know the Enchanted Castle—it doth stand°
Upon a Rock on the Border of a Lake
Nested in Trees, which all do seem to shake
From some old Magic like Urganda's sword.°
O Phœbus that I had thy sacred word 30
To shew this Castle in fair dreaming wise
Unto my friend, while sick and ill he lies.

You know it well enough, where it doth seem
A mossy place, a Merlin's Hall, a dream.
You know the clear lake, and the little Isles,
The Mountains blue, and cold near neighbour rills—
All which elsewhere are but half animate
Here do they look alive to love and hate;
To smiles and frowns; they seem a lifted mound
Above some giant, pulsing underground. 40

Part of the building was a chosen See
Built by a banish'd santon of Chaldee:°
The other part two thousand years from him
Was built by Cuthbert de Saint Aldebrim;°
Then there's a little wing, far from the sun,
Built by a Lapland Witch turn'd maudlin nun—°
And many other juts of aged stone
Founded with many a mason-devil's groan.

The doors all look as if they oped themselves,
The windows as if latch'd by fays and elves— 50
And from them comes a silver flash of light
As from the Westward of a summer's night;
Or like a beauteous woman's large blue eyes
Gone mad through olden songs and Poesies—

See what is coming from the distance dim!
A golden galley all in silken trim!
Three rows of oars are lightening moment-whiles
Into the verdurous bosoms of those Isles.
Towards the shade under the Castle Wall
It comes in silence—now 'tis hidden all. 60
The clarion sounds; and from a postern grate
An echo of sweet music doth create
A fear in the poor herdsman who doth bring
His beasts to trouble the enchanted spring:
He tells of the sweet music and the spot
To all his friends, and they believe him not.

O that our dreamings all of sleep or wake
Would all their colours from the sunset take:
From something of material sublime,
Rather than shadow our own Soul's daytime 70
In the dark void of Night. For in the world
We jostle—but my flag is not unfurl'd°
On the Admiral staff—and to philosophize
I dare not yet!—Oh never will the prize,
High reason, and the lore of good and ill
Be my award. Things cannot to the will
Be settled, but they tease us out of thought.
Or is it that Imagination brought
Beyond its proper bound, yet still confined,—
Lost in a sort of Purgatory blind, 80
Cannot refer to any standard law
Of either earth or heaven?—It is a flaw
In happiness to see beyond our bourn—
It forces us in Summer skies to mourn:
It spoils the singing of the Nightingale.

Dear Reynolds. I have a mysterious tale
And cannot speak it. The first page I read
Upon a Lampit Rock of green sea weed°
Among the breakers—'Twas a quiet Eve;
The rocks were silent—the wide sea did weave 90
An untumultuous fringe of silver foam
Along the flat brown sand. I was at home,

And should have been most happy—but I saw
Too far into the sea; where every maw
The greater on the less feeds evermore:—
But I saw too distinct into the core
Of an eternal fierce destruction,
And so from Happiness I far was gone.
Still am I sick of it: and though to-day
I've gathered young spring-leaves, and flowers gay 100
Of Periwinkle and wild strawberry,
Still do I that most fierce destruction see,
The shark at savage prey—the hawk at pounce,
The gentle Robin, like a pard or ounce,
Ravening a worm—Away ye horrid moods,
Moods of one's mind! You know I hate them well,°
You know I'd sooner be a clapping bell
To some Kamschatkan missionary church,°
Than with these horrid moods be left in lurch—
Do you get health—and Tom the same—I'll dance, 110
And from detested moods in new Romance
Take refuge—Of bad lines a Centaine dose°
Is sure enough—and so 'here follows prose.'—°

Isabella;
or,
The Pot of Basil

A STORY FROM BOCCACCIO

I

Fair Isabel, poor simple Isabel!
  Lorenzo, a young palmer in Love's eye!
They could not in the self-same mansion dwell
  Without some stir of heart, some malady;
They could not sit at meals but feel how well
  It soothed each to be the other by;
They could not, sure, beneath the same roof sleep
But to each other dream, and nightly weep.

2

With every morn their love grew tenderer,
With every eve deeper and tenderer still; 10
He might not in house, field, or garden stir,
But her full shape would all his seeing fill;
And his continual voice was pleasanter
To her, than noise of trees or hidden rill;
Her lute-string gave an echo of his name,
She spoilt her half-done broidery with the same.

3

He knew whose gentle hand was at the latch,
Before the door had given her to his eyes;
And from her chamber-window he would catch
Her beauty farther than the falcon spies; 20
And constant as her vespers would he watch,
Because her face was turn'd to the same skies;
And with sick longing all the night outwear,
To hear her morning-step upon the stair.

4

A whole long month of May in this sad plight
Made their cheeks paler by the break of June:
'To-morrow will I bow to my delight,
To-morrow will I ask my lady's boon.'—
'O may I never see another night,
Lorenzo, if thy lips breathe not love's tune.'— 30
So spake they to their pillows; but, alas,
Honeyless days and days did he let pass;

5

Until sweet Isabella's untouch'd cheek
Fell sick within the rose's just domain,
Fell thin as a young mother's, who doth seek
By every lull to cool her infant's pain:
'How ill she is,' said he, 'I may not speak,
And yet I will, and tell my love all plain:
If looks speak love-laws, I will drink her tears,
And at the least 'twill startle off her cares.' 40

6

So said he one fair morning, and all day
  His heart beat awfully against his side;
And to his heart he inwardly did pray
  For power to speak; but still the ruddy tide°
Stifled his voice, and puls'd resolve away—
  Fever'd his high conceit of such a bride,
Yet brought him to the meekness of a child:
Alas! when passion is both meek and wild!

7

So once more he had wak'd and anguished
  A dreary night of love and misery, 50
If Isabel's quick eye had not been wed
  To every symbol on his forehead high;
She saw it waxing very pale and dead,
  And straight all flush'd; so, lisped tenderly,
'Lorenzo!'—here she ceas'd her timid quest,
But in her tone and look he read the rest.

8

'O Isabella, I can half perceive
  That I may speak my grief into thine ear;
If thou didst ever any thing believe,
  Believe how I love thee, believe how near 60
My soul is to its doom: I would not grieve
  Thy hand by unwelcome pressing, would not fear°
Thine eyes by gazing; but I cannot live
Another night, and not my passion shrive.

9

'Love! thou art leading me from wintry cold,
  Lady! thou leadest me to summer clime,
And I must taste the blossoms that unfold
  In its ripe warmth this gracious morning time.'
So said, his erewhile timid lips grew bold,
  And poesied with hers in dewy rhyme: 70
Great bliss was with them, and great happiness
Grew, like a lusty flower in June's caress.

10

Parting they seem'd to tread upon the air,
  Twin roses by the zephyr blown apart
Only to meet again more close, and share
  The inward fragrance of each other's heart.
She, to her chamber gone, a ditty fair
  Sang, of delicious love and honey'd dart;
He with light steps went up a western hill,
And bade the sun farewell, and joy'd his fill. 80

11

All close they met again, before the dusk
  Had taken from the stars its pleasant veil,
All close they met, all eves, before the dusk
  Had taken from the stars its pleasant veil,
Close in a bower of hyacinth and musk,
  Unknown of any, free from whispering tale.
Ah! better had it been for ever so,
Than idle ears should pleasure in their woe.

12

Were they unhappy then?—It cannot be—
  Too many tears for lovers have been shed, 90
Too many sighs give we to them in fee,
  Too much of pity after they are dead,
Too many doleful stories do we see,
  Whose matter in bright gold were best be read;
Except in such a page where Theseus' spouse°
Over the pathless waves towards him bows.

13

But, for the general award of love,
  The little sweet doth kill much bitterness;
Though Dido silent is in under-grove,°
  And Isabella's was a great distress, 100
Though young Lorenzo in warm Indian clove
  Was not embalm'd, this truth is not the less—
Even bees, the little almsmen of spring-bowers,
Know there is richest juice in poison-flowers.

14

With her two brothers this fair lady dwelt,°
  Enriched from ancestral merchandize,
And for them many a weary hand did swelt°
  In torched mines and noisy factories,
And many once proud-quiver'd loins did melt
  In blood from stinging whip;—with hollow eyes 110
Many all day in dazzling river stood,
To take the rich-ored driftings of the flood.

15

For them the Ceylon diver held his breath,°
  And went all naked to the hungry shark;
For them his ears gush'd blood; for them in death
  The seal on the cold ice with piteous bark
Lay full of darts; for them alone did seethe
  A thousand men in troubles wide and dark:
Half-ignorant, they turn'd an easy wheel,
That set sharp racks at work, to pinch and peel. 120

16

Why were they proud? Because their marble founts
  Gush'd with more pride than do a wretch's tears?—
Why were they proud? Because fair orange-mounts°
  Were of more soft ascent than lazar stairs?—°
Why were they proud? Because red-lin'd accounts°
  Were richer than the songs of Grecian years?—°
Why were they proud? again we ask aloud,
Why in the name of Glory were they proud?

17

Yet were these Florentines as self-retired
  In hungry pride and gainful cowardice, 130
As two close Hebrews in that land inspired,°
  Paled in and vineyarded from beggar-spies;
The hawks of ship-mast forests—the untired
  And pannier'd mules for ducats and old lies—
Quick cat's-paws on the generous stray-away,—
Great wits in Spanish, Tuscan, and Malay.

18

How was it these same ledger-men could spy
  Fair Isabella in her downy nest?
How could they find out in Lorenzo's eye
  A straying from his toil? Hot Egypt's pest° 140
Into their vision covetous and sly!
  How could these money-bags see east and west?—
Yet so they did—and every dealer fair
Must see behind, as doth the hunted hare.

19

O eloquent and famed Boccaccio!
  Of thee we now should ask forgiving boon,
And of thy spicy myrtles as they blow,
  And of thy roses amorous of the moon,
And of thy lilies, that do paler grow
  Now they can no more hear thy ghittern's tune,° 150
For venturing syllables that ill beseem
The quiet glooms of such a piteous theme.

20

Grant thou a pardon here, and then the tale
  Shall move on soberly, as it is meet;
There is no other crime, no mad assail
  To make old prose in modern rhyme more sweet:
But it is done—succeed the verse or fail—
  To honour thee, and thy gone spirit greet;
To stead thee as a verse in English tongue,
An echo of thee in the north-wind sung. 160

21

These brethren having found by many signs°
  What love Lorenzo for their sister had,
And how she lov'd him too, each unconfines
  His bitter thoughts to other, well nigh mad
That he, the servant of their trade designs,
  Should in their sister's love be blithe and glad,
When 'twas their plan to coax her by degrees
To some high noble and his olive-trees.

22

And many a jealous conference had they,
  And many times they bit their lips alone, 170
Before they fix'd upon a surest way
  To make the youngster for his crime atone;
And at the last, these men of cruel clay
  Cut Mercy with a sharp knife to the bone;
For they resolved in some forest dim
To kill Lorenzo, and there bury him.

23

So on a pleasant morning, as he leant
  Into the sun-rise, o'er the balustrade
Of the garden-terrace, towards him they bent
  Their footing through the dews; and to him said, 180
'You seem there in the quiet of content,
  Lorenzo, and we are most loth to invade
Calm speculation; but if you are wise,
Bestride your steed while cold is in the skies.

24

'To-day we purpose, ay, this hour we mount
  To spur three leagues towards the Apennine;
Come down, we pray thee, ere the hot sun count
  His dewy rosary on the eglantine.'
Lorenzo, courteously as he was wont,
  Bow'd a fair greeting to these serpents' whine; 190
And went in haste, to get in readiness,
With belt, and spur, and bracing huntsman's dress.

25

And as he to the court-yard pass'd along,
  Each third step did he pause, and listen'd oft
If he could hear his lady's matin-song,
  Or the light whisper of her footstep soft;
And as he thus over his passion hung,
  He heard a laugh full musical aloft;
When, looking up, he saw her features bright
Smile through an in-door lattice, all delight. 200

26

'Love, Isabel!' said he, 'I was in pain
  Lest I should miss to bid thee a good morrow:
Ah! what if I should lose thee, when so fain
  I am to stifle all the heavy sorrow
Of a poor three hours' absence? but we'll gain
  Out of the amorous dark what day doth borrow.
Good bye! I'll soon be back.'—'Good bye!' said she:—
And as he went she chanted merrily.

27

So the two brothers and their murder'd man°
  Rode past fair Florence, to where Arno's stream 210
Gurgles through straiten'd banks, and still doth fan
  Itself with dancing bulrush, and the bream
Keeps head against the freshets. Sick and wan
  The brothers' faces in the ford did seem,
Lorenzo's flush with love.—They pass'd the water
Into a forest quiet for the slaughter.

28

There was Lorenzo slain and buried in,
  There in that forest did his great love cease;
Ah! when a soul doth thus its freedom win,
  It aches in loneliness—is ill at peace 220
As the break-covert blood-hounds of such sin:
  They dipp'd their swords in the water, and did tease
Their horses homeward, with convulsed spur,
Each richer by his being a murderer.

29

They told their sister how, with sudden speed,
  Lorenzo had ta'en ship for foreign lands,
Because of some great urgency and need
  In their affairs, requiring trusty hands.
Poor Girl! put on thy stifling widow's weed,
  And 'scape at once from Hope's accursed bands; 230
To-day thou wilt not see him, nor to-morrow,
And the next day will be a day of sorrow.

30

She weeps alone for pleasures not to be;
  Sorely she wept until the night came on,
And then, instead of love, O misery!
  She brooded o'er the luxury alone:
His image in the dusk she seem'd to see,
  And to the silence made a gentle moan,
Spreading her perfect arms upon the air,
And on her couch low murmuring 'Where? O where?' 240

31

But Selfishness, Love's cousin, held not long
  Its fiery vigil in her single breast;
She fretted for the golden hour, and hung
  Upon the time with feverish unrest—
Not long—for soon into her heart a throng
  Of higher occupants, a richer zest,
Came tragic; passion not to be subdued,
And sorrow for her love in travels rude.

32

In the mid days of autumn, on their eves
  The breath of Winter comes from far away, 250
And the sick west continually bereaves
  Of some gold tinge, and plays a roundelay
Of death among the bushes and the leaves,
  To make all bare before he dares to stray
From his north cavern. So sweet Isabel
By gradual decay from beauty fell,

33

Because Lorenzo came not. Oftentimes
  She ask'd her brothers, with an eye all pale,
Striving to be itself, what dungeon climes
  Could keep him off so long? They spake a tale 260
Time after time, to quiet her. Their crimes
  Came on them, like a smoke from Hinnom's vale;°
And every night in dreams they groan'd aloud,
To see their sister in her snowy shroud.

34

And she had died in drowsy ignorance,
  But for a thing more deadly dark than all;
It came like a fierce potion, drunk by chance,
  Which saves a sick man from the feather'd pall
For some few gasping moments; like a lance,
  Waking an Indian from his cloudy hall 270
With cruel pierce, and bringing him again
Sense of the gnawing fire at heart and brain.

35

It was a vision.—In the drowsy gloom,
  The dull of midnight, at her couch's foot
Lorenzo stood, and wept: the forest tomb
  Had marr'd his glossy hair which once could shoot
Lustre into the sun, and put cold doom
  Upon his lips, and taken the soft lute
From his lorn voice, and past his loamed ears
Had made a miry channel for his tears. 280

36

Strange sound it was, when the pale shadow spake;
  For there was striving, in its piteous tongue,
To speak as when on earth it was awake,
  And Isabella on its music hung:
Languor there was in it, and tremulous shake,
  As in a palsied Druid's harp unstrung;
And through it moan'd a ghostly under-song,
Like hoarse night-gusts sepulchral briars among.

37

Its eyes, though wild, were still all dewy bright
  With love, and kept all phantom fear aloof 290
From the poor girl by magic of their light,
  The while it did unthread the horrid woof
Of the late darken'd time,—the murderous spite
  Of pride and avarice,—the dark pine roof
In the forest,—and the sodden turfed dell,
Where, without any word, from stabs he fell.

38

Saying moreover, 'Isabel, my sweet!
  Red whortle-berries droop above my head,
And a large flint-stone weighs upon my feet;
  Around me beeches and high chestnuts shed 300
Their leaves and prickly nuts; a sheep-fold bleat
  Comes from beyond the river to my bed:
Go, shed one tear upon my heather-bloom,
And it shall comfort me within the tomb.

39

'I am a shadow now, alas! alas!
  Upon the skirts of human-nature dwelling
Alone: I chant alone the holy mass,
  While little sounds of life are round me knelling,
And glossy bees at noon do fieldward pass,
  And many a chapel bell the hour is telling, 310
Paining me through: those sounds grow strange to me,
And thou art distant in Humanity.

40

'I know what was, I feel full well what is,
  And I should rage, if spirits could go mad;
Though I forget the taste of earthly bliss,
  That paleness warms my grave, as though I had
A Seraph chosen from the bright abyss
  To be my spouse: thy paleness makes me glad;
Thy beauty grows upon me, and I feel
A greater love through all my essence steal.' 320

41

The Spirit mourn'd 'Adieu!'—dissolv'd, and left°
  The atom darkness in a slow turmoil;
As when of healthful midnight sleep bereft,
  Thinking on rugged hours and fruitless toil,
We put our eyes into a pillowy cleft,
  And see the spangly gloom froth up and boil:
It made sad Isabella's eyelids ache,
And in the dawn she started up awake;

42

'Ha! ha!' said she, 'I knew not this hard life,
  I thought the worst was simple misery; 330
I thought some Fate with pleasure or with strife
  Portion'd us—happy days, or else to die;
But there is crime—a brother's bloody knife!
  Sweet Spirit, thou hast school'd my infancy:
I'll visit thee for this, and kiss thine eyes,
And greet thee morn and even in the skies.'

43

When the full morning came, she had devised
  How she might secret to the forest hie;
How she might find the clay, so dearly prized,
  And sing to it one latest lullaby; 340
How her short absence might be unsurmised,
  While she the inmost of the dream would try.
Resolv'd, she took with her an aged nurse,
And went into that dismal forest-hearse.°

44

See, as they creep along the river side,
  How she doth whisper to that aged Dame,
And, after looking round the champaign wide,
  Shows her a knife.—'What feverous hectic flame
Burns in thee, child?—What good can thee betide,
  That thou should'st smile again?'—The evening came, 350
And they had found Lorenzo's earthy bed;
The flint was there, the berries at his head.

45

Who hath not loiter'd in a green church-yard,
  And let his spirit, like a demon-mole,
Work through the clayey soil and gravel hard,
  To see scull, coffin'd bones, and funeral stole;
Pitying each form that hungry Death hath marr'd,
  And filling it once more with human soul?
Ah! this is holiday to what was felt
When Isabella by Lorenzo knelt. 360

46

She gaz'd into the fresh-thrown mould, as though
  One glance did fully all its secrets tell;
Clearly she saw, as other eyes would know
  Pale limbs at bottom of a crystal well;
Upon the murderous spot she seem'd to grow,
  Like to a native lily of the dell:
Then with her knife, all sudden, she began
To dig more fervently than misers can.

47

Soon she turn'd up a soiled glove, whereon
  Her silk had play'd in purple phantasies,° 370
She kiss'd it with a lip more chill than stone,
  And put it in her bosom, where it dries
And freezes utterly unto the bone
  Those dainties made to still an infant's cries:°
Then 'gan she work again; nor stay'd her care,
But to throw back at times her veiling hair.

48

That old nurse stood beside her wondering,
  Until her heart felt pity to the core
At sight of such a dismal labouring,
  And so she kneeled, with her locks all hoar, 380
And put her lean hands to the horrid thing:°
  Three hours they labour'd at this travail sore;
At last they felt the kernel of the grave,
And Isabella did not stamp and rave.

49

Ah! wherefore all this wormy circumstance?
  Why linger at the yawning tomb so long?
O for the gentleness of old Romance,
  The simple plaining of a minstrel's song!
Fair reader, at the old tale take a glance,
  For here, in truth, it doth not well belong 390
To speak:—O turn thee to the very tale,
And taste the music of that vision pale.

50

With duller steel than the Perséan sword°
  They cut away no formless monster's head,
But one, whose gentleness did well accord
  With death, as life. The ancient harps have said,°
Love never dies, but lives, immortal Lord:
  If love impersonate was ever dead,
Pale Isabella kiss'd it, and low moan'd.
'Twas love; cold,—dead indeed, but not dethroned. 400

51

In anxious secrecy they took it home,
  And then the prize was all for Isabel:
She calm'd its wild hair with a golden comb,
  And all around each eye's sepulchral cell
Pointed each fringed lash; the smeared loam
  With tears, as chilly as a dripping well,
She drench'd away:—and still she comb'd, and kept
Sighing all day—and still she kiss'd, and wept.

52

Then in a silken scarf,—sweet with the dews
  Of precious flowers pluck'd in Araby, 410
And divine liquids come with odorous ooze
  Through the cold serpent-pipe refreshfully,—
She wrapp'd it up; and for its tomb did choose
  A garden-pot, wherein she laid it by,
And cover'd it with mould, and o'er it set
Sweet Basil, which her tears kept ever wet.

53

And she forgot the stars, the moon, and sun,
  And she forgot the blue above the trees,
And she forgot the dells where waters run,
  And she forgot the chilly autumn breeze; 420
She had no knowledge when the day was done,
  And the new morn she saw not: but in peace
Hung over her sweet Basil evermore,
And moisten'd it with tears unto the core.

54

And so she ever fed it with thin tears,
  Whence thick, and green, and beautiful it grew,
So that it smelt more balmy than its peers
  Of Basil-tufts in Florence; for it drew
Nurture besides, and life, from human fears,
  From the fast mouldering head there shut from view: 430
So that the jewel, safely casketed,
Came forth, and in perfumed leafits spread.

55

O Melancholy, linger here awhile!
  O Music, Music, breathe despondingly!
O Echo, Echo, from some sombre isle,
  Unknown, Lethean, sigh to us—O sigh!
Spirits in grief, lift up your heads, and smile;
  Lift up your heads, sweet Spirits, heavily,
And make a pale light in your cypress glooms,
Tinting with silver wan your marble tombs. 440

56

Moan hither, all ye syllables of woe,
  From the deep throat of sad Melpomene!
Through bronzed lyre in tragic order go,
  And touch the strings into a mystery;
Sound mournfully upon the winds and low;
  For simple Isabel is soon to be
Among the dead: She withers, like a palm
Cut by an Indian for its juicy balm.

57

O leave the palm to wither by itself;
  Let not quick Winter chill its dying hour!— 450
It may not be—those Baälites of pelf,°
  Her brethren, noted the continual shower
From her dead eyes; and many a curious elf,
  Among her kindred, wonder'd that such dower
Of youth and beauty should be thrown aside
By one mark'd out to be a Noble's bride.

58

And, furthermore, her brethren wonder'd much
  Why she sat drooping by the Basil green,
And why it flourish'd, as by magic touch;
  Greatly they wonder'd what the thing might mean: 460
They could not surely give belief, that such
  A very nothing would have power to wean
Her from her own fair youth, and pleasures gay,
And even remembrance of her love's delay.

59

Therefore they watch'd a time when they might sift
  This hidden whim; and long they watch'd in vain;
For seldom did she go to chapel-shrift,
  And seldom felt she any hunger-pain;
And when she left, she hurried back, as swift
  As bird on wing to breast its eggs again; 470
And, patient as a hen-bird, sat her there
Beside her Basil, weeping through her hair.

60

Yet they contriv'd to steal the Basil-pot,
  And to examine it in secret place:
The thing was vile with green and livid spot,
  And yet they knew it was Lorenzo's face:
The guerdon of their murder they had got,
  And so left Florence in a moment's space,
Never to turn again.—Away they went,
With blood upon their heads, to banishment. 480

61

O Melancholy, turn thine eyes away!
  O Music, Music, breathe despondingly!
O Echo, Echo, on some other day,
  From isles Lethean, sigh to us—O sigh!
Spirits of grief, sing not your 'Well-a-way!'
  For Isabel, sweet Isabel, will die;
Will die a death too lone and incomplete,
Now they have ta'en away her Basil sweet.

62

Piteous she look'd on dead and senseless things,
  Asking for her lost Basil amorously; 490
And with melodious chuckle in the strings
  Of her lorn voice, she oftentimes would cry
After the Pilgrim in his wanderings,
  To ask him where her Basil was; and why
T'was hid from her: 'For cruel 'tis,' said she,
'To steal my Basil-pot away from me.'

63

And so she pined, and so she died forlorn,
  Imploring for her Basil to the last.
No heart was there in Florence but did mourn
  In pity of her love, so overcast. 500
And a sad ditty of this story born
  From mouth to mouth through all the country pass'd:
Still is the burthen sung—'O cruelty,
To steal my Basil-pot away from me!'

'Mother of Hermes! and still youthful Maia!'

Mother of Hermes! and still youthful Maia!
    May I sing to thee
As thou wast hymned on the shores of Baiæ?°
    Or may I woo thee
In earlier Sicilian? or thy smiles°
Seek as they once were sought, in Grecian isles,
By Bards who died content in pleasant sward,
Leaving great verse unto a little clan?
O give me their old vigour, and unheard,
Save of the quiet Primrose, and the span 10
    Of Heaven, and few ears
Rounded by thee my song should die away
    Content as theirs
Rich in the simple worship of a day.—

'Give me your patience Sister while I frame'

Give me your patience Sister while I frame
Exact in Capitals your golden name:
Or sue the fair Apollo and he will
Rouse from his heavy slumber and instill
Great love in me for thee and Poesy.
Imagine not the greatest mastery
And Kingdom over all the Realms of verse
Nears more to Heaven in aught than when we nurse
And surety give to love and Brotherhood.

Anthropopagi in Othello's mood,° 10
Ulysses stormed, and his enchanted belt
Glow with the Muse, but they are never felt
Unbosom'd so and so eternal made,
Such tender incense is their Laurel shade,
To all the regent sisters of the Nine,
As this poor offering to you sister mine.

Kind Sister! aye, this third name says you are;
Enchanted has it been the Lord knows where.
And may it taste to you like good old wine
Take you to real happiness and give 20
Sons daughters and a home like honied hive.

'Sweet sweet is the greeting of eyes'

Sweet sweet is the greeting of eyes,
And sweet is the voice in its greeting,
When Adieux have grown old and goodbyes
Fade away where old time is retreating—

Warm the nerve of a welcoming hand
And earnest a kiss on the Brow,
When we meet over sea and o'er Land
Where furrows are new to the Plough.

On visiting the Tomb of Burns

The Town, the churchyard, and the setting sun,
The Clouds, the trees, the rounded hills all seem
Though beautiful, cold—strange—as in a dream,
I dreamed long ago, now new begun
The shortlived, paly summer is but won
From winter's ague, for one hour's gleam;
Though saphire warm, their stars do never beam,
All is cold Beauty; pain is never done
For who has mind to relish Minos-wise,°
The real of Beauty, free from that dead hue 10
Fickly imagination and sick pride°
Cast wan upon it! Burns! with honour due°
I have oft honoured thee. Great shadow, hide
Thy face, I sin against thy native skies.

'Old Meg she was a Gipsey'

Old Meg she was a Gipsey
  And liv'd upon the Moors;
Her bed it was the brown heath turf,
  And her house was out of doors—
Her apples were swart blackberries,
  Her currants pods o' Broom,
Her wine was dew o' the wild white rose,
  Her book a churchyard tomb—
Her brothers were the craggy hills,
  Her sisters larchen trees— 10
Alone with her great family
  She liv'd as she did please.
No Breakfast had she many a morn,
  No dinner many a noon;
And 'stead of supper she would stare
  Full hard against the Moon—
But every Morn, of wood bine fresh
  She made her garlanding;
And every night the dark glen Yew
  She wove and she would sing— 20

And with her fingers old and brown
  She plaited Mats o' Rushes,
And gave them to the Cottagers
  She met among the Bushes—
Old Meg was brave as Margaret Queen°
  And tall as Amazon:
An old red blanket cloak she wore
  A chip hat had she on—°
God rest her aged bones somewhere
  She died full long agone! 30

'There was a naughty Boy'

There was a naughty Boy
  A naughty boy was he
He would not stop at home
  He could not quiet be—
    He took
    In his Knapsack
    A Book
    Full of vowels
    And a shirt
    With some towels— 10
    A slight cap
    For night cap—
    A hair brush
    Comb ditto
    New Stockings
    For old ones
    Would split O!
    This Knapsack
    Tight at 's back
    He revetted close° 20
    And follow'd his Nose
    To the North
    To the North
And follow'd his nose
    To the North—

There was a naughty boy
  And a naughty boy was he
For nothing would he do
  But scribble poetry—
    He took 30
    An inkstand
    In his hand
    And a Pen
    Big as ten
    In the other
    And away
    In a Pother
    He ran
    To the mountains
    And fountains 40
    And ghostes
    And Postes
    And witches
    And ditches
    And wrote
    In his coat
    When the weather
    Was cool
    Fear of gout
    And without 50
    When the weather
    Was warm—
    Och the charm
    When we choose
    To follow one's nose
    To the north
    To the north
To follow one's nose to the north!

There was a naughty boy
  And a naughty boy was he 60
He kept little fishes
  In washing tubs three
    In spite
    Of the might

Of the Maid
Nor affraid
Of his Granny-good—
He often would
Hurly burly
Get up early 70
And go
By hook or crook
To the brook
And bring home
Miller's thumb°
Tittle bat
Not over fat
Minnows small
As the stall
Of a glove 80
Not above
The size
Of a nice
Little Baby's
Little finger—
O he made
'T was his trade
Of Fish a pretty kettle
A kettle—A kettle
Of Fish a pretty kettle 90
A kettle!

There was a naughty Boy
And a naughty Boy was he
He ran away to Scotland
The people for to see—
There he found
That the ground
Was as hard
That a yard
Was as long, 100
That a song
Was as merry,
That a cherry

Was as red—
That lead
Was as weighty
That fourscore
Was as eighty
That a door
Was as wooden 110
As in england—
So he stood in
His shoes
And he wonderd
He wonderd
He stood in his
Shoes and he wonder'd—

'Ah! ken ye what I met the day'

Ah! ken ye what I met the day
  Out oure the Mountains
A coming down by craggis grey
  An' mossie fountains
Ah goud hair'd Marie yeve I pray
  Ane minute's guessing—
For that I met upon the way
  Is past expressing—
As I stood where a rocky brig
  A torrent crosses 10
I spied upon a misty rig
  A troup o' Horses—
And as they trotted down the glen
  I sped to meet them
To see if I might know the Men,
  To stop and greet them.
First Willie on his sleek mare came
  At canting gallop
His long hair rustled like a flame
  On board a shallop— 20

Then came his brother Rab and then
  Young Peggy's Mither
And Peggy too—adown the glen
  They went togither—
I saw her wrappit in her hood
  Fra wind and raining—
Her cheek was flush wi' timid blood
  'Twixt growth and waning—
She turn'd her dazed head full oft
  For thence her Brithers 30
Came riding with her Bridegroom soft
  An' mony ithers.
Young Tam came up an' eyed me quick
  With reddened cheek:
Braw Tam was daffed like a chick,
  He could na speak—
Ah Marie they are all gane hame
  Through blustring weather
An' every heart is full on flame
  An' light as feather. 40
Ah! Marie they are all gone hame
  Fra happy wedding,
Whilst I—Ah is it not a shame?
  Sad tears am shedding—

Sonnet to Ailsa Rock

Hearken, thou craggy ocean pyramid!
  Give answer from thy voice, the sea fowls' screams!
  When were thy shoulders mantled in huge streams?
When, from the sun, was thy broad forehead hid?
How long is't since the mighty powers bid
  Thee heave to airy sleep from fathom dreams?
  Sleep in the lap of thunder or sunbeams,
  Or when grey clouds are thy cold coverlid.

Thou answer'st not, for thou art dead asleep;
  Thy life is but two dead eternities—

The last in air, the former in the deep;
  First with the whales, last with the eagle-skies—
Drown'd wast thou till an earthquake made thee steep,
  Another cannot wake thy giant size.

'This mortal body of a thousand days'

This mortal body of a thousand days
  Now fills, O Burns, a space in thine own room,
Where thou didst dream alone on budded bays,
  Happy and thoughtless of thy day of doom!
My pulse is warm with thine old Barley-bree,
  My head is light with pledging a great soul,
My eyes are wandering, and I cannot see,
  Fancy is dead and drunken at its goal;
Yet can I stamp my foot upon thy floor,
  Yet can I ope thy window-sash to find 10
The meadow thou hast tramped o'er and o'er,—
  Yet can I think of thee till thought is blind,—
Yet can I gulp a bumper to thy name,—
O smile among the shades, for this is fame!

'There is a joy in footing slow across a silent plain'

There is a joy in footing slow across a silent plain
Where Patriot Battle has been fought when Glory had the gain;
There is a pleasure on the heath where Druids old have been,
Where Mantles grey have rustled by and swept the nettles
  green:
There is a joy in every spot, made known by times of old,
New to the feet, although the tale a hundred times be told:
There is a deeper joy than all, more solemn in the heart,
More parching to the tongue than all, of more divine a smart,
When weary feet forget themselves upon a pleasant turf,
Upon hot sand, or flinty road, or Sea shore iron scurf, 10

Toward the Castle or the Cot where long ago was born
One who was great through mortal days and died of fame
unshorn.
Light Hether-bells may tremble then, but they are far away;
Woodlark may sing from sandy fern,—the Sun may hear his
Lay;
Runnels may kiss the grass on shelves and shallows clear
But their low voices are not heard though come on travels
drear;
Bloodred the sun may set behind black mountain peaks;
Blue tides may sluice and drench their time in Caves and weedy
creeks;
Eagles may seem to sleep wing wide upon the Air;
Ring doves may fly convuls'd across to some high cedar'd lair; 20
But the forgotten eye is still fast wedded to the ground—
As Palmer's that with weariness mid desert shrine hath found.
At such a time the Soul's a Child, in Childhood is the brain
Forgotten is the worldly heart—alone, it beats in vain—
Aye if a Madman could have leave to pass a healthful day,
To tell his forehead's swoon and faint when first began decay,
He might make tremble many a Man whose Spirit had gone
forth
To find a Bard's low Cradle place about the silent north.
Scanty the hour and few the steps beyond the Bourn of Care,
Beyond the sweet and bitter world—beyond it unaware; 30
Scanty the hour and few the steps because a longer stay
Would bar return and make a Man forget his mortal way.
O horrible! to lose the sight of well remember'd face,
Of Brother's eyes, of Sister's Brow, constant to every place;
Filling the Air as on we move with Portraiture intense
More warm than those heroic tints that fill a Painter's sense,
When Shapes of old come striding by and visages of old,
Locks shining black, hair scanty grey and passions manifold.
No, no that horror cannot be—for at the Cable's length
Man feels the gentle Anchor pull and gladdens in its 40
strength—
One hour half idiot he stands by mossy waterfall,
But in the very next he reads his Soul's memorial:
He reads it on the Mountain's height where chance he may sit
down
Upon rough marble diadem, that Hill's eternal crown.

Yet be the Anchor e'er so fast, room is there for a prayer
That Man may never loose his Mind on Mountains bleak and
  bare;
That he may stray league after League some great Birthplace to
  find,
And keep his vision clear from speck, his inward sight
  unblind—

'All gentle folks who owe a grudge'

All gentle folks who owe a grudge
  To any living thing
Open your ears and stay your trudge
  Whilst I in dudgeon sing—

The gad fly he hath stung me sore
  O may he ne'er sting you!
But we have many a horrid bore
  He may sting black and blue.

Has any here an old grey Mare
  With three Legs all her store?
O put it to her Buttocks bare
  And straight she'll run on four.

Has any here a Lawyer suit
  Of 1743?
Take Lawyer's nose and put it to't
  And you the end will see—

Is there a Man in Parliament
  Dum founder'd in his speech?
O let his neighbour make a rent
  And put one in his breech— 20

O Lowther how much better thou°
  Hadst figur'd t' other day.
When to the folks thou madst a bow
  And hadst no more to say.

If lucky gad fly had but ta'en
His seat upon thine A—e
And put thee to a little pain
To save thee from a worse.

Better than Southey it had been°
Better than Mr. D——° 30
Better than Wordsworth too I ween
Better than Mr. V——.

Forgive me pray good people all
For deviating so
In spirit sure I had a call—
And now I on will go—

Has any here a daughter fair
Too fond of reading novels
Too apt to fall in love with care
And charming Mister Lovels?° 40

O put a gadfly to that thing
She keeps so white and pert
I mean the finger for the ring
And it will breed a Wert—

Has any here a pious spouse
Who seven times a day
Scolds as King David pray'd; to chouse°
And have her holy way?

O let a Gadfly's little sting
Persuade her sacred tongue 50
That noises are a common thing
But that her bell has rung.

And as this is the summum bo-
Num of all conquering
I leave withouten wordes mo'
The Gadfly's little sting—

'Not Aladin magian'

Not Aladin magian
Ever such a work began,
Not the Wizard of the Dee°
Ever such a dream could see;
Not St. John in Patmos' isle°
In the passion of his toil
When he saw the churches seven
Golden aisled built up in heaven
Gazed at such a rugged wonder.
As I stood its roofing under 10
Lo! I saw one sleeping there
On the marble cold and bare
While the surges washed his feet
And his garments white did beat
Drench'd about the sombre rocks,
On his neck his well-grown locks
Lifted dry above the Main
Were upon the curl again—
'What is this and what art thou?'
Whisper'd I and touch'd his brow. 20
'What art thou and what is this?'
Whisper'd I and strove to kiss
The Spirit's hand to wake his eyes.
Up he started in a thrice.°
'I am Lycidas,' said he,°
'Fam'd in funeral Minstrelsy—
This was architected thus
By the great Oceanus,
Here his mighty waters play
Hollow Organs all the day, 30
Here by turns his dolphins all
Finny palmers great and small°
Come to pay devotion due—
Each a mouth of pearls must strew;
Many a Mortal of these days
Dares to pass our sacred ways,
Dares to touch audaciously
This Cathedral of the Sea—

I have been the Pontif priest°
Where the Waters never rest, 40
Where a fledgy sea bird choir
Soars for ever—holy fire
I have hid from Mortal Man.
Proteus is my Sacristan.
But the stupid eye of Mortal
Hath pass'd beyond the Rocky portal
So for ever will I leave
Such a taint and soon unweave
All the magic of the place—
'Tis now free to stupid face, 50
To cutters and to fashion boats,°
To cravats and to Petticoats.
The great Sea shall war it down
For its fame shall not be blown
At every farthing quadrille dance.'
So saying with a Spirit's glance
He dived—

'Read me a Lesson muse, and speak it loud'

Read me a Lesson muse, and speak it loud
  Upon the top of Nevis blind in Mist!
I look into the Chasms and a Shroud
  Vaprous doth hide them; just so much I wist
Mankind do know of Hell: I look o'erhead
  And there is sullen Mist; even so much
Mankind can tell of Heaven: Mist is spread
  Before the Earth beneath me—even such,
Even so vague is Man's sight of himself.
  Here are the craggy Stones beneath my feet; 10
Thus much I know, that a poor witless elf
  I tread on them; that all my eye doth meet
    Is mist and Crag—not only on this height
    But in the world of thought and mental might—

'Upon my Life Sir Nevis I am piqued'

MRS. C——

Upon my Life Sir Nevis I am piqued
That I have so far panted tugg'd and reek'd°
To do an honor to your old bald pate
And now am sitting on you just to bate,°
Without your paying me one compliment.
Alas 'tis so with all, when our intent
Is plain, and in the eye of all Mankind
We fair ones show a preference, too blind!
You Gentlemen immediately turn tail.
O let me then my hapless fate bewail! 10
Ungrateful Baldpate have I not disdain'd
The pleasant Valleys—have I not mad brain'd
Deserted all my Pickles and preserves
My China closet too—with wretched Nerves
To boot—say wretched ingrate have I not
Left my soft cushion chair and caudle pot.
'Tis true I had no corns—no! thank the fates
My Shoemaker was always Mr. Bates.
And if not Mr. Bates why I'm not old!
Still dumb ungrateful Nevis—still so cold! 20

(Here the lady took some more whiskey and was putting even more to her lips when she dashed it to the Ground for the Mountain began to grumble which continued for a few Minutes before he thus began,)
thus began,

BEN NEVIS

What whining bit of tongue and Mouth thus dares
Disturb my Slumber of a thousand years—
Even so long my sleep has been secure
And to be so awaked I'll not endure.
Oh pain—for since the Eagle's earliest scream
I've had a damned confounded ugly dream
A Nightmare sure—What Madam was it you?
It cannot be! My old eyes are not true!

*A domestic of Ben's

Red-Crag*, my Spectacles! Now let me see!
Good Heavens Lady how the gemini° 30
Did you get here? O I shall split my Sides!
I shall earthquake—°

MRS. C——

Sweet Nevis do not quake, for though I love
Your honest Countenance all things above
Truly I should not like to be convey'd
So far into your Bosom—gentle Maid
Loves not too rough a treatment, gentle sir
Pray thee be calm and do not quake nor stir—
No not a Stone or I shall go in fits—

BEN NEVIS

I must—I shall—I meet not such tit bits 40
I meet not such sweet creatures every day
By my old night cap night cap night and day
I must have one sweet Buss—I must and shall!°
Red-Crag!—What Madam can you then repent
Of all the toil and vigour you have spent
To see Ben Nevis and to touch his nose?
Red-Crag I say! O I must have you close!
Red-Crag, there lies beneath my farthest toe
A vein of Sulphur—go dear Red-Crag go—
And rub your flinty back against it—budge! 50
Dear Madam I must kiss you, faith I must!
I must embrace you with my dearest gust!°

*another domestic of Ben's

Block-head*, d'ye hear—Block-head, I'll make her feel
There lies beneath my east leg's northern heel
A cave of young earth dragons—well my boy
Go thither quick and so complete my joy—
Take you a bundle of the largest pines
And where the sun on fiercest Phosphor shines
Fire them and ram them in the Dragon's nest
Then will the dragons fry and fizz their best 60
Until ten thousand, now no bigger than
Poor Aligators poor things of one span,
Will each one swell to twice ten times the size
Of northern whale—then for the tender prize—

The moment then—for then will Red-Crag rub
His flinty back and I shall kiss and snub
And press my dainty morsel to my breast
Block-head make haste!

O Muses weep the rest—
The Lady fainted and he thought her dead
So pulled the clouds again about his head 70
And went to sleep again—soon she was rous'd
By her affrighted Servants—next day hous'd
Safe on the lowly ground she bless'd her fate
That fainting fit was not delayed too late—

On Some Skulls in Beauley Abbey, near Inverness

I shed no tears;°
Deep thought, or awful vision, I had none;
By thousand petty fancies I was crossed.

WORDSWORTH

And mock'd the dead bones that lay scatter'd by.

SHAKSPEARE

I

In silent barren Synod met
Within these roofless walls, *where yet*
The shafted arch and carved fret
Cling to the Ruin,
The Brethren's Skulls mourn, dewy wet,
Their Creed's undoing.°

2

The mitred ones of Nice and Trent
Were not so tongue-tied,—no, they went
Hot to their Councils, scarce content
With Orthodoxy; 10
But ye, poor tongueless things, were meant
To speak by proxy.

3

Your Chronicles no more exist,
Since Knox, the Revolutionist,
Destroy'd the work of every fist
That scrawl'd black letter;
Well! I'm a Craniologist,
And may do better.

4

This skull-cap wore the cowl from sloth,
Or discontent, perhaps from both; 20
And yet one day, against his oath,
He tried escaping,
For men, tho' idle, may be loth
To live on gaping.

5

A Toper this! he plied his glass
More strictly than he said the Mass,
And lov'd to see a tempting Lass
Come to confession,
Letting her absolution pass
O'er fresh transgression. 30

6

This crawl'd through life in feebleness,
Boasting he never knew excess,
Cursing those crimes he scarce could guess,
Or feel but faintly,
With prayers that Heaven would cease to bless
Men so unsaintly.

7

Here's a true Churchman! he'd affect
Much charity, and ne'er neglect
To pray for mercy on th' elect,
But thought no evil 40
In sending Heathen, Turk, and Sect
All to the Devil!

8

Poor Skull, thy fingers set ablaze,°
With silver Saint in golden rays,
The holy Missal; thou did'st craze
'Mid bead and spangle,
While others pass'd their idle days
In coil and wrangle.

9

Long time this sconce a helmet wore,
But sickness smites the conscience sore; 50
He broke his sword, and hither bore
His gear and plunder,
Took to the cowl,—then rav'd and swore
At his damn'd blunder!

10

This lily colour'd skull, with all
The teeth complete, so white and small,
Belong'd to one whose early pall
A lover shaded;
He died ere Superstition's gall
His heart invaded. 60

11

Ha! here is 'undivulged crime'!°
Despair forbad his soul to climb
Beyond this world, this mortal time
Of fever'd sadness,
Until their Monkish Pantomime
Dazzled his madness!

12

A younger brother this! a man
Aspiring as a Tartar Khan,
But, curb'd and baffled, he began
The trade of frightening; 70
It smack'd of power!—and here he ran
To deal Heaven's lightening.

13

This ideot-skull belong'd to one,
A buried Miser's only son,
Who penitent, ere he'd begun
To taste of pleasure,
And, hoping Heaven's dread wrath to shun,
Gave Hell his treasure.

14

Here is the forehead of an Ape,
A robber's mark,—and near the nape 80
That bone, fie on't, bears just the shape
Of carnal passion;
Ah! he was one for theft and rape,
In Monkish fashion!

15

This was the Porter!—he could sing,
Or dance, or play, do any thing,
And what the Friars bade him bring,
They ne'er were balk'd of;
Matters not worth remembering,
And seldom talk'd of. 90

16

Enough! why need I further pore?
This corner holds at least a score,
And yonder twice as many more
Of Reverend Brothers;
'Tis the same story o'er and o'er,—
They're like the others!

Nature withheld Cassandra in the Skies

Nature withheld Cassandra in the Skies
For meet adornment a full thousand years;
She took their cream of Beauty, fairest dies
And shaped and tinted her above all peers;

Love meanwhile held her dearly with his wings
  And underneath their shadow charm'd her eyes
To such a richness, that the cloudy Kings
  Of high Olympus utter'd slavish sighs—
When I beheld her on the Earth descend
  My heart began to burn—and only pains 10
They were my pleasures—they my sad Life's end—
  Love pour'd her Beauty into my warm veins—

' 'Tis "the witching time of night" '

'Tis 'the witching time of night'
Orbed is the Moon and bright
And the Stars they glisten, glisten
Seeming with bright eyes to listen—
For what listen they?
For a song and for a charm
See they glisten in alarm
And the Moon is waxing warm
To hear what I shall say.
Moon keep wide thy golden ears 10
Hearken Stars, and hearken Spheres
Hearken thou eternal Sky
I sing an infant's lullaby,
A pretty Lullaby!
Listen, Listen, listen, listen
Glisten, glisten, glisten, glisten
And hear my lullaby!
Though the Rushes that will make
Its cradle still are in the lake:
Though the linnen that will be° 20
Its swathe is on the cotton tree;
Though the wollen that will keep
It warm, is on the silly sheep;
Listen Stars' light, listen, listen
Glisten, Glisten, glisten, glisten
And hear my lullaby!
Child! I see thee! Child I've found thee
Midst of the quiet all around thee!

Child I see thee! Child I spy thee
And thy mother sweet is nigh thee! 30
Child I know thee! Child no more
But a Poet *ever*more.
See, See the Lyre, the Lyre
In a flame of fire
Upon the little cradle's top
Flaring, flaring, flaring.
Past the eyesight's bearing—
Awake it from its sleep
And see if it can keep
Its eyes upon the blaze. 40
Amaze, Amaze!
It stares, it stares, it stares
It dares what no one dares
It lifts its little hand into the flame
Unharm'd, and on the strings
Paddles a little tune and sings
With dumb endeavour sweetly!
Bard art thou completely!
Little Child
O' the western wild 50
Bard art thou completely!—
Sweetly, with dumb endeavour,—
A Poet now or never!
Little Child
O' the western wild
A Poet now or never!

'And what is Love?—It is a doll dress'd up'

And what is Love?—It is a doll dress'd up
For idleness to cosset, nurse, and dandle;
A thing of soft misnomers, so divine
That silly youth doth think to make itself
Divine by loving, and so goes on
Yawning and doating a whole summer long,
Till Miss's comb is made a pearl tiara,
And common Wellingtons turn Romeo boots;°

Till Cleopatra lives at Number Seven,
And Anthony resides in Brunswick Square. 10
Fools! if some passions high have warm'd the world,
If Queens and Soldiers have play'd high for hearts,
It is no reason why such agonies
Should be more common than the growth of weeds.
Fools! make me whole again that weighty pearl°
The Queen of Œgypt melted, and I'll say
That ye may love in spite of beaver hats.°

Fragment: 'Welcome joy, and welcome sorrow'

Under the flag
Of each his faction, they to battle bring
Their embryo atoms.

MILTON

Welcome joy, and welcome sorrow,
  Lethe's weed, and Hermes' feather,°
Come to-day, and come to-morrow,
  I do love you both together!
  I love to mark sad faces in fair weather,
And hear a merry laugh amid the thunder;
  Fair and foul I love together;
Meadows sweet where flames burn under;
And a giggle at a wonder;
Visage sage at Pantomime; 10
Funeral and steeple-chime;
Infant playing with a skull;°
Morning fair and storm-wreck'd hull;
Night-shade with the woodbine kissing;
Serpents in red roses hissing;
Cleopatra, regal drest,
With the aspics at her breast;
Dancing Music, Music sad,
Both together, sane and mad;
Muses bright and Muses pale; 20
Sombre Saturn, Momus hale,

Laugh and sigh, and laugh again,
Oh! the sweetness of the pain!°
Muses bright and Muses pale,
Bare your faces of the veil,
Let me see, and let me write
Of the day, and of the night,
Both together,—let me slake
All my thirst for sweet heart-ache!
Let my bower be of Yew, 30
Interwreath'd with Myrtles new,
Pines, and Lime-trees full in bloom,
And my couch a low grass tomb.°

Fragment: 'Where's the Poet? Show him! show him!'

Where's the Poet? Show him! show him!
Muses nine, that I may know him!
'Tis the man, who with a man
  Is an equal, be he King,
Or poorest of the beggar-clan,
  Or any other wondrous thing
A man may be 'twixt ape and Plato;
  'Tis the man who with a bird,°
Wren or eagle, finds his way to
  All its instincts;—he hath heard 10
The Lion's roaring, and can tell
  What his horny throat expresseth;
And to him the Tiger's yell
  Comes articulate, and presseth
On his ear like mother-tongue;

To Homer

Standing aloof in giant ignorance,°
  Of thee I hear and of the Cyclades,
As one who sits ashore and longs perchance
  To visit dolphin-coral in deep seas,

So wast thou blind;—but then the veil was rent,
  For Jove uncurtain'd Heaven to let thee live,
And Neptune made for thee a spumy tent,
  And Pan made sing for thee his forest-hive;
Aye on the shores of darkness there is light,
  And precipices show untrodden green, 10
There is a budding morrow in midnight,
  There is a triple sight in blindness keen;°
Such seeing hadst thou, as it once befel
To Dian, Queen of Earth, and Heaven, and Hell.

Hyperion: A Fragment

BOOK I

Deep in the shady sadness of a vale°
Far sunken from the healthy breath of morn,
Far from the fiery noon, and eve's one star,
Sat gray-hair'd Saturn, quiet as a stone,
Still as the silence round about his lair;
Forest on forest hung above his head
Like cloud on cloud. No stir of air was there,
Not so much life as on a summer's day
Robs not one light seed from the feather'd grass,°
But where the dead leaf fell, there did it rest. 10
A stream went voiceless by, still deadened more
By reason of his fallen divinity
Spreading a shade: the Naiad 'mid her reeds
Press'd her cold finger closer to her lips.

  Along the margin-sand large foot-marks went,
No further than to where his feet had stray'd,
And slept there since. Upon the sodden ground
His old right hand lay nerveless, listless, dead,
Unsceptred; and his realmless eyes were closed;
While his bow'd head seem'd list'ning to the Earth, 20
His ancient mother, for some comfort yet.

  It seem'd no force could wake him from his place;
But there came one, who with a kindred hand

Touch'd his wide shoulders, after bending low
With reverence, though to one who knew it not.
She was a Goddess of the infant world;
By her in stature the tall Amazon
Had stood a pigmy's height: she would have ta'en
Achilles by the hair and bent his neck;
Or with a finger stay'd Ixion's wheel. 30
Her face was large as that of Memphian sphinx,°
Pedestal'd haply in a palace court,
When sages look'd to Egypt for their lore.
But oh! how unlike marble was that face:
How beautiful, if sorrow had not made
Sorrow more beautiful than Beauty's self.
There was a listening fear in her regard,
As if calamity had but begun;
As if the vanward clouds of evil days
Had spent their malice, and the sullen rear 40
Was with its stored thunder labouring up.
One hand she press'd upon that aching spot
Where beats the human heart, as if just there,
Though an immortal, she felt cruel pain:
The other upon Saturn's bended neck
She laid, and to the level of his ear
Leaning with parted lips, some words she spake
In solemn tenour and deep organ tone:
Some mourning words, which in our feeble tongue
Would come in these like accents; O how frail 50
To that large utterance of the early Gods!
'Saturn, look up!—though wherefore, poor old King?
I have no comfort for thee, no not one:
I cannot say, "O wherefore sleepest thou?"
For heaven is parted from thee, and the earth
Knows thee not, thus afflicted, for a God;
And ocean too, with all its solemn noise,
Has from thy sceptre pass'd; and all the air
Is emptied of thine hoary majesty.
Thy thunder, conscious of the new command, 60
Rumbles reluctant o'er our fallen house;°
And thy sharp lightning in unpractised hands
Scorches and burns our once serene domain.
O aching time! O moments big as years!

All as ye pass swell out the monstrous truth,
And press it so upon our weary griefs
That unbelief has not a space to breathe.
Saturn, sleep on:—O thoughtless, why did I
Thus violate thy slumbrous solitude?
Why should I ope thy melancholy eyes? 70
Saturn, sleep on! while at thy feet I weep.'

As when, upon a tranced summer-night,
Those green-rob'd senators of mighty woods,
Tall oaks, branch-charmed by the earnest stars,
Dream, and so dream all night without a stir,
Save from one gradual solitary gust
Which comes upon the silence, and dies off,
As if the ebbing air had but one wave;
So came these words and went; the while in tears
She touch'd her fair large forehead to the ground, 80
Just where her falling hair might be outspread
A soft and silken mat for Saturn's feet.
One moon, with alteration slow, had shed
Her silver seasons four upon the night,
And still these two were postured motionless,
Like natural sculpture in cathedral cavern;°
The frozen God still couchant on the earth,
And the sad Goddess weeping at his feet:
Until at length old Saturn lifted up
His faded eyes, and saw his Kingdom gone, 90
And all the gloom and sorrow of the place,
And that fair kneeling Goddess; and then spake,
As with a palsied tongue, and while his beard
Shook horrid with such aspen-malady:
'O tender spouse of gold Hyperion,
Thea, I feel thee ere I see thy face;
Look up, and let me see our doom in it;
Look up, and tell me if this feeble shape
Is Saturn's; tell me, if thou hear'st the voice
Of Saturn; tell me, if this wrinkling brow, 100
Naked and bare of its great diadem,
Peers like the front of Saturn. Who had power
To make me desolate? whence came the strength?
How was it nurtur'd to such bursting forth,

While Fate seem'd strangled in my nervous grasp?
But it is so; and I am smother'd up,
And buried from all godlike exercise
Of influence benign on planets pale,
Of admonitions to the winds and seas,
Of peaceful sway above man's harvesting, 110
And all those acts which Deity supreme
Doth ease its heart of love in.—I am gone
Away from my own bosom: I have left
My strong identity, my real self,
Somewhere between the throne, and where I sit
Here on this spot of earth. Search, Thea, search!
Open thine eyes eterne, and sphere them round
Upon all space: space starr'd, and lorn of light;
Space region'd with life-air; and barren void;
Spaces of fire, and all the yawn of hell.— 120
Search, Thea, search! and tell me, if thou seest
A certain shape or shadow, making way
With wings or chariot fierce to repossess
A heaven he lost erewhile: it must—it must
Be of ripe progress—Saturn must be King.
Yes, there must be a golden victory;
There must be Gods thrown down, and trumpets blown
Of triumph calm, and hymns of festival
Upon the gold clouds metropolitan,°
Voices of soft proclaim, and silver stir 130
Of strings in hollow shells; and there shall be
Beautiful things made new, for the surprise
Of the sky-children; I will give command:
Thea! Thea! Thea! where is Saturn?'

This passion lifted him upon his feet,
And made his hands to struggle in the air,
His Druid locks to shake and ooze with sweat,°
His eyes to fever out, his voice to cease.
He stood, and heard not Thea's sobbing deep;
A little time, and then again he snatch'd 140
Utterance thus.—'But cannot I create?
Cannot I form? Cannot I fashion forth
Another world, another universe,
To overbear and crumble this to nought?

Where is another Chaos? Where?'—That word
Found way unto Olympus, and made quake
The rebel three.—Thea was startled up,°
And in her bearing was a sort of hope,
As thus she quick-voic'd spake, yet full of awe.

'This cheers our fallen house: come to our friends, 150
O Saturn! come away, and give them heart;
I know the covert, for thence came I hither.'
Thus brief; then with beseeching eyes she went
With backward footing through the shade a space:
He follow'd, and she turn'd to lead the way
Through aged boughs, that yielded like the mist
Which eagles cleave upmounting from their nest.

Meanwhile in other realms big tears were shed,
More sorrow like to this, and such like woe,
Too huge for mortal tongue or pen of scribe: 160
The Titans fierce, self-hid, or prison-bound,
Groan'd for the old allegiance once more,
And listen'd in sharp pain for Saturn's voice.
But one of the whole mammoth-brood still kept
His sov'reignty, and rule, and majesty;—
Blazing Hyperion on his orbed fire
Still sat, still snuff'd the incense, teeming up°
From man to the sun's God; yet unsecure:
For as among us mortals omens drear
Fright and perplex, so also shuddered he— 170
Not at dog's howl, or gloom-bird's hated screech,
Or the familiar visiting of one
Upon the first toll of his passing-bell,
Or prophesyings of the midnight lamp;
But horrors, portion'd to a giant nerve,
Oft made Hyperion ache. His palace bright
Bastion'd with pyramids of glowing gold,
And touch'd with shade of bronzed obelisks,
Glar'd a blood-red through all its thousand courts,
Arches, and domes, and fiery galleries; 180
And all its curtains of Aurorian clouds
Flush'd angerly: while sometimes eagle's wings,
Unseen before by Gods or wondering men,

Darken'd the place; and neighing steeds were heard,
Not heard before by Gods or wondering men.
Also, when he would taste the spicy wreaths
Of incense, breath'd aloft from sacred hills,
Instead of sweets, his ample palate took
Savour of poisonous brass and metal sick:
And so, when harbour'd in the sleepy west, 190
After the full completion of fair day,—
For rest divine upon exalted couch
And slumber in the arms of melody,
He pac'd away the pleasant hours of ease
With stride colossal, on from hall to hall;
While far within each aisle and deep recess,°
His winged minions in close clusters stood,
Amaz'd and full of fear; like anxious men
Who on wide plains gather in panting troops,
When earthquakes jar their battlements and towers. 200
Even now, while Saturn, rous'd from icy trance,
Went step for step with Thea through the woods,
Hyperion, leaving twilight in the rear,
Came slope upon the threshold of the west;
Then, as was wont, his palace-door flew ope
In smoothest silence, save what solemn tubes,
Blown by the serious Zephyrs, gave of sweet
And wandering sounds, slow-breathed melodies;
And like a rose in vermeil tint and shape,
In fragrance soft, and coolness to the eye, 210
That inlet to severe magnificence
Stood full blown, for the God to enter in.

He enter'd, but he enter'd full of wrath;
His flaming robes stream'd out beyond his heels,
And gave a roar, as if of earthly fire,
That scar'd away the meek ethereal Hours
And made their dove-wings tremble. On he flared,
From stately nave to nave, from vault to vault,
Through bowers of fragrant and enwreathed light,
And diamond-paved lustrous long arcades, 220
Until he reach'd the great main cupola;
There standing fierce beneath, he stampt his foot,
And from the basements deep to the high towers

Jarr'd his own golden region; and before
The quavering thunder thereupon had ceas'd,
His voice leapt out, despite of godlike curb,
To this result: 'O dreams of day and night!
O monstrous forms! O effigies of pain!
O spectres busy in a cold, cold gloom!
O lank-eared Phantoms of black-weeded pools! 230
Why do I know ye? why have I seen ye? why
Is my eternal essence thus distraught
To see and to behold these horrors new?
Saturn is fallen, am I too to fall?
Am I to leave this haven of my rest,
This cradle of my glory, this soft clime,
This calm luxuriance of blissful light,
These crystalline pavilions, and pure fanes,
Of all my lucent empire? It is left
Deserted, void, nor any haunt of mine. 240
The blaze, the splendor, and the symmetry,
I cannot see—but darkness, death and darkness.
Even here, into my centre of repose,
The shady visions come to domineer,
Insult, and blind, and stifle up my pomp.—
Fall!—No, by Tellus and her briny robes!
Over the fiery frontier of my realms
I will advance a terrible right arm
Shall scare that infant thunderer, rebel Jove,
And bid old Saturn take his throne again.'— 250
He spake, and ceas'd, the while a heavier threat
Held struggle with his throat but came not forth;°
For as in theatres of crowded men
Hubbub increases more they call out 'Hush!'
So at Hyperion's words the Phantoms pale
Bestirr'd themselves, thrice horrible and cold;
And from the mirror'd level where he stood
A mist arose, as from a scummy marsh.
At this, through all his bulk an agony°
Crept gradual, from the feet unto the crown, 260
Like a lithe serpent vast and muscular
Making slow way, with head and neck convuls'd
From over-strained might. Releas'd, he fled
To the eastern gates, and full six dewy hours

Before the dawn in season due should blush,
He breath'd fierce breath against the sleepy portals,
Clear'd them of heavy vapours, burst them wide
Suddenly on the ocean's chilly streams.
The planet orb of fire, whereon he rode
Each day from east to west the heavens through, 270
Spun round in sable curtaining of clouds;
Not therefore veiled quite, blindfold, and hid,
But ever and anon the glancing spheres,
Circles, and arcs, and broad-belting colure,°
Glow'd through, and wrought upon the muffling dark
Sweet-shaped lightnings from the nadir deep
Up to the zenith,—hieroglyphics old,
Which sages and keen-eyed astrologers
Then living on the earth, with labouring thought
Won from the gaze of many centuries: 280
Now lost, save what we find on remnants huge°
Of stone, or marble swart; their import gone,
Their wisdom long since fled.—Two wings this orb
Possess'd for glory, two fair argent wings,
Ever exalted at the God's approach:
And now, from forth the gloom their plumes immense
Rose, one by one, till all outspreaded were;
While still the dazzling globe maintain'd eclipse,
Awaiting for Hyperion's command.
Fain would he have commanded, fain took throne 290
And bid the day begin, if but for change.
He might not:—No, though a primeval God:
The sacred seasons might not be disturb'd.
Therefore the operations of the dawn
Stay'd in their birth, even as here 'tis told.
Those silver wings expanded sisterly,
Eager to sail their orb; the porches wide
Open'd upon the dusk demesnes of night;
And the bright Titan, phrenzied with new woes,
Unus'd to bend, by hard compulsion bent 300
His spirit to the sorrow of the time;
And all along a dismal rack of clouds,
Upon the boundaries of day and night,
He stretch'd himself in grief and radiance faint.
There as he lay, the Heaven with its stars

Look'd down on him with pity, and the voice
Of Cœlus, from the universal space,
Thus whisper'd low and solemn in his ear.
'O brightest of my children dear, earth-born
And sky-engendered, Son of Mysteries 310
All unrevealed even to the powers
Which met at thy creating; at whose joys
And palpitations sweet, and pleasures soft,
I, Cœlus, wonder, how they came and whence;
And at the fruits thereof what shapes they be,
Distinct, and visible; symbols divine,
Manifestations of that beauteous life
Diffus'd unseen throughout eternal space:
Of these new-form'd art thou, oh brightest child!
Of these, thy brethren and the Goddesses! 320
There is sad feud among ye, and rebellion
Of son against his sire. I saw him fall,
I saw my first-born tumbled from his throne!°
To me his arms were spread, to me his voice
Found way from forth the thunders round his head!
Pale wox I, and in vapours hid my face.
Art thou, too, near such doom? vague fear there is:
For I have seen my sons most unlike Gods.
Divine ye were created, and divine
In sad demeanour, solemn, undisturb'd, 330
Unruffled, like high Gods, ye liv'd and ruled:
Now I behold in you fear, hope, and wrath;
Actions of rage and passion; even as
I see them, on the mortal world beneath,
In men who die.—This is the grief, O Son!
Sad sign of ruin, sudden dismay, and fall!
Yet do thou strive; as thou art capable,
As thou canst move about, an evident God;
And canst oppose to each malignant hour
Ethereal presence:—I am but a voice; 340
My life is but the life of winds and tides,
No more than winds and tides can I avail:—
But thou canst.—Be thou therefore in the van
Of circumstance; yea, seize the arrow's barb
Before the tense string murmur.—To the earth!
For there thou wilt find Saturn, and his woes.

Meantime I will keep watch on thy bright sun,
And of thy seasons be a careful nurse.'—
Ere half this region-whisper had come down,
Hyperion arose, and on the stars 350
Lifted his curved lids, and kept them wide
Until it ceas'd; and still he kept them wide:
And still they were the same bright, patient stars.
Then with a slow incline of his broad breast,°
Like to a diver in the pearly seas,
Forward he stoop'd over the airy shore,
And plung'd all noiseless into the deep night.

BOOK II

Just at the self-same beat of Time's wide wings
Hyperion slid into the rustled air,
And Saturn gain'd with Thea that sad place
Where Cybele and the bruised Titans mourn'd.
It was a den where no insulting light
Could glimmer on their tears; where their own groans
They felt, but heard not, for the solid roar°
Of thunderous waterfalls and torrents hoarse,
Pouring a constant bulk, uncertain where.
Crag jutting forth to crag, and rocks that seem'd 10
Ever as if just rising from a sleep,
Forehead to forehead held their monstrous horns;
And thus in thousand hugest phantasies
Made a fit roofing to this nest of woe.
Instead of thrones, hard flint they sat upon,
Couches of rugged stone, and slaty ridge
Stubborn'd with iron. All were not assembled:
Some chain'd in torture, and some wandering.
Cœus, and Gyges, and Briareüs,
Typhon, and Dolor, and Porphyrion, 20
With many more, the brawniest in assault,
Were pent in regions of laborious breath;
Dungeon'd in opaque element, to keep
Their clenched teeth still clench'd, and all their limbs
Lock'd up like veins of metal, crampt and screw'd;
Without a motion, save of their big hearts
Heaving in pain, and horribly convuls'd

With sanguine feverous boiling gurge of pulse.
Mnemosyne was straying in the world;
Far from her moon had Phœbe wandered; 30
And many else were free to roam abroad,
But for the main, here found they covert drear.
Scarce images of life, one here, one there,°
Lay vast and edgeways; like a dismal cirque
Of Druid stones, upon a forlorn moor,
When the chill rain begins at shut of eve,
In dull November, and their chancel vault,
The Heaven itself, is blinded throughout night.
Each one kept shroud, nor to his neighbour gave°
Or word, or look, or action of despair. 40
Creüs was one; his ponderous iron mace
Lay by him, and a shatter'd rib of rock
Told of his rage, ere he thus sank and pined.
Iäpetus another; in his grasp,
A serpent's plashy neck; its barbed tongue
Squeez'd from the gorge, and all its uncurl'd length
Dead; and because the creature could not spit
Its poison in the eyes of conquering Jove.
Next Cottus: prone he lay, chin uppermost,
As though in pain; for still upon the flint 50
He ground severe his skull, with open mouth
And eyes at horrid working. Nearest him
Asia, born of most enormous Caf,
Who cost her mother Tellus keener pangs,
Though feminine, than any of her sons:
More thought than woe was in her dusky face,
For she was prophesying of her glory;
And in her wide imagination stood
Palm-shaded temples, and high rival fanes,
By Oxus or in Ganges' sacred isles. 60
Even as Hope upon her anchor leans,
So leant she, not so fair, upon a tusk
Shed from the broadest of her elephants.
Above her, on a crag's uneasy shelve,
Upon his elbow rais'd, all prostrate else,
Shadow'd Enceladus; once tame and mild
As grazing ox unworried in the meads;
Now tiger-passion'd, lion-thoughted, wroth,

He meditated, plotted, and even now
Was hurling mountains in that second war, 70
Not long delay'd, that scar'd the younger Gods
To hide themselves in forms of beast and bird.
Not far hence Atlas; and beside him prone
Phorcus, the sire of Gorgons. Neighbour'd close
Oceanus, and Tethys, in whose lap
Sobb'd Clymene among her tangled hair.
In midst of all lay Themis, at the feet
Of Ops the queen all clouded round from sight;
No shape distinguishable, more than when
Thick night confounds the pine-tops with the clouds: 80
And many else whose names may not be told.
For when the Muse's wings are air-ward spread,
Who shall delay her flight? And she must chaunt
Of Saturn, and his guide, who now had climb'd
With damp and slippery footing from a depth
More horrid still. Above a sombre cliff
Their heads appear'd, and up their stature grew
Till on the level height their steps found ease:
Then Thea spread abroad her trembling arms
Upon the precincts of this nest of pain, 90
And sidelong fix'd her eye on Saturn's face:
There saw she direst strife; the supreme God
At war with all the frailty of grief,
Of rage, of fear, anxiety, revenge,
Remorse, spleen, hope, but most of all despair.
Against these plagues he strove in vain; for Fate
Had pour'd a mortal oil upon his head,
A disanointing poison: so that Thea,
Affrighted, kept her still, and let him pass
First onwards in, among the fallen tribe. 100

As with us mortal men, the laden heart
Is persecuted more, and fever'd more,
When it is nighing to the mournful house
Where other hearts are sick of the same bruise;
So Saturn, as he walk'd into the midst,
Felt faint, and would have sunk among the rest,
But that he met Enceladus's eye,
Whose mightiness, and awe of him, at once

Came like an inspiration; and he shouted,
'Titans, behold your God!' at which some groan'd; 110
Some started on their feet; some also shouted;
Some wept, some wail'd, all bow'd with reverence;
And Ops, uplifting her black folded veil,
Show'd her pale cheeks, and all her forehead wan,
Her eye-brows thin and jet, and hollow eyes.
There is a roaring in the bleak-grown pines
When Winter lifts his voice; there is a noise
Among immortals when a God gives sign,
With hushing finger, how he means to load
His tongue with the full weight of utterless thought, 120
With thunder, and with music, and with pomp:
Such noise is like the roar of bleak-grown pines;
Which, when it ceases in this mountain'd world,
No other sound succeeds; but ceasing here,
Among these fallen, Saturn's voice therefrom
Grew up like organ, that begins anew
Its strain, when other harmonies, stopt short,
Leave the dinn'd air vibrating silverly.
Thus grew it up—'Not in my own sad breast,
Which is its own great judge and searcher out, 130
Can I find reason why ye should be thus:
Not in the legends of the first of days,
Studied from that old spirit-leaved book
Which starry Uranus with finger bright
Sav'd from the shores of darkness, when the waves
Low-ebb'd still hid it up in shallow gloom;—
And the which book ye know I ever kept
For my firm-based footstool:—Ah, infirm!
Not there, nor in sign, symbol, or portent
Of element, earth, water, air, and fire,— 140
At war, at peace, or inter-quarreling
One against one, or two, or three, or all
Each several one against the other three,
As fire with air loud warring when rain-floods
Drown both, and press them both against earth's face,
Where, finding sulphur, a quadruple wrath
Unhinges the poor world;—not in that strife,
Wherefrom I take strange lore, and read it deep,
Can I find reason why ye should be thus:

No, no-where can unriddle, though I search, 150
And pore on Nature's universal scroll
Even to swooning, why ye, Divinities,
The first-born of all shap'd and palpable Gods,
Should cower beneath what, in comparison,
Is untremendous might. Yet ye are here,
O'erwhelm'd, and spurn'd, and batter'd, ye are here!
O Titans, shall I say 'Arise!'—Ye groan:
Shall I say 'Crouch!'—Ye groan. What can I then?
O Heaven wide! O unseen parent dear!
What can I? Tell me, all ye brethren Gods, 160
How we can war, how engine our great wrath!°
O speak your counsel now, for Saturn's ear
Is all a-hunger'd. Thou, Oceanus,
Ponderest high and deep; and in thy face
I see, astonied, that severe content
Which comes of thought and musing: give us help!'

So ended Saturn; and the God of the Sea,
Sophist and sage, from no Athenian grove,°
But cogitation in his watery shades,
Arose, with locks not oozy, and began,° 170
In murmurs, which his first-endeavouring tongue
Caught infant-like from the far-foamed sands.
'O ye, whom wrath consumes! who, passion-stung,
Writhe at defeat, and nurse your agonies!
Shut up your senses, stifle up your ears,
My voice is not a bellows unto ire.
Yet listen, ye who will, whilst I bring proof
How ye, perforce, must be content to stoop:
And in the proof much comfort will I give,
If ye will take that comfort in its truth. 180
We fall by course of Nature's law, not force
Of thunder, or of Jove. Great Saturn, thou
Hast sifted well the atom-universe;
But for this reason, that thou art the King,
And only blind from sheer supremacy,
One avenue was shaded from thine eyes,
Through which I wandered to eternal truth.
And first, as thou wast not the first of powers,
So art thou not the last; it cannot be:

Thou art not the beginning nor the end. 190
From Chaos and parental Darkness came
Light, the first fruits of that intestine broil,
That sullen ferment, which for wondrous ends
Was ripening in itself. The ripe hour came,
And with it Light, and Light, engendering
Upon its own producer, forthwith touch'd
The whole enormous matter into life.
Upon that very hour, our parentage,
The Heavens and the Earth, were manifest:
Then thou first-born, and we the giant-race, 200
Found ourselves ruling new and beauteous realms.
Now comes the pain of truth, to whom 'tis pain;
O folly! for to bear all naked truths,
And to envisage circumstance, all calm,
That is the top of sovereignty. Mark well!
As Heaven and Earth are fairer, fairer far
Than Chaos and blank Darkness, though once chiefs;
And as we show beyond that Heaven and Earth
In form and shape compact and beautiful,
In will, in action free, companionship, 210
And thousand other signs of purer life;
So on our heels a fresh perfection treads,
A power more strong in beauty, born of us
And fated to excel us, as we pass
In glory that old Darkness: nor are we
Thereby more conquer'd, than by us the rule
Of shapeless Chaos. Say, doth the dull soil
Quarrel with the proud forests it hath fed,
And feedeth still, more comely than itself?
Can it deny the chiefdom of green groves? 220
Or shall the tree be envious of the dove
Because it cooeth, and hath snowy wings
To wander wherewithal and find its joys?
We are such forest-trees, and our fair boughs
Have bred forth, not pale solitary doves,
But eagles, golden-feather'd, who do tower
Above us in their beauty, and must reign
In right thereof; for 'tis the eternal law
That first in beauty should be first in might:
Yea, by that law, another race may drive 230

Our conquerors to mourn as we do now.
Have ye beheld the young God of the Seas,°
My dispossessor? Have ye seen his face?
Have ye beheld his chariot, foam'd along
By noble winged creatures he hath made?
I saw him on the calmed waters scud,
With such a glow of beauty in his eyes,
That it enforc'd me to bid sad farewell
To all my empire: farewell sad I took,
And hither came, to see how dolorous fate 240
Had wrought upon ye; and how I might best
Give consolation in this woe extreme.
Receive the truth, and let it be your balm.'

Whether through poz'd conviction, or disdain,
They guarded silence, when Oceanus
Left murmuring, what deepest thought can tell?
But so it was, none answer'd for a space,
Save one whom none regarded, Clymene;
And yet she answer'd not, only complain'd,
With hectic lips, and eyes up-looking mild, 250
Thus wording timidly among the fierce:
'O Father, I am here the simplest voice,
And all my knowledge is that joy is gone,
And this thing woe crept in among our hearts,
There to remain for ever, as I fear:
I would not bode of evil, if I thought
So weak a creature could turn off the help
Which by just right should come of mighty Gods;
Yet let me tell my sorrow, let me tell
Of what I heard, and how it made me weep, 260
And know that we had parted from all hope.
I stood upon a shore, a pleasant shore,
Where a sweet clime was breathed from a land
Of fragrance, quietness, and trees, and flowers.
Full of calm joy it was, as I of grief;
Too full of joy and soft delicious warmth;
So that I felt a movement in my heart
To chide, and to reproach that solitude
With songs of misery, music of our woes;
And sat me down, and took a mouthed shell 270

And murmur'd into it, and made melody—
O melody no more! for while I sang,
And with poor skill let pass into the breeze
The dull shell's echo, from a bowery strand
Just opposite, an island of the sea,
There came enchantment with the shifting wind,
That did both drown and keep alive my ears.
I threw my shell away upon the sand,
And a wave fill'd it, as my sense was fill'd
With that new blissful golden melody.° 280
A living death was in each gush of sounds,
Each family of rapturous hurried notes,
That fell, one after one, yet all at once,
Like pearl beads dropping sudden from their string:
And then another, then another strain,
Each like a dove leaving its olive perch,
With music wing'd instead of silent plumes,
To hover round my head, and make me sick
Of joy and grief at once. Grief overcame,
And I was stopping up my frantic ears, 290
When, past all hindrance of my trembling hands,
A voice came sweeter, sweeter than all tune,
And still it cried, 'Apollo! young Apollo!
The morning-bright Apollo! young Apollo!'
I fled, it follow'd me, and cried 'Apollo!'
O Father, and O Brethren, had ye felt
Those pains of mine; O Saturn, hadst thou felt,
Ye would not call this too indulged tongue
Presumptuous, in thus venturing to be heard.'

So far her voice flow'd on, like timorous brook 300
That, lingering along a pebbled coast,
Doth fear to meet the sea: but sea it met,
And shudder'd; for the overwhelming voice
Of huge Enceladus swallow'd it in wrath:
The ponderous syllables, like sullen waves
In the half-glutted hollows of reef-rocks,
Came booming thus, while still upon his arm
He lean'd; not rising, from supreme contempt.
'Or shall we listen to the over-wise,
Or to the over-foolish, Giant-Gods? 310

Not thunderbolt on thunderbolt, till all
That rebel Jove's whole armoury were spent,
Not world on world upon these shoulders piled,
Could agonize me more than baby-words
In midst of this dethronement horrible.
Speak! roar! shout! yell! ye sleepy Titans all.
Do ye forget the blows, the buffets vile?
Are ye not smitten by a youngling arm?
Dost thou forget, sham Monarch of the Waves,
Thy scalding in the seas? What, have I rous'd 320
Your spleens with so few simple words as these?
O joy! for now I see ye are not lost:
O joy! for now I see a thousand eyes
Wide glaring for revenge!'—As this he said,
He lifted up his stature vast, and stood,
Still without intermission speaking thus:
'Now ye are flames, I'll tell you how to burn,
And purge the ether of our enemies;
How to feed fierce the crooked stings of fire,
And singe away the swollen clouds of Jove, 330
Stifling that puny essence in its tent.
O let him feel the evil he hath done;
For though I scorn Oceanus's lore,
Much pain have I for more than loss of realms:
The days of peace and slumberous calm are fled;
Those days, all innocent of scathing war,
When all the fair Existences of heaven
Came open-eyed to guess what we would speak:—
That was before our brows were taught to frown,
Before our lips knew else but solemn sounds; 340
That was before we knew the winged thing,°
Victory, might be lost, or might be won.
And be ye mindful that Hyperion,
Our brightest brother, still is undisgraced—
Hyperion, lo! his radiance is here!'

All eyes were on Enceladus's face,
And they beheld, while still Hyperion's name
Flew from his lips up to the vaulted rocks,
A pallid gleam across his features stern:
Not savage, for he saw full many a God 350

Wroth as himself. He look'd upon them all,
And in each face he saw a gleam of light,
But splendider in Saturn's, whose hoar locks
Shone like the bubbling foam about a keel
When the prow sweeps into a midnight cove.
In pale and silver silence they remain'd,
Till suddenly a splendour, like the morn,
Pervaded all the beetling gloomy steeps,
All the sad spaces of oblivion,
And every gulf, and every chasm old, 360
And every height, and every sullen depth,
Voiceless, or hoarse with loud tormented streams:
And all the everlasting cataracts,
And all the headlong torrents far and near,
Mantled before in darkness and huge shade,
Now saw the light and made it terrible.
It was Hyperion:—a granite peak
His bright feet touch'd, and there he stay'd to view
The misery his brilliance had betray'd
To the most hateful seeing of itself. 370
Golden his hair of short Numidian curl,
Regal his shape majestic, a vast shade
In midst of his own brightness, like the bulk
Of Memnon's image at the set of sun°
To one who travels from the dusking East:
Sighs, too, as mournful as that Memnon's harp
He utter'd, while his hands contemplative
He press'd together, and in silence stood.
Despondence seiz'd again the fallen Gods
At sight of the dejected King of Day, 380
And many hid their faces from the light:
But fierce Enceladus sent forth his eyes
Among the brotherhood; and, at their glare,
Uprose Iäpetus, and Creüs too,
And Phorcus, sea-born, and together strode
To where he towered on his eminence.
There those four shouted forth old Saturn's name;
Hyperion from the peak loud answered, 'Saturn!'
Saturn sat near the Mother of the Gods,
In whose face was no joy, though all the Gods 390
Gave from their hollow throats the name of 'Saturn!'

BOOK III

Thus in alternate uproar and sad peace,
Amazed were those Titans utterly.
O leave them, Muse! O leave them to their woes;
For thou art weak to sing such tumults dire:
A solitary sorrow best befits
Thy lips, and antheming a lonely grief.
Leave them, O Muse! for thou anon wilt find
Many a fallen old Divinity
Wandering in vain about bewildered shores.
Meantime touch piously the Delphic harp, 10
And not a wind of heaven but will breathe
In aid soft warble from the Dorian flute;°
For lo! 'tis for the Father of all verse.°
Flush every thing that hath a vermeil hue,
Let the rose glow intense and warm the air,
And let the clouds of even and of morn
Float in voluptuous fleeces o'er the hills;
Let the red wine within the goblet boil,
Cold as a bubbling well; let faint-lipp'd shells,
On sands, or in great deeps, vermilion turn 20
Through all their labyrinths; and let the maid
Blush keenly, as with some warm kiss surpris'd.
Chief isle of the embowered Cyclades,
Rejoice, O Delos, with thine olives green,
And poplars, and lawn-shading palms, and beech,
In which the Zephyr breathes the loudest song,
And hazels thick, dark-stemm'd beneath the shade:
Apollo is once more the golden theme!
Where was he, when the Giant of the Sun°
Stood bright, amid the sorrow of his peers? 30
Together had he left his mother fair°
And his twin-sister sleeping in their bower,
And in the morning twilight wandered forth
Beside the osiers of a rivulet,
Full ankle-deep in lilies of the vale.
The nightingale had ceas'd, and a few stars
Were lingering in the heavens, while the thrush
Began calm-throated. Throughout all the isle

There was no covert, no retired cave
Unhaunted by the murmurous noise of waves, 40
Though scarcely heard in many a green recess.
He listen'd, and he wept, and his bright tears
Went trickling down the golden bow he held.
Thus with half-shut suffused eyes he stood,
While from beneath some cumbrous boughs hard by
With solemn step an awful Goddess came,°
And there was purport in her looks for him,
Which he with eager guess began to read
Perplex'd, the while melodiously he said:
'How cam'st thou over the unfooted sea? 50
Or hath that antique mien and robed form
Mov'd in these vales invisible till now?
Sure I have heard those vestments sweeping o'er
The fallen leaves, when I have sat alone
In cool mid-forest. Surely I have traced
The rustle of those ample skirts about
These grassy solitudes, and seen the flowers
Lift up their heads, as still the whisper pass'd.
Goddess! I have beheld those eyes before,
And their eternal calm, and all that face, 60
Or I have dream'd.'—'Yes,' said the supreme shape,
'Thou hast dream'd of me; and awaking up
Didst find a lyre all golden by thy side,
Whose strings touch'd by thy fingers, all the vast
Unwearied ear of the whole universe
Listen'd in pain and pleasure at the birth
Of such new tuneful wonder. Is't not strange
That thou shouldst weep, so gifted? Tell me, youth,
What sorrow thou canst feel; for I am sad
When thou dost shed a tear: explain thy griefs 70
To one who in this lonely isle hath been
The watcher of thy sleep and hours of life,
From the young day when first thy infant hand
Pluck'd witless the weak flowers, till thine arm
Could bend that bow heroic to all times.
Show thy heart's secret to an ancient Power
Who hath forsaken old and sacred thrones°
For prophecies of thee, and for the sake
Of loveliness new born.'—Apollo then,

With sudden scrutiny and gloomless eyes, 80
Thus answer'd, while his white melodious throat
Throbb'd with the syllables.—'Mnemosyne!°
Thy name is on my tongue, I know not how;
Why should I tell thee what thou so well seest?
Why should I strive to show what from thy lips
Would come no mystery? For me, dark, dark,°
And painful vile oblivion seals my eyes:
I strive to search wherefore I am so sad,
Until a melancholy numbs my limbs;
And then upon the grass I sit, and moan, 90
Like one who once had wings.—O why should I
Feel curs'd and thwarted, when the liegeless air
Yields to my step aspirant? why should I
Spurn the green turf as hateful to my feet?
Goddess benign, point forth some unknown thing:
Are there not other regions than this isle?
What are the stars? There is the sun, the sun!
And the most patient brilliance of the moon!
And stars by thousands! Point me out the way
To any one particular beauteous star, 100
And I will flit into it with my lyre,
And make its silvery splendour pant with bliss.
I have heard the cloudy thunder: Where is power?
Whose hand, whose essence, what divinity
Makes this alarum in the elements,
While I here idle listen on the shores
In fearless yet in aching ignorance?
O tell me, lonely Goddess, by thy harp,
That waileth every morn and eventide,
Tell me why thus I rave, about these groves! 110
Mute thou remainest—Mute! yet I can read
A wondrous lesson in thy silent face:
Knowledge enormous makes a God of me.°
Names, deeds, gray legends, dire events, rebellions,
Majesties, sovran voices, agonies,
Creations and destroyings, all at once
Pour into the wide hollows of my brain,
And deify me, as if some blithe wine
Or bright elixir peerless I had drunk,
And so become immortal.'—Thus the God, 120

While his enkindled eyes, with level glance
Beneath his white soft temples, stedfast kept
Trembling with light upon Mnemosyne.
Soon wild commotions shook him, and made flush
All the immortal fairness of his limbs;
Most like the struggle at the gate of death;
Or liker still to one who should take leave
Of pale immortal death, and with a pang
As hot as death's is chill, with fierce convulse
Die into life: so young Apollo anguish'd: 130
His very hair, his golden tresses famed
Kept undulation round his eager neck.
During the pain Mnemosyne upheld
Her arms as one who prophesied.—At length
Apollo shriek'd;—and lo! from all his limbs
Celestial°

Fancy

Ever let the Fancy roam,
Pleasure never is at home:
At a touch sweet Pleasure melteth,
Like to bubbles when rain pelteth;
Then let winged Fancy wander
Through the thought still spread beyond her:
Open wide the mind's cage-door,
She'll dart forth, and cloudward soar.
O sweet Fancy! let her loose;
Summer's joys are spoilt by use, 10
And the enjoying of the Spring
Fades as does its blossoming;
Autumn's red-lipp'd fruitage too,
Blushing through the mist and dew,
Cloys with tasting: What do then?
Sit thee by the ingle, when
The sear faggot blazes bright,
Spirit of a winter's night;

When the soundless earth is muffled,
And the caked snow is shuffled 20
From the ploughboy's heavy shoon;°
When the Night doth meet the Noon
In a dark conspiracy
To banish Even from her sky.
Sit thee there, and send abroad,
With a mind self-overaw'd,
Fancy, high-commission'd:—send her!
She has vassals to attend her:
She will bring, in spite of frost,
Beauties that the earth hath lost; 30
She will bring thee, all together,
All delights of summer weather;
All the buds and bells of May,
From dewy sward or thorny spray;
All the heaped Autumn's wealth,
With a still, mysterious stealth:
She will mix these pleasures up
Like three fit wines in a cup,
And thou shalt quaff it:—thou shalt hear
Distant harvest-carols clear; 40
Rustle of the reaped corn;
Sweet birds antheming the morn:
And, in the same moment—hark!
'Tis the early April lark,
Or the rooks, with busy caw,
Foraging for sticks and straw.
Thou shalt, at one glance, behold
The daisy and the marigold;
White-plum'd lilies, and the first
Hedge-grown primrose that hath burst; 50
Shaded hyacinth, alway
Sapphire queen of the mid-May;
And every leaf, and every flower
Pearled with the self-same shower.
Thou shalt see the field-mouse peep
Meagre from its celled sleep;
And the snake all winter-thin
Cast on sunny bank its skin;
Freckled nest-eggs thou shalt see

Hatching in the hawthorn-tree, 60
When the hen-bird's wing doth rest
Quiet on her mossy nest;
Then the hurry and alarm
When the bee-hive casts its swarm;
Acorns ripe down-pattering,
While the autumn breezes sing.

Oh, sweet Fancy! let her loose;
Every thing is spoilt by use:
Where's the cheek that doth not fade,
Too much gaz'd at? Where's the maid 70
Whose lip mature is ever new?
Where's the eye, however blue,
Doth not weary? Where's the face
One would meet in every place?
Where's the voice, however soft,
One would hear so very oft?
At a touch sweet Pleasure melteth
Like to bubbles when rain pelteth.
Let, then, winged Fancy find
Thee a mistress to thy mind: 80
Dulcet-eyed as Ceres' daughter,°
Ere the God of Torment taught her
How to frown and how to chide;
With a waist and with a side
White as Hebe's, when her zone
Slipt its golden clasp, and down
Fell her kirtle to her feet,
While she held the goblet sweet,
And Jove grew languid.—Break the mesh
Of the Fancy's silken leash; 90
Quickly break her prison-string
And such joys as these she'll bring.—
Let the winged Fancy roam,
Pleasure never is at home.

Ode: 'Bards of Passion and of Mirth'

Bards of Passion and of Mirth,
Ye have left your souls on earth!
Have ye souls in heaven too,
Double-lived in regions new?
Yes, and those of heaven commune
With the spheres of sun and moon;
With the noise of fountains wond'rous,
And the parle of voices thund'rous;
With the whisper of heaven's trees
And one another, in soft ease 10
Seated on Elysian lawns
Brows'd by none but Dian's fawns;
Underneath large blue-bells tented,
Where the daisies are rose-scented,
And the rose herself has got
Perfume which on earth is not;
Where the nightingale doth sing
Not a senseless, tranced thing,
But divine melodious truth;
Philosophic numbers smooth; 20
Tales and golden histories
Of heaven and its mysteries.

Thus ye live on high, and then°
On the earth ye live again;
And the souls ye left behind you
Teach us, here, the way to find you,
Where your other souls are joying,
Never slumber'd, never cloying.
Here, your earth-born souls still speak
To mortals, of their little week; 30
Of their sorrows and delights;
Of their passions and their spites;
Of their glory and their shame;
What doth strengthen and what maim.
Thus ye teach us, every day,
Wisdom, though fled far away.

Bards of Passion and of Mirth,
Ye have left your souls on earth!
Ye have souls in heaven too,
Double-lived in regions new! 40

'I had a dove and the sweet dove died'

I had a dove and the sweet dove died,
And I have thought it died of grieving:
O what could it mourn for? it was tied
With a silken thread of my own hand's weaving.
Sweet little red-feet why did you die?
Why would you leave me—sweet dove why?
You lived alone on the forest tree—
Why pretty thing could you not live with me?
I kiss'd you oft, and I gave you white peas—
Why not live sweetly as in the green trees? 10

Faery Song: 'Ah! woe is me! poor Silver-wing!'

Ah! woe is me! poor Silver-wing!
That I must chaunt thy Lady's dirge,
And death to this fair haunt of spring
Of melody, and streams of flowery verge,—
Poor Silver-wing! Ah! woe is me!
That I must see
These blossoms snow upon thy lady's pall!
Go, pretty Page, and in her ear
Whisper that the hour is near!
Softly tell her not to fear 10
Such calm favonian burial!°
Go, pretty Page, and soothly tell,—
The blossoms hang by a melting spell,
And fall they must, ere a star wink thrice
Upon her closed eyes,
That now in vain are weeping their last tears,
At sweet life leaving, and these arbours green,—
Rich dowry from the Spirit of the spheres,—
Alas! poor Queen!

Song: 'Hush, hush, tread softly, hush, hush my dear'

1

Hush, hush, tread softly, hush, hush my dear,
All the house is asleep, but we know very well
That the jealous, the jealous old Baldpate may hear,
Though you've padded his night-cap, O sweet Isabel.
Though your feet are more light than a fairy's feet,
Who dances on bubbles where brooklets meet.
Hush hush, tread softly, hush hush, my dear,
For less than a nothing the jealous can hear:

2

No leaf doth tremble, no ripple is there
On the river—All's still, and the night's sleepy eye 10
Closes up, and forgets all its Lethean care
Charmed to death by the drone of the humming may fly.
And the moon, whether prudish or complaisant,
Hath fled to her bower, well knowing I want
No light in the darkness, no torch in the gloom;
But my Isabel's eyes and her lips pulped with bloom.

3

Lift the latch, ah gently! ah tenderly, sweet
We are dead, if that latchet gives one little chink:
Well done, now those lips and a flowery seat—
The old man may dream and the planets may wink;° 20
The shut rose shall dream of our loves and awake
Full blown and such warmth for the morning take;
The stockdove shall hatch her soft brace and shall coo,
While I kiss to the melody aching all through.

The Eve of St. Agnes

1

St. Agnes' Eve—Ah, bitter chill it was!
The owl, for all his feathers, was a-cold;
The hare limp'd trembling through the frozen grass,
And silent was the flock in woolly fold:

Numb were the Beadsman's fingers, while he told
His rosary, and while his frosted breath,
Like pious incense from a censer old,
Seem'd taking flight for heaven, without a death,
Past the sweet Virgin's picture, while his prayer he saith.

2

His prayer he saith, this patient, holy man; 10
Then takes his lamp, and riseth from his knees,
And back returneth, meagre, barefoot, wan,
Along the chapel aisle by slow degrees:
The sculptur'd dead, on each side, seem to freeze,°
Emprison'd in black, purgatorial rails:
Knights, ladies, praying in dumb orat'ries,
He passeth by; and his weak spirit fails
To think how they may ache in icy hoods and mails.

3

Northward he turneth through a little door,
And scarce three steps, ere Music's golden tongue 20
Flatter'd to tears this aged man and poor;
But no—already had his deathbell rung;
The joys of all his life were said and sung:
His was harsh penance on St. Agnes' Eve:
Another way he went, and soon among
Rough ashes sat he for his soul's reprieve,
And all night kept awake, for sinners' sake to grieve.

4

That ancient Beadsman heard the prelude soft;
And so it chanc'd, for many a door was wide,
From hurry to and fro. Soon, up aloft, 30
The silver, snarling trumpets 'gan to chide:
The level chambers, ready with their pride,
Were glowing to receive a thousand guests:
The carved angels, ever eager-eyed,
Star'd, where upon their heads the cornice rests,
With hair blown back, and wings put cross-wise on their
breasts.

5

At length burst in the argent revelry,
With plume, tiara, and all rich array,
Numerous as shadows haunting fairily
The brain, new stuff'd, in youth, with triumphs gay 40
Of old romance. These let us wish away,
And turn, sole-thoughted, to one Lady there,
Whose heart had brooded, all that wintry day,
On love, and wing'd St. Agnes' saintly care,
As she had heard old dames full many times declare.

6

They told her how, upon St. Agnes' Eve,
Young virgins might have visions of delight,
And soft adorings from their loves receive
Upon the honey'd middle of the night,
If ceremonies due they did aright; 50
As, supperless to bed they must retire,
And couch supine their beauties, lily white;
Nor look behind, nor sideways, but require
Of Heaven with upward eyes for all that they desire.

7

Full of this whim was thoughtful Madeline:
The music, yearning like a God in pain,
She scarcely heard: her maiden eyes divine,
Fix'd on the floor, saw many a sweeping train
Pass by—she heeded not at all: in vain
Came many a tiptoe, amorous cavalier, 60
And back retir'd, not cool'd by high disdain;
But she saw not: her heart was otherwhere:
She sigh'd for Agnes' dreams, the sweetest of the year.

8

She danc'd along with vague, regardless eyes,
Anxious her lips, her breathing quick and short:
The hallow'd hour was near at hand: she sighs
Amid the timbrels, and the throng'd resort
Of whisperers in anger, or in sport;

'Mid looks of love, defiance, hate, and scorn,
Hoodwink'd with faery fancy; all amort,° 70
Save to St. Agnes and her lambs unshorn,°
And all the bliss to be before to-morrow morn.

9

So, purposing each moment to retire,
She linger'd still. Meantime, across the moors,
Had come young Porphyro, with heart on fire
For Madeline. Beside the portal doors,
Buttress'd from moonlight, stands he, and implores
All saints to give him sight of Madeline,
But for one moment in the tedious hours,
That he might gaze and worship all unseen; 80
Perchance speak, kneel, touch, kiss—in sooth such things have been.

10

He ventures in: let no buzz'd whisper tell:
All eyes be muffled, or a hundred swords
Will storm his heart, Love's fev'rous citadel:
For him, those chambers held barbarian hordes,
Hyena foemen, and hot-blooded lords,
Whose very dogs would execrations howl
Against his lineage: not one breast affords
Him any mercy, in that mansion foul,
Save one old beldame, weak in body and in soul. 90

11

Ah, happy chance! the aged creature came,
Shuffling along with ivory-headed wand,
To where he stood, hid from the torch's flame,
Behind a broad hall-pillar, far beyond
The sound of merriment and chorus bland:
He startled her; but soon she knew his face,
And grasp'd his fingers in her palsied hand,
Saying, 'Mercy, Porphyro! hie thee from this place;
They are all here to-night, the whole blood-thirsty race!

12

'Get hence! get hence! there's dwarfish Hildebrand; 100
He had a fever late, and in the fit
He cursed thee and thine, both house and land:
Then there's that old Lord Maurice, not a whit
More tame for his gray hairs—Alas me! flit!
Flit like a ghost away.'—'Ah, Gossip dear,
We're safe enough; here in this arm-chair sit,
And tell me how'—'Good Saints! not here, not here;
Follow me, child, or else these stones will be thy bier.'

13

He follow'd through a lowly arched way,
Brushing the cobwebs with his lofty plume, 110
And as she mutter'd 'Well-a—well-a-day!'
He found him in a little moonlight room,
Pale, lattic'd, chill, and silent as a tomb.
'Now tell me where is Madeline,' said he,
'O tell me, Angela, by the holy loom
Which none but secret sisterhood may see,
When they St. Agnes' wool are weaving piously.'

14

'St. Agnes! Ah! it is St. Agnes' Eve—
Yet men will murder upon holy days:
Thou must hold water in a witch's sieve, 120
And be liege-lord of all the Elves and Fays,
To venture so: it fills me with amaze
To see thee, Porphyro!—St. Agnes' Eve!
God's help! my lady fair the conjuror plays
This very night: good angels her deceive!
But let me laugh awhile, I've mickle time to grieve.'

15

Feebly she laugheth in the languid moon,
While Porphyro upon her face doth look,
Like puzzled urchin on an aged crone
Who keepeth clos'd a wond'rous riddle-book, 130

As spectacled she sits in chimney nook.
But soon his eyes grew brilliant, when she told
His lady's purpose; and he scarce could brook
Tears, at the thought of those enchantments cold,
And Madeline asleep in lap of legends old.

16

Sudden a thought came like a full-blown rose,
Flushing his brow, and in his pained heart
Made purple riot: then doth he propose
A stratagem, that makes the beldame start:
'A cruel man and impious thou art: 140
Sweet lady, let her pray, and sleep, and dream
Alone with her good angels, far apart
From wicked men like thee. Go, go!—I deem
Thou canst not surely be the same that thou didst seem.'

17

'I will not harm her, by all saints I swear,'
Quoth Porphyro: 'O may I ne'er find grace
When my weak voice shall whisper its last prayer,
If one of her soft ringlets I displace,
Or look with ruffian passion in her face:
Good Angela, believe me by these tears; 150
Or I will, even in a moment's space,
Awake, with horrid shout, my foemen's ears,
And beard them, though they be more fang'd than wolves and bears.'

18

'Ah! why wilt thou affright a feeble soul?
A poor, weak, palsy-stricken, churchyard thing,
Whose passing-bell may ere the midnight toll;
Whose prayers for thee, each morn and evening,
Were never miss'd.'—Thus plaining, doth she bring
A gentler speech from burning Porphyro;
So woful, and of such deep sorrowing, 160
That Angela gives promise she will do
Whatever he shall wish, betide her weal or woe.

19

Which was, to lead him, in close secrecy,
Even to Madeline's chamber, and there hide
Him in a closet, of such privacy
That he might see her beauty unespied,
And win perhaps that night a peerless bride,
While legion'd fairies pac'd the coverlet,
And pale enchantment held her sleepy-eyed.
Never on such a night have lovers met, 170
Since Merlin paid his Demon all the monstrous debt.°

20

'It shall be as thou wishest,' said the Dame:
'All cates and dainties shall be stored there°
Quickly on this feast-night: by the tambour frame
Her own lute thou wilt see: no time to spare,
For I am slow and feeble, and scarce dare
On such a catering trust my dizzy head.
Wait here, my child, with patience; kneel in prayer
The while: Ah! thou must needs the lady wed,
Or may I never leave my grave among the dead.' 180

21

So saying, she hobbled off with busy fear.
The lover's endless minutes slowly pass'd;
The dame return'd, and whisper'd in his ear
To follow her; with aged eyes aghast
From fright of dim espial. Safe at last,
Through many a dusky gallery, they gain
The maiden's chamber, silken, hush'd, and chaste;
Where Porphyro took covert, pleas'd amain.
His poor guide hurried back with agues in her brain.

22

Her falt'ring hand upon the balustrade, 190
Old Angela was feeling for the stair,
When Madeline, St. Agnes' charmed maid,
Rose, like a mission'd spirit, unaware:

With silver taper's light, and pious care,
She turn'd, and down the aged gossip led
To a safe level matting. Now prepare,
Young Porphyro, for gazing on that bed;
She comes, she comes again, like ring-dove fray'd and fled.

23

Out went the taper as she hurried in;
Its little smoke, in pallid moonshine, died: 200
She clos'd the door, she panted, all akin
To spirits of the air, and visions wide:
No uttered syllable, or, woe betide!
But to her heart, her heart was voluble,
Paining with eloquence her balmy side;
As though a tongueless nightingale should swell°
Her throat in vain, and die, heart-stifled, in her dell.

24

A casement high and triple-arch'd there was,°
All garlanded with carven imag'ries
Of fruits, and flowers, and bunches of knot-grass, 210
And diamonded with panes of quaint device,
Innumerable of stains and splendid dyes,
As are the tiger-moth's deep-damask'd wings;
And in the midst, 'mong thousand heraldries,
And twilight saints, and dim emblazonings,
A shielded scutcheon blush'd with blood of queens and kings.

25

Full on this casement shone the wintry moon,
And threw warm gules on Madeline's fair breast,
As down she knelt for heaven's grace and boon;
Rose-bloom fell on her hands, together prest, 220
And on her silver cross soft amethyst,
And on her hair a glory, like a saint:
She seem'd a splendid angel, newly drest,
Save wings, for heaven:—Porphyro grew faint:
She knelt, so pure a thing, so free from mortal taint.

26

Anon his heart revives: her vespers done,
Of all its wreathed pearls her hair she frees;
Unclasps her warmed jewels one by one;
Loosens her fragrant boddice; by degrees
Her rich attire creeps rustling to her knees: 230
Half-hidden, like a mermaid in sea-weed,
Pensive awhile she dreams awake, and sees,
In fancy, fair St. Agnes in her bed,
But dares not look behind, or all the charm is fled.

27

Soon, trembling in her soft and chilly nest,
In sort of wakeful swoon, perplex'd she lay,
Until the poppied warmth of sleep oppress'd
Her soothed limbs, and soul fatigued away;
Flown, like a thought, until the morrow-day;
Blissfully haven'd both from joy and pain; 240
Clasp'd like a missal where swart Paynims pray;°
Blinded alike from sunshine and from rain,
As though a rose should shut, and be a bud again.

28

Stol'n to this paradise, and so entranced,
Porphyro gazed upon her empty dress,
And listen'd to her breathing, if it chanced
To wake into a slumberous tenderness;
Which when he heard, that minute did he bless,
And breath'd himself: then from the closet crept,
Noiseless as fear in a wide wilderness, 250
And over the hush'd carpet, silent, stept,
And 'tween the curtains peep'd, where, lo!—how fast she slept.

29

Then by the bed-side, where the faded moon
Made a dim, silver twilight, soft he set
A table, and, half anguish'd, threw thereon
A cloth of woven crimson, gold, and jet:—

O for some drowsy Morphean amulet!°
The boisterous, midnight, festive clarion,
The kettle-drum, and far-heard clarionet,
Affray his ears, though but in dying tone:— 260
The hall door shuts again, and all the noise is gone.°

30

And still she slept an azure-lidded sleep,
In blanched linen, smooth, and lavender'd,
While he from forth the closet brought a heap
Of candied apple, quince, and plum, and gourd;
With jellies soother than the creamy curd,°
And lucent syrops, tinct with cinnamon;°
Manna and dates, in argosy transferr'd
From Fez; and spiced dainties, every one,
From silken Samarcand to cedar'd Lebanon. 270

31

These delicates he heap'd with glowing hand
On golden dishes and in baskets bright
Of wreathed silver: sumptuous they stand
In the retired quiet of the night,
Filling the chilly room with perfume light.—
'And now, my love, my seraph fair, awake!
Thou art my heaven, and I thine eremite:
Open thine eyes, for meek St. Agnes' sake,
Or I shall drowse beside thee, so my soul doth ache.'

32

Thus whispering, his warm, unnerved arm° 280
Sank in her pillow. Shaded was her dream
By the dusk curtains:—'twas a midnight charm
Impossible to melt as iced stream:
The lustrous salvers in the moonlight gleam;
Broad golden fringe upon the carpet lies:
It seem'd he never, never could redeem
From such a stedfast spell his lady's eyes;
So mus'd awhile, entoil'd in woofed phantasies.

33

Awakening up, he took her hollow lute,—
Tumultuous,—and, in chords that tenderest be, 290
He play'd an ancient ditty, long since mute,
In Provence call'd, 'La belle dame sans mercy':°
Close to her ear touching the melody;—
Wherewith disturb'd, she utter'd a soft moan:
He ceased—she panted quick—and suddenly
Her blue affrayed eyes wide open shone:
Upon his knees he sank, pale as smooth-sculptured stone.

34

Her eyes were open, but she still beheld,
Now wide awake, the vision of her sleep:
There was a painful change, that nigh expell'd 300
The blisses of her dream so pure and deep
At which fair Madeline began to weep,
And moan forth witless words with many a sigh;
While still her gaze on Porphyro would keep;
Who knelt, with joined hands and piteous eye,
Fearing to move or speak, she look'd so dreamingly.

35

'Ah, Porphyro!' said she, 'but even now
Thy voice was at sweet tremble in mine ear,
Made tuneable with every sweetest vow;
And those sad eyes were spiritual and clear: 310
How chang'd thou art! how pallid, chill, and drear!
Give me that voice again, my Porphyro,
Those looks immortal, those complainings dear!
Oh leave me not in this eternal woe,
For if thou diest, my Love, I know not where to go.'

36

Beyond a mortal man impassion'd far
At these voluptuous accents, he arose,
Ethereal, flush'd, and like a throbbing star
Seen mid the sapphire heaven's deep repose;

Into her dream he melted, as the rose 320
Blendeth its odour with the violet,—
Solution sweet: meantime the frost-wind blows
Like Love's alarum pattering the sharp sleet
Against the window-panes; St. Agnes' moon hath set.

37

'Tis dark: quick pattereth the flaw-blown sleet:°
'This is no dream, my bride, my Madeline!'
'Tis dark: the iced gusts still rave and beat:
'No dream, alas! alas! and woe is mine!
Porphyro will leave me here to fade and pine.—
Cruel! what traitor could thee hither bring? 330
I curse not, for my heart is lost in thine,
Though thou forsakest a deceived thing;—
A dove forlorn and lost with sick unpruned wing.'

38

'My Madeline! sweet dreamer! lovely bride!
Say, may I be for aye thy vassal blest?
Thy beauty's shield, heart-shap'd and vermeil dyed?
Ah, silver shrine, here will I take my rest
After so many hours of toil and quest,
A famish'd pilgrim,—saved by miracle.
Though I have found, I will not rob thy nest 340
Saving of thy sweet self; if thou think'st well
To trust, fair Madeline, to no rude infidel.

39

'Hark! 'tis an elfin-storm from faery land,
Of haggard seeming, but a boon indeed:°
Arise—arise! the morning is at hand;—
The bloated wassaillers will never heed:—
Let us away, my love, with happy speed;
There are no ears to hear, or eyes to see,—
Drown'd all in Rhenish and the sleepy mead:
Awake! arise! my love, and fearless be, 350
For o'er the southern moors I have a home for thee.'

40

She hurried at his words, beset with fears,
For there were sleeping dragons all around,
At glaring watch, perhaps, with ready spears—
Down the wide stairs a darkling way they found.—°
In all the house was heard no human sound.
A chain-droop'd lamp was flickering by each door;
The arras, rich with horseman, hawk, and hound,
Flutter'd in the besieging wind's uproar;
And the long carpets rose along the gusty floor. 360

41

They glide, like phantoms, into the wide hall;
Like phantoms, to the iron porch, they glide;
Where lay the Porter, in uneasy sprawl,
With a huge empty flaggon by his side:
The wakeful bloodhound rose, and shook his hide,
But his sagacious eye an inmate owns:
By one, and one, the bolts full easy slide:—
The chains lie silent on the footworn stones;—
The key turns, and the door upon its hinges groans.

42

And they are gone: ay, ages long ago 370
These lovers fled away into the storm.
That night the Baron dreamt of many a woe,
And all his warrior-guests, with shade and form
Of witch, and demon, and large coffin-worm,
Were long be-nightmar'd. Angela the old
Died palsy-twitch'd, with meagre face deform;
The Beadsman, after thousand aves told,
For aye unsought for slept among his ashes cold.

The Eve of St. Mark

Upon a sabbath day it fell,
Twice holy was the sabbath bell,
That call'd the folk to evening prayer—
The City streets were clean and fair
From wholesome drench of April rains
And on the western window panes
The chilly sunset faintly told
Of unmatur'd green vallies cold,
Of the green thorny bloomless hedge,
Of rivers new with springtide sedge, 10
Of Primroses by shelter'd rills
And daisies on the aguish hills—
Twice holy was the sabbath bell:
The silent Streets were crowded well
With staid and pious companies
Warm from their fireside orat'ries
And moving with demurest air
To even song and vesper prayer.
Each arched porch and entry low
Was fill'd with patient folk and slow, 20
With whispers hush and shuffling feet
While play'd the organs loud and sweet—

The Bells had ceas'd, the prayers begun
And Bertha had not yet half done:
A curious volume patch'd and torn
That all day long from earliest morn
Had taken captive her two eyes
Among its golden broideries—°
Perplex'd her with a thousand things—
The Stars of heaven and angels' wings, 30
Martyrs in a fiery blaze—
Azure saints mid silver rays,
Aaron's breastplate, and the seven°
Candlesticks John saw in heaven—
The winged Lion of St. Mark
And the covenantal Ark

With its many mysteries,
Cherubim and golden Mice.°

Bertha was a maiden fair°
Dwelling in the old Minster Square; 40
From her fireside she could see
Sidelong its rich antiquity—
Far as the Bishop's garden wall
Where Sycamores and elm trees tall
Full leav'd the forest had outstript—
By no sharp north wind ever nipt
So shelter'd by the mighty pile—
Bertha arose and read awhile
With forehead 'gainst the window pane—
Again she tried and then again 50
Until the dusk eve left her dark
Upon the Legend of St. Mark.
From pleated lawn-frill fine and thin
She lifted up her soft warm chin,
With aching neck and swimming eyes
And dazed with saintly imageries.

All was gloom, and silent all
Save now and then the still footfall
Of one returning townwards late—
Past the echoing minster gate— 60
The clamorous daws that all the day
Above tree tops and towers play
Pair by pair had gone to rest,
Each in its ancient belfry nest
Where asleep they fall betimes
To musick of the drowsy chimes.
All was silent—all was gloom
Abroad and in the homely room—
Down she sat poor cheated soul
And struck a Lamp from the dismal coal, 70
Leaned forward, with bright drooping hair
And slant book full against the glare.
Her shadow in uneasy guize
Hover'd about a giant size
On ceiling beam and old oak chair,

The Parrot's cage and pannel square
And the warm angled winter screen
On which were many monsters seen
Call'd Doves of Siam, Lima Mice°
And legless birds of Paradise,° 80
Macaw and tender av'davat°
And silken furr'd angora cat—
Untir'd she read; her shadow still
Glower'd about as it would fill
The Room with wildest forms and shades
As though some ghostly Queens of spades
Had come to mock behind her back—
And dance, and ruffle their garments black.
Untir'd she read the Legend page
Of holy Mark from youth to age; 90
On Land, on Seas, in pagan-chains,
Rejoicing for his many pains—
Sometimes the learned Eremite
With golden star, or dagger bright
Referr'd to pious poesies
Written in smallest crowquill size
Beneath the text; and thus the rhyme
Was parcel'd out from time to time:°
—'Als writith he of swevenis
Men han beforne they wake in bliss, 100
Whanne thate hir friendes thinke hem bound
In crimpid shroude farre under grounde;
And how a litling child mote be
A sainte er its nativitie;
Gif that the modre (god her blesse)
Kepen in solitarinesse,
And kissen devoute the holy croce.
Of Goddis love and Sathan's force
He writith; and thinges many mo:
Of swiche thinges I may not shew; 110
Bot I must tellen verilie
Somdel of Saintè Cicilie;
And chieflie whate he auctorethe
Of Saintè Markis life and dethe.'

At length her constant eyelids come

Upon the fervent Martyrdom;
Then lastly to his holy shrine°
Exalt amid the tapers' shine
At Venice—

'Gif ye wol stonden hardie wight'

Gif ye wol stonden hardie wight—
Amiddes of the blacke night—
Righte in the churche porch, pardie
Ye wol behold a companie
Approuchen thee full dolorouse
For sooth to sain from everich house
Be it in City or village
Wol come the Phantom and image
Of ilka gent and ilka carle
Whom coldè Deathè hath in parle 10
And wol some day that very year
Touchen with foulè venime spear
And sadly do them all to die—
Hem all shalt thou see verilie
And everichon shall by the pass
All who must die that year Alas—

'Why did I laugh tonight? No voice will tell'

Why did I laugh tonight? No voice will tell:
  No God, no Demon of severe response
Deigns to reply from heaven or from Hell.—
  Then to my human heart I turn at once—
Heart! thou and I are here sad and alone;
  Say, wherefore did I laugh? O mortal pain!
O Darkness! Darkness! ever must I moan
  To question Heaven and Hell and Heart in vain!

Why did I laugh? I know this being's lease°
  My fancy to its utmost blisses spreads: 10
Yet could I on this very midnight cease,
  And the world's gaudy ensigns see in shreds.
Verse, fame and Beauty are intense indeed
But Death intenser—Death is Life's high meed.

'When they were come unto the Faery's Court'

When they were come unto the Faery's Court
They rang—no one at home—all gone to sport
And dance and kiss and love as faeries do
For Faeries be as humans lovers true—
Amid the woods they were so lone and wild
Where even the Robin feels himself exil'd
And where the very brooks as if affraid
Hurry along to some less magic shade.
'No one at home!' the fretful princess cry'd,
'And all for nothing such a drery ride 10
And all for nothing my new diamond cross
No one to see my Persian feathers toss°
No one to see my Ape, my Dwarf, my Fool
Or how I pace my Otaheitan mule—°
Ape, Dwarf and Fool, why stand you gaping there?
Burst the door open, quick—or I declare
I'll switch you soundly and in pieces tear.'
The Dwarf began to tremble and the Ape
Star'd at the Fool, the Fool was all agape
The Princess grasp'd her switch but just in time 20
The dwarf with piteous face began to rhyme.
'O mighty Princess did you ne'er hear tell
What your poor servants know but too too well?
Know you the three "great crimes" in faery land?
The first alas! poor Dwarf I understand—
I made a whipstock of a faery's wand
The next is snoring in their company,
The next, the last the direst of the three
Is making free when they are not at home.
I was a Prince—a baby prince—my doom 30

You see, I made a whipstock of a wand—
My top has henceforth slept in faery land.
He was a Prince, the Fool a grown up Prince
But he has never been a king's son since
He fell a snoring at a faery Ball—
Your poor Ape was a Prince, and he poor thing
Picklock'd a faery's boudoir—now no King
But ape—so pray your highness stay awhile
'Tis sooth indeed. We know it to our sorrow—
Persist and *you* may be an ape tomorrow—' 40
While the Dwarf spake the Princess all for spite
Peel'd the brown hazel twig to lilly white,
Clench'd her small teeth, and held her lips apart
Try'd to look unconcern'd with beating heart
They saw her highness had made up her mind
A quavering like the reeds before the wind—
And they had had it, but O happy chance
The Ape for very fear began to dance
And grinn'd as all his ugliness did ache—
She staid her vixen fingers for his sake, 50
He was so very ugly: then she took
Her pocket mirror and began to look
First at herself and at him and then
She smil'd at her own beauteous face again.
Yet for all this—for all her pretty face—
She took it in her head to see the place.
Women gain little from experience
Either in Lovers, husbands or expence—
The more the beauty, the more fortune too;
Beauty before the wide world never knew— 60
So each Fair reasons—tho' it oft miscarries.
She thought *her* pretty face would please the faeries.
'My darling Ape I won't whip you to-day
Give me the Picklock sirrah and go play—'
They all three wept—but counsel was as vain
As crying cup biddy to drops of rain—°
Yet lingeringly did the sad Ape forth draw°
The Picklock from the Pocket in his Jaw.
The Princess took it and dismounting straight
Tripp'd in blue silver'd slippers to the gate 70
And touch'd the wards; the Door full courteously°

Opened—she enter'd with her servants three—
Again it clos'd and there was nothing seen
But the Mule grasing on the herbage green.

End of Canto xii

Canto the xiii

The Mule no sooner saw himself alone
Than he pricked up his Ears—and said, 'Well done,
At least unhappy Prince I may be free—
No more a Princess shall side saddle me.
O king of Otaheitè—tho' a Mule—
"Aye every inch a king"—tho' "Fortune's fool—"° 80
Well done—for by what Mr. Dwarfy said
I would not give a sixpence for her head.'
Even as he spake he trotted in high glee
To the knotty side of an old Pollard tree
And rubbed his sides against the mossed bark
Till his Girths burst and left him naked stark
Except his Bridle—how get rid of that
Buckled and tied with many a twist and plait?
At last it struck him to pretend to sleep
And then the thievish Monkies down would creep 90
And filch the unpleasant trammels quite away.
No sooner thought of than adown he lay
Shamm'd a good snore—the Monkey-men descended
And whom they thought to injure they befriended.
They hung his Bridle on a topmost bough
And off he went run, trot, or any how—

'He is to weet a melancholy Carle'

He is to weet a melancholy Carle
Thin in the waist, with bushy head of hair
As hath the seeded thistle when in parle
It holds the Zephyr ere it sendeth fair
Its light balloons into the summer air;

Thereto his beard had not begun to bloom
No brush had touch'd his chin or razor sheer
No care had touch'd his cheek with mortal doom
But new he was and bright as scarf from Persian loom—

Ne cared he for wine, or half and half° 10
Ne cared he for fish or flesh or fowl
And sauces held he worthless as the chaff;
He 'sdeign'd the swine-herd at the wassail bowl
Ne with lewd ribbalds sat he cheek by jowl,°
Ne with sly Lemans in the scorner's chair°
But after water brooks this Pilgrim's soul°
Panted, and all his food was woodland air
Though he would ofttimes feast on gilliflowers rare—

The slang of cities in no wise he knew—
Tipping the wink to him was heathen Greek— 20
He sipp'd no olden Tom or ruin blue°
Or nantz, or cheery brandy drank full meek°
By many a Damsel hoarse and rouge of cheek
Nor did he know each aged Watchman's beat—
Nor in obscured purlieus would he seek
For curled Jewesses with ankles neat°
Who as they walk abroad make tinkling with their feet—

A dream, after reading Dante's Episode of Paolo and Francesca

As Hermes once took to his feathers light,
When lulled Argus, baffled, swoon'd and slept,
So on a Delphic reed my idle spright
So play'd, so charm'd, so conquer'd, so bereft
The dragon-world of all its hundred eyes; 5
And, seeing it asleep, so fled away—
Not unto Ida with its snow-cold skies,
Nor unto Tempe, where Jove griev'd a day;

But to that second circle of sad hell,
  Where 'mid the gust, the world-wind, and the flaw° 10
Of rain and hailstones, lovers need not tell
  Their sorrows. Pale were the sweet lips I saw,
Pale were the lips I kiss'd, and fair the form
I floated with about that melancholy storm.

La belle dame sans merci

O what can ail thee knight at arms
  Alone and palely loitering?
The sedge has withered from the Lake
  And no birds sing!

O what can ail thee knight at arms
  So haggard and so woe begone?
The squirrel's granary is full
  And the harvest's done.

I see a lilly on thy brow
  With anguish moist and fever dew, 10
And on thy cheeks a fading rose
  Fast withereth too—

I met a Lady in the Meads
  Full beautiful, a faery's child
Her hair was long, her foot was light
  And her eyes were wild—

I made a Garland for her head,
  And bracelets too, and fragrant Zone:
She look'd at me as she did love
  And made sweet moan— 20

I set her on my pacing steed
  And nothing else saw all day long
For sidelong would she bend and sing
  A faery's song—

She found me roots of relish sweet
  And honey wild and manna dew
And sure in language strange she said
  'I love thee true'—

She took me to her elfin grot
  And there she wept and sigh'd full sore 30
And there I shut her wild wild eyes
  With kisses four.

And there she lulled me asleep
  And there I dream'd—Ah Woe betide!
The latest dream I ever dreamt
  On the cold hill side.

I saw pale kings and Princes too
  Pale warriors, death pale were they all;
They cried 'La belle dame sans merci
  Thee hath in thrall.' 40

I saw their starv'd lips in the gloam
  With horrid warning gaped wide
And I awoke and found me here
  On the cold hill's side

And this is why I sojourn here
  Alone and palely loitering;
Though the sedge is wither'd from the Lake
  And no birds sing——

Song of four Fairies: Fire, Air, Earth, and Water

SALAMANDER, ZEPHYR, DUSKETHA, AND BREAMA

SALAMANDER

Happy, happy glowing fire!

ZEPHYR

Fragrant air! Delicious light!

DUSKETHA

Let me to my glooms retire!

BREAMA

I to green-weed rivers bright!

SALAMANDER

Happy, happy glowing fire,
Dazzling bowers of soft retire,
Ever let my nourish'd wing,
Like a Bat's, still wandering,
Nimbly fan your fiery spaces,
Spirit sole in deadly places; 10
In unhaunted roar and blaze,
Open eyes that never daze:
Let me see the myriad shapes
Of men, and beasts, and fish, and apes,
Portray'd in many a fiery den
And wrought by spumy bitumen
On the deep intenser roof,
Arched every way aloof;
Let me breathe upon their skies,
And anger their live tapestries; 20
Free from cold and every care
Of chilly rain, and shivering air.

ZEPHYR

Spirit of Fire—away, away!
Or your very roundelay
Will sear my plumage newly budded
From its quilled sheath, and studded
With the self same dews that fell
On the May-grown Asphodel.
Spirit of fire—away, away!

BREAMA

Spirit of fire—away, away! 30
Zephyr, blue-eyed faery turn
And see my cool sedge-buried urn,
Where it rests its mossy Brim

'Mid water mint and Cresses dim;
And the flowers in sweet troubles
Lift their eyes above the bubbles,
Like our Queen when she would please
To sleep and Oberon will tease.
Love me, blue-eyed Faery true
Soothly I am sick for you. 40

ZEPHYR

Gentle Breama! by the first
Violet young nature nurst,
I will bathe myself with thee
So you sometime follow me
To my home, far far in west,
Beyond the nimble-wheeled quest
Of the golden-presenc'd Sun.
Come with me, o'er tops of trees,
To my fragrant Pallaces,
Where they ever floating are 50
Beneath the cherish of a Star
Call'd Vesper, who with silver veil
Ever hides his brilliance pale,
Ever gently drows'd doth keep
Twilight for the Fays to sleep.
Fear not that your watry hair
Will thirst in drouthy ringlets there;
Clouds of stored summer rains
Thou shalt taste, before the stains
From the mountain soil they take, 60
And too unlucent for thee make.
I love thee, chrystal Fairy true;
Sooth I am as sick for you!

SALAMANDER

Out, ye aguish Fairies, out!
Chilly Lovers, what a rout
Keep ye with your frozen breath,
Colder than the mortal death.
Adder-eyed Dusketha, speak,
Shall we leave these and go seek
In the Earth's wide entrails old 70

Couches warm as theirs is cold?
O for a fiery-gloom and thee
Dusketha, so enchantingly
Freckle-wing'd, and lizard-sided!

DUSKETHA

By thee, Sprite, will I be guided!
I care not for cold or heat;
Frost or flame, or sparks, or sleet
To my essence are the same;
But I honor more the flame.
Sprite of Fire! I follow thee 80
Wheresoever it may be;
To the torrid spouts and fountains
Underneath earth-quaked mountains;
Or, at thy supreme desire,
Touch the very pulse of fire
With my bare unlidded eyes.

SALAMANDER

Sweet Dusketh'! Paradise!
Off, ye icy spirits, fly,
Frosty creatures of the Sky!

DUSKETHA

Breathe upon them fiery Sprite! 90

ZEPHYR AND BREAMA

Away, away to our delight!

SALAMANDER

Go feed on icicles, while we
Bedded in tongued flames will be.

DUSKETHA

Lead me to those fevrous glooms
Sprite of Fire!

BREAMA

Me to the blooms,
Blue eyed Zephyr, of those Flowers

Far in the west where the May-cloud lowers
And the beams of still Vesper when winds are all wist
Are shed through the rain and the milder mist
And twilight your floating Bowers.° 100

Sonnet to Sleep

O soft embalmer of the still midnight,
Shutting with careful fingers and benign
Our gloom-pleas'd eyes, embower'd from the light,
Enshaded in forgetfulness divine:
O soothest sleep! if so it please thee, close,
In midst of this thine hymn my willing eyes,
Or wait the Amen ere thy poppy throws
Around my bed its lulling charities.
Then save me or the passed day will shine
Upon my pillow breeding many woes: 10
Save me from curious conscience that still hoards
Its strength for darkness, burrowing like the mole;
Turn the Key deftly in the oiled wards
And seal the hushed Casket of my soul—

Ode to Psyche

O Goddess! hear these tuneless numbers, wrung
By sweet enforcement and remembrance dear,
And pardon that thy secrets should be sung
Even into thine own soft-conched ear:
Surely I dreamt to-day, or did I see
The winged Psyche with awaken'd eyes?
I wander'd in a forest thoughtlessly,
And, on the sudden, fainting with surprise,
Saw two fair creatures, couched side by side
In deepest grass, beneath the whisp'ring roof 10
Of leaves and trembled blossoms, where there ran
A brooklet, scarce espied:

'Mid hush'd, cool-rooted flowers, fragrant-eyed,
  Blue, silver-white, and budded Tyrian,°
They lay calm-breathing on the bedded grass;
  Their arms embraced, and their pinions too;
  Their lips touch'd not, but had not bade adieu
As if disjoined by soft-handed slumber,
And ready still past kisses to outnumber
  At tender eye-dawn of aurorean love: 20
    The winged boy I knew;°
  But who wast thou, O happy, happy dove?
    His Psyche true!

O latest born and loveliest vision far
  Of all Olympus' faded hierarchy!
Fairer than Phœbe's sapphire-region'd star,
  Or Vesper, amorous glow-worm of the sky;
Fairer than these, though temple thou hast none,°
    Nor altar heap'd with flowers;
Nor virgin-choir to make delicious moan 30
    Upon the midnight hours;
No voice, no lute, no pipe, no incense sweet
  From chain-swung censer teeming;
No shrine, no grove, no oracle, no heat
  Of pale-mouth'd prophet dreaming.

O brightest! though too late for antique vows,
  Too, too late for the fond believing lyre,
When holy were the haunted forest boughs,
  Holy the air, the water, and the fire;
Yet even in these days so far retir'd 40
  From happy pieties, thy lucent fans,°
  Fluttering among the faint Olympians,
I see, and sing, by my own eyes inspired.
So let me be thy choir, and make a moan
    Upon the midnight hours;
Thy voice, thy lute, thy pipe, thy incense sweet
  From swinged censer teeming;
Thy shrine, thy grove, thy oracle, thy heat
  Of pale-mouth'd prophet dreaming.

Yes, I will be thy priest, and build a fane 50
In some untrodden region of my mind,
Where branched thoughts, new grown with pleasant pain,
Instead of pines shall murmur in the wind:
Far, far around shall those dark-cluster'd trees°
Fledge the wild-ridged mountains steep by steep;
And there by zephyrs, streams, and birds, and bees,
The moss-lain Dryads shall be lull'd to sleep;
And in the midst of this wide quietness
A rosy sanctuary will I dress
With the wreath'd trellis of a working brain, 60
With buds, and bells, and stars without a name,
With all the gardener Fancy e'er could feign,
Who breeding flowers, will never breed the same:
And there shall be for thee all soft delight
That shadowy thought can win,°
A bright torch, and a casement ope at night,
To let the warm Love in!

On Fame

Fame like a wayward girl will still be coy
To those who woo her with too slavish knees
But makes surrender to some thoughtless boy
And dotes the more upon a heart at ease—
She is a Gipsey will not speak to those
Who have not learnt to be content without her,
A Jilt whose ear was never whisper'd close
Who think they scandal her who talk about her—
A very Gipsey is she Nilus born,
Sister in law to jealous Potiphar.—° 10
Ye lovesick Bards, repay her scorn for scorn.
Ye lovelorn Artists madmen that ye are,
Make your best bow to her and bid adieu
Then if she likes it she will follow you—

On Fame

You cannot eat your cake and have it too.—PROVERB

How fever'd is that Man who cannot look
   Upon his mortal days with temperate blood,
Who vexes all the leaves of his Life's book
   And robs his fair name of its maidenhood;
It is as if the rose should pluck herself
   Or the ripe plum finger its misty bloom;
As if a clear Lake meddling with itself
   Should cloud its pureness with a muddy gloom.
But the rose leaves herself upon the Briar
For winds to kiss and grateful Bees to feed 10
And the ripe plum still wears its dim attire—
   The undisturbed Lake has crystal space—
   Why then should Man teasing the world for grace°
Spoil his salvation by a fierce miscreed?

'If by dull rhymes our English must be chain'd'

If by dull rhymes our English must be chain'd,
   And, like Andromeda, the Sonnet sweet
      Fetter'd, in spite of pained loveliness;°
Let us find out, if we must be constrain'd,
   Sandals more interwoven and complete
To fit the naked foot of Poesy;
      Let us inspect the Lyre, and weigh the stress
Of every chord, and see what may be gain'd
   By ear industrious, and attention meet;
      Misers of sound and syllable, no less 10
Than Midas of his coinage, let us be
   Jealous of dead leaves in the bay wreath crown,
So, if we may not let the Muse be free,
   She will be bound with garlands of her own.

Two or three Posies

Two or three Posies
With two or three simples
Two or three Noses
With two or three pimples,
Two or three wise men
And two or three ninnys
Two or three purses
And two or three guineas
Two or three raps
At two or three doors 10
Two or three naps
Of two or three hours—
Two or three Cats
And two or three mice
Two or three sprats
At a very great price—
Two or three sandies
And two or three tabbies
Two or three dandies—
And two Mrs.— mum!° 20
Two or three Smiles
And two or three frowns
Two or three Miles
To two or three towns
Two or three pegs
For two or three bonnets
Two or three dove's eggs
To hatch into sonnets—

Ode on Indolence

They toil not, neither do they spin.°

I

One morn before me were three figures seen,
With bowed necks, and joined hands, side-faced;
And one behind the other stepp'd serene,
In placid sandals, and in white robes graced:
They pass'd, like figures on a marble Urn,
When shifted round to see the other side;
They came again; as when the Urn once more
Is shifted round, the first seen Shades return;
And they were strange to me, as may betide
With Vases, to one deep in Phidian Lore.° 10

2

How is it, Shadows, that I knew ye not?
How came ye muffled in so hush a Masque?
Was it a silent deep-disguised plot
To steal away, and leave without a task
My idle days? Ripe was the drowsy hour;
The blissful cloud of summer-indolence
Benumb'd my eyes; my pulse grew less and less;
Pain had no sting, and pleasure's wreath no flower.
O, why did ye not melt, and leave my sense
Unhaunted quite of all but—nothingness? 20

3

A third time pass'd they by, and, passing, turn'd
Each one the face a moment whiles to me;
Then faded, and to follow them I burn'd
And ached for wings, because I knew the three:
The first was a fair Maid, and Love her name;
The second was Ambition, pale of cheek,
And ever watchful with fatigued eye;
The last, whom I love more, the more of blame
Is heap'd upon her, Maiden most unmeek,—
I knew to be my demon Poesy. 30

4

They faded, and, forsooth! I wanted wings:
  O folly! What is love? and where is it?
And for that poor Ambition—it springs
  From a man's little heart's short fever-fit;
For Poesy!—no,—she has not a joy,—
  At least for me,—so sweet as drowsy noons,
    And evenings steep'd in honied indolence;
O, for an age so shelter'd from annoy,
  That I may never know how change the moons,
    Or hear the voice of busy common-sense! 40

5

A third time came they by;—alas! wherefore?
  My sleep had been embroider'd with dim dreams;
My soul had been a lawn besprinkled o'er°
  With flowers, and stirring shades, and baffled beams:
The morn was clouded, but no shower fell,
  Though in her lids hung the sweet tears of May;
    The open casement press'd a new-leaved vine,
  Let in the budding warmth and throstle's lay;
O Shadows! 'twas a time to bid farewell!
    Upon your skirts had fallen no tears of mine. 50

6

So, ye three Ghosts, adieu! Ye cannot raise
  My head cool-bedded in the flowery grass;
For I would not be dieted with praise,
  A pet-lamb in a sentimental Farce!°
Fade softly from my eyes, and be once more
  In masque-like figures on the dreamy Urn;
    Farewell! I yet have visions for the night,
And for the day faint visions there is store;
    Vanish, ye Phantoms, from my idle spright,
Into the clouds, and never more return! 60

'Shed no tear—O shed no tear'

Shed no tear—O shed no tear
The Flower will bloom another year—
Weep no more—O weep no more—
Young buds sleep in the root's white core—
Dry your eyes—O dry your eyes
For I was taught in Paradise
To ease my breast of Melodies—
Shed no tear—

Over head—look over head
'Mong the blossoms white and red— 10
Look up, look up—I flutter now
On this flush pomgranate bow—
See me 'tis this silvery bill
Ever cures the good man's ill—
Shed no tear—O shed no tear
The flower will bloom another year
Adieu—Adieu—I fly adieu
I vanish in the heaven's blue—
Adieu Adieu—

Ode to a Nightingale

I

My heart aches, and a drowsy numbness pains°
My sense, as though of hemlock I had drunk,°
Or emptied some dull opiate to the drains
One minute past, and Lethe-wards had sunk:
'Tis not through envy of thy happy lot,
But being too happy in thine happiness,—
That thou, light-wingèd Dryad of the trees,
In some melodious plot
Of beechen green, and shadows numberless,
Singest of summer in full-throated ease. 10

2

O, for a draught of vintage! that hath been
Cool'd a long age in the deep-delved earth,
Tasting of Flora and the country green,
Dance, and Provençal song, and sunburnt mirth!
O for a beaker full of the warm South,°
Full of the true, the blushful Hippocrene,
With beaded bubbles winking at the brim,
And purple-stained mouth;
That I might drink, and leave the world unseen,
And with thee fade away into the forest dim: 20

3

Fade far away, dissolve, and quite forget
What thou among the leaves hast never known,
The weariness, the fever, and the fret
Here, where men sit and hear each other groan;
Where palsy shakes a few, sad, last gray hairs,
Where youth grows pale, and spectre-thin, and dies;°
Where but to think is to be full of sorrow
And leaden-eyed despairs,
Where Beauty cannot keep her lustrous eyes,
Or new Love pine at them beyond tomorrow. 30

4

Away! away! for I will fly to thee,
Not charioted by Bacchus and his pards,
But on the viewless wings of Poesy,
Though the dull brain perplexes and retards:
Already with thee! tender is the night,
And haply the Queen-Moon is on her throne,
Cluster'd around by all her starry Fays;
But here there is no light,
Save what from heaven is with the breezes blown
Through verdurous glooms and winding mossy ways. 40

5

I cannot see what flowers are at my feet,
Nor what soft incense hangs upon the boughs,
But, in embalmed darkness, guess each sweet
Wherewith the seasonable month endows
The grass, the thicket, and the fruit-tree wild;
White hawthorn, and the pastoral eglantine;°
Fast fading violets cover'd up in leaves;
And mid-May's eldest child,
The coming musk-rose, full of dewy wine,
The murmurous haunt of flies on summer eves. 50

6

Darkling I listen; and, for many a time°
I have been half in love with easeful Death,
Call'd him soft names in many a mused rhyme,
To take into the air my quiet breath;
Now more than ever seems it rich to die,
To cease upon the midnight with no pain,
While thou art pouring forth thy soul abroad
In such an ecstasy!
Still wouldst thou sing, and I have ears in vain—
To thy high requiem become a sod.° 60

7

Thou wast not born for death, immortal Bird!
No hungry generations tread thee down;
The voice I hear this passing night was heard
In ancient days by emperor and clown:
Perhaps the self-same song that found a path°
Through the sad heart of Ruth, when, sick for home,
She stood in tears amid the alien corn;
The same that oft-times hath
Charm'd magic casements, opening on the foam
Of perilous seas, in faery lands forlorn. 70

8

Forlorn! the very word is like a bell
   To toll me back from thee to my sole self!
Adieu! the fancy cannot cheat so well
   As she is fam'd to do, deceiving elf.
Adieu! adieu! thy plaintive anthem fades
   Past the near meadows, over the still stream,
      Up the hill-side; and now 'tis buried deep
         In the next valley-glades:
   Was it a vision, or a waking dream?
      Fled is that music:—Do I wake or sleep? 80

Ode on a Grecian Urn

I

Thou still unravish'd bride of quietness,
   Thou foster-child of silence and slow time,
Sylvan historian, who canst thus express
   A flowery tale more sweetly than our rhyme:
What leaf-fring'd legend haunts about thy shape
   Of deities or mortals, or of both,
      In Tempe or the dales of Arcady?
   What men or gods are these? What maidens loth?
What mad pursuit? What struggle to escape?
      What pipes and timbrels? What wild ecstasy? 10

2

Heard melodies are sweet, but those unheard
   Are sweeter; therefore, ye soft pipes, play on;
Not to the sensual ear, but, more endear'd,
   Pipe to the spirit ditties of no tone:
Fair youth, beneath the trees, thou canst not leave
   Thy song, nor ever can those trees be bare;
      Bold Lover, never, never canst thou kiss,
Though winning near the goal—yet, do not grieve;
   She cannot fade, though thou hast not thy bliss,
      For ever wilt thou love, and she be fair! 20

3

Ah, happy, happy boughs! that cannot shed
  Your leaves, nor ever bid the Spring adieu;
And, happy melodist, unwearied,
  For ever piping songs for ever new;
More happy love! more happy, happy love!
  For ever warm and still to be enjoy'd,
    For ever panting, and for ever young;
All breathing human passion far above,°
  That leaves a heart high-sorrowful and cloy'd,
    A burning forehead, and a parching tongue. 30

4

Who are these coming to the sacrifice?
  To what green altar, O mysterious priest,
Lead'st thou that heifer lowing at the skies,
  And all her silken flanks with garlands drest?
What little town by river or sea shore,
  Or mountain-built with peaceful citadel,
    Is emptied of this folk, this pious morn?
And, little town, thy streets for evermore
  Will silent be; and not a soul to tell
    Why thou art desolate, can e'er return. 40

5

O Attic shape! Fair attitude! with brede°
  Of marble men and maidens overwrought,
With forest branches and the trodden weed;
  Thou, silent form, dost tease us out of thought
As doth eternity: Cold Pastoral!
  When old age shall this generation waste,
    Thou shalt remain, in midst of other woe
Than ours, a friend to man, to whom thou say'st,°
  'Beauty is truth, truth beauty,'—that is all°
    Ye know on earth, and all ye need to know. 50

Ode on Melancholy

1

No, no, go not to Lethe, neither twist
  Wolf's-bane, tight-rooted, for its poisonous wine;
Nor suffer thy pale forehead to be kiss'd
  By nightshade, ruby grape of Proserpine;
Make not your rosary of yew-berries,
  Nor let the beetle, nor the death-moth be°
    Your mournful Psyche, nor the downy owl
A partner in your sorrow's mysteries;
  For shade to shade will come too drowsily,
    And drown the wakeful anguish of the soul. 10

2

But when the melancholy fit shall fall
  Sudden from heaven like a weeping cloud,
That fosters the droop-headed flowers all,
  And hides the green hill in an April shroud;
Then glut thy sorrow on a morning rose,
  Or on the rainbow of the salt sand-wave,
    Or on the wealth of globed peonies;
Or if thy mistress some rich anger shows,
  Emprison her soft hand, and let her rave,
    And feed deep, deep upon her peerless eyes. 20

3

She dwells with Beauty—Beauty that must die;
  And Joy, whose hand is ever at his lips
Bidding adieu; and aching Pleasure nigh,
  Turning to poison while the bee-mouth sips:
Ay, in the very temple of Delight
  Veil'd Melancholy has her sovran shrine,
    Though seen of none save him whose strenuous tongue
Can burst Joy's grape against his palate fine;
  His soul shall taste the sadness of her might,
    And be among her cloudy trophies hung. 30

The Fall of Hyperion: A Dream

CANTO I

Fanatics have their dreams, wherewith they weave
A paradise for a sect; the savage too
From forth the loftiest fashion of his sleep
Guesses at Heaven: pity these have not
Trac'd upon vellum or wild Indian leaf
The shadows of melodious utterance.
But bare of laurel they live, dream and die;
For Poesy alone can tell her dreams,
With the fine spell of words alone can save
Imagination from the sable charm 10
And dumb enchantment. Who alive can say
'Thou art no Poet; may'st not tell thy dreams'?
Since every man whose soul is not a clod
Hath visions, and would speak, if he had lov'd
And been well nurtured in his mother tongue.
Whether the dream now purposed to rehearse
Be poet's or Fanatic's will be known
When this warm scribe my hand is in the grave.

Methought I stood where trees of every clime,
Palm, Myrtle, oak, and sycamore, and beech, 20
With plantane, and spice blossoms, made a screen;
In neighbourhood of fountains, by the noise
Soft showering in mine ears; and, by the touch
Of scent, not far from roses. Turning round,
I saw an arbour with a drooping roof
Of trellis vines, and bells, and larger blooms,
Like floral-censers swinging light in air;
Before its wreathed doorway, on a mound
Of moss, was spread a feast of summer fruits,
Which nearer seen, seem'd refuse of a meal 30
By Angel tasted, or our Mother Eve;°
For empty shells were scattered on the grass,
And grape stalks but half bare, and remnants more,
Sweet smelling, whose pure kinds I could not know.

Still was more plenty than the fabled horn
Thrice emptied could pour forth, at banqueting
For Proserpine return'd to her own fields,
Where the white heifers low. And appetite
More yearning than on earth I ever felt
Growing within, I ate deliciously; 40
And, after not long, thirsted, for thereby
Stood a cool vessel of transparent juice,
Sipp'd by the wander'd bee, the which I took,
And, pledging all the Mortals of the World,
And all the dead whose names are in our lips,
Drank. That full draught is parent of my theme.
No Asian poppy, nor Elixir fine
Of the soon fading jealous Caliphat;°
No poison gender'd in close Monkish cell
To thin the scarlet conclave of old Men, 50
Could so have rapt unwilling life away.
Among the fragrant husks and berries crush'd,
Upon the grass I struggled hard against
The domineering potion; but in vain:
The cloudy swoon came on, and down I sunk
Like a Silenus on an antique vase.°
How long I slumber'd 'tis a chance to guess.
When sense of life return'd, I started up
As if with wings; but the fair trees were gone,
The mossy mound and arbour were no more; 60
I look'd around upon the carved sides
Of an old sanctuary with roof august,
Builded so high, it seem'd that filmed clouds
Might spread beneath, as o'er the stars of heaven;
So old the place was, I remembered none
The like upon the earth; what I had seen
Of grey Cathedrals, buttress'd walls, rent towers,
The superannuations of sunk realms,
Or Nature's Rocks toil'd hard in waves and winds,
Seem'd but the faulture of decrepit things° 70
To that eternal domed Monument.
Upon the marble at my feet there lay
Store of strange vessels, and large draperies,
Which needs had been of dyed asbestus wove,°
Or in that place the moth could not corrupt,°

So white the linen; so, in some, distinct
Ran imageries from a sombre loom.
All in a mingled heap confus'd there lay
Robes, golden tongs, censer, and chafing dish.
Girdles, and chains, and holy jewelries— 80

Turning from these with awe, once more I rais'd
My eyes to fathom the space every way;
The embossed roof, the silent massy range
Of columns north and South, ending in mist
Of nothing, then to Eastward, where black gates
Were shut against the sunrise evermore.
Then to the West I look'd, and saw far off
An Image, huge of feature as a cloud,
At level of whose feet an altar slept,
To be approach'd on either side by steps, 90
And marble balustrade, and patient travail
To count with toil the innumerable degrees.
Towards the altar sober-pac'd I went,
Repressing haste, as too unholy there;
And, coming nearer, saw beside the shrine
One minist'ring; and there arose a flame.
When in mid-May the sickening East Wind°
Shifts sudden to the South, the small warm rain
Melts out the frozen incense from all flowers,
And fills the air with so much pleasant health 100
That even the dying man forgets his shroud;
Even so that lofty sacrificial fire,
Sending forth Maian incense, spread around
Forgetfulness of every thing but bliss,
And clouded all the altar with soft smoke,
From whose white fragrant curtains thus I heard
Language pronounc'd. 'If thou canst not ascend
These steps, die on that marble where thou art.
Thy flesh, near cousin to the common dust,
Will parch for lack of nutriment—thy bones 110
Will wither in few years, and vanish so
That not the quickest eye could find a grain
Of what thou now art on that pavement cold.
The sands of thy short life are spent this hour,
And no hand in the Universe can turn

Thy hour glass, if these gummed leaves be burnt
Ere thou canst mount up these immortal steps.'
I heard, I look'd: two senses both at once
So fine, so subtle, felt the tyranny
Of that fierce threat, and the hard task proposed. 120
Prodigious seem'd the toil, the leaves were yet
Burning,—when suddenly a palsied chill
Struck from the paved level up my limbs,
And was ascending quick to put cold grasp
Upon those streams that pulse beside the throat:
I shriek'd; and the sharp anguish of my shriek
Stung my own ears—I strove hard to escape
The numbness; strove to gain the lowest step.
Slow, heavy, deadly was my pace: the cold
Grew stifling, suffocating, at the heart; 130
And when I clasp'd my hands I felt them not.
One minute before death, my iced foot touch'd
The lowest stair; and as it touch'd, life seem'd
To pour in at the toes: I mounted up,
As once fair Angels on a ladder flew
From the green turf to heaven.—'Holy Power,'
Cried I, approaching near the horned shrine,°
'What am I that should so be sav'd from death?
What am I, that another death come not
To choak my utterance sacrilegious here?' 140
Then said the veiled shadow—'Thou hast felt
What 'tis to die and live again before
Thy fated hour. That thou hadst power to do so
Is thy own safety; thou hast dated on°
Thy doom.'—'High Prophetess,' said I, 'purge off
Benign, if so it please thee, my mind's film'—
'None can usurp this height,' return'd that shade,
'But those to whom the miseries of the world
Are misery, and will not let them rest.
All else who find a haven in the world, 150
Where they may thoughtless sleep away their days,
If by a chance into this fane they come,
Rot on the pavement where thou rotted'st half.'—
'Are there not thousands in the world,' said I,
Encourag'd by the sooth voice of the shade,
'Who love their fellows even to the death;

Who feel the giant agony of the world;°
And more, like slaves to poor humanity,
Labour for mortal good? I sure should see
Other men here: but I am here alone.' 160
'They whom thou spak'st of are no vision'ries,'
Rejoin'd that voice—'They are no dreamers weak,
They seek no wonder but the human face;
No music but a happy-noted voice—
They come not here, they have no thought to come—
And thou art here, for thou art less than they—
What benefit canst thou do, or all thy tribe
To the great World? Thou art a dreaming thing;
A fever of thyself—think of the Earth;
What bliss even in hope is there for thee? 170
What haven? Every creature hath its home;
Every sole man hath days of joy and pain,
Whether his labours be sublime or low—
The pain alone; the joy alone; distinct:
Only the dreamer venoms all his days,
Bearing more woe than all his Sins deserve.
Therefore, that happiness be somewhat shar'd,
Such things as thou art are admitted oft
Into like gardens thou didst pass erewhile,
And suffer'd in these Temples; for that cause 180
Thou standest safe beneath this statue's knees.'
'That I am favored for unworthiness,
By such propitious parley medicin'd
In sickness not ignoble, I rejoice,
Aye, and could weep for love of such award.'
So answer'd I, continuing, 'If it please
Majestic shadow, tell me: sure not all°
Those melodies sung into the world's ear
Are useless: sure a poet is a sage;
A humanist, Physician to all Men. 190
That I am none I feel, as Vultures feel
They are no birds when Eagles are abroad.
What am I then? Thou spakest of my tribe:
What tribe?'—The tall shade veil'd in drooping white
Then spake, so much more earnest, that the breath
Mov'd the thin linen folds that drooping hung
About a golden censer from the hand

Pendent.—'Art thou not of the dreamer tribe?
The poet and the dreamer are distinct,
Diverse, sheer opposite, antipodes. 200
The one pours out a balm upon the world,
The other vexes it.' Then shouted I
Spite of myself, and with a Pythia's spleen,
'Apollo! faded, far flown Apollo!
Where is thy misty pestilence to creep°
Into the dwellings, through the door crannies,
Of all mock lyrists, large self worshipers,°
And careless Hectorers in proud bad verse.
Though I breathe death with them it will be life
To see them sprawl before me into graves. 210
Majestic shadow, tell me where I am:
Whose altar this; for whom this incense curls:
What Image this, whose face I cannot see,
For the broad marble knees; and who thou art,
Of accent feminine, so courteous.'
Then the tall shade in drooping linens veil'd
Spake out, so much more earnest, that her breath
Stirr'd the thin folds of gauze that drooping hung
About a golden censer from her hand
Pendent; and by her voice I knew she shed 220
Long treasured tears. 'This temple sad and lone
Is all spar'd from the thunder of a war°
Foughten long since by giant hierarchy
Against rebellion: this old image here,
Whose carved features wrinkled as he fell,
Is Saturn's; I, Moneta, left supreme°
Sole Priestess of his desolation.'—
I had no words to answer; for my tongue,
Useless, could find about its roofed home
No syllable of a fit Majesty 230
To make rejoinder to Moneta's mourn.
There was a silence while the altar's blaze
Was fainting for sweet food: I look'd thereon
And on the paved floor, where nigh were pil'd
Faggots of cinnamon, and many heaps
Of other crisped spice-wood—then again
I look'd upon the altar and its horns
Whiten'd with ashes, and its lang'rous flame,

And then upon the offerings again;
And so by turns—till sad Moneta cried, 240
'The sacrifice is done, but not the less,
Will I be kind to thee for thy good will.
My power, which to me is still a curse,
Shall be to thee a wonder; for the scenes
Still swooning vivid through my globed brain
With an electral changing misery
Thou shalt with those dull mortal eyes behold,
Free from all pain, if wonder pain thee not.'
As near as an immortal's sphered words°
Could to a Mother's soften, were these last: 250
But yet I had a terror of her robes,
And chiefly of the veils, that from her brow
Hung pale, and curtain'd her in mysteries
That made my heart too small to hold its blood.
This saw that Goddess, and with sacred hand
Parted the veils. Then saw I a wan face,
Not pin'd by human sorrows, but bright blanch'd
By an immortal sickness which kills not;
It works a constant change, which happy death
Can put no end to; deathwards progressing 260
To no death was that visage; it had pass'd
The lily and the snow; and beyond these
I must not think now, though I saw that face—
But for her eyes I should have fled away.
They held me back, with a benignant light,
Soft mitigated by divinest lids
Half closed, and visionless entire they seem'd
Of all external things—they saw me not,
But in blank splendor beam'd like the mild moon,
Who comforts those she sees not, who knows not 270
What eyes are upward cast. As I had found
A grain of gold upon a mountain's side,
And twing'd with avarice strain'd out my eyes
To search its sullen entrails rich with ore,
So at the view of sad Moneta's brow,
I ached to see what things the hollow brain
Behind enwombed: what high tragedy
In the dark secret Chambers of her skull
Was acting, that could give so dread a stress

To her cold lips, and fill with such a light 280
Her planetary eyes; and touch her voice
With such a sorrow—'Shade of Memory!'
'By all the gloom hung round thy fallen house,
By this last Temple, by the golden age,
By great Apollo, thy dear foster child,
And by thy self, forlorn divinity,
The pale Omega of a wither'd race,
Let me behold, according as thou said'st,
What in thy brain so ferments to and fro.'— 290
No sooner had this conjuration pass'd
My devout Lips, than side by side we stood.
(Like a stunt bramble by a solemn Pine)
Deep in the shady sadness of a vale,°
Far sunken from the healthy breath of morn,
Far from the fiery noon, and Eve's one star.
Onward I look'd beneath the gloomy boughs,
And saw, what first I thought an Image huge.
Like to the Image pedestal'd so high
In Saturn's Temple. Then Moneta's voice 300
Came brief upon mine ear,—'So Saturn sat
When he had lost his realms'—Whereon there grew
A power within me of enormous ken,
To see as a God sees, and take the depth
Of things as nimbly as the outward eye
Can size and shape pervade. The lofty theme
At those few words hung vast before my mind,
With half unravel'd web. I set myself
Upon an Eagle's watch, that I might see,
And seeing ne'er forget. No stir of life 310
Was in this shrouded vale, not so much air
As in the zoning of a Summer's day
Robs not one light seed from the feather'd grass,
But where the dead leaf fell there did it rest:
A stream went voiceless by, still deaden'd more
By reason of the fallen Divinity
Spreading more shade: the Naiad mid her reeds
Press'd her cold finger closer to her lips.
Along the margin sand large footmarks went
No farther than to where old Saturn's feet 320
Had rested, and there slept, how long a sleep!

Degraded, cold, upon the sodden ground
His old right hand lay nerveless, listless, dead,
Unsceptred; and his realmless eyes were clos'd,
While his bow'd head seem'd listening to the Earth,
His antient mother, for some comfort yet.

It seemed no force could wake him from his place;
But there came one who with a kindred hand
Touch'd his wide shoulders, after bending low
With reverence, though to one who knew it not. 330
Then came the griev'd voice of Mnemosyne,
And griev'd I hearken'd. 'That divinity
Whom thou saw'st step from yon forlornest wood,
And with slow pace approach our fallen King,
Is Thea, softest-natur'd of our Brood.'
I mark'd the goddess in fair statuary
Surpassing wan Moneta by the head,
And in her sorrow nearer woman's tears.
There was a listening fear in her regard,
As if calamity had but begun; 340
As if the vanward clouds of evil days
Had spent their malice, and the sullen rear
Was with its stored thunder labouring up.
One hand she press'd upon that aching spot
Where beats the human heart; as if just there
Though an immortal, she felt cruel pain;
The other upon Saturn's bended neck
She laid, and to the level of his hollow ear
Leaning, with parted lips, some words she spake
In solemn tenor and deep organ tune; 350
Some mourning words, which in our feeble tongue
Would come in this-like accenting; how frail
To that large utterance of the early Gods!—
'Saturn! look up—and for what, poor lost King?
I have no comfort for thee, no—not one—
I cannot cry, *Wherefore thus sleepest thou?*
For heaven is parted from thee, and the earth
Knows thee not, so afflicted, for a God;
And Ocean too, with all its solemn noise,
Has from thy sceptre pass'd and all the air 360
Is emptied of thine hoary Majesty.

Thy thunder, captious at the new command,
Rumbles reluctant o'er our fallen house;
And thy sharp lightning in unpracticed hands
Scorches and burns our once serene domain.
With such remorseless speed still come new woes
That unbelief has not a space to breathe.
Saturn, sleep on:—Me thoughtless, why should I
Thus violate thy slumbrous solitude?
Why should I ope thy melancholy eyes? 370
Saturn, sleep on, while at thy feet I weep.—

As when, upon a tranced Summer Night,
Forests, branch-charmed by the earnest stars,
Dream, and so dream all night, without a noise,
Save from one gradual solitary gust,
Swelling upon the silence; dying off;
As if the ebbing air had but one wave;
So came these words, and went; the while in tears
She press'd her fair large forehead to the earth,
Just where her fallen hair might spread in curls, 380
A soft and silken mat for Saturn's feet.
Long, long, those two were postured motionless,°
Like sculpture builded up upon the grave
Of their own power. A long awful time
I look'd upon them; still they were the same;
The frozen God still bending to the Earth,
And the sad Goddess weeping at his feet.
Moneta silent. Without stay or prop
But my own weak mortality, I bore
The load of this eternal quietude, 390
The unchanging gloom, and the three fixed shapes
Ponderous upon my senses a whole Moon.
For by my burning brain I measured sure
Her silver seasons shedded on the night
And every day by day methought I grew
More gaunt and ghostly—Oftentimes I pray'd
Intense, that Death would take me from the Vale
And all its burthens—Gasping with despair
Of change, hour after hour I curs'd myself:
Until old Saturn rais'd his faded eyes, 400
And look'd around, and saw his Kingdom gone,

And all the gloom and sorrow of the place,
And that fair kneeling Goddess at his feet.
As the moist scent of flowers, and grass, and leaves
Fills forest dells with a pervading air
Known to the woodland nostril, so the words
Of Saturn fill'd the mossy glooms around,
Even to the hollows of time-eaten oaks,
And to the windings in the foxes' hole,
With sad low tones, while thus he spake, and sent 410
Strange musings to the solitary Pan.°

'Moan, brethren, moan; for we are swallow'd up
And buried from all godlike exercise
Of influence benign on planets pale,
And peaceful sway above man's harvesting,
And all those acts which Deity supreme
Doth ease its heart of love in. Moan and wail.
Moan, brethren, moan; for lo! the rebel spheres
Spin round, the stars their antient courses keep,
Clouds still with shadowy moisture haunt the earth, 420
Still suck their fill of light from Sun and Moon,
Still buds the tree, and still the sea-shores murmur.
There is no death in all the universe
No smell of Death—there shall be death—Moan, moan,
Moan, Cybele, moan, for thy pernicious babes
Have chang'd a God into a shaking Palsy.
Moan, brethren, moan; for I have no strength left,
Weak as the reed—weak—feeble as my voice—
O, O, the pain, the pain of feebleness.
Moan, moan; for still I thaw—or give me help: 430
Throw down those Imps and give me victory.
Let me hear other groans, and trumpets blown
Of triumph calm, and hymns of festival
From the gold peaks of Heaven's high piled clouds;
Voices of soft proclaim, and silver stir
Of strings in hollow shells; and let there be
Beautiful things made new for the surprize
Of the sky children'—So he feebly ceas'd,
With such a poor and sickly sounding pause,
Methought I heard some old Man of the earth 440
Bewailing earthly loss; nor could my eyes

And ears act with that pleasant unison of sense
Which marries sweet sound with the grace of form,
And dolourous accent from a tragic harp
With large limb'd visions—More I scrutinized:
Still fix'd he sat beneath the sable trees,
Whose arms spread straggling in wild serpent forms,
With leaves all hush'd: his awful presence there
(Now all was silent) gave a deadly lie
To what I erewhile heard: only his lips 450
Trembled amid the white curls of his beard.
They told the truth, though, round, the snowy locks
Hung nobly, as upon the face of heaven
A midday fleece of clouds. Thœa arose
And stretch'd her white arm through the hollow dark,
Pointing some whither: whereat he too rose
Like a vast giant seen by men at sea
To grow pale from the waves at dull midnight.
They melted from my sight into the woods:
Ere I could turn, Moneta cried—'These twain 460
Are speeding to the families of grief,
Where roof'd in by black rocks they waste in pain
And darkness for no hope.'—And she spake on,
As ye may read who can unwearied pass
Onward from the Antichamber of this dream,°
Where even at the open doors awhile
I must delay, and glean my memory
Of her high phrase: perhaps no further dare.—

CANTO II

'Mortal, that thou may'st understand aright,°
I humanize my sayings to thine ear,
Making comparisons of earthly things;
Or thou might'st better listen to the wind,
Whose language is to thee a barren noise,
Though it blows legend-laden through the trees—
In melancholy realms big tears are shed,
More sorrow like to this, and such-like woe,
Too huge for mortal tongue, or pen of scribe.
The Titans fierce, self-hid, or prison-bound, 10
Groan for the old allegiance once more,

Listening in their doom for Saturn's voice.
But one of our whole eagle-brood still keeps
His sov'reignty, and Rule, and Majesty;
Blazing Hyperion on his orbed fire
Still sits, still snuffs the incense teeming up
From man to the Sun's God: yet unsecure,
For as upon the Earth dire prodigies
Fright and perplex, so also shudders he:
Nor at dog's howl, or gloom-bird's Even screech, 20
Or the familiar visitings of one
Upon the first toll of his passing bell:
But horrors portion'd to a giant nerve
Make great Hyperion ache. His palace bright,
Bastion'd with pyramids of glowing gold,
And touch'd with shade of bronzed obelisks,
Glares a blood red through all the thousand Courts,
Arches, and domes, and fiery galeries:
And all its curtains of Aurorian clouds
Flush angerly: when he would taste the wreaths 30
Of incense breath'd aloft from sacred hills,
Instead of sweets, his ample palate takes
Savour of poisonous brass and metals sick.
Wherefore when harbour'd in the sleepy West,
After the full completion of fair day,
For rest divine upon exalted couch
And slumber in the arms of melody,
He paces through the pleasant hours of ease,
With strides colossal, on from Hall to Hall;
While, far within each aisle and deep recess, 40
His winged minions in close clusters stand
Amaz'd, and full of fear; like anxious men
Who on a wide plain gather in sad troops,
When earthquakes jar their battlements and towers.
Even now, while Saturn, rous'd from icy trance
Goes, step for step, with Thea from yon woods,
Hyperion, leaving twilight in the rear,
Is sloping to the threshold of the west.—
Thither we tend.'—Now in clear light I stood,
Reliev'd from the dusk vale. Mnemosyne° 50
Was sitting on a square edg'd polish'd stone,
That in its lucid depth reflected pure

Her priestess-garments. My quick eyes ran on
From stately nave to nave, from vault to vault,
Through bowers of fragrant and enwreathed light,
And diamond paved lustrous long arcades.
Anon rush'd by the bright Hyperion;
His flaming robes stream'd out beyond his heels,
And gave a roar, as if of earthly fire,
That scar'd away the meek ethereal hours 60
And made their dove-wings tremble: on he flared—

Lamia

PART I

Upon a time, before the faery broods°
Drove Nymph and Satyr from the prosperous woods,
Before King Oberon's bright diadem,
Sceptre, and mantle, clasp'd with dewy gem,
Frighted away the Dryads and the Fauns
From rushes green, and brakes, and cowslip'd lawns,
The ever-smitten Hermes empty left
His golden throne, bent warm on amorous theft:
From high Olympus had he stolen light,
On this side of Jove's clouds, to escape the sight 10
Of his great summoner, and made retreat°
Into a forest on the shores of Crete.
For somewhere in that sacred island dwelt
A nymph, to whom all hoofed Satyrs knelt;
At whose white feet the languid Tritons poured
Pearls, while on land they wither'd and adored.
Fast by the springs where she to bathe was wont,
And in those meads where sometime she might haunt,
Were strewn rich gifts, unknown to any Muse,
Though Fancy's casket were unlock'd to choose. 20
Ah, what a world of love was at her feet!
So Hermes thought, and a celestial heat
Burnt from his winged heels to either ear,
That from a whiteness, as the lily clear,
Blush'd into roses 'mid his golden hair,
Fallen in jealous curls about his shoulders bare.

From vale to vale, from wood to wood, he flew,
Breathing upon the flowers his passion new,
And wound with many a river to its head,
To find where this sweet nymph prepar'd her secret bed: 30
In vain; the sweet nymph might nowhere be found,
And so he rested, on the lonely ground,
Pensive, and full of painful jealousies
Of the Wood-Gods, and even the very trees.
There as he stood, he heard a mournful voice,

Such as once heard, in gentle heart, destroys°
All pain but pity: thus the lone voice spake:
'When from this wreathed tomb shall I awake!
When move in a sweet body fit for life,
And love, and pleasure, and the ruddy strife 40
Of hearts and lips! Ah, miserable me!'
The God, dove-footed, glided silently
Round bush and tree, soft-brushing, in his speed,
The taller grasses and full-flowering weed,
Until he found a palpitating snake,
Bright, and cirque-couchant in a dusky brake.°

She was a gordian shape of dazzling hue,
Vermilion-spotted, golden, green, and blue;
Striped like a zebra, freckled like a pard,
Eyed like a peacock, and all crimson barr'd; 50
And full of silver moons, that, as she breathed,
Dissolv'd, or brighter shone, or interwreathed
Their lustres with the gloomier tapestries—
So rainbow-sided, touch'd with miseries,
She seem'd, at once, some penanced lady elf,°
Some demon's mistress, or the demon's self.
Upon her crest she wore a wannish fire
Sprinkled with stars, like Ariadne's tiar:
Her head was serpent, but ah, bitter-sweet!
She had a woman's mouth with all its pearls complete: 60
And for her eyes: what could such eyes do there
But weep, and weep, that they were born so fair?
As Proserpine still weeps for her Sicilian air.
Her throat was serpent, but the words she spake
Came, as through bubbling honey, for Love's sake,
And thus; while Hermes on his pinions lay,
Like a stoop'd falcon ere he takes his prey.

'Fair Hermes, crown'd with feathers, fluttering light,
I had a splendid dream of thee last night:
I saw thee sitting, on a throne of gold, 70
Among the Gods, upon Olympus old,
The only sad one; for thou didst not hear
The soft, lute-finger'd Muses chaunting clear,
Nor even Apollo when he sang alone,

Deaf to his throbbing throat's long, long melodious moan.
I dreamt I saw thee, robed in purple flakes,
Break amorous through the clouds, as morning breaks,
And, swiftly as a bright Phœbean dart,
Strike for the Cretan isle; and here thou art!
Too gentle Hermes, hast thou found the maid?' 80
Whereat the star of Lethe not delay'd°
His rosy eloquence, and thus inquired:
'Thou smooth-lipp'd serpent, surely high inspired!
Thou beauteous wreath, with melancholy eyes,
Possess whatever bliss thou canst devise,
Telling me only where my nymph is fled,—
Where she doth breathe!' 'Bright planet, thou hast said,'
Return'd the snake, 'but seal with oaths, fair God!'
'I swear,' said Hermes, 'by my serpent rod,
And by thine eyes, and by thy starry crown!' 90
Light flew his earnest words, among the blossoms blown.
Then thus again the brilliance feminine:
'Too frail of heart! for this lost nymph of thine,
Free as the air, invisibly, she strays
About these thornless wilds; her pleasant days
She tastes unseen; unseen her nimble feet
Leave traces in the grass and flowers sweet;
From weary tendrils, and bow'd branches green,
She plucks the fruit unseen, she bathes unseen:
And by my power is her beauty veil'd 100
To keep it unaffronted, unassail'd
By the love-glances of unlovely eyes,
Of Satyrs, Fauns, and blear'd Silenus' sighs.
Pale grew her immortality, for woe
Of all these lovers, and she grieved so
I took compassion on her, bade her steep
Her hair in weïrd syrops, that would keep
Her loveliness invisible, yet free
To wander as she loves, in liberty.
Thou shalt behold her, Hermes, thou alone, 110
If thou wilt, as thou swearest, grant my boon!'
Then, once again, the charmed God began
An oath, and through the serpent's ears it ran
Warm, tremulous, devout, psalterian.°
Ravish'd, she lifted her Circean head,°

Blush'd a live damask, and swift-lisping said,
'I was a woman, let me have once more
A woman's shape, and charming as before.
I love a youth of Corinth—O the bliss!
Give me my woman's form, and place me where he is. 120
Stoop, Hermes, let me breathe upon thy brow,
And thou shalt see thy sweet nymph even now.'
The God on half-shut feathers sank serene,
She breath'd upon his eyes, and swift was seen
Of both the guarded nymph near-smiling on the green.
It was no dream; or say a dream it was,°
Real are the dreams of Gods, and smoothly pass
Their pleasures in a long immortal dream.
One warm, flush'd moment, hovering, it might seem
Dash'd by the wood-nymph's beauty, so he burn'd; 130
Then, lighting on the printless verdure, turn'd
To the swoon'd serpent, and with languid arm,
Delicate, put to proof the lythe Caducean charm.
So done, upon the nymph his eyes he bent
Full of adoring tears and blandishment,
And towards her stept: she, like a moon in wane,
Faded before him, cower'd, nor could restrain
Her fearful sobs, self-folding like a flower
That faints into itself at evening hour:
But the God fostering her chilled hand, 140
She felt the warmth, her eyelids open'd bland,
And, like new flowers at morning song of bees,
Bloom'd, and gave up her honey to the lees.
Into the green-recessed woods they flew;
Nor grew they pale, as mortal lovers do.

Left to herself, the serpent now began°
To change; her elfin blood in madness ran,
Her mouth foam'd, and the grass, therewith besprent,
Wither'd at dew so sweet and virulent;
Her eyes in torture fix'd, and anguish drear, 150
Hot, glaz'd, and wide, with lid-lashes all sear,
Flash'd phosphor and sharp sparks, without one cooling tear.
The colours all inflam'd throughout her train,
She writh'd about, convuls'd with scarlet pain:
A deep volcanian yellow took the place

Of all her milder-mooned body's grace;
And, as the lava ravishes the mèad,
Spoilt all her silver mail, and golden brede;
Made gloom of all her frecklings, streaks and bars,
Eclips'd her crescents, and lick'd up her stars: 160
So that, in moments few, she was undrest
Of all her sapphires, greens, and amethyst,
And rubious-argent: of all these bereft,°
Nothing but pain and ugliness were left.
Still shone her crown; that vanish'd, also she
Melted and disappear'd as suddenly;
And in the air, her new voice luting soft,
Cried, 'Lycius! gentle Lycius!'—Borne aloft°
With the bright mists about the mountains hoar
These words dissolv'd: Crete's forests heard no more. 170

Whither fled Lamia, now a lady bright,
A full-born beauty new and exquisite?
She fled into that valley they pass o'er
Who go to Corinth from Cenchreas' shore;°
And rested at the foot of those wild hills,
The rugged founts of the Peræan rills,
And of that other ridge whose barren back
Stretches, with all its mist and cloudy rack,
South-westward to Cleone. There she stood°
About a young bird's flutter from a wood, 180
Fair, on a sloping green of mossy tread,
By a clear pool, wherein she passioned
To see herself escap'd from so sore ills,
While her robes flaunted with the daffodils.

Ah, happy Lycius!—for she was a maid
More beautiful than ever twisted braid,
Or sigh'd, or blush'd, or on spring-flowered lea
Spread a green kirtle to the minstrelsy:
A virgin purest lipp'd, yet in the lore
Of love deep learned to the red heart's core: 190
Not one hour old, yet of sciential brain
To unperplex bliss from its neighbour pain;
Define their pettish limits, and estrange
Their points of contact, and swift counterchange;

Intrigue with the specious chaos, and dispart
Its most ambiguous atoms with sure art;
As though in Cupid's college she had spent
Sweet days a lovely graduate, still unshent,
And kept his rosy terms in idle languishment.

Why this fair creature chose so fairily 200
By the wayside to linger, we shall see;
But first 'tis fit to tell how she could muse
And dream, when in the serpent prison-house,
Of all she list, strange or magnificent:
How, ever, where she will'd, her spirit went;
Whether to faint Elysium, or where
Down through tress-lifting waves the Nereids fair
Wind into Thetis' bower by many a pearly stair;
Or where in God Bacchus drains his cups divine,
Stretch'd out, at ease, beneath a glutinous pine; 210
Or where in Pluto's gardens palatine
Mulciber's columns gleam in far piazzian line.
And sometimes into cities she would send
Her dream, with feast and rioting to blend;
And once, while among mortals dreaming thus.
She saw the young Corinthian Lycius
Charioting foremost in the envious race,
Like a young Jove with calm uneager face,
And fell into a swooning love of him.
Now on the moth-time of that evening dim 220
He would return that way, as well she knew,
To Corinth from the shore; for freshly blew
The eastern soft wind, and his galley now
Grated the quaystones with her brazen prow
In port Cenchreas, from Egina isle
Fresh anchor'd; whither he had been awhile
To sacrifice to Jove, whose temple there
Waits with high marble doors for blood and incense rare.
Jove heard his vows, and better'd his desire;
For by some freakful chance he made retire 230
From his companions, and set forth to walk,
Perhaps grown wearied of their Corinth talk:
Over the solitary hills he fared,
Thoughtless at first, but ere eve's star appeared

His phantasy was lost, where reason fades,
In the calm'd twilight of Platonic shades.
Lamia beheld him coming, near, more near—
Close to her passing, in indifference drear,
His silent sandals swept the mossy green;
So neighbour'd to him, and yet so unseen 240
She stood: he pass'd, shut up in mysteries,
His mind wrapp'd like his mantle, while her eyes
Follow'd his steps, and her neck regal white
Turn'd—syllabling thus, 'Ah, Lycius bright,
And will you leave me on the hills alone?
Lycius, look back! and be some pity shown.'
He did; not with cold wonder fearingly,
But Orpheus-like at an Eurydice;
For so delicious were the words she sung,
It seem'd he had lov'd them a whole summer long: 250
And soon his eyes had drunk her beauty up,
Leaving no drop in the bewildering cup,
And still the cup was full,—while he, afraid
Lest she should vanish ere his lip had paid
Due adoration, thus began to adore;
Her soft look growing coy, she saw his chain so sure:°
'Leave thee alone! Look back! Ah, Goddess, see
Whether my eyes can ever turn from thee!
For pity do not this sad heart belie—
Even as thou vanishest so I shall die. 260
Stay! though a Naiad of the rivers, stay!
To thy far wishes will thy streams obey:
Stay! though the greenest woods be thy domain,
Alone they can drink up the morning rain:
Though a descended Pleiad, will not one
Of thine harmonious sisters keep in tune
Thy spheres, and as thy silver proxy shine?
So sweetly to these ravish'd ears of mine
Came thy sweet greeting, that if thou shouldst fade
Thy memory will waste me to a shade:— 270
For pity do not melt!'—'If I should stay,'
Said Lamia, 'here, upon this floor of clay,
And pain my steps upon these flowers too rough,
What canst thou say or do of charm enough
To dull the nice remembrance of my home?

Thou canst not ask me with thee here to roam
Over these hills and vales, where no joy is,—
Empty of immortality and bliss!
Thou art a scholar, Lycius, and must know°
That finer spirits cannot breathe below 280
In human climes, and live: Alas! poor youth,
What taste of purer air hast thou to soothe
My essence? What serener palaces,
Where I may all my many senses please,°
And by mysterious sleights a hundred thirsts appease?
It cannot be—Adieu!' So said, she rose
Tiptoe with white arms spread. He, sick to lose
The amorous promise of her lone complain,
Swoon'd, murmuring of love, and pale with pain.
The cruel lady, without any show 290
Of sorrow for her tender favourite's woe,
But rather, if her eyes could brighter be,
With brighter eyes and slow amenity,
Put her new lips to his, and gave afresh
The life she had so tangled in her mesh:
And as he from once trance was wakening
Into another, she began to sing,
Happy in beauty, life, and love, and every thing,
A song of love, too sweet for earthly lyres,
While, like held breath, the stars drew in their panting fires. 300
And then she whisper'd in such trembling tone,
As those who, safe together met alone
For the first time through many anguish'd days,
Use other speech than looks; bidding him raise
His drooping head, and clear his soul of doubt,
For that she was a woman, and without
Any more subtle fluid in her veins
Than throbbing blood, and that the self-same pains
Inhabited her frail-strung heart as his.
And next she wonder'd how his eyes could miss 310
Her face so long in Corinth, where, she said,
She dwelt but half retir'd, and there had led
Days happy as the gold coin could invent
Without the aid of love; yet in content
Till she saw him, as once she pass'd him by,
Where 'gainst a column he leant thoughtfully

At Venus' temple porch, 'mid baskets heap'd
Of amorous herbs and flowers, newly reap'd
Late on that eve, as 'twas the night before
The Adonian feast; whereof she saw no more,° 320
But wept alone those days, for why should she adore?
Lycius from death awoke into amaze,
To see her still, and singing so sweet lays;
Then from amaze into delight he fell
To hear her whisper woman's lore so well;
And every word she spake entic'd him on
To unperplex'd delight and pleasure known.
Let the mad poets say whate'er they please
Of the sweets of Fairies, Peris, Goddesses,°
There is not such a treat among them all, 330
Haunters of cavern, lake, and waterfall,
As a real woman, lineal indeed
From Pyrrha's pebbles or old Adam's seed.
Thus gentle Lamia judg'd, and judg'd aright,°
That Lycius could not love in half a fright,
So threw the goddess off, and won his heart
More pleasantly by playing woman's part,
With no more awe than what her beauty gave,
That, while it smote, still guaranteed to save.
Lycius to all made eloquent reply, 340
Marrying to every word a twinborn sigh;
And last, pointing to Corinth, ask'd her sweet,
If 'twas too far that night for her soft feet.
The way was short, for Lamia's eagerness
Made, by a spell, the triple league decrease
To a few paces; not at all surmised
By blinded Lycius, so in her comprized.°
They pass'd the city gates, he knew not how,
So noiseless, and he never thought to know.

As men talk in a dream, so Corinth all, 350
Throughout her palaces imperial,
And all her populous streets and temples lewd,
Mutter'd, like tempest in the distance brew'd,
To the wide-spreaded night above her towers.
Men, women, rich and poor, in the cool hours,
Shuffled their sandals o'er the pavement white,

Companion'd or alone; while many a light
Flared, here and there, from wealthy festivals,
And threw their moving shadows on the walls,
Or found them cluster'd in the corniced shade 360
Of some arch'd temple door, or dusky colonnade.

Muffling his face, of greeting friends in fear,
Her fingers he press'd hard, as one came near
With curl'd gray beard, sharp eyes, and smooth bald crown,
Slow-stepp'd, and robed in philosophic gown:
Lycius shrank closer, as they met and past,
Into his mantle, adding wings to haste,
While hurried Lamia trembled: 'Ah,' said he,
'Why do you shudder, love, so ruefully?
Why does your tender palm dissolve in dew?'— 370
'I'm wearied,' said fair Lamia: 'tell me who
Is that old man? I cannot bring to mind
His features:—Lycius! wherefore did you blind
Yourself from his quick eyes?' Lycius replied,
''Tis Apollonius sage, my trusty guide'°
And good instructor; but to-night he seems
The ghost of folly haunting my sweet dreams.'

While yet he spake they had arrived before
A pillar'd porch, with lofty portal door,
Where hung a silver lamp,whose phosphor glow 380
Reflected in the slabbed steps below,
Mild as a star in water; for so new,
And so unsullied was the marble hue,
So through the crystal polish, liquid fine,
Ran the dark veins, that none but feet divine
Could e'er have touch'd there. Sounds Æolian
Breath'd from the hinges, as the ample span
Of the wide doors disclos'd a place unknown°
Some time to any, but those two alone,
And a few Persian mutes, who that same year 390
Were seen about the markets: none knew where
They could inhabit; the most curious
Were foil'd, who watch'd to trace them to their house:
And but the flitter-winged verse must tell,
For truth's sake, what woe afterwards befel,

'Twould humour many a heart to leave them thus,
Shut from the busy world, of more incredulous.

PART II

Love in a hut, with water and a crust,
Is—Love, forgive us!—cinders, ashes, dust;
Love in a palace is perhaps at last
More grievous torment that a hermit's fast:—
That is a doubtful tale from faery land,
Hard for the non-elect to understand.
Had Lycius liv'd to hand his story down,
He might have given the moral a fresh frown,
Or clench'd it quite: but too short was their bliss
To breed distrust and hate, that make the soft voice hiss. 10
Besides, there, nightly, with terrific glare,
Love, jealous grown of so complete a pair,
Hover'd and buzz'd his wings, with fearful roar,
Above the lintel of their chamber door,
And down the passage cast a glow upon the floor.

For all this came a ruin: side by side
They were enthroned, in the even tide,
Upon a couch, near to a curtaining
Whose airy texture, from a golden string,
Floated into the room, and let appear 20
Unveil'd the summer heaven, blue and clear,
Betwixt two marble shafts:—there they reposed,
Where use had made it sweet, with eyelids closed,
Saving a tythe which love still open kept,
That they might see each other while they almost slept;
When from the slope side of a suburb hill,
Deafening the swallow's twitter, came a thrill
Of trumpets—Lycius started—the sounds fled,
But left a thought, a buzzing in his head.
For the first time, since first he harbour'd in 30
That purple-lined palace of sweet sin,
His spirit pass'd beyond its golden bourn
Into the noisy world almost forsworn.
The lady, ever watchful, penetrant,
Saw this with pain, so arguing a want

Of something more, more than her empery
Of joys; and she began to moan and sigh
Because he mused beyond her, knowing well
That but a moment's thought is passion's passing bell.
'Why do you sigh, fair creature?' whisper'd he: 40
'Why do you think?' return'd she tenderly:
'You have deserted me;—where am I now?
Not in your heart while care weighs on your brow:
No, no, you have dismiss'd me; and I go
From your breast houseless: ay, it must be so.'
He answer'd, bending to her open eyes,
Where he was mirror'd small in paradise,
'My silver planet, both of eve and morn!
Why will you plead yourself so sad forlorn,
While I am striving how to fill my heart 50
With deeper crimson, and a double smart?
How to entangle, trammel up and snare
Your soul in mine, and labyrinth you there
Like the hid scent in an unbudded rose?
Ay, a sweet kiss—you see your mighty woes.
My thoughts! shall I unveil them? Listen then!
What mortal hath a prize, that other men
May be confounded and abash'd withal,
But lets it sometimes pace abroad majestical,
And triumph, as in thee I should rejoice 60
Amid the hoarse alarm of Corinth's voice.
Let my foes choke, and my friends shout afar,
While through the thronged streets your bridal car
Wheels round its dazzling spokes.'—The lady's cheek
Trembled; she nothing said, but, pale and meek,
Arose and knelt before him, wept a rain
Of sorrows at his words; at last with pain
Beseeching him, the while his hand she wrung,
To change his purpose. He thereat was stung,
Perverse, with stronger fancy to reclaim 70
Her wild and timid nature to his aim:
Besides, for all his love, in self despite,
Against his better self, he took delight
Luxurious in her sorrows, soft and new.
His passion, cruel grown, took on a hue
Fierce and sanguineous as 'twas possible

In one whose brow had no dark veins to swell.
Fine was the mitigated fury, like
Apollo's presence when in act to strike
The serpent—Ha, the serpent! certes, she° 80
Was none. She burnt, she lov'd the tyranny,°
And, all subdued, consented to the hour
When to the bridal he should lead his paramour.
Whispering in midnight silence, said the youth,
'Sure some sweet name thou hast, though, by my truth,
I have not ask'd it, ever thinking thee
Not mortal, but of heavenly progeny,
As still I do. Hast any mortal name,
Fit appellation for this dazzling frame?
Or friends or kinsfolk on the citied earth, 90
To share our marriage feast and nuptial mirth?'
'I have no friends,' said Lamia, 'no, not one;
My presence in wide Corinth hardly known:
My parents' bones are in their dusty urns
Sepulchred, where no kindled incense burns,
Seeing all their luckless race are dead, save me,
And I neglect the holy rite for thee.
Even as you list invite your many guests;
But if, as now it seems, your vision rests
With any pleasure on me, do not bid 100
Old Apollonius—from him keep me hid.'
Lycius, perplex'd at words so blind and blank,
Made close inquiry; from whose touch she shrank,
Feigning a sleep; and he to the dull shade
Of deep sleep in a moment was betray'd.

It was the custom then to bring away
The bride from home at blushing shut of day,
Veil'd, in a chariot, heralded along
By strewn flowers, torches, and a marriage song,
With other pageants: but this fair unknown 110
Had not a friend. So being left alone,
(Lycius was gone to summon all his kin)
And knowing surely she could never win
His foolish heart from its mad pompousness,
She set herself, high-thoughted, how to dress
The misery in fit magnificence.

She did so, but 'tis doubtful how and whence
Came, and who were her subtle servitors.
About the halls, and to and from the doors,
There was a noise of wings, till in short space 120
The glowing banquet-room shone with wide-arched grace.
A haunting music, sole perhaps and lone°
Supportress of the faery-roof, made moan
Throughout, as fearful the whole charm might fade.
Fresh carved cedar, mimicking a glade
Of palm and plantain, met from either side,
High in the midst, in honour of the bride:
Two palms and then two plantains, and so on,
From either side their stems branch'd one to one
All down the aisled place; and beneath all 130
There ran a stream of lamps straight on from wall to wall.
So canopied, lay an untasted feast
Teeming with odours. Lamia, regal drest,
Silently paced about, and as she went,
In pale contented sort of discontent,
Mission'd her viewless servants to enrich°
The fretted splendour of each nook and niche.
Between the tree-stems, marbled plain at first,
Came jasper pannels; then, anon, there burst
Forth creeping imagery of slighter trees, 140
And with the larger wove in small intricacies.
Approving all, she faded at self-will,
And shut the chamber up, close, hush'd and still,
Complete and ready for the revels rude,
When dreadful guests would come to spoil her solitude.

The day appear'd, and all the gossip rout.
O senseless Lycius! Madman! wherefore flout
The silent-blessing fate, warm cloister'd hours,
And show to common eyes these secret bowers?
The herd approach'd; each guest, with busy brain, 150
Arriving at the portal, gaz'd amain,
And enter'd marveling: for they knew the street,
Remember'd it from childhood all complete
Without a gap, yet ne'er before had seen
That royal porch, that high-built fair demesne;
So in they hurried all, maz'd, curious and keen:

Save one, who look'd thereon with eye severe,
And with calm-planted steps walk'd in austere;
'Twas Apollonius: something too he laugh'd,
As though some knotty problem, that had daft° 160
His patient thought, had now begun to thaw,
And solve and melt:—'twas just as he foresaw.

He met within the murmurous vestibule
His young disciple. ''Tis no common rule,
Lycius,' said he, 'for uninvited guest
To force himself upon you, and infest
With an unbidden presence the bright throng
Of younger friends; yet must I do this wrong,
And you forgive me.' Lycius blush'd, and led
The old man through the inner doors broad-spread; 170
With reconciling words and courteous mien
Turning into sweet milk the sophist's spleen.

Of wealthy lustre was the banquet-room,
Fill'd with pervading brilliance and perfume:
Before each lucid pannel fuming stood
A censer fed with myrrh and spiced wood,
Each by a sacred tripod held aloft,
Whose slender feet wide-swerv'd upon the soft
Wool-woofed carpets: fifty wreaths of smoke
From fifty censers their light voyage took 180
To the high roof, still mimick'd as they rose°
Along the mirror'd walls by twin-clouds odorous.
Twelve sphered tables, by silk seats insphered,
High as the level of a man's breast rear'd
On libbard's paws, upheld the heavy gold
Of cups and goblets, and the store thrice told
Of Ceres' horn, and, in huge vessels, wine
Come from the gloomy tun with merry shine.
Thus loaded with a feast the tables stood,
Each shrining in the midst the image of a God. 190

When in an antichamber every guest
Had felt the cold full sponge to pleasure press'd,
By minist'ring slaves, upon his hands and feet,
And fragrant oils with ceremony meet

Pour'd on his hair, they all mov'd to the feast
In white robes, and themselves in order placed
Around the silken couches, wondering
Whence all this mighty cost and blaze of wealth could spring.

Soft went the music the soft air along,°
While fluent Greek a vowel'd undersong 200
Kept up among the guests, discoursing low
At first, for scarcely was the wine at flow;
But when the happy vintage touch'd their brains,
Louder they talk, and louder come the strains
Of powerful instruments:—the gorgeous dyes,
The space, the splendour of the draperies,
The roof of awful richness, nectarous cheer,
Beautiful slaves, and Lamia's self, appear,
Now, when the wine has done its rosy deed,
And every soul from human trammels freed, 210
No more so strange; for merry wine, sweet wine,
Will make Elysian shades not too fair, too divine.

Soon was God Bacchus at meridian height;
Flush'd were their cheeks, and bright eyes double bright:
Garlands of every green, and every scent
From vales deflower'd, or forest-trees branch-rent,
In baskets of bright osier'd gold were brought
High as the handles heap'd, to suit the thought
Of every guest; that each, as he did please,
Might fancy-fit his brows, silk-pillow'd at his ease. 220

What wreath for Lamia? What for Lycius?
What for the sage, old Apollonius?
Upon her aching forehead be there hung
The leaves of willow and of adder's tongue;
And for the youth, quick, let us strip for him
The thyrsus, that his watching eyes may swim°
Into forgetfulness; and, for the sage,
Let spear-grass and the spiteful thistle wage
War on his temples. Do not all charms fly°
At the mere touch of cold philosophy? 230
There was an awful rainbow once in heaven:
We know her woof, her texture; she is given

In the dull catalogue of common things.
Philosophy will clip an Angel's wings,
Conquer all mysteries by rule and line,
Empty the haunted air, and gnomed mine—°
Unweave a rainbow, as it erewhile made
The tender-person'd Lamia melt into a shade.

By her glad Lycius sitting, in chief place,
Scarce saw in all the room another face, 240
Till, checking his love trance, a cup he took
Full brimm'd, and opposite sent forth a look
'Cross the broad table, to beseech a glance
From his old teacher's wrinkled countenance,
And pledge him. The bald-head philosopher
Had fix'd his eye, without a twinkle or stir
Full on the alarmed beauty of the bride,
Brow-beating her fair form, and troubling her sweet pride.
Lycius then press'd her hand, with devout touch,
As pale it lay upon the rosy couch: 250
'Twas icy, and the cold ran through his veins;
Then sudden it grew hot, and all the pains
Of an unnatural heat shot to his heart.
'Lamia, what means this? Wherefore dost thou start?
Know'st thou that man?' Poor Lamia answer'd not.
He gaz'd into her eyes, and not a jot
Own'd they the lovelorn piteous appeal:
More, more he gaz'd: his human senses reel:
Some hungry spell that loveliness absorbs;
There was no recognition in those orbs. 260
'Lamia!' he cried—and no soft-toned reply.
The many heard, and the loud revelry
Grew hush; the stately music no more breathes;
The myrtle sicken'd in a thousand wreaths.
By faint degrees, voice, lute, and pleasure ceased;
A deadly silence step by step increased,
Until it seem'd a horrid presence there,
And not a man but felt the terror in his hair.
'Lamia!' he shriek'd; and nothing but the shriek
With its sad echo did the silence break. 270
'Begone, foul dream!' he cried, gazing again
In the bride's face, where now no azure vein

Wander'd on fair-spaced temples; no soft bloom
Misted the cheek; no passion to illume
The deep-recessed vision:—all was blight;
Lamia, no longer fair, there sat a deadly white.
'Shut, shut those juggling eyes, thou ruthless man!
Turn them aside, wretch! or the righteous ban
Of all the Gods, whose dreadful images
Here represent their shadowy presences, 280
May pierce them on the sudden with the thorn
Of painful blindness; leaving thee forlorn,
In trembling dotage to the feeblest fright
Of conscience, for their long offended might,
For all thine impious proud-heart sophistries,
Unlawful magic, and enticing lies.
Corinthians! look upon that gray-beard wretch!
Mark how, possess'd, his lashless eyelids stretch
Around his demon eyes! Corinthians, see!
My sweet bride withers at their potency.' 290
'Fool!' said the sophist, in an under-tone
Gruff with contempt; which a death-nighing moan
From Lycius answer'd, as heart-struck and lost,
He sank supine beside the aching ghost.
'Fool! Fool!' repeated he, while his eyes still
Relented not, nor mov'd; 'from every ill
Of life have I preserv'd thee to this day,
And shall I see thee made a serpent's prey?'
Then Lamia breath'd death breath; the sophist's eye,
Like a sharp spear, went through her utterly, 300
Keen, cruel, perceant, stinging: she, as well
As her weak hand could any meaning tell,
Motion'd him to be silent; vainly so,
He look'd and look'd again a level—No!
'A Serpent!' echoed he; no sooner said,
Than with a frightful scream she vanished:
And Lycius' arms were empty of delight,
As were his limbs of life, from that same night.
On the high couch he lay!—his friends came round—
Supported him—no pulse, or breath they found, 310
And, in its marriage robe, the heavy body wound.*

*'Philostratus, in his fourth book *de Vita Apollonii*, hath a memorable instance in this kind, which I may not omit, of one Menippus Lycius, a young man

twenty-five years of age, that going betwixt Cenchreas and Corinth, met such a phantasm in the habit of a fair gentlewoman, which taking him by the hand, carried him home to her house, in the suburbs of Corinth, and told him she was a Phœnician by birth, and if he would tarry with her, he should hear her sing and play, and drink such wine as never any drank, and no man should molest him; but she, being fair and lovely, would live and die with him, that was fair and lovely to behold. The young man, a philosopher, otherwise staid and discreet, able to moderate his passions, though not this of love, tarried with her a while to his great content, and at last married her, to whose wedding, amongst other guests, came Apollonius; who, by some probable conjectures, found her out to be a serpent, a lamia; and that all her furniture was, like Tantalus' gold, described by Homer, no substance but mere illusions. When she saw herself descried, she wept, and desired Apollonius to be silent, but he would not be moved, and thereupon she, plate, house, and all that was in it, vanished in an instant: many thousands took notice of this fact, for it was done in the midst of Greece.'

Burton's 'Anatomy of Melancholy.' Part 3. Sect. 2. Memb. 1. Subs. 1.

'Pensive they sit, and roll their languid eyes'

Pensive they sit, and roll their languid eyes
Nibble their tosts, and cool their tea with sighs,
Or else forget the purpose of the night
Forget their tea—forget their appetite.
See with cross'd arms they sit—ah hapless crew
The fire is going out, and no one rings
For coals, and therefore no coals Betty brings.
A Fly is in the milk pot—must he die
Circled by a humane society?°
No no there Mr. Werter takes his spoon° 10
Inverts it—dips the handle and lo, soon
The little struggler sav'd from perils dark
Across the teaboard draws a long wet mark.
Romeo! Arise! take Snuffers by the handle°
There's a large Cauliflower in each candle.°
A winding-sheet—Ah me! I must away°
To No. 7 just beyond the Circus gay.
'Alas my friend! your Coat sits very well:
Where may your Taylor live?' 'I may not tell—

O pardon me—I'm absent now and then. 20
Where *might* my Taylor live?—I say again
I cannot tell. Let me no more be teas'd—
He lives in Wapping *might* live where he pleas'd."°

To Autumn

1

Season of mists and mellow fruitfulness,
Close bosom-friend of the maturing sun;
Conspiring with him how to load and bless
With fruit the vines that round the thatch-eves run;
To bend with apples the moss'd cottage-trees,
And fill all fruit with ripeness to the core;
To swell the gourd, and plump the hazel shells
With a sweet kernel; to set budding more,
And still more, later flowers for the bees,
Until they think warm days will never cease, 10
For Summer has o'er-brimm'd their clammy cells.

2

Who hath not seen thee oft amid thy store?
Sometimes whoever seeks abroad may find
Thee sitting careless on a granary floor,
Thy hair soft-lifted by the winnowing wind;
Or on a half-reap'd furrow sound asleep,
Drows'd with the fume of poppies while thy hook
Spares the next swath and all its twined flowers:
And sometimes like a gleaner thou dost keep
Steady thy laden head across a brook; 20
Or by a cyder-press, with patient look,
Thou watchest the last oozings hours by hours.

3

Where are the songs of Spring? Ay, where are they?
Think not of them, thou hast thy music too,—
While barred clouds bloom the soft-dying day,°
And touch the stubble-plains with rosy hue;°

Then in a wailful choir the small gnats mourn
  Among the river sallows, borne aloft
    Or sinking as the light wind lives or dies;
And full-grown lambs loud bleat from hilly bourn; 30
  Hedge-crickets sing; and now with treble soft
  The red-breast whistles from a garden-croft;
    And gathering swallows twitter in the skies.

'Bright Star, would I were stedfast as thou art'

Bright Star, would I were stedfast as thou art—°
  Not in lone splendor hung aloft the night,
And watching, with eternal lids apart,
  Like nature's patient, sleepless Eremite,
The moving waters at their priestlike task
  Of pure ablution round earth's human shores,
Or gazing on the new soft-fallen masque
  Of snow upon the mountains and the moors—
No—yet still stedfast, still unchangeable
  Pillow'd upon my fair love's ripening breast, 10
To feel for ever its soft swell and fall,
  Awake for ever in a sweet unrest,
Still, still to hear her tender-taken breath,
And so live ever—or else swoon to death—

On Coaches

from 'THE JEALOUSIES'

I

Eban, untempted by the Pastry-Cooks,
(Of Pastry he got store within the Palace)
With hasty steps, wrapp'd cloak, and solemn looks,
Incognito upon his errand sallies,
His smelling-bottle ready for the allies;

He pass'd the Hurdy-gurdies with disdain,
Vowing he'd have them sent on board the gallies;
Just as he made his vow, it 'gan to rain,
Therefore he call'd a coach, and bade it drive amain.

2

'I'll pull the string,' said he, and further said,° 10
'Polluted Jarvey! Ah, thou filthy hack!°
Whose springs of life are all dried up and dead,
Whose linsey-wolsey lining hangs all slack,°
Whose rug is straw, whose wholeness is a crack:
And evermore thy steps go clatter-clitter;
Whose glass once up can never be got back,
Who prov'st, with jolting arguments and bitter,
That 'tis of vile no-use to travel in a litter.

3

'Thou inconvenience! thou hungry crop°
For all corn! thou snail-creeper to and fro, 20
Who while thou goest ever seem'st to stop,
And fiddle-faddle standest while you go;
I' the morning, freighted with a weight of woe,
Unto some Lazar-house thou journiest,
And in the evening tak'st a double row
Of dowdies, for some dance or party drest,
Besides the goods meanwhile thou movest east and west.

4

'By thy ungallant bearing and sad mien,
An inch appears the utmost thou couldst budge;
Yet at the slightest nod, or hint, or sign, 30
Round to the curb-stone patient does thou trudge,
School'd in a beckon, learned in a nudge,
A dull-eyed Argus watching for a fare;
Quiet and plodding thou dost bear no grudge
To whisking Tilburies, or Phaetons rare,°
Curricles, or Mail-coaches, swift beyond compare.'

5

Philosophising thus, he pull'd the check,
And bade the Coachman wheel to such a street,
Who turning much his body, more his neck,
Louted full low, and hoarsely did him greet.° 40

'The day is gone, and all its sweets are gone'

The day is gone, and all its sweets are gone!
Sweet voice, sweet lips, soft hand, and softer breast,
Warm breath, light whisper, tender semi-tone,
Bright eyes, accomplish'd shape, and lang'rous waist!
Faded the flower and all its budded charms,
Faded the sight of beauty from my eyes,
Faded the shape of beauty from my arms,
Faded the voice, warmth, whiteness, paradise,
Vanish'd unseasonably at shut of eve,
When the dusk Holiday—or Holinight— 10
Of fragrant curtain'd Love begins to weave
The woof of darkness, thick, for his delight;
But, as I've read Love's Missal through to-day,
He'll let me sleep, seeing I fast and pray.

'What can I do to drive away'

What can I do to drive away
Remembrance from my eyes? for they have seen,
Aye, an hour ago, my brilliant Queen!
Touch has a memory. O say, love, say,
What can I do to kill it and be free
In my old liberty?
When every fair one that I saw was fair,
Enough to catch me in but half a snare,
Not keep me there:

When, howe'er poor or particolour'd things, 10
My muse had wings,
And ever ready was to take her course
Whither I bent her force,
Unintellectual, yet divine to me;—
Divine, I say!—What sea-bird o'er the sea
Is a philosopher the while he goes
Winging along where the great water throes?

How shall I do
To get anew
Those moulted feathers, and so mount once more 20
Above, above
The reach of fluttering Love,
And make him cower lowly while I soar?
Shall I gulp wine? No, that is vulgarism.
A heresy and schism,
Foisted into the canon law of love;—
No,—wine is only sweet to happy men;
More dismal cares
Seize on me unawares,—
Where shall I learn to get my peace again? 30
To banish thoughts of that most hateful land,
Dungeoner of my friends, that wicked strand
Where they were wreck'd and live a wrecked life;°
That monstrous region, whose dull rivers pour,
Ever from their sordid urns unto the shore,
Unown'd of any weedy-haired gods;
Whose winds, all zephyrless, hold scourging rods,
Iced in the great lakes, to afflict mankind;
Whose rank-grown forests, frosted, black, and blind,
Would fright a Dryad; whose harsh herbaged meads 40
Make lean and lank the starv'd ox while he feeds;
There flowers have no scent, birds no sweet song,°
And great unerring Nature once seems wrong.

O, for some sunny spell
To dissipate the shadows of this hell!
Say they are gone,—with the new dawning light
Steps forth my lady bright!

O, let me once more rest
My soul upon that dazzling breast!
Let once again these aching arms be placed, 50
The tender gaolers of thy waist!
And let me feel that warm breath here and there
To spread a rapture in my very hair,—
O, the sweetness of the pain!
Give me those lips again!
Enough! Enough! it is enough for me
To dream of thee!

'I cry your mercy—pity—love!—aye, love'

I cry your mercy—pity—love!—aye, love,
Merciful love that tantalises not,
One-thoughted, never wand'ring, guileless love,
Unmask'd, and being seen—without a blot!
O, let me have thee whole,—all,—all—be mine!
That shape, that fairness, that sweet minor zest
Of love, your kiss, those hands, those eyes divine,
That warm, white, lucent, million-pleasured breast,—
Yourself—your soul—in pity give me all,
Withold no atom's atom or I die, 10
Or living on perhaps, your wretched thrall,
Forget, in the midst of idle misery,
Life's purposes,—the palate of my mind
Losing its gust, and my ambition blind.

To Fanny

Physician Nature! let my spirit blood!°
O ease my heart of verse and let me rest;
Throw me upon thy Tripod, till the flood
Of stifling numbers ebbs from my full breast.
A Theme! a Theme! Great Nature! give a theme;
Let me begin my dream.
I come—I see Thee, as Thou standest there,
Beckon me out into the wintry air.°

Ah! dearest Love, sweet home of all my fears
  And hopes and joys and panting miseries,— 10
To-night, if I may guess, thy beauty wears
    A smile of such delight,
    As brilliant and as bright,
As when with ravished, aching, vassal eyes,
    Lost in a soft amaze,
    I gaze, I gaze!

Who now, with greedy looks, eats up my feast?
  What stare outfaces now my silver moon!
Ah! keep that hand unravished at the least;
    Let, let the amorous burn— 20
    But, prithee, do not turn
  The current of your Heart from me so soon:
    O save, in charity,
    The quickest pulse for me.

Save it for me, sweet love! though music breathe
  Voluptuous visions into the warm air,
Though swimming through the dance's dangerous wreath;°
    Be like an April day,
    Smiling and cold and gay,
  A temperate lily, temperate as fair; 30
    Then, Heaven! there will be
    A warmer June for me.

Why this—you'll say—my Fanny!—is not true;
  Put your soft hand upon your snowy side,
Where the heart beats: confess—'tis nothing new—
    Must not a woman be
    A feather on the sea,
  Swayed to and fro by every wind and tide?
    Of as uncertain speed
    As blow-ball from the mead?° 40

I know it—and to know it is despair
  To one who loves you as I love sweet Fanny,
Whose heart goes fluttering for you every where,
    Nor when away you roam,
    Dare keep its wretched home:

Love, Love alone, has pains severe and many;
Then, Loveliest! keep me free,
From torturing jealousy.

Ah! if you prize my subdued soul above
The poor, the fading, brief, pride of an hour: 50
Let none profane my Holy See of Love,
Or with a rude hand break
The sacramental cake:
Let none else touch the just new-budded flower;
If not—may my eyes close
Love on their last repose!

'This living hand, now warm and capable'

This living hand, now warm and capable°
Of earnest grasping, would, if it were cold
And in the icy silence of the tomb,
So haunt thy days and chill thy dreaming nights
That thou would wish thine own heart dry of blood
So in my veins red life might stream again,
And thou be conscience-calm'd—see here it is
I hold it towards you—

'In after time a Sage of mickle lore'

In after time a Sage of mickle lore,
Yclep'd Typographus, the Giant took°
And did refit his limbs as heretofore,
And made him read in many a learned book,
And into many a lively legend look;
Thereby in goodly themes so training him,
That all his brutishness he quite forsook,
When meeting Artegall and Talus grim,
The one he struck stone blind, the other's eyes wox dim.

'Whenne Alexandre the Conqueroure'

Whenne Alexandre the Conqueroure was wayfayringe in y^e londe of Inde, there mette hym a damoselle of marveillouse beautie slepynge uponne the herbys and flourys. He colde ne loke uponne her withouten grete plesance, and he was welle nighe loste in wondrement. Her forme was everyche whytte lyke ye fayrest carvynge of Quene Cythere, onlie thatte y^t was swellyd and blushyd wyth warmthe and lyffe wythalle.

Her forhed was as whytte as ys the snowe whyche ye talle hed of a Norwegian pyne stelythe from ye northerne wynde. One of her fayre hondes was yplaced thereonne, and thus whytte wyth whytte was ymyngld as ye gode Arthure saythe, lyke whytest lylys yspredde on whyttest snowe; and her bryght eyne whenne she them oped, sparklyd lyke Hesperus through an evenynge cloude.

They were yclosed yn slepe, save that two slauntynge raies shotte to her mouthe, and were theyre bathyd yn swetenesse, as whenne by chaunce ye moone fyndeth a banke of violettes and droppethe thereonne ye silverie dewe.

The authoure was goynge onne withouthen descrybynge ye ladye's breste, whenne lo, a genyus appearyd—'Cuthberte,' sayeth he, 'an thou canst not descrybe ye ladye's breste, and fynde a simile thereunto, I forbyde thee to proceede yn thy romaunt.' Thys, I kennd fulle welle, far surpassyd my feble powres, and forthwythe I was fayne to droppe my quille.

Keats's Marginalia to the Shakespeare Folio

A MIDSUMMER NIGHT'S DREAM

These are the forgeries of iealousie,
And never since the middle Summers spring
Met we on hil, in dale, forrest, or mead,
By paued fountaine, or by rushie brooke,
Or in the beached margent of the sea,
To dance our ringlets to the whistling Winde,
But with thy braules thou hast disturb'd our sport.

[II. i. 81–7]

There is something exquisitely rich and luxurious in Titania's saying 'since the middle summer's spring' as if Bowers were not exuberant and covert enough for fairy sports untill their second sprouting—which is surely the most bounteous overwhelming of all Nature's goodnesses. She steps forth benignly in the spring and her conduct is so gracious that by degrees all things are becoming happy under her wings and nestle against her bosom; she feels this Love & gratitude too much to remain selfsame, and unable to contain herself buds forth the overflowings of her heart about the middle summer—O Shakespeare thy ways are but just searchable! The thing is a piece of profound verdure.

TROYLUS AND CRESSIDA

> I have (as when the Sunne doth light a-scorne)
> Buried this sigh, in wrinkle of a smile:
>
> [I. i. 39–40]

I have not read this copy much and yet have had time to find many faults—however 'tis certain that the Commentators have contrived to twist many beautiful passages into common places as they have done with respect to 'a scorn' which they have hocus pocus'd in 'a storm'° thereby destroying the depth of the simile—taking away all the surrounding Atmosphere of Imagery and leaving a bare and unapt picture. Now however beautiful a Comparison may be for a bare aptness—Shakespeare is seldom guilty of one—he could not be content to the 'sun lighting a Storm,' but he gives us Apollo in the act of drawing back his head and forcing a smile upon the world—'the Sun doth light as scorn.'

> PANDARUS —But to prooue to you that Hellen loues him, she came and puts me her white hand to his clouen chin.
> CRESSIDA — —Iuno haue mercy, how came it clauen? [I. ii. 128–31]

A most delicate touch—Juno being the Goddess of Childbirth.

> Sith euery action that hath gone before,
> Whereof we haue Record, Triall did draw
> Bias and thwart, not answering the ayme:
> And that vnbodied figure of the thought
> That gaue't surmised shape.
>
> [I. iii. 13–17]

The Genius of Shakespeare was an inate universality—wherefore he had the utmost atchievement of human intellect prostrate beneath his indolent and kingly gaze—He could do easily Man's utmost. His plans of tasks to come were not of this world—if what he purposed to do here after would not in his own Idea 'answer the aim' how tremendous must have been his Conception of Ultimates.

> Blunt wedges rive hard knots: the seeded Pride
> That hath to this maturity blowne up
> In ranke Achilles, must or now be cropt,
> Or shedding breed a Nursery of like euil
> To ouer-bulke vs all.
>
> [I. iii. 316–20]

'Blowne up' &c. One's very breath while leaning over these Pages is held for fear of blowing this line away—as easily as the gentlest breeze Robs dandelions of their fleecy Crowns.°

> Sweet, rouse your selfe; and the weake wanton Cupid
> Shall from your necke unloose his amorous fould,
> And like a dew drop from the Lyons mane,
> Be shooke to ayrie ayre.°
>
> [III. iii. 223–6]

Where fore should this ayrie be left out?

KING LEAR

GON. —You see how full of changes his age is, the obseruation we haue made of it hath been little. He alwaies lou'd our Sister most; and with what poor iudgment he hath now cast her off, appeares too grossely.

REG. 'Tis the infirmity of his age; yet he hath euer but slenderly knowne himselfe.

GON. The best and soundest of his time hath bin but rash, then must we looke from his age, to receiue not alone the imperfections of long ingraffed condition, but therewithall the vnruly way-wardnesse that infirm and cholericke yeares bring with them.

REG. Such vnconstant starts are we like to haue from him, as this of Kent's banishment. [I. i. 291–303]

How finely is the brief of Lear's character sketched in this conference—from this point does Shakespeare spur him out to the mighty grapple—'the seeded pride that hath to this maturity blowne up' Shakespeare doth scatter abroad on the winds of Passion, where the germs take buoyant root in stormy Air, suck lightning sap, and become voiced dragons—self-will and pride and wrath are taken at a rebound by his giant hand and mounted to the Clouds—there to remain and thunder evermore—

... though she's as like this, as a Crabbe's like an Apple, ... [I. v. 15–16]

'Thy fifty yet doth double five and twenty.'

REG. —Was he not companion with the riotous Knights
That tended vpon my Father?

GLO. —I know not Madam, 'tis too bad, too bad.

BAST. —Yes Madam, he was of that consort. [II. i. 96–8]

This bye-writing is more marvellous than the whole ripped up contents of Pernambuca—or any buca whatever—on the earth or in the waters under the earth—

Keats's Marginalia to Paradise Lost

[on the title-page] The Genius of Milton, more particularly in respect to its span in immensity, calculated him, by a sort of birthright, for such an 'argument' as the paradise lost—he had an exquisite passion for what is properly, in the sense of ease and pleasure poetical Luxury—and with that it appears to me he would fain have been content if he could so doing have preserved his self respect and feel of duty perform'd—but there was working in him as it were that same sort of thing as operates in the great world to the end of a Prophecy's being accomplish'd: therefore he devoted himself rather to the Ardours than the pleasures of Song, solacing himself at intervals with cups of old wine—and those are with some exceptions the finest parts of the Poem. With some exceptions—for the spirit of mounting and adventure can never be unfruitful or unrewarded—had he not broken through the clouds which envellope so deliciously the Elysian fields of Verse and committed himself to the Extreme we never should have seen Satan as described—

'But his face
Deep Scars of thunder had entrench'd,' &c.°

[above 'The Argument'] There is a greatness which the Paradise Lost possesses over every other Poem—*the Magnitude of Contrast* and that is softened by the contrast being ungrotesque to a degree—Heaven moves on like music throughout—Hell is also peopled with angels—it also moves on like music not grating and harsh but like a grand accompaniment in the Base to Heaven.

[above Book I] There is always a great charm in the openings of great Poems, more particularly where the action begins—that of Dante's Hell—of Hamlet. the first step must be heroic and full of power and nothing can be more impressive and shaded than the commencement of the action here '*Round he throws his baleful eyes*'

> But his doom
> Reserv'd him to more wrath; for now the thought
> Both of lost happiness and lasting pain
> Torments him; round he throws his baleful eyes,
> That witness'd huge affliction and dismay
> Mix'd with obdurate pride and stedfast hate:
> At once, as far as Angels' ken, he views
> The dismal situation waste and wild;
> A dungeon horrible on all sides round
> As one great furnace flamed, yet from those flames
> No light, but rather darkness visible
> Serv'd only to discover sights of woe,
> Regions of sorrow, doleful shades, where peace
> And rest can never dwell; hope never comes
> That comes to all; but torture without end
> Still urges, and a fiery deluge, fed
> With ever-burning sulphur unconsum'd:
> Such place eternal Justice had prepared
> For those rebellious, here their prison ordain'd
> In utter darkness, and their portion set
> As far removed from God and light of Heaven,
> As from the centre thrice to the utmost pole.
> Oh how unlike the place from whence they fell!
>
> [i. 53–75]

One of the most mysterious of semi-speculations is, one would suppose, that of one Mind's imagining into another. Things may be described by a Man's self in parts so as to make a grand whole which that Man himself would scarcely inform to its excess. A Poet can seldom have justice done to his imagination—for men are as distinct in their conceptions of material shadowings as they are in matters of spiritual understanding—it can scarcely be conceived how Milton's Blindness might pervade° the magnitude of his conceptions as a bat in a large gothic vault—

> He call'd so loud, that all the hollow deep
> Of Hell resounded. Princes, Potentates,
> Warriors, the flower of Heaven, once yours, now lost,
> If such astonishment as this can seize
> Eternal Spirits; or have ye chosen this place
> After the toil of battle to repose

Your wearied virtue, for the ease you find
To slumber here, as in the vales of Heaven?

[i. 314–21]

There is a cool pleasure in the very sound of vale—The english word is of the happiest chance. Milton has put vales in heaven and hell with the very utter affection and yearning of a great Poet—It is a sort of delphic Abstraction, a beautiful thing made more beautiful by being reflected and put in a Mist—the next mention of Vale is one of the most pathetic in the whole range of Poetry.

Others more mild
retreated in a silent Valley &c.°

How much of the charm is in the Valley!—

but he, his wonted pride
Soon recollecting, with high words, that bore
Semblance of worth not substance, gently raised
Their fainting courage, and dispell'd their fears.
Then straight commands that at the warlike sound
Of trumpets loud and clarions be uprear'd
His mighty standard: that proud honour claim'd
Azazel as his right, a Cherub tall;
Who forthwith from the glittering staff unfurl'd
The imperial ensign, which full high advanced
Shone like a meteor streaming to the wind,
With gems and golden lustre rich emblazed,
Seraphic arms and trophies; all the while
Sonorous metal blowing martial sounds:
At which the universal host up-sent
A shout, that tore Hell's concave, and beyond
Frighted the reign of Chaos and old Night.
All in a moment through the gloom were seen
Ten thousand banners rise into the air
With orient colours waving: with them rose
A forest huge of spears, and thronging helms
Appear'd, and serried shields in thick array
Of depth immeasurable: anon they move
In perfect phalanx to the Dorian mood
Of flutes and soft recorders; such as raised
To height of noblest temper heroes old
Arming to battle, and instead of rage
Deliberate valour breath'd, firm and unmoved
With dread of death to flight or foul retreat;

Nor wanting power to mitigate and swage
With solemn touches, troubled thoughts, and chase
Anguish, and doubt, and fear, and sorrow, and pain,
From mortal or immortal minds. Thus they
Breathing united force with fixed thought
Moved on in silence to soft pipes, that charm'd
Their painful steps o'er the burnt soil; and now
Advanced in view they stand, a horrid front
Of dreadful length and dazzling arms, in guise
Of warriors old with order'd spear and shield,
Awaiting what command their mighty chief
Had to impose.

[i. 527–67]

The light and shade—the sort of black brightness—the ebon diamonding—the ethiop Immortality—the sorrow, the pain. the sad-sweet Melody—the Phalanges of Spirits so depressed as to be 'uplifted beyond hope'—the short mitigation of Misery—the thousand Melancholies and Magnificences of this Page—leaves no room for anything to be said thereon but: 'so it is—'

his form had not yet lost
All her original brightness, nor appear'd
Less than Arch-Angel ruin'd, and the excess
Of glory obscured; as when the sun new risen
Looks through the horizontal misty air
Shorn of his beams: or from behind the moon
In dim eclipse disastrous twilight sheds
On half the nations, and with fear of change
Perplexes monarchs. Darken'd so, yet shone
Above them all the Arch-Angel:

[i. 591–600]

How noble and collected an indignation against Kings, '*and for fear of change perplexes Monarchs*' &c. His very wishing should have had power to pull that feeble animal Charles from his bloody throne. 'The evil days' had come to him—he hit the new System of things a mighty mental blow—the exertion must have had or is yet to have some sequences—

Anon out of the earth a fabric huge
Rose like an exhalation, with the sound
Of dulcet symphonies and voices sweet,
Built like a temple, where pilasters round

Were set, and Doric pillars overlaid
With golden architrave; nor did there want
Cornice or frieze, with bossy sculptures graven;
The roof was fretted gold. Not Babylon,
Nor great Alcairo such magnificence
Equall'd in all their glories, to inshrine
Belus or Serapis their Gods, or seat
Their kings, when Egypt with Assyria strove
To wealth and luxury. The ascending pile
Stood fix'd her stately height; and straight the doors
Opening their brazen folds, discover, wide
Within, her ample spaces, o'er the smooth
And level pavement: from the arched roof
Pendent by subtle magic many a row
Of starry lamps and blazing cressets, fed
With Naptha and Asphaltus, yielded light
As from a sky.

[i. 710–30]

What creates the intense pleasure of not knowing? A sense of independence, of power from the fancy's creating a world of its own by the sense of probabilities. We have read the Arabian Nights and hear there are thousands of those sort of Romances lost—we imagine after them—but not their realities if we had them nor our fancies in their strength can go further than this Pandemonium—

'Straight the doors opening' &c.
'rose like an exhalation'—

Others, with vast Typhoean rage more fell
Rend up both rocks and hills, and ride the air
In whirlwind; Hell scarce holds the wild uproar.
As when Alcides from Œchalia crown'd
With conquest, felt the envenom'd robe, and tore
Through pain up by the roots Thessalian pines,
And Lichas from the top of Œta threw
Into the Euboic sea. Others more mild,
Retreated in a silent valley, sing
With notes angelical to many a harp
Their own heroic deeds and hapless fall
By doom of battle; and complain that fate
Free Virtue should inthrall to force or chance.
Their song was partial, but the harmony
(What could it less when Spirits immortal sing?)

Suspended Hell, and took with ravishment
The thronging audience. In discourse more sweet
(For eloquence the soul, song charms the sense,)
Others apart sat on a hill retired,
In thoughts more elevate, and reason'd high
On providence, foreknowledge, will, and fate,
Fix'd fate, free will, foreknowledge absolute,
And found no end, in wandering mazes lost.

[ii. 539–61]

Milton is godlike in the sublime pathetic. In Demons, fallen Angels, and Monsters the delicacies of passion, living in and from their immortality, is of the most softening and dissolving nature. It is carried to the utmost here—'Others more mild'—nothing can express the sensation one feels at '*Their song was partial*' &c. Examples of this nature are divine to the utmost in other poets—in Caliban '*Sometimes a thousand twangling instruments*' &c. In Theocritus—Polyphemus—and Homer's Hymn to Pan where Mercury is represented as taking his '*homely fac'd*'° to heaven. There are numerous other instances° in Milton—where Satan's progeny is called his '*daughter dear*,' and where this same Sin, a female, and with a feminine instinct for the showy and martial is in pain least death should sully his bright arms, '*nor vainly hope to be invulnerable in those bright arms*.' Another instance is '*pensive I sat* alone.' We need not mention '*Tears such as Angels weep*.'

but thou
Revisitst not these eyes, that roll in vain
To find thy piercing ray, and find no dawn;
So thick a drop serene hath quench'd their orbs,
Or dim suffusion veiled. Yet not the more
Cease I to wander where the Muses haunt
Clear spring, or shady grove, or sunny hill,
Smit with the love of sacred song; but chief
Thee, Sion, and the flowery brooks beneath
That wash thy hallow'd feet, and warbling flow,
Nightly I visit: nor sometimes forget
Those other two equall'd with me in fate,
So were I equall'd with them in renown,
Blind Thamyris, and blind Maeonides,
And Tiresias and Phineus prophets old.
Then feed on thoughts, that voluntary move
Harmonious numbers; as the wakeful bird
Sings darkling, and in shadiest covert hid

Tunes her nocturnal note. Thus with the year
Seasons return, but not to me returns
Day, or the sweet approach of even or morn,
Or sight of vernal bloom, or summer's rose,
Or flocks, or herds, or human face divine;
But cloud in stead, and ever-during dark
Surrounds me, from the cheerful ways of men
Cut off, and for the book of knowledge fair
Presented with a universal blank
Of nature's works to me expunged and rased,
And wisdom at one entrance quite shut out.
So much the rather thou, celestial Light,
Shine inward, and the mind through all her powers
Irradiate; there plant eyes, all mist from thence
Purge and disperse, that I may see and tell
Of things invisible to mortal sight.
Now had the Almighty Father from above,
From the pure empyrean where he sits
High throned above all height, bent down his eye,
His own works and their works at once to view.

[iii. 22–59]

The management of this Poem is Apollonian.° Satan first '*throws round his baleful eyes*' then awakes his legions, he consults, he sets forward on his voyage—and just as he is getting to the end of it we see the Great God and our first parent, and that same satan all brought in one's vision—we have the invocation to light before we mount to heaven—we breathe more freely—we feel the great Author's consolations coming thick upon him at a time when he complains most—we are getting ripe for diversity—the immediate topic of the Poem opens with a grand Perspective of all concerned.

Thus while God spake, ambrosial Fragrance fill'd
All Heaven, and in the blessed Spirits elect
Sense of new joy ineffable diffused:

[iii. 135–7]

Hell is finer than this—

A violent cross wind from either coast
†Blows them traverse ten thousand leagues awry
Into the devious air;

[iii. 487–9]

†This part of its sound is unaccountably expressive of the description.

Here matter new to gaze the Devil met
Undazzled, far and wide his eye commands,
For sight no obstacle found here, nor shade,
But all sunshine, as when his beams at noon
Culminate from the Equator,

[iii. 613–17]

A Spirits eye . . .

[above Book iv] A friend of mine says this Book has the finest opening of any—the point of time is gigantically critical—the wax is melted, the seal is about to be applied—and Milton breaks out '*O for that warring voice*' &c. There is moreover an opportunity for a Grandeur of Tenderness—the opportunity is not lost. Nothing can be higher—No thing so more than delphic—

Not that fair field
Of Enna, where Proserpine gathering flowers,
Herself a fairer flower, by gloomy Dis
Was gather'd, which cost Ceres all that pain
To seek her through the world;

[iv. 268–72]

There are two specimins of a very extraordinary beauty in the Paradise Lost, they are of a nature as far as I have read, unexampled elsewhere—they are entirely distinct from the brief pathos of Dante—and they are not to be found even in Shakespeare—they are according to the great prerogative of poetry better described in themselves than by a volume the one is in this fol—'*which cost Ceres all that pain*'—the other is that ending '*Nor could the Muse defend her son*'—they appear exclusively Miltonic without the Shadow of another mind ancient or modern—

reluctant flames, the sign
Of wrath awaked;

[vi. 58–9]

'Reluctant'° with its original and modern meaning combined and woven together, with all its Shades of signification has a powerful effect—

Meanwhile the tepid caves, and fens, and shores,
Their brood as numerous hatch, from the egg that soon

Bursting with kindly rupture, forth disclosed
Their callow young, but feather'd soon and fledge
They summ'd their pens, and, soaring the air sublime
With clang despised the ground, under a cloud
In prospect;

[vii. 417–23]

Milton in every instance pursues his imagination to the utmost—he is 'sagacious of his Quarry,' he sees Beauty on the wing, pounces upon it and gorges it to the producing his essential verse. 'So from the root the springs lighten the green stalk,'° &c. But in no instance is this sort of perseverence more exemplified than in what may be called his *Stationing or statuary*: He is not content with simple description, he must station,—thus here, we not only see how the Birds '*with clang despised the ground*,' but we see them '*under a cloud in prospect*' So we see Adam *Fair indeed and tall—under a plantane*—and so we see Satan '*disfigured—on the Assyrian Mount*' This last with all its accompaniments, and keeping in mind the Theory of Spirits' eyes and the simile of Gallilio, has a dramatic vastness and solemnity fit and worthy to hold one amazed in the midst of this Paradise Lost—

Me, of these
Nor skill'd nor studious, higher argument
Remains, sufficient of itself to raise
That name, unless an age too late, or cold
Climate, or years, damp my intended wing
Depress'd; and much they may, if all be mine,
*Not hers who brings it nightly to my ear.

[ix. 41–7]

Had not Shakespeare liv'd?

So saying, through each thicket, dark or dry,
Like a black mist low creeping, he held on
His midnight search, where soonest he might find
The serpent: him fast sleeping soon he found
In labyrinth of many a round self-roll'd,
His head the midst, well stored with subtle wiles.
Not yet in horrid shade or dismal den,
Nor nocent yet; but, on the grassy herb
Fearless, unfear'd he slept: in at his mouth
The Devil enter'd, and his brutal sense,
In heart or head, possessing, soon inspired

With act intelligential; but his sleep
Disturb'd not, waiting close the approach of morn.
[ix. 179–91]

Satan having entered the Serpent, and inform'd his brutal sense—might seem sufficient—but Milton goes on '*but his sleep disturb'd not.*' Whose spirit does not ache° at the smothering and confinement—the unwilling stillness—the '*waiting close*'? Whose head is not dizzy at the prosiable° speculations of satan in the serpent prison—no passage of poetry ever can give a greater pain of suffocations—

Mr. Kean

'In our unimaginative days,'—*Habeas Corpus'd*° as we are, out of all wonder, uncertainty and fear;—in these fireside, delicate, gilded days,—these days of sickly safety and comfort, we feel very grateful to Mr. Kean for giving us some excitement by his old passion in one of the old plays. He is a relict of romance;—a Posthumous ray of chivalry, and always seems just arrived from the camp of Charlemagne. In Richard he is his sword's dear cousin; in Hamlet his footing is germain to the platform. In Macbeth his eye laughs siege to scorn; in Othello he is welcome to Cyprus. In Timon he is of the palace—of Athens—of the woods, and is worthy to sleep in a grave 'which once a day with its embossed froth, the turbulent surge doth cover.'° For all these was he greeted with enthusiasm on his re-appearance in Richard; for all these, his sickness will ever be a public misfortune. His return was full of power. He is not the man to 'bate a jot.' On Thursday evening, he acted *Luke* in *Riches*,° as far as the stage will admit, to perfection. In the hypocritical self-possession, in the caution, and afterwards the pride, cruelty and avarice, Luke appears to us a man incapable of imagining to the extreme hienousness of crimes. To him, they are mere magic-lantern horrors. He is at no trouble to deaden his conscience.

Mr. Kean's two characters of this week, comprising as they do, the utmost of quiet and turbulence, invite us to say a few words on his acting in general. We have done this before, but we do it again without remorse. Amid his numerous excellencies, the one which at this moment most weighs upon us, is the elegance, gracefulness and music of elocution. A melodious passage in poetry is full of pleasures both sensual and spiritual. The spiritual is felt when the very letters and points of charactered language show like the hieroglyphics of beauty:—the mysterious signs of an immortal freemasonry! 'A thing to dream of, not to tell!'° The sensual life of verse springs warm from the lips of Kean, and to one learned in Shakespearean hieroglyphics°,—learned in the spiritual portion of those lines to which Kean adds a sensual grandeur: his tongue must seem to have robbed 'the hybla bees, and left them honeyless.'° There is an indescribable gusto in his voice, by which we feel that the utterer is thinking of the past and future, while speaking of the instant. When he says in Othello, 'put up your bright swords, for the dew will rust them,'° we feel that his throat had commanded where swords were as thick as reeds. From eternal risk, he speaks as though his body were unassailable. Again, his exclamation of 'blood, blood, blood!' is direful and slaughterous to the deepest degree, the very words appear stained and gory. His nature hangs over them, making a prophetic repast. His voice is loosed on them, like the wild dog on the savage relics of an eastern conflict; and we can distinctly hear it 'gorging, and growling o'er carcase and limb.'° In Richard, 'Be stirring with the lark to-morrow, gentle Norfolk!'° comes from him, as through the morning atmosphere, towards which he yearns. We could cite a volume of such immortal scraps, and dote upon them with our remarks; but as an end must come, we will content ourselves with a single syllable. It is in those lines of impatience to the night who 'like a foul and ugly witch, doth limp so tediously away.'° Surely this intense power of anatomizing the passion of every syllable—of taking to himself the wings of verse, is the means by which he becomes a storm with such fiery decision; and by which, with a still deeper charm, he 'does his spiriting gently.'° Other actors are continually thinking of their sum-total effect throughout a play. Kean delivers himself up to the instant feeling, without the shadow of a thought about any thing else. He feels his being as deeply as Wordsworth, or any of our intellectual monopolists. From all his comrades he stands alone, reminding us of him, whom Dante has so finely described in his Hell:

'And sole apart retir'd, the Soldan fierce!'°

Although so many times he has lost the Battle of Bosworth Field, we can easily conceive him really expectant of victory, and a different termination of the piece. Yet we are as moths about a candle, in speaking of this great man. 'Great, let us call him, for he conquered us!'° We will say no more. Kean! Kean! have a carefulness of thy health, an in-nursed respect for thy own genius, a pity for us in these cold and enfeebling times! Cheer us a little in the failure of our days! for romance lives but in books. The goblin is driven from the heath, and the rainbow is robbed of its mystery!°

Rejected Preface to Endymion

Endymion
a Romance
by John Keats

The stretched metre of an antique song—
Shakespeare's Sonnets

Inscribed
with every feeling of pride and regret,
and with 'a bowed mind,'
To the memory of
The most english of Poets except Shakespeare,
Thomas Chatterton—

Preface

In a great nation, the work of an individual is of so little importance; his pleadings and excuses are so uninteresting; his 'way of life' such a nothing that a preface seems a sort of impertinent bow to Strangers who care nothing about it.

A preface however should be down in so many words; and such a one that by an eye glance over the type, the Reader may catch an idea of an Author's modesty, and non opinion of himself—which I sincerely hope may be seen in the few lines I have to write, notwithstanding certain proverbs of many ages' old which men find a great pleasure in receiving for gospel.

About a twelvemonth since, I published a little book of verses; it was read by some dozen of my friends who lik'd it; and some dozen whom I was unaquainted with, who did not. Now when a dozen human beings, are at words with another dozen, it becomes a matter of anxiety to side with one's friends—more especially when excited thereto by a great love of Poetry.

I fought under disadvantages. Before I began I had no inward feel of being able to finish; and as I proceeded my steps were all uncertain. So this Poem must rather be considerd as an endeavour than a thing accomplish'd; a poor prologue to what, if I live, I humbly hope to do. In duty to the Public I should have kept it back for a year or two, knowing it to be so faulty: but I really cannot do so:—by repetition my favorite Passages sound vapid in my ears, and I would rather redeem myself with a new Poem should this one be found of any interest.

I have to apologise to the lovers of Simplicity for touching the spell of Loveliness that hung about Endymion: if any of my lines plead for me with such people I shall be proud.

It has been too much the fashion of late to consider men biggotted and adicted to every word that may chance to escape their lips: now I here declare that I have not any particular affection for any particular phrase, word or letter in the whole affair. I have written to please myself and in hopes to please others, and for a love of fame; if I neither please myself, nor others not get fame, of what consequence is Phraseology?

I would fain escape the bickerings that all Works not exactly in chime bring upon their begetters:—but this is not fair to expect, there must be conversation of some sort and to object shows a Man's consequence.— In case of a London drizzle or a scotch Mist, the following quotation from Marston may perhaps stead me as an umbrella for an hour or so: 'let it be the Curtesy of my peruser rather to pity my self hindering labours than to malice me'°

One word more.—for we cannot help seeing our own affairs in every point of view.—Should any one call my dedication to Chatterton affected I answer as followeth:

'Were I dead Sir I should like a Book dedicated to me'—

Teignmouth March 19th 1818—

Letter to C. C. Clarke, 9 October 1816

Wednesday Octr 9th—

My dear Sir,

The busy time has just gone by, and I can now devote any time you may mention to the pleasure of seeing Mr Hunt—'t will be an Era in my existence—I am anxious too to see the Author of the Sonnet to the Sun,° for it is no mean gratification to become acquainted with Men who in their admiration of Poetry do not jumble together Shakspeare and Darwin°—I have coppied out a sheet or two of Verses which I composed some time ago,

worst
and find so much to blame in them that the best part will go into the fire—those to G. Mathew I will suffer to meet the eye of Mr H. notwithstanding that the Muse is so frequently mentioned—I here sinned in the face of Heaven even while remembering what, I think, Horace says, 'never presume to make a God appear but for an Action worthy of a God'.° From a few Words of yours when last I saw you, I have no doubt but that you have something in your Portfolio which I should by rights see—I will put you in Mind of it—Although the Borough is a beastly place in dirt, turnings and windings; yet No 8 Dean Street is not difficult to find; and if you would run the Gauntlet over London Bridge, take the first turning to the left and then the first to the right and moreover knock at my door which is nearly opposite a Meeting,° you would do one a Charity which as St Paul saith is the father of all the Virtues—At all events let me hear from you soon—I say at all events not excepting the Gout in your fingers—

Yours' sincerely
JOHN KEATS—

Letter to J. H. Reynolds, 17, 18 April 1817

Carisbrooke April 17th

My dear Reynolds,

Ever since I wrote to my Brothers from Southampton I have been in a taking, and at this moment I am about to become settled. for I have unpacked my books, put them into a snug corner—pinned up Haydon—Mary Queen of Scotts, and Milton with his daughters in a row. In the passage I found a head of Shakspeare which I had not before seen—It is most likely the same that George spoke so well of; for I like it extremely—Well—this head I have hung over my Books, just above the three in a row, having first discarded a french Ambassador—Now this alone is a good morning's work—Yesterday I went to Shanklin, which occasioned a great debate in my mind whether I should live there or at Carisbrooke. Shanklin is a most beautiful place—sloping wood and meadow ground reaches round the Chine, which is a cleft between the Cliffs of the depth of nearly 300 feet at least. This cleft is filled with trees & bushes in the narrow part; and as it widens becomes bare, if it were not for primroses on one side, which spread to the very verge of the Sea, and some fishermen's huts on the other, perched midway in the Ballustrades of beautiful green Hedges along their steps down to the sands.—But the sea, Jack, the sea°—the little waterfall—then the white cliff—then St Catherine's Hill—'the sheep in the meadows, the cows in the corn.'—Then, why are you at Carisbrooke? say you—Because, in the first place, I shod be at twice the Expense and three times the inconvenience—next that from here I can see your continent—from a little hill close by, the whole north Angle of the Isle of Wight, with the water between us. In the 3d place, I see Carisbrooke Castle from my window, and have found several delightful wood-alleys, and copses, and quick freshes°—As for Primroses—the Island ought to be called Primrose Island: that is, if the nation of Cowslips agree thereto, of which there are diverse Clans just beginning to lift up their heads and if an how the Rain holds whereby that is Birds eyes abate—another reason of my fixing is that I am more in reach of the places around me—I intend to walk over the island east—West—North South—I have not seen many specimens of Ruins—I dont think however I shall ever see one to surpass Carisbrooke Castle. The trench is o'ergrown with the smoothest turf, and the walls with ivy—The Keep within side is one Bower of ivy—a Colony of Jackdaws have been there many years—I dare say I have seen many a descendant of some old cawer who peeped through the Bars at Charles the first, when he was there in Confinement.° On the road from Cowes to Newport I saw some extensive Barracks which disgusted me extremely with Government for placing such a Nest of Debauchery in so beautiful a place—I asked a man on the Coach about this—and he said that the people had been spoiled—In the room where I slept at Newport I found this on the Window 'O Isle spoilt by the Mil*a*tary'—I must in honesty however confess that I did not feel very sorry at the idea of the Women being a little profligate—The Wind is in a sulky fit, and I feel that it would be no bad thing to be the favorite of some Fairy, who would give one the power of seeing how our Friends got on, at a Distance—I should like, of all Loves, a sketch of you and Tom and George in ink which Haydon will do if you tell him how I want them—From want of regular rest, I have been rather *narvus*—and the passage in Lear—'Do you not hear the Sea?"°—has haunted me intensely.

[*A draft of the Sonnet 'On the Sea' follows*]

April 18th

Will you have the goodness to do this? Borrow a Botanical Dictionary—turn to the words Laurel and Prunus show the explanations to your sisters and Mrs Dilk and without more ado let them send me the Cups Basket and Books they trifled and put off and off while I was in Town—ask them what they can say for themselves—ask Mrs Dilk wherefore she does so distress me—Let me know how Jane has her health—the Weather is unfavorable for her—Tell George and Tom to write.—I'll tell you what—On the 23rd was Shakespeare born—now If I should receive a Letter from you and another from my Brothers on that day 'twould be a parlous good thing—Whenever you write say a Word or two on some Passage in Shakespeare that may have come rather new to you; which must be continually happening, notwithstandg that we read the same Play forty times—for instance, the following, from the Tempest, never struck me so forcibly as at present,

'Urchins
Shall, for that vast of Night that they may work,
All exercise on thee—'

How can I help bringing to your mind the Line—

In the dark backward and abysm of time—°

I find that I cannot exist without poetry—without eternal poetry—half the day will not do—the whole of it—I began with a little, but habit has made me a Leviathan—I had become all in a Tremble from not having written any thing of late—the Sonnet over leaf did me some good. I slept the better last night for it—this Morning, however, I am nearly as bad again—Just now I opened Spencer, and the first Lines I saw were these.—

'The noble Heart that harbors vertuous thought,
And is with Child of glorious great intent,
Can never rest, until it forth have brought
Th' eternal Brood of Glory excellent—'°

Let me know particularly about Haydon; ask him to write to me about Hunt, if it be only ten lines—I hope all is well—I shall forthwith begin my Endymion, which I hope I shall have got some way into by the time you come, when we will read our verses in a delightful place I have set my heart upon near the Castle—Give my Love to your Sisters severally—To George and Tom—Remember me to Rice Mr & Mrs Dilk and all we know.——

Your sincere Friend
JOHN KEATS.

Direct J. Keats Mrs Cook's new Village
Carisbrooke

Letter to Leigh Hunt, 10 May 1817

Margate May 10th

My dear Hunt,

The little Gentleman that sometimes lurks in a gossips bowl ought to have come in very likeness of a *coasted* crab° and choaked me outright for not having answered your Letter ere this—however you must not suppose that I was in Town to receive it; no, it followed me to the isle of Wight and I got it just as I was going to pack up for Margate, for reasons which you anon shall hear. On arriving at this treeless affair I wrote to my Brother George to request C. C .C. to do the thing you wot of respecting Rimini;° and George tells me he has undertaken it with great Pleasure; so I hope there has been an understanding between you for many Proofs—C. C. C. is well acquainted with Bensley. Now why did you not send the key of your Cupboard which I know was full of Papers? We would have lock'd them all in a trunk together with those you told me to destroy; which indeed I did not do for fear of demolishing Receipts. There not being a more unpleasant thing in the world (saving a thousand and one others) than to pay a Bill twice. Mind you—Old Wood's a very Varmant—sharded in Covetousness—And now I am upon a horrid subject—what a horrid one you were upon last Sunday° and well you handled it. The last Examiner was Battering Ram against Christianity—Blasphemy—Tertullian—Erasmus—Sr Philip Sidney. And then the dreadful Petzelians and their expiation by Blood—and do Christians shudder at the same thing in a Newspaper which the attribute to their God in its most aggravated form? What is to be the end of this?—I must mention Hazlitt's Southey—O that he had left out the grey hairs!—Or that they had been in any other Paper not concluding with such a Thunderclap—that sentence° about making a Page of the feelings of a whole life appears to me like a Whale's back in the Sea of Prose. I ought to have said a word on Shakspeare's Christianity—there are two, which I have not looked over with you, touching the thing: the one for, the other against: That in favor is in Measure for Measure Act. 2. S. 2 Isab. Alas! Alas!

> Why all the Souls that were; were forfeit once
> And he that might the vantage best have took,
> Found out the Remedy—

That against is in Twelfth Night. Act. 3. S 2. Maria—'for there is no Christian, that means to be saved by believing rightly, can ever believe such impossible Passages of grossness!' Before I come to the Nymphs° I must get through all disagreeables—I went to the Isle of Wight—thought so much about Poetry so long together that I could not get to sleep at night—and moreover, I know not how it was, I could not get wholesome food—By this means in a Week or so I became not over capable in my upper Stories, and set off pell mell for Margate, at least 150 Miles—because forsooth I fancied that I should like my old Lodging here, and could contrive to do without Trees. Another thing I was too much in Solitude, and consequently was obliged to be in continual burning of thought as an only resource. However Tom is with me at present and we are very comfortable. We intend though to get among some Trees. How have you got on among them? How are the Nymphs? I suppose they have led you a fine dance—Where are you now—In Judea, Cappadocia, or the Parts of Lybia about Cyrene, Strangers from Heaven, Hues and Prototypes—I wager you have given given several new turns to the old saying 'Now the Maid was fair and pleasant to look on' as well as mad a little variation in 'once upon a time' perhaps too you have rather varied 'thus endeth the first Lesson' I hope you have made a Horse shoe business of—'unsuperfluous lift' 'faint Bowers' and fibrous roots. I vow that I have been down in the Mouth lately at this Work. These last two day however I have felt more confident—I have asked myself so often why I should be a Poet more than other Men,—seeing how great a thing it is,—how great things are to be gained by it—What a thing to be in the Mouth of Fame—that at last the Idea has grown so monstrously beyond my seeming Power of attainment that the other day I nearly consented with myself to drop into a Phæton—yet 't is a disgrace to fail even in a huge attempt, and at this moment I drive the thought from me. I began my Poem° about a Fortnight since and have done some every day except travelling ones—Perhaps I may have done a good deal for the time but it appears such a Pin's Point to me that I will not coppy any out—When I consider that so many of these Pin points go to form a Bodkin point (God send I end not my Life with a bare Bodkin, in its modern sense) and that it requires a thousand bodkins to make a Spear bright enough to throw any light to posterity—I see that nothing but continual uphill Journeying? Now is there any thing more unpleasant (it may come among the thousand and one) than to be so journeying and miss the Goal at last—But I intend to whistle all these cogitations into the Sea where I hope they will breed Storms violent enough to block up all exit from Russia. Does Shelley° go on telling strange Stories of the Death of kings? Tell him there are strange Stories of the death of Poets—some have died before they were conceived 'how do you make that out Master Vellum' Does Mrs S— cut Bread and Butter as neatly as ever? Tell her to procure some fatal Scissars and cut the thread of Life of all to be disappointed Poets. Does Mrs Hunt tear linen in half as straight as ever? Tell her to tear from the book of Life all blank Leaves. Remember me to them all—to Miss Kent° and the little ones all—

Your sincere friend
JOHN KEATS alias JUNKETS—

You shall know where we move—

Letter to B. R. Haydon, 10, 11 May 1817

Margate Saturday Eve

My dear Haydon,

> Let Fame, which all hunt after in their Lives,
> Live register'd upon our brazen tombs,
> And so grace us in the disgrace of death:
> When spite of cormorant devouring time
> The endeavour of this present breath may buy
> That Honor which shall bate his Scythe's keen edge
> And make us heirs of all eternity.°

To think that I have no right to couple myself with you in this speech would be death to me so I have e'en written it—and I pray God that our brazen Tombs be nigh neighbors. It cannot be long first the endeavor of this present breath will soon be over—and yet it is as well to breathe freely during our sojourn—it is as well if you have not been teased with that Money affair—that bill-pestilence. However I must think that difficulties nerve the Spirit of a Man—they make our Prime Objects a Refuge as well as a Passion. The Trumpet of Fame is as a tower of Strength the ambitious bloweth it and is safe—I suppose by your telling me not to give way to forebodings George has mentioned to you what I have lately said in my Letters to him—truth is I have been in such a state of Mind as to read over my Lines and hate them. I am 'one that gathers Samphire dreadful trade'° the Cliff of Poesy Towers above me—yet when, Tom who meets with some of Pope's Homer in Plutarch's Lives reads some of those to me they seem like Mice to mine. I read and write about eight hours a day. There is an old saying 'well begun is half done'—'t is a bad one. I would use instead—'Not begun at all 'till half done' so according to that I have not begun my Poem and consequently (a priori) can say nothing about it. Thank God! I do begin arduously where I leave off, notwithstanding occasional depressions: and I hope for the support of a High Power while I clime this little eminence and especially in my Years of more momentous Labor. I remember your saying that you had notions of a good Genius presiding over you—I have of late had the same thought. for things which I do half at Random are afterwards confirmed by my judgment in a dozen features of Propriety—Is it too daring to Fancy Shakspeare this Presider? When in the Isle of Wight I met with a Shakspeare° in the Passage of the House at which I lodged—it comes nearer to my idea of him than any I have seen—I was but there a Week yet the old Woman made me take it with me though I went off in a hurry—Do you not think this is ominous of good? I am glad you say every Man of great Views is at times tormented as I am—

Sunday Aft. This Morning I received a letter from George by which it appears that Money Troubles are to follow us up for some time to come perhaps for always—these vexations are a great hindrance to one—they are not like Envy and detraction stimulants to further exertion as being immediately relative and reflected on at the same time with the prime object—but rather like a nettle leaf or two in your bed. So now I revoke my Promise of finishing my Poem by the Autumn which I should have done had I gone on as I have done—but I cannot write while my spirit is fevered in a contrary direction and I am now sure of having plenty of it this Summer—At this moment I am in no enviable Situation—I feel that I am not in a Mood to write any to day; and it appears that the loss of it is the beginning of all sorts of irregularities. I am extremely glad that a time must come when every thing will leave not a wrack behind. You tell me never to despair—I wish it was as easy for me to observe the saying—truth is I have a horrid Morbidity of Temperament which has shown itself at intervals—it is I have no doubt the greatest Enemy and stumbling block I have to fear—I may even say that it is likely to be the cause of my disappointment. How ever every ill has its share of good—this very bane would at any time enable me to look with an obstinate eye on the Devil Himself—ay to be as proud of being the lowest of the human race as Alfred could be in being of the highest.° I feel confident I should have been a rebel Angel had the opportunity been mine. I am very sure that you do love me as your own Brother—I have seen it in your continual anxiety for me—and I assure you that your wellfare and fame is and will be a chief pleasure to me all my Life.

I know no one but you who can be fully sensible of the turmoil and anxiety, the sacrifice of all what is called comfort the readiness to Measure time by what is done and to die in 6 hours could plans be brought to conclusions.—the looking upon the Sun the Moon the Stars, the Earth and its contents as materials to form greater things—that is to say ethereal things—but here I am talking like a Madman greater things that our Creator himself made!! I wrote to Hunt yesterday—scarcely know what I said in it—I could not talk about Poetry in the way I should have liked for I was not in humor with either his or mine. His self delusions are very lamentable they have inticed him into a Situation which I should be less eager after than that of a galley Slave—what you observe thereon is very true must be in time. Perhaps it is a self delusion to say so—but I think I could not be deceived in the Manner that Hunt is—may I die tomorrow if I am to be. There is no greater Sin after the 7 deadly than to flatter oneself into an idea of being a great Poet—or one of those beings who are privileged to wear out their Lives in the pursuit of Honor—how comfortable a feel it is that such a Crime must bring its heavy Penalty? That if one be a Selfdeluder accounts will be balanced? I am glad you are hard at Work—'t will now soon be done—I long to see Wordsworth's as well as to have mine in:° but I would rather not show my face in Town till the end of the Year—if that will be time enough—if not I shall be disappointed if you do not write for me even when you think best—I never quite despair and I read Shakspeare—indeed I shall I think never read any other Book much—Now this might lead me into a long Confab but I desist. I am very near Agreeing with Hazlit that Shakspeare is enough for us—By the by what a tremendous Southean Article his last was—I wish he had left out 'grey hairs' It was very gratifying to meet your remarks of the Manuscript°—I was reading Anthony and Cleopat when I got the Paper and there are several Passages applicable to the events you commentate. You say that he arrived by degrees, and not by any single Struggle to the height of his ambition—and that his Life had been as common in particulars as other Mens—Shakspeare makes Enobarb say—Where's Antony Eros—He's walking in the garden—thus: *and spurns the rush that lies* before him, cries fool, Lepidus! In the same scene we find: 'let determined things to destiny hold unbewailed their way'. Dolabella says of Antony's Messenger

'An argument that he is pluck'd when hither
He sends so poor a pinion of his wing'—Then again,
Eno—'I see Men's Judgments are

A parcel of their fortunes; and things outward
Do draw the inward quality after them,
To suffer all alike'—The following applies well to Bertram°
'Yet he that can endure
To follow with allegience a fallen Lord,
Does conquer him that did his Master conquer,
And earns a place i' the story"°

But how differently does Buonap bear his fate from Antony! 'T is good too that the Duke of Wellington has a good Word or so in the Examiner. A Man ought to have the Fame he deserves—and I begin to think that detracting from him as well as from Wordsworth is the same thing. I wish he had a little more taste—and did not in that respect 'deal in Lieutenantry' You should have heard from me before this—but in the first place I did not like to do so before I had got a little way in the 1st Book and in the next as G. told me you were going to write I delayed till I had heard from you—Give my Respects the next time you write to the North° and also to John Hunt°—Remember me to Reynolds and tell him to write, Ay, and when you sent Westward tell your Sister that I mentioned her in this—So now in the Name of Shakespeare Raphael and all our Saints I commend you to the care of heaven!

Your everlasting friend
JOHN KEATS

Letter to Jane and Mariane Reynolds, 14 September 1817

Oxford Sunday Evening

My dear Jane,

You are such a literal translator that I shall some day amuse myself with looking over some foreign sentences and imagining how you would render them into english. This is an age for typical curiosities and I would advise you, as a good speculation, to study Hebrew and astonish the world with a figurative version in our native tongue. 'The Mountains skipping like Rams and the little Hills like Lambs"° you will leave as far behind as the Hare did the Tortoise. It must be so or you would never have thought that I really meant you would like to pro and con about those Honey combs—no, I had no such idea, or if I had 'twoud be only to tease you a little for Love. So now let me put down in black and white briefly my sentiments thereon. Imprimis—I sincerely believe that Imogen is the finest Creature; and that I should have been disappointed at hearing you prefer Juliet. Item Yet I feel such a yearning towards Juliet and that I would rather follow her into Pandemonium than Imogen into Paradize—heartily wishing myself a Romeo to be worthy of her and to hear the Devils quote the old Proverb—'Birds of a feather flock together'—Amen. Now let us turn to the sea shore. Believe me, my dear Jane it is a great Happiness to me that you are in this finest part of the year, winning a little enjoyment from the hard World—in truth the great Elements we know of are no mean Comforters—the open Sky sits upon our senses like a Sapphire Crown—the Air is our Robe of State—the Earth is our throne and the Sea a mighty Minstrell playing before it—able like David's Harp to charm the evil spirit from such Creatures as I am—able like Ariel's to make such a one as you forget almost the tempest-cares of life. I have found in the Ocean's Musick—varying (though selfsame) more than the passion of Timotheus,° an enjoyment not to be put into words and 'though inland far I be'° I now hear the voice most audibly while pleasing myself in the Idea of your Sensations. Marianne is getting well apace and if you have a few trees and a little Harvesting about you I'll snap my fingers in Lucifer's eye. I hope you bathe too—if you do not I earnestly recommend it—bathe thrice a Week and let us have no more sitting up next Winter. Which is the best of Shakespeare's Plays?—I mean in what mood and with what accompenament do you like the Sea best? It is very fine in the morning when the Sun

> 'opening on Neptune with fair blessed beams
> Turns into yellow gold his salt sea streams'°

and superb when

> The sun from meridian height,
> Illumines the depth of the sea—
> and the fishes beginning to sweat
> Cry damn it how hot we shall be°

and gorgeous when the fair planet hastens—'to his home within the western foam'° but dont you think there is something extremely fine after sunset when there are a few white Clouds about and a few stars blinking—when the the waters are ebbing and the Horison a Mystery? This state of things has been so fulfelling to me that I am anxious to hear whether it is a favorite with you—so when you and Marrianne club your Letter to me put in a word or to about it—I am glad that you will spend a little time with the Dilks—tell Dilk that it would be perhaps as well if he left a Pheasant or Partridge alive here and there to keep up a supply of Game for next season—tell him to reign in if possible all the Nimrod of his disposition, he being a mighty hunter befor the Lord°—of the Manor. Tell him to shoot far and not have at the poor devils in a furrow—when they are flying he may fire and nobody will be the wiser. Give my sincerest Respects to Mrs Dilk saying that I have not forgiven myself for not having got her the little Box of Medicine I promised for her after dinner flushings. and that had I remained at Hampstead I would have made precious havoc with her house and furniture—drawn a great harrow over her garden—poisoned Boxer—eaten her Cloathes pegs,—fried her Cabbages fricaceed (how is it spelt?) her radishes—ragouted her Onions—belaboured her beet root—outstripped her Scarlet Runners—parlez vou'd with her french Beans—devoured her Mignon or Mignonette—metamorphosed her Bell handles—splintered her looking glasses—bullock'd at her cups and saucers—agonized her decanters—put old Philips to pickle in the Brine tub—disorganized her Piano—dislocated her Candlesticks—emptied her wine bins in a fit of despair—turned out her maid to Grass and Astonished Brown—whose Letter to her on these events I would rather see than the original copy of the Book of Genesis. Should you see Mr. W. D.° remember me° to him—and to little Robinson Crusoe—and to Mr. Snook—Poor Bailey scarcely ever well has gone to bed very so so, and pleased that I am writing to you. To your Brother John (whom henceforth I shall consider as mine) and to you my dear firends Marrianne and Jane I shall ever feel grateful for having made known to me so real a fellow as Bailey. He delights me in the Selfish and (please god) the disenterrested part of my disposition. If the old Poets have any pleasure in looking down at the Enjoyers of their Works, their eyes must bend with double satisfaction upon him—I sit as at a feast when he is over them and pray that if after my death any of my Labours should be worth saving, they may have as 'honest a Chronicler'° as Bailey. Out of this his Enthusiasm in his own pursuit and for all good things is of an exalted kind—worthy a more healthful frame and an untorn Spirit. He must have happy years to come—he shall not die by God—A Letter from John the other day was a chief Happiness to me. I made a little mistake when just now I talked of being far inland: how can that be when Endymion and I are at the bottom of the Sea? Whence I hope to bring him in safety before you leave the Sea Side and if I can so contrive it you shall be greeted by him on the Sands and he shall tell you all his adventures: which at having finished he shall thus proceed. 'My dear Ladies, favorites of my gentle Mistress, how ever my friend Keats may have teazed and vexed you believe me he loves you not the less—for instance I am deep in his favor and yet he has been hawling me through the Earth and Sea with unrelenting Perseverence—I know for all this that he is mightily fond of me, by his contriving me all sorts of pleasures—nor is this the least fair Ladies—this one of meeting you on desart Shore and greeting you in his Name—He sends you moreover this little scroll'—

'My dear Girls,

I send you per favor of Endymion the assurance of my esteem of you and my utmost wishes for you Health and Pleasure—being ever—Your affectionate

Brother. JOHN KEATS—

George and Tom are well—

(Remberences to little Britain)

Letter to Benjamin Bailey, 8 October 1817

Hampstead Oct[r] Wednesday

My dear Bailey,

After a tolerable journey I went from Coach to Coach to as far as Hampstead where I found my Brothers—the next Morning finding myself tolerably well I went to Lambs Conduit Street° and delivered your Parcel—Jane and Marianne were greatly improved Marianne especially she has no unhealthy plumpness in the face—but she comes me healthy and angular to the Chin—I did not see John I was extremely sorry to hear that poor Rice after having had capital Health During his tour, was very ill. I dare say you have heard from him. From No. 19 I went to Hunt's and Haydon's who live now neighbours.° Shelley was there—I know nothing about any thing in this part of the world—every Body seems at Loggerheads. There's Hunt infatuated—theres Haydon's Picture in statu quo. There's Hunt walks up and down his painting room criticising every head most unmercifully—There's Horace Smith tired of Hunt. 'The web of our Life is of mingled Yarn'° Haydon having removed entirely from Marlborough street Crips must direct his Letter to Lisson Grove North Paddington. Yesterday Morning while I was at Brown's in came Reynolds—he was pretty bobbish we had a pleasant day—but he would walk home at night that cursed cold distance. M[rs] Bentley's children are making a horrid row—whereby I regret I cannot be transported to your Room to write to you. I am quite disgusted with literary Men and will never know another except Wordsworth—no not even Byron—Here is an instance of the friendships of such—Haydon and Hunt have known each other many years—now they live pour ainsi dire jealous Neighbours. Haydon says to me Keats dont show your Lines to Hunt on any account or he will have done half for you—so it appears Hunt wishes it to be thought. When he met Reynolds in the Theatre John told him that I was getting on to the completion of 4000 Lines. Ah! says Hunt, had it not been for me they would have been 7000! If he will say this to Reynolds what would he to other People? Haydon received a Letter a little while back on this subject from some Lady—which contains a caution to me through him on this subject—Now is not all this a most paultry thing to think about? You may see the whole of the case by the following extract from a Letter I wrote to George in the spring 'As to what you say about my being a poet, I can retun no answer but by saying that the high Idea I have of poetical fame makes me think I see it towering to high above me. At any rate I have no right to talk until Endymion is finished—it will be a test, a trial of my Powers of Imagination and chiefly of my invention which is a rare thing indeed—by which I must make 4000 Lines of one bare circumstance and fill them with Poetry; and when I consider that this is a great task, and that when done it will take me but a dozen paces towards the Temple of Fame—it makes me say—God forbid that I should be without such a task! I have heard Hunt say and may be asked—why endeavour after a long Poem? To which I should answer—Do not the Lovers of Poetry like to have a little Region to wander in where they may pick and choose, and in which the images are so numerous that many are forgotten and found new in a second Reading: which may be food for a Week's stroll in the Summer? Do not they like this better than what they can read through before Mrs Williams comes down stairs? a Morning work at most. Besides a long Poem is a test of Invention which I take to be the Polar Star of Poetry, as Fancy is the Sails, and Imagination the Rudder. Did our great Poets ever write short Pieces? I mean in the shape of Tales—This same invention seems indeed of late Years to have been forgotten as a Poetical excellence. But enough of this, I put on no Laurels till I shall have finished Endymion, and I hope Apollo is not angered at my having made a Mockery at him at Hunts'° You see Bailey how independant my writing has been—Hunts dissuasion was of no avail—I refused to visit Shelley, that I might have my own unfetterd scope—and after all I shall have the Reputation of Hunt's elevé—His corrections and amputations will by the knowing ones be trased in the Poem—This is to be sure the vexation of a day—nor would I say so many Words about it to any but those whom I know to have my wellfare and Reputation at Heart—Haydon promised to give directions for those Casts and you may expect to see them soon—with as many Letters You will soon hear the dinning of Bells—never mind you and Gleg will defy the foul fiend—But do not sacrifice your health to Books do take it kindly and not so voraciously. I am certain if you are your own Physician your stomach will resume its proper strength and then what great Benefits will follow. My Sister wrote a Letter to me which I think must be at y[e] post office Ax Will° to see. My Brothers kindest remembrances to you—we are going to dine at Brown's where I have some hopes of meeting Reynolds. The little Mercury° I have taken has corrected the Poison and improved my Health—though I feel from my employment that I shall never be again secure in Robustness—would that you were as well as

your sincere friend & brother
JOHN KEATS

The Dilks are expected to day—

Letter to Benjamin Bailey, 3 November 1817

Monday—Hampstead

My dear Bailey,

Before I received your Letter I had heard of your disappointment—an unlook'd for piece of villainy.° I am glad to hear there was an hindrance to your speaking your Mind to the Bishop: for all may go straight yet—as to being ordained—but the disgust consequent cannot pass away in a hurry—it must be shocking to find in a sacred Profession such barefaced oppression and impertinence—The Stations and Grandeurs of the World have taken it into their heads that they cannot commit themselves towards and inferior in rank—but is not the impertinence from one above to one below more wretchedly mean than from the low to the high? There is something so nauseous in self-willed yawning impudence in the shape of conscience—it sinks the Bishop of Lincoln into a smashed frog putrifying: that a rebel against common decency should escape the Pillory! That a mitre should cover a Man guilty of the most coxcombical, tyranical and indolent impertinence! I repeat this word for the offence appears to me most especially *impertinent*—and a very serious return would be the Rod—Yet doth he sit in his Palace. Such is this World—and we live—you have surely in a continual struggle against the suffocation of accidents—we must bear (and my Spleen is mad at the thought thereof) the Proud Mans Contumely—O for a recourse somewhat human independant of the great Consolations of Religion and undepraved Sensations. of the Beautiful. the poetical in all things—O for a Remedy against such wrongs within the pale of the World! Should not those things be pure enjoyment should they stand the chance of being contaminated by being called in as antagonists to Bishops? Would not earthly things do? By Heavens my dear Bailey, I know you have a spice of what I mean—you can set me and have set it in all the rubs that may befal me you have I know a sort of Pride which would kick the Devil on the Jaw Bone and make him drunk with the kick—There is nothing so balmy to a soul imbittered as yours must be, as Pride—When we look at the Heavens we cannot be proud—but shall stocks and stones be impertinent and say it does not become us to kick them? At this Moment I take your hand let us walk up yon Mountain of common sense now if our Pride be vainglorious such a support woud fail—yet you feel firm footing—now look beneath at that parcel of knaves and fools. Many a Mitre is moving among them. I cannot express how I despise the Man who would wrong or be impertinent to you—The thought that we are mortal makes us groan I will speak of something else or my Spleen will get higher and higher—and I am not a bearer of the two egded Sword. I hope you will recieve an answer from Haydon soon—if not Pride! Pride! Pride! I have received no more subscription°—but shall soon have a full health Liberty and leisure to give a good part of my time to him—I will certainly be in time for him—We have promised him one year let that have elapsed and then do as we think proper. If I did not know how impossible it is, I should say 'do not at this time of disappointments disturb yourself about others'—There has been a flaming attack upon Hunt in the Endinburgh Magazine°—I never read any thing so virulent—accusing him of the greatest Crimes—depreciating his Wife his Poetry—his Habits—his company, his Conversation—These Philipics are to come out in Numbers—calld 'the Cockney School of Poetry' There has been but one Number published—that on Hunt to which they have prefixed a Motto from one Cornelius Webb Poetaster—who unfortunately was of our Party occasionally at Hampstead and took it into his head to write the following—something about—'we'll talk on Wordsworth Byron—a theme we never tire on' and so forth till he comes to Hunt and Keats. In the Motto they have put Hunt and Keats in large Letters—I have no doubt that the second Number was intended for me: but have hopes of its non appearance from the following advertisement in last Sunday's Examiner. 'To Z. The writer of the Article signed Z in Blackwood's Edinburgh magazine for October 1817 is invited to send his address to the printer of the Examiner, in order that Justice may be executed of the proper person' I dont mind the thing much—but if he should go to such lenghts with me as he has done with Hunt I must infalibly call him to an account—if he be a human being and appears in Squares and Theatres where we might possibly meet—I dont relish his abuse Yesterday Rice and I were at Reynolds—John was to be articled tomrow I suppose by this time it is done. Jane was much better—At one time or other I will do you a Pleasure and the Poets a little Justice—but it ought to be in a Poem of greater moment than Endymion—I will do it some day—I have seen two Letters of a little Story Reynolds is writing—I wish he would keep at it—Here is the song I enclosed to Jane if you can make it out in this cross wise writing.

[*There follows a draft of* Endymion *iv. 146–81*]

O that I had Orpheus lute—and was able to charm away all your Griefs and Cares—but all my power is a Mite—amid all you troubles I shall ever be—

your sincere and affectionate friend
JOHN KEATS

My brothers remembrances to you
Give my respects to Gleig and Whitehead

Letter to Benjamin Bailey, 22 November 1817

My dear Bailey,

I will get over the first part of this (*un*said) Letter as soon as possible for it relates to the affair of poor Crips—To a Man of your nature, such a Letter as Haydon's must have been extremely cutting—What occasions the greater part of the World's Quarrels? simply this, two Minds meet and do not understand each other time enough to praevent any shock or surprise at the conduct of either party—As soon as I had known Haydon three days I had got enough of his character not to have been surpised at such a Letter as he has hurt you with. Nor when I knew it was it a principle with me to drop his acquaintance although with you it would have been an imperious feeling. I wish you knew all that I think about Genius and the Heart—and yet I think you are thoroughly acquainted with my innermost breast in that respect or you could not have known me even thus long and still hold me worthy to be your dear friend. In passing however I must say of one thing that has pressed upon me lately and encreased my Humility and capability of submission and that is this truth—Men of Genius are great as certain ethereal Chemicals operating on the Mass of neutral intellect—by they have not any individuality, any determined Character. I would call the top and head of those who have a proper self Men of Power—°

But I am running my head into a Subject which I am certain I could not do justice to under five years study and 3 vols octavo—and moreover long to be talking about the Imagination—so my dear Bailey do not think of this unpleasant affair if possible—do not—I defy any ham to come of it—I defy—I'll shall write to Crips this Week and requet him to tell me all his goings on from time to time by Letter wherever I may be—it will all go on well—so dont because you have suddenly discover'd a Coldness in Haydon suffer yourself to be teased. Do not my dear fellow. O I wish I was as certain of the end of all your troubles as that of your momentary start about the authenticity of the Imagination. I am certain of nothing but of the holiness of the Heart's affections and the truth of Imagination—What the imagination seizes as Beauty must be truth—whether it existed before or not—for I have the same Idea of all our Passions as of Love they are all in their sublime, creative of essential Beauty—In a Word, you may know my favorite Speculation by my first Book° and the little song I sent in my last—which is a representation from the fancy of the probable mode of operating in these Matters—The Imagination may be compared to Adam's dream—he awoke and found it truth.° I am the more zealous in this affair, because I have never yet been able to perceive how any thing can be known for truth by consequitive reasoning—and yet it must be—Can it be that even the greatest Philosopher ever arrived at his goal without putting aside numerous objections—However it may be, O for a Life of Sensations rather than of Thoughts! It is 'a Vision in the form of Youth' a Shadow of reality to come—and this consideration has further convinced me for it has come as auxiliary to another favorite Speculation of mine, that we shall enjoy ourselves here after by having what we called happiness on Earth repeated in a finer tone and so repeated—And yet such a fate can only befall those who delight in sensation rather than hunger as you do after Truth—Adam's dream will do here and seems to be a conviction that Imagination and its empyreal reflection is the same as human Life and its spiritual repetition. But as I was saying—the simple imaginative Mind may have its rewards in the repetion of its own silent Working coming continually on the spirit with a fine suddenness—to compare great things with small—have you never by being surprised with an old Melody—in a delicious place—by a delicious voice, felt over again your very speculations and surmises at the time it first operated on your soul—do you not remember forming to yourself the singer's face more beautiful that it was possible and yet with the elevation of the Moment you did not think so—even then you were mounted on the Wings of Imagination so high—that the Prototype must be here after—that delicious face you will see—What a time! I am continually running away from the subject—sure this cannot be exactly the case with a complex Mind—one that is imaginative and at the same time careful of its fruits—who would exist partly on sensation partly on thought—to whom it is necessary that years should bring the philosophic Mind°—such an one I consider your's and therefore it is necessary to your eternal Happiness that you not only drink this old Wine of Heaven which I shall call the redigestion of our most ethereal Musings on Earth; but also increase in knowledge and know all things. I am glad to hear you are in a fair Way for Easter—you will soon get through your unpleasant reading and then!—but the world is full of troubles and I have not much reason to think myself pesterd with many—I think Jane or Marianne has a better opinion of me than I deserve—for really and truly I do not think my Brothers illness connected with mine—you know more of the real Cause than they do—nor have I any chance of being rack'd as you have been—you perhaps at one time thought there was such a thing as Worldly Happiness to be arrived at, at certain periods of time marked out—you have of necessity from your disposition been thus led away—I scarcely remember counting upon any Happiness—I look not for it if it be not in the present hour—nothing startles me beyond the Moment. The setting sun will always set me to rights—or if a Sparrow come before my Window I take part in its existince and pick about the Gravel. The first thing that strikes me on hearing a Misfortune having befalled another is this. 'Well it cannot be helped.—he will have the pleasure of trying the resources of his spirit, and I beg now my dear Bailey that hereafter should you observe any thing cold in me not to put it to the account of heartlessness but abstraction—for I assure you I sometimes feel not the influence of a Passion or Affection during a whole week—and so long this sometimes continues I begin to suspect myself and the genuiness of my feelings at other times—thinking them a few barren Tragedy-tears—My Brother Tom is much improved—he is going to Devonshire—whither I shall follow him—at present I am just arrived at Dorking to change the Scene—change the Air and give me a spur to wind up my Poem, of which there are wanting 500 Lines. I should have been here a day sooner but the Reynoldses persuaded me to spop in Town to meet your friend Christie°—There were Rice and Martin—we talked about Ghosts—I will have some talk with Taylor and let you know—when please God I come down a Christmas—I will find that Examiner if possible. My best regards to Gleig—My Brothers to you and Mrs Bentley

Your affectionate friend
JOHN KEATS

I want to say much more to you—a few hints will set me going
Direct Burford Bridge near dorking

Letter to J. H. Reynolds, 22 November 1817

Saturday

My Dear Reynolds,

There are two things which tease me here—one of them Crips—and the other that I cannot go with Tom into Devonshire—however I hope to do my duty to myself in a week or so; and then Ill try what I can do for my neighbour—now is not this virtuous? on returning to Town—Ill damn all Idleness—indeed, in superabundance of employment, I must not be content to run here and there on little two penny errands—but turn Rakehell i e go a *making* or Bailey will think me just as great a Promise keeper as *he* thinks you—for my self I do not,—and do not remember above one Complaint against you for matter o' that—Bailey writes so abominable a hand, to give his Letter a fair reading requires a little time; so I had not seen when I saw you last, his invitation to Oxford at Christmas—I'll go with you—You know how poorly Rice was—I do not think it was all corporeal—bodily pain was not used to keep him silent. Ill tell you what; he was hurt at what your Sisters said about his joking with your Mother he was, soothly to sain—It will all blow over. God knows, my Dear Reynolds, I should not talk any sorrow to you—you must have enough vexations—so I won't any more. If I ever start a rueful subject in a Letter to you—blow me! Why dont you—Now I was a going to ask a very silly Question neither you nor any body else could answer, under a folio, or at least a Pamphlet—you shall judge—Why dont you, as I do, look unconcerned at what may be called more particularly Heart-vexations? They never surprize me—lord! a man should have the fine point of his soul taken off to become fit for this world—I like this place° very much—There is Hill & Dale and a little River—I went up Box hill this Evening after the Moon—you a' seen the Moon—came down—and wrote some lines. Whenever I am separated from you, and not engaged in a continued Poem—every Letter shall bring you a lyric—but I am too anxious for you to enjoy the whole, to send you a particle. One of the three Books I have with me is Shakespear's Poems: I neer found so many beauties in the sonnets—they seem to be full of fine things said unintentionally—in the intensity of working out conceits—Is this to be borne? Hark ye!

> When lofty trees I see barren of leaves
> Which erst from heat did canopy the herd,
> And Summer's green all girded up in sheaves,
> Borne on the bier with white and bristly beard.°

He has left nothing to say about nothing or any thing: for look at Snails, you know what he says about Snails, you know where he talks about 'cockled snails'—well, in one of these sonnets, he says—the chap slips into—no! I lie! this is in the Venus and Adonis: the Simile brought it to my Mind.

> Audi—As the snail, whose tender horns being hit,
> Shrinks back into his shelly cave with pain,
> And there all smothered up in shade doth sit,
> Long after fearing to put forth again:
> So at his blody view her eyes are fled,
> Into the deep dark Cabins of her head.°

He overwhelms a genuine Lover of Poesy will all manner of abuse, talking about—

> 'a poets rage
> And stretched metre of an antique song'°—

Which by the by will be a capital Motto for my Poem—wont it?—He speaks too of 'Time's antique pen'—and 'aprils first born flowers'—and 'deaths eternal cold'—By the Whim King! I'll give you a Stanza, because it is not material in connection and when I wrote it I wanted you to—give your vote, pro or con.—

[Endymion *iv. 581–90 follows*]

Now I hope I shall not fall off in the winding up,—as the Woman said to the rounce°—I mean up and down. I see there is an advertizement in the chronicle to Poets—he is so overloaded with poems on the late Princess.°—I suppose you do not lack—send me a few—lend me thy hand to laugh a little—send me a little pullet sperm, a few finch eggs—and remember me to each of our Card playing Club—when you die you will all be turned into Dice, and be put in pawn with the Devil—for Cards they crumple up like any King—I mean John in the stage play what pertains Prince Arthur—I rest

Your affectionate friend
JOHN KEATS

Give my love to both houses—hinc atque illinc.°

Letter to George and Tom Keats, 21, ?27 December 1817

Hampstead Sunday
22 December 1818

My dear Brothers

I must crave your pardon for not having written ere this & &° I saw Kean return to the public in Richard III, & finely he did it, & at the request of Reynolds I went to criticise his Luke in Riches°—the critique is in todays champion, which I send you with the Examiner in which you will find very proper lamentation on the obsoletion of christmas Gambols & pastimes: but it was mixed up with so much egotism of that drivelling nature that pleasure is entirely lost. Hone the publisher's trial, you must find very amusing; & as Englishmen very encouraging—his *Not Guilty* is a thing, which not to have been, would have dulled still more Liberty's Emblazoning—Lord Ellenborough has been paid in his own coin—Wooler & Hone have done us an essential service°—I have had two very pleasant evenings with Dilke yesterday & today; & am at this moment just come from him & feel in the humour to go on with this, began in the morning, & from which he came to fetch me. I spent Friday evening with Wells & went the next morning to see *Death on the Pale horse.*° It is a wonderful picture, when West's age is considered;

But there is nothing to be intense upon; no women one feels mad to kiss; no face swelling into reality. the excellence of every Art is its intensity, capable of making all disagreeables evaporate, from their being in close relationship with Beauty & Truth—Examine King Lear & you will find this examplified throughout; but in this picture we have unpleasantness without any momentous depth of speculation excited, in which to bury its repulsiveness—The picture is larger than Christ rejected°—I dined with Haydon the sunday after you left, & had a very pleasant day, I dined too (for I have been out too much lately) with Horace Smith & met his two brothers with Hill & Kingston & one Du Bois, they only served to convince me, how superior humour is to wit in respect to enjoyment—These men say things which make one start, without making one feel, they are all alike; their manners are alike; they all know fashionables; they have a mannerism in their very eating & drinking, in their mere handling a Decanter—They talked of Kean & his low company—Would I were with that company instead of yours said I to myself! I know such like acquaintance will never do for me & yet I am going to Reynolds, on wednesday—Brown & Dilke walked with me & back from the Christmas pantomime. I had not a dispute but a disquisition with Dilke, on various subjects; several things dovetailed in my mind, & at once it struck me, what quality went to form a Man of Achievement especially in Literature & which Shakespeare posessed so enormously—I mean *Negative Capability*, that is when man is capable of being in uncertainties, Mysteries, doubts, without any irritable reaching after fact & reason—Coleridge, for instance, would let go by a fine isolated verisimilitude caught from the Penetralium of mystery, from being incapable of remaining content with half knowledge. This pursued through Volumes would perhaps take us no further than this, that with a great poet the sense of Beauty overcomes every other consideration, or rather obliterates all consideration.

Shelley's poem° is out & there are words about its being objected too, as much as Queen Mab was. Poor Shelley I think he has his Quota of good qualities, in sooth la!! Write soon to your most sincere friend & affectionate Brother

JOHN

Letter to B. R. Haydon, 23 January 1818

Friday 23rd

My dear Haydon,

I have a complete fellow-feeling with you in this business°—so much so that it would be as well to wait for a choice out of *Hyperion*—when that Poem is done there will be a wide range for you—in Endymion I think you may have many bits of the deep and sentimental cast—the nature of *Hyperion* will lead me to treat it in a more naked and grecian Manner—and the march of passion and endeavour will be undeviating—and one great contrast between them will be—that the Hero of the written tale being mortal is led on, like Buonaparte, by circumstance; whereas the Apollo in Hyperion being a fore-seeing God will shape his actions like one. But I am counting &c.

Your proposal pleases me—and, believe me, I would not have my Head in the shop windows from any hand but yours—no by Apelles! I will write Taylor and you shall hear from me

Your's ever JOHN KEATS

To Benjamin Bailey, 23 January 1818

My dear Bailey, Friday Jany 23rd

Twelve days have pass'd since your last reached me—what has gone through the myriads of human Minds since the 12th we talk of the immense number of Books, the Volumes ranged thousands by thousands—but perhaps more goes through the human intelligence in 12 days than ever was written. How has that unfortunate Family° lived through the twelve? One saying of your's I shall never forget—you may not recollect it—it being perhaps said when you were looking on the surface and seeming of Humanity alone, without a thought of the past or the future—or the deeps of good and evil—you were at the moment estranged from speculation and I think you have arguments ready for the Man who would utter it to you—this is a formidable preface for a simple thing—merely you said; '*Why should Woman suffer?*' Aye. Why should she? 'By heavens I'd coin my very Soul and drop my Blood for Drachmas."°! These things are, and he who feels how incompetent the most skyey Knight errantry its to heal this bruised fairness is like a sensitive leaf on the hot hand of thought. Your tearing, my dear friend,

a spiritless and gloomy Letter up to rewrite to me is what I shall never forget—it was to me a real thing. Things have happen'd lately of great Perplexity—You must have heard of them—Reynolds and Haydon retorting and recrimminating—and parting for ever—the same thing has happened between Haydon and Hunt—It is unfortunate—Men should bear with each other—there lives not the Man who may not be cut up, aye hashed to pieces on his weakest side. The best of Men have but a portion of good in them—a kind of spiritual yeast in their frames which creates the ferment of existence—by which a Man is propell'd to act and strive and buffet with Circumstance. The sure way Bailey, is first to know a Man's faults, and then be passive, if after that he insensibly draws you towards him then you have no Power to break the link. Before I felt interested in either Reynolds or Haydon—I was well read in their faults yet knowing them I have been cementing gradually with both—I have an affection for them both for reasons almost opposite—and to both must I of necessity cling—supported always by the hope that when a little time—a few years shall have tried me more fully in their esteem I may be able to bring them together—the time must come because they have both hearts—and they will recollect the best parts of each other when this gust is overblown. I had a Message from you through a Letter to Jane I think about Cripps—there can be no idea of binding° till a sufficient sum is sure for him—and even then the thing should be maturely consider'd by all his helpers. I shall try my luck upon as many fat-purses as I can meet with—Cripps is improving very fast—I have the greater hopes of him because he is so slow in devellopment—a Man of great executing Powers at 20—with a look and a speech almost stupid is sure to do something. I have just look'd through the second side of your Letter—I feel a great content at it. I was at Hunt's the other day, and he surprised me with a real authenticated Lock of *Milton's Hair*. I know you would like what I wrote thereon—so here it is—*as they say of a Sheep in a Nursery* Book
On seeing a Lock of Milton's Hair—

[*A draft of 'Lines on seeing a Lock of Milton's hair' follows*]

This I did at Hunt's at his request—perhaps I should have done something better alone and at home—I have sent my first book to the Press—and this afternoon shall begin preparing the second—my visit to you will be a great spur to quicken the Proceeding—I have not had your Sermon returned—I long to make it the subject of a Letter to you—What do they say at Oxford?
I trust you and Gleig pass much fine time together. Remember me to

him and Whitehead. My Brother Tom is getting stronger but his Spitting of blood continues—I sat down to read King Lear yesterday, and felt the greatness of the thing up to the writing of a Sonnet preparatory thereto—in my next you shall have it There were some miserable reports of Rice's health—I went and lo! Master Jemmy had been to the play the night before and was out at the time—he always comes on his Legs like a Cat—I have seen a good deal of Wordsworth. Hazlitt is lecturing on Poetry at the Surry institution—I shall be there next Tuesday.

Your most affectionate Friend
JOHN KEATS—

Letter to George and Tom Keats, 23, 24 January 1818

Friday, 23^d January 1818

My dear Brothers.
I was thinking what hindered me from writing so long, for I have many things to say to you & know not where to begin. It shall be upon a thing most interesting to you my Poem. Well! I have given the 1st book to Taylor; he seemed more than satisfied with it, & to my surprise proposed publishing it in Quarto if Haydon would make a drawing of some event therein, for a Frontispeice. I called on Haydon, he said he would do anything I liked, but said he would rather paint a finished picture, from it, which he seems eager to do; this in a year or two will be a glorious thing for us; & it will be, for Haydon is struck with the 1st Book. I left Haydon & the next day received a letter from him, proposing to make, as he says, with all his might, a finished chalk sketch of my head, to be engraved in the first style & put at the head of my Poem, saying at the same time he had never done the thing for any human being, & that it must have considerable effect as he will put the name to it—I begin to day to copy my 2nd Book 'thus far into the bowels of the Land'°—You shall hear whether it will be Quarto or non Quarto, picture or non Picture. Leigh Hunt I showed my 1st Book to, he allows it not much merit as a whole; says it is unnatural & made ten objections to it in the mere skimming over. He says the conversation is unnatural & too high-flown for the Brother & Sister. Says it should be simple forgetting do ye mind, that they are both overshadowed by a Supernatural Power, & of force could not speak like Franchesca in the Rimini. He must first prove that Caliban's poetry is unnatural,—This

with me completely over-turns his objections—the fact is he & Shelley are hurt & perhaps justly, at my not having showed them the affair officiously & from several hints I have had they appear much disposed to dissect & anatomize, any trip or slip I may have made. But whose afraid Ay! Tom! demme if I am. I went last tuesday, an hour too late, to Hazlitt's Lecture on poetry, got there just as they were coming out, when all these pounced upon me. Hazlitt, John Hunt & son, Wells, Bewick, all the Landseers, Bob Harris, Rox of the Burrough° Aye & more; the Landseers enquired after you particularly—I know not whether Wordsworth has left town—But sunday I dined with Hazlitt & Haydon, also that I took Haslam with me—I dined with Brown lately. Dilke having taken the Champion, Theatricals was obliged to be in Town. Fanny has returned to Walthamstow—Mr Abbey appeared very glum, the last time I went to see her, & said in an indirect way, that I had no business there—Rice has been ill, but has been mending much lately—I think a little change has taken place in my intellect lately—I cannot bear to be uninterested or unemployed, I, who for so long a time, have been addicted to passiveness—Nothing is finer for the purposes of great productions, than a very gradual ripening of the intellectual powers—As an instance of this—observe—I sat down yesterday to read King Lear once again the thing appeared to demand the prologue of a Sonnet, I wrote it & began to read—(I know you would like to see it)

[*A draft of 'On Sitting Down to Read* King Lear *Once Again' follows*]

So you see I am getting at it, with a sort of determination & strength, though verily I do not feel it at this moment—this is my fourth letter this morning & feel rather tired & my head rather swimming—so I will leave it open till tomorrow's post.——

I am in the habit of taking my papers to Dilkes & copying there; so I chat & proceed at the same time. I have been there at my work this evening, & the walk over the Heath takes off all sleep, so I will even proceed with you—I left off short in my last, just as I began an account of a private theatrical—Well it was of the lowest order, all greasy & oily, insomuch that if they had lived in olden times, when signs were hung over the doors; the only appropriate one for that oily place would have been—a guttered Candle—they played John Bull The Review. & it was to conclude with Bombastes Furioso°—I saw from a Box the 1st Act of John Bull, then I went to Drury & did not return till it was over; when by Wells' interest we got behind the scenes. there was not a yard wide all the way round for actors, scene shifters & interlopers to move in; for

'Note Bene' the Green Room was under the stage & there was I threatened over & over again to be turned out by the oily scene shifters—there did I hear a little painted Trollop own, very candidly, that she had failed in Mary, with a 'damned if she'd play a serious part again, as long as she lived,' & at the same time she was habited as the Quaker in the Review—there was a quarrel & a fat good natured looking girl in soldiers Clothes wished she had only been a man for Tom's sake—One fellow began a song but an unlucky finger-point from the Gallery sent him off like a shot, One chap was dressed to kill for the King in Bombastes. & stood at the edge of the scene in the very sweat of anxiety to show himself, but Alas the thing was not played. the sweetest morsel of the night moreover was, that the musicians began pegging & fagging away at an overture—never did you see faces more in earnest, three times did they play it over, dropping all kinds of correctness & still did not the curtain draw up—Well then they went into a country-dance then into a region they well knew, into their old boonsome Pot-house. & then to see how pompous o' the sudden they turned; how they looked about, & chatted; how they did not care a Damn; was a great treat—I hope I have not tired you by this filling up of the dash in my last,—Constable the Bookseller has offered Reynolds ten gineas a sheet to write for his magazine. it is an Edinburgh one which, Blackwoods started up in opposition to. Hunt said he was nearly sure that the 'Cockney School' was written by Scott,° so you are right Tom!—There are no more little bits of news I can remember at present I remain

My dear Brothers Your very affectionate Brother

JOHN

Letter to John Taylor, 30 January 1818

My dear Taylor, Friday

These Lines, as they now stand, about Happiness have rung in my ears like a 'chime a mending'.°see here,

Behold
Wherein Lies happiness Pœona? fold—

This appears to me the very contrary of blessed. I hope this will appear to you more elegible.

[Endymion *i. 777–81 follows*]

You must indulge me by putting this in for setting aside the badness of the other, such a preface is necessary to the Subject. The whole thing must I think have appeared to you, who are a consequitive Man, as a thing almost of mere words—but I assure you that when I wrote it, it was a regular stepping of the Imagination towards a Truth. My having written that Argument will perhaps be of the greatest Service to me of any thing I ever did—It set before me at once the gradations of Happiness even like a kind of Pleasure Thermometer—and is my first step towards the chief Attempt in the Drama—the playing of different Natures with Joy and Sorrow.

Do me this favor and believe Me, Your sincere friend

JOHN KEATS

I hope your next Work° will be of a more general Interest—I sppose you cogitate a little about it now and then.

Letter to J. H. Reynolds, 3 February 1818

Hampstead Tuesday.

My dear Reynolds,

I thank you for your dish of Filberts—Would I could get a basket of them by way of desert every day for the sum of two pence— Would we were a sort of ethereal Pigs, & turn'd loose to feed upon spiritual Mast° & Acorns—which would be merely being a squirrel & feed upon filberts. for what is a squirrel but an airy pig, or a filbert but a sort of archangelical acorn. About the nuts being worth cracking, all I can say is that where there are a throng of delightful Images ready drawn simplicity is the only thing. the first is the best on account of the first line, and the 'arrow—foil'd of its antler'd food'—and moreover (and this is the only word or two I find fault with, the more because I have had so much reason to shun it as a quicksand) the last has 'tender and true'—We must cut this, and not be rattlesnaked itno any more of the like—It may be said that we ought to read our Contemporaries. that Wordsworth &c should have their due from us. but for the sake of a few fine imaginative or domestic passages, are we to be bullied into a certain Philosophy engendered in the whims of an Egotist—Every man has his speculations, but every man does not brood and peacock over them till he makes a false coinage and deceives himself—Many a man can travel to the very bourne of Heaven, and yet want confidence to put down his half-seeing. Sancho° will invent a Journey heavenward as well as any body.

We hate poetry that has a palpable design upon us—and if we do not agree, seems to put its hand in its breeches pocket. Poetry should be great & unobtrusive, a thing which enters into one's soul, and does not startle it or amaze it with itself but with its subject.—How beautiful are the retired flowers! how would they lose their beauty were they to throng into the highway crying out, 'admire me I am a violet! dote upon me I am a primrose!'° Modern poets differ from the Elizabethans in this. Each of the moderns like an Elector of Hanover governs his petty state, & knows how many straws are swept daily from the Causeways in all his dominions & has a continual itching that all the Housewives should have their coppers well scoured: the antients were Emperors of vast Provinces, they had only heard of the remote ones and scarcely cared to visit them.—I will cut all this—I will have no more of Wordsworth or Hunt in particular—Why should we be of the tribe of Manasseh when we can wander with Esau?° why should we kick against the Pricks, when we can walk on Roses? Why should we be owls, when we can be Eagles? Why be teased with 'nice Eyed wagtails,"° when we have in sight 'the Cherub Contemplation'?°—Why with Wordsworths 'Matthew with a bough of wilding in his hand' when we can have Jacques 'under an oak &c"°—The secret of the Bough of Wilding will run through your head faster than I can write it—Old Matthew spoke to him some years ago on some nothing, & because he happens in an Evening Walk to imagine the figure of the old man—he must stamp it down in black & white, and it is henceforth sacred—I don't mean to deny Wordsworth's grandeur & Hunt's merit, but I mean to say we need not be teazed with grandeur & merit—when we can have them uncontaminated & unobtrusive. Let us have the old Poets, & robin Hood Your letter and its sonnets gave me more pleasure than will the 4th Book of Childe Harold & the whole of any body's life & opinions. In return for your dish of filberts, I have gathered a few Catkins,° I hope they'll look pretty.

To J. H. R. In answer to his Robin Hood Sonnets.

[*'Robin Hood' follows*]

I hope you will like them they are at least written in the Spirit of Outlawry.—Here are the Mermaid lines

[*'Lines on the Mermaid Tavern' follows*]

I will call on you at 4 tomorrow, and we will trudge together for it is not the thing to be a stranger in the Land of Harpsicols.° I hope also to bring you my 2[d] book—In the hope that these Scribblings will be some amusement for you this Evening—I remain copying on the Hill

Y[r] sincere friend and Coscribbler
JOHN KEATS

Letter to J. H. Reynolds, 19 February 1818

My dear Reynolds,

I have an idea that a Man might pass a very pleasant life in this manner—let him on any certain day read a certain Page of full Poesy or distilled Prose and let him wander with it, and muse upon it, and reflect from it, and bring home to it, and prophesy upon it, and dream upon it—untill it becomes stale—but when will it do so? Never—When Man has arrived at a certain ripeness in intellect any one grand and spiritual passage serves him as a starting post towards all 'the two-and thirty Pallaces'° How happy is such a voyage of conception! what delicious diligent Indolence! A doze upon a Sofa does not hinder it, and a napp upon Clover engenders ethereal finger-pointings—the prattle of a child gives it wings, and the converse of middle age a strength to beat them—a strain of musick conducts to 'an odd angle of the Isle'° and when the leaves whisper it puts a 'girdle round the earth'.° Nor will this sparing touch of noble Books be any irreverance to their Writers—for perhaps the honors paid by Man to Man are trifles in comparison to the Benefit done by great Works to the 'Spirit and pulse of good'° by their mere passive existence. Memory should not be called Knowledge—Many have original Minds who do not think it—they are led away by Custom—Now it appears to me that almost any Man may like the Spider spin from his own inwards his own airy Citadel—the points of leaves and twigs on which the Spider begins her work are few and she fills the Air with a beautiful circuiting: man should be content with as few points to tip with the fine Webb of his Soul and weave a tapestry empyrean—full of Symbols for his spiritual eye, of softness for his spiritual touch, of space for his wandering of distinctness for his Luxury—But the Minds of Mortals are so different and bent on such diverse Journeys that it may at first appear impossible for any common taste and fellowship to exist between two or three under these suppositions. It is however quite the contrary. Minds would leave each other in contrary directions, traverse each other in Numberless points, and all last greet each other at the Journeys end—A old Man and a child would talk together and the old Man be led on his Path, and the child left thinking—Man should not dispute or assert but whisper results to his neighbour, and thus by every germ of Spirit sucking the Sap from mould ethereal every human might become great, and Humanity instead of being a wide heath of Furse and Briars with here and there a remote Oak or Pine, would become a grand democracy of Forest Trees. It has been an old Comparison for our urging on—the Bee hive—however it seems to me that we should rather be the flower than the Bee—for it is a false notion that more is gained by receiving than giving—no the receiver and the giver are equal in their benefits—The flower I doubt not receives a fair guerdon from the Bee—its leaves blush deeper in the next spring—and who shall say between Man and Woman which is the most delighted? Now it is more noble to sit like Jove that to fly like Mercury—let us not therefore go hurrying about and collecting honey-bee like, buzzing here and there impatiently from a knowledge of what is to be arrived at: but let us open our leaves like a flower and be passive and receptive—budding patiently under the eye of Apollo and taking hints from evey noble insect that favors us with a visit—sap will be given us for Meat and dew for drink—I was led into these thoughts, my dear Reynolds, by the beauty of the morning operating on a sense of Idleness—I have not read any Books—the Morning said I was right—I had no Idea but of the Morning and the Thrush said I was right—seeming to say—

[*'O thou whose face hath felt the Winter's wind' follows*]

Now I am sensible all this is a mere sophistication double however it may neighbour to any truths, to excuse my own indolence—so I will not deceive myself that Man should be equal with jove—but think himself very well off as a sort of scullion-Mercury or even a humble Bee—It is not matter whether I am right or wrong either one way or another, if there is sufficient to lift a little time from your Shoulders.

Your affectionate friend
JOHN KEATS—

Letter to John Taylor, 27 February 1818

Hampstead 27 Feby–

My dear Taylor,

Your alteration strikes me as being a great improvement—the page looks much better. And now I will attend to the Punctuations you speak of—the comma should be at *soberly*, and in the other passage the comma should follow *quiet*,.° I am extremely indebted to you for this attention and also for your after admonitions—It is a sorry thing for me that any one should have to overcome Prejudices in reading my Verses—that affects me more than any hypercriticism on any particular Passage. In *Endymion* I have most likely but moved into the Go-cart° from the leading strings. In Poetry I have a few Axioms, and you will see how far I am from their Centre. 1st I think Poetry should surprise by a fine excess and not by Singularity—it should strike the Reader as a wording of his own highest thoughts, and appear almost a Remembrance—2nd Its touches of Beauty should never be half way therby making the reader breathless instead of content: the rise, the progress, the setting of imagery should like the Sun come natural natural too him—shine over him and set soberly although in magnificence leaving him in the Luxury of twilight—but it is easier to think what Poetry should be than to write it—and this leads me on to another axiom. That if Poetry comes not as naturally as the Leaves to a tree it had better not come at all. However it may be with me I cannot help looking into new countries with 'O for a Muse of fire to ascend!'°—If Endymion serves me as a Pioneer perhaps I ought to be content. I have great reason to be content, for thank God I can read and perhaps understand Shakspeare to his depths, and I have I am sure many friends, who, if I fail, will attribute any change in my Life and Temper to Humbleness rather than to Pride—to a cowering under the Wings of great Poets rather than to a Bitterness that I am not appreciated. I am anxious to get Endymion printed that I may forget it and proceed. I have coppied the 3rd Book and have begun the 4th. On running my Eye over the Proofs—I saw one Mistake I will notice it presently and also any others if there be any—There should be no comma in 'the raft branch down sweeping from a tall Ash top' —I have besides made one or two alterations and also altered the 13 Line Page 32 to make sense of it as you will see. I will take care the Printer shall not trip up my Heels—There should be no dash after Dryope in the Line 'Dryope's lone lulling of her Child'. Remember me to Percy Street.°

Your sincere and oblig^d friend
JOHN KEATS—

P.S. You shall have a short *Preface* in good time—

Letter to Benjamin Bailey, 13 March 1818

My dear Bailey, Teignmouth Friday

When a poor devil is drowning, it is said he comes thrice to the surface, ere he makes his final sink if however, even at the third rise, he can manage to catch hold of a piece of weed or rock, he stands a fair chance,—as I hope I do now, of being saved. I have sunk twice in our Correspondence, have risen twice and been too idle, or something worse, to extricate myself—I have sunk the third time and just now risen again at this two of the Clock P.M. and saved myself from utter perdition—by beginning this, all drench'd as I am and fresh from the Water—and I would rather endure the present inconvenience of a Wet Jacket, than you should keep a laced one in store for me. Why did I not stop at Oxford in my Way?—How can you ask such a Question? Why did I not promise to do so? Did I not in a Letter to you make a promise to do so? Then how can you be so unreasonable as to ask me why I did not? This is the thing—(for I have been rubbing up my invention; trying several sleights—I first polish'd a cold, felt it in my fingers tried it on the table, but could not pocket it: I tried Chilblains, Rheumatism, Gout, tight Boots, nothing of that sort would do, so this is, as I was going to say, the thing.—I have a Letter from Tom saying how much better he had got, and thinking he had better stop—I went down to prevent his coming up—Will not this do? Turn it which way you like—it is selvaged all round—I have used it these three last days to keep out the abominable Devonshire Weather—by the by you may say what you will of devonshire: the thuth is, it is a splashy, rainy, misty snowy, foggy, haily floody, muddy, slipshod County—the hills are very beautiful, when you get a sight of 'em—the Primroses are out, but then you are in—the Cliffs are of a fine deep Colour, but then the Clouds are continually vieing with them—The Women like your London People in a sort of negative way—because the native men are the poorest creatures in England—because Government never have thought it worth while to send a recruiting party among them. When I think of Wordsworth's Sonnet 'Vanguard of Liberty! ye Men of Kent!' the degenerated race about me are Pulvis Ipecac. Simplex° a strong dose—Were I a Corsair I'd make a descent on the South Coast of Devon, if I did not run the chance of having Cowardice imputed to me: as for the Men they'd run away into the methodist meeting houses, and the Women would be glad of it—Had England been a large devonshire we should not have won the Battle of Waterloo—There are knotted oaks—there are lusty rivulets there are Meadows such as are not—there are vallies of femminine Climate—but there are no thews and Sinews—Moor's Almanack is here a curiosity—Arms Neck and shoulders may at least be seen there, and The Ladies read it as some out of the way romance—Such a quelling Power have these thoughts over me, that I fancy the very Air of a deteriorating quality—I fancy the flowers, all precocious, have an Acrasian° spell about them—I feel able to beat off the devonshire waves like soap froth—I think it well for the honor of Brittain that Julius Cæsar did not first land in this County—A Devonshirer standing on his native hills is not a distinct object—he does not show against the light—a wolf or two would dispossess him. I like, I love England, I like its strong Men—Give me a 'long brown plain' for my Morning so I may meet with some of Edmond Iron side's desendants—Give me a barren mould so I may meet with some shadowing of Alfred in the shape of a Gipsey, a Huntsman or as Shepherd. Scenery is fine—but human nature is finer—The Sward is richer for the tread of a real, nervous, english foot—the eagles nest is finer for the Mountaineer has look'd into it—Are these facts or prejudices? Whatever they are, for them I shall never be able to relish entirely any devonshire scenery—Homer is very fine, Achilles is fine, Diomed is fine, Shakspeare is fine, Hamlet is fine, Lear is fine, but dwindled englishmen are not fine—Where too the Women are so passable, and have such english names, such as Ophelia, Cordelia &—that they should have such Paramours or rather Imparamours—As for them I cannot, in thought help wishing as did the cruel Emperour, that they had but one head and I might cut it off to deliver them from any horrible Courtesy they may do their undeserving Countrymen—I wonder I meet with no born Monsters—O Devonshire, last night I thought the Moon had dwindled in heaven—I have never had your Sermon from Wordsworth but Mrs Dilke lent it me—You know my ideas about Religion—I do not think myself more in the right than other people and that nothing in this world is proveable. I wish I could enter into all your feelings on the subject merely for one short 10 Minutes and give you a Page or two to your liking. I am sometimes so very sceptical as to think Poetry itself a mere Jack a lanthern to amuse whoever may chance to be struck with its brilliance—As Tradesmen say every thing is worth what it will fetch, so probably every mental pursuit takes its reality and worth from the ardour of the pursuer—being in itself a nothing—Ehtereal thing may at least be thus real, divided under three heads—Things real—things semireal—and no things—Things real—such as existences of Sun Moon & Stars and passages of Shakspeare—Things semireal such as Love, the Clouds &c which require a greeting of the Spirit to make them wholly exist—and Nothings which are made Great and dignified by an ardent pursuit—Which by the by stamps the burgundy mark on the bottles of our Minds, insomuch as they are able to *'consecrate whate'er they look upon'*° I have written a Sonnet here of a somewhat collateral nature—so don't imagine it an a propos des bottes.

[*'The Human Seasons' follows*]

Aye this may be carried—but what am I talking of—it is an old maxim of mine and of course must be well known that evey point of thought is the centre of an intellectual world—the two uppermost thoughts in a Man's mind are the two poles of his World he revolves on them and every thing is southward or northward to him through their means—We take but three steps from feathers to iron. Now my dear fellow I must once for all tell you I have not one Idea of the truth of any of my speculations—I shall never be a Reasoner because I care not to be in the right, when retired from bickering and in a proper philosophical temper—So you must not stare if in any future letter I endeavour to prove that Apollo as he had a cat gut string to his Lyre used a cats' paw as a Pecten°—and further from said Pecten's reiterated and continual teasing came the term Hen peck'd. My Brother Tom desires to be remember'd to you—he has just this moment had a spitting of blood poor fellow—Remember me to Greig° and Whitehed—

Your affectionate friend
JOHN KEATS—

Letter to J. H. Reynolds, 14 March 1818

Teignmouth Saturday

Dear Reynolds,

I escaped being blown over and blown under & trees & house being toppled on me.—I have since hearing of Brown's accident had an aversion to a dose of parapet.° and being also a lover of antiquities I would sooner have a harmless piece of herculaneum° sent me quietly as a present, than ever so modern a chimney pot tumbled onto my head—Being agog to see some Devonshire, I would have taken a walk the first day, but the rain wo^{d} not let me; and the second, but the rain wo^{d} not let me; and the third; but the rain forbade it—Ditto 4 ditto 5—So I made up my Mind to stop in doors, and catch a sight flying between the showers; and behold I saw a pretty valley—pretty cliffs, pretty Brooks, pretty Meadows, pretty trees, both standing as they were created, and blown down as they are uncreated—The green is beautiful, as they say, and pity it is that it is amphibious—mais! but alas! the flowers here wait as naturally for the rain twice a day as the Muscles do for the Tide.—so we look upon a brook in these parts as you look upon a dash° in your Country—there must be something to support this, aye fog, hail, snow rain—Mist—blanketing up three parts of the year—This devonshire is like Lydia Languish,° very entertaining when at smiles, but cursedly subject to sympathetic moisture. You have the sensation of walking under one great Lamplighter: and you cant go on the other side of the ladder to keep your frock clean, and cosset your superstition. Buy a girdle—put a pebble in your Mouth—loosen your Braces—for I am going among Scenery whence I intend to tip you the Damosel Radcliffe°—I'll cavern you, and grotto you, and waterfall you, and wood you, and water you, and immense-rock you, and tremendous sound you, and solitude you. Ill make a lodgment on your glacis by a row of Pines, and storm your covered way with bramble Bushes. Ill have at you with hip and haw smallshot, and cannonade you with Shingles—Ill be witty upon salt fish, and impede your cavalry with clotted cream. But ah Coward! to talk at this rate to a sick man, or I hope to one that was sick—for I hope by this you stand on your right foot.—If you are not—that's all,—I intend to cut all sick people if they do not make up their minds to cut sickness—a fellow to whom I have a complete aversion, and who strange to say is harboured and countenanced in several houses where I visit—he is sitting now quite impudent between me and Tom—He insults me at poor Jem Rice's—and you have seated him before now between us at the Theatre—where I thought he look'd with a longing eye at poor Kean. I shall say, once for all, to my friends generally and severally, cut that fellow, or I cut you—I went to the Theatre here the other night, which I forgot to tell George, and got insulted, which I ought to remember to forget to tell any Body; for I did not fight, and as yet have had no redress—'Lie thou there, sweetheart!' I wrote to Bailey yesterday, obliged to speak in a high way, and a damme who's affraid—for I had owed him so long; however, he shall see I will be better in future. Is he in Town yet? I have directed to Oxford as the better chance. I have copied my fourth Book, and shall write the preface soon. I wish it was all done; for I want to forget it and make my mind free for something new—Atkins the Coachman, Bartlet the Surgeon, Simmons the Barber, and the Girls over at the Bonnet shop say we shall now have a Month of seasonable Weather. warm, witty, and full of invention—Write to me and tell me you are well or thereabouts, or by the holy Beaucœur,—which I suppose is the virgin Mary, or the repented Magdalen, (beautiful name, that Magdalen) Ill take to my Wings and fly away to any where but old or Nova Scotia—I wish I had a little innocent bit of Metaphysic in my head, to criss-cross this letter: but you know a favorite tune is hardest to be remembered when one wants it most and you, I know, have long ere this taken it for granted that I never have any speculations without assocating you in them, where they are of a pleasant nature and you know enough to me to tell the places where I haunt most, so that if you think for five minutes after having read this you will find it a long letter and see written in the Air above you,

Your most affectionate friend
JOHN KEATS

Remember me to all. Tom's remembrances to you.

Letter to B. R. Haydon, 21 March 1818

My dear Haydon— Teignmouth Saturd—Morn

In sooth, I hope you are not too sanguine about that seal°—in sooth I hope it is not Brumidgeum—in double sooth I hope it is his—and in tripple sooth I hope I shall have an impression. Such a piece of intelligence came doubly welcome to me while in your own County and in your own hand—not but I have blown up said County for its urinal qualifications—the 6 first days I was here it did nothing but rain and at that time having to write to a friend I gave Devonshire a good blowing up—it has been fine for about three days and I was coming round a bit; but to day it rains again—with me the County is yet upon its good behaviour—I have enjoyed the most delightful Walks these three fine days beautiful enough to make me content here all the summer could I stay.

[*'For there's Bishop's Teign' follows*]

Here's some doggrel for you—Perhaps you would like a bit of B—hrell—°

[*'Where be ye going you Devon maid' follows*]

I know not if this rhyming fit has done any thing—it will be safe with you if worthy to put among my Lyrics How does the Work go on? I should like to bring out my Dentatus° at the time your Epic makes its appearance. I expect to have my Mind soon clear for something new. Tom has been much worse: but is now getting better.—his remembrances to you—I think of seeing the dart° and Plymouth—but I dont know. It has as yet been a Mystery to me how and when Wordsworth went—I cant help thinking he has returned to his Shell—with his beautiful Wife and his enchanting Sister. It is a great Pity that People should by associating themselves with the finest things, spoil them. Hunt has damned Hampstead and Masks and Sonnets and italian tales—Wordsworth has damned the lakes—Millman° has damned the old drama—West has damned—wholesale—Peacock has damned sattire Ollier has damn'd Music—Hazlitt has damned the bigotted and the blue-stockined how durst the Man?! he is your only good damner and if ever I am damn'd—I shoul'nt like him to damn me. It will not be long ere I see you, but I thought I would just give you a line out of Devon—

Your's affectionately
JOHN KEATS

Rember me to all we know

Letter to James Rice, 24 March 1818

Teignmouth Tuesday,

My dear Rice,

Being in the midst of your favourite Devon, I should not by rights, pen one word but it should contain a vast portion of Wit, Wisdom, and learning—for I have heard that Milton ere he wrote his Answer to Salmasius° came into these parts, and for on whole Month, rolled himself, for three whole hours in a certain meadow hard by us—where the mark of his nose at equidistances is still shown. The exhibitor of said Meadow further saith that after these rollings, not a nettle sprang up in all the seven acres for seven years and that from said time a new sort of plant was made from the white thorn, of a thornless nature very much used by the Bucks of the present day to rap their Boots withall—This accout made me very naturally suppose that the nettles and thorns etherealized by the Scholars rotatory motion and gardner'd in his head, thence flew after a new fermentation against the luckless Salmasius and accasioned his well known and unhappy end. What a happy thing it would be if we could settle our thoughts, make our minds up on any matter in five Minutes and remain content—that is to build a sort of mental Cottage of feelings quiet and pleasant—to have a sort of Philosophical Back Garden, and cheerful holiday-keeping front one—but Alas! this never can be: for as the material Cottager knows there are such places as france and Italy and the Andes and the Burning Mountains—so the spiritual Cottager has knowledge of the terra semi incognita of things unearthly; and cannot for his Life, keep in the check rein—Or I should stop here quiet and comfortable in my theory of Nettles. You will see however I am obliged to run wild, being attracted by the Loadstone Concatenation. No sooner had I settle the notty point of Salmasius than the Devil put this whim into my head in the likeness of one of Pythagora's questionings 'Did Milton do more good or ham to the world? He wrote let me infom you (for I have it from a friend, who had it of—) he wrote Lycidas, Comus, Paradise Lost and other Poems, with much delectable prose—he was moreover an active friend to Man all his Life and has been since his death. Very good—but my dear fellow I must let you know that as there is ever the same quantity of matter constituting this habitable globe—as the ocean notwithstanding the enormous changes and revolutions taking place in some or other of its demesnes—notwithstanding Waterspouts whirpools and mighty Rivers emptying themselves into it, it still is made up of the same bulk—nor ever varies the number of its Atoms—And as a certain bulk of Water was instituted at the Creation—so very likely a certain portion of intellect was spun forth into the thin Air for the Brains of Man to prey upon it—You will see my drift without any unnecessary parenthesis. That which is contained in the Pacific can't° lie in the hollow of the Caspian—that which was in Miltons head could not find Room in Charles the seconds—he like a Moon attracted Intellect to its flow—it has not ebbd yet—but has left the shore pebble all bare—I mean all Bucks Authors of Hengist and Castlereaghs° of the present day—who without Miltons gormandizing might have been all wise Men—Now for as much as—I was very peedisposed to a Country I had heard you speak so highly of, I took particular notice of every thing during my journey and have bought some folio asses skin for Memorandums—I have seen evey thing but the wind—and that they say becomes visible by taking a dose of Acorns or sleeping on night in a hog trough with your tail to the Sow Sow west. Some of the little Barmaids look'd at me as if I knew Jem Rice—but when I took a glass of Brandy they were quite convinced. One asked whether you preserved a secret she gave you on the nail—another how my buttons of your Coat were buttoned in general—I told her it used to be four—but since you had become acquainted with one Martin you had reduced it to three and had been turning this third one in your Mind—and would do so with finger and thumb only you had taken to snuff—I have met with a Brace or twain of little Long heads—not a kit o' the german°—all in the neatest little dresses, and avoiding all the puddes—but very fond of peppermint drops, laming ducks, and seeing little Girls affairs. Well I cant tell! I hope you are showing poor Reynolds the way to get well—send me a good account of him and if I can I'll send you one of Tom—Oh! for a day and all well! I went yesterday to dawlish fair—

[*'Over the hill and over the dale' follows*]

Tom's Remembrances and mine to all—

Your sincere friend
JOHN KEATS

Letter to B. R. Haydon, 8 April 1818

Wednesday—

My dear Haydon,

I am glad you were pleased with my nonsense° and if it so happen that the humour takes me when I have set down to prose to you I will not gainsay it. I should be (god forgive me) ready to swear because I cannot make use of you assistance in going through Devon if I was not in my own Mind determined to visit it thoroughly at some more favorable time of the year. But now Tom (who is getting greatly better) is anxious to be in Town therefore I put off my threading the County. I purpose within a Month to put my knapsack at my back and make a pedestrian tour through the North of England, and part of Scotland—to make a sort of Prologue to the Life I intend to pursue—that is to write, to study and to see all Europe at the lowest expence. I will clamber through the Clouds and exist. I will get such an accumulation of stupendous recollolections that as I walk through the suburbs of London I may not see them—I will stand upon Mount Blanc and remember this coming Summer when I intend to straddle ben Lomond—with my Soul!—galligaskins° are out of the Question—I am nearer myself to hear your Christ is being tinted into immortality—Believe me Haydon your picture is a part of myself—I have ever been too sensible of the labyrinthian path to eminence in Art (judging from Poetry) ever to think I understood the emphasis of Painting. The innumerable compositions and decompositions° which take place between the intellect and its thousand materials before it arrives at that trembling delicate and snail-horn perception of Beauty—I know not you many havens of intenseness—nor ever can know them—but for this I hope not you atchieve is lost upon me: for when a Schoolboy the abstract Idea I had of an heroic painting—was what I cannot describe I saw it somewhat sideways large prominent round and colour'd with magnificence—somewhat like the feel I have of Anthony and Cleopatra. Or of Alcibiades, leaning on his Crimson Couch in his Galley, his broad shoulders imperceptibly heaving with the Sea—That° passage in Shakspeare is finer than this

'see how the surly Warwick mans the Wall'°

I like your consignment of Corneille—that's the humor of it—They shall be called your Posthumous Works. I don't understand you bit of Italian.° I hope she will awake from her dream and flourish fair—my respects to her—The Hedges by this time are beginng to leaf—Cats are becoming more vociferous— young Ladies that wear Watches are always looking at them—Women about forty five think the Season very back ward—Lady's Mares have but half an allowance of food—It rains here again, has been doing so for three days—however as I told you I'll take a trial in June July or August next year—

I am affraid Wordsworth went rather huff'd out of Town—I am sorry for it. he cannot expect his fireside Divan to be infallible he cannot expect but that every Man of worth is as proud as himself. O that he had not fit with a Warrener that is din'd at Kingston's.° I shall be in town in about a fortnight and then we will have a day or so now and then before I set out on my northern expedition—we will have no more abominable Rows—for they leave one in a fearful silence having settled the Methodists let us be rational—not upon compulsion—no if it will out let it—but I will not play the Basoon any more deliberately—Remember me to Hazlitt, and Bewick—Your affectionate friend

JOHN KEATS

Letter to J. H. Reynolds, 9 April 1818

My Dear Reynolds. Th[y] Morn[g]

Since you all agree that the thing° is bad, it must be so— though I am not aware there is any thing like Hunt in it, (and if there is, it is my natural way, and I have something in common with Hunt) look it over again and examine into the motives, the seeds from which any one sentence sprung—I have not the slightest feel of humility towards the Public—or to any thing in existence,—but the eternal Being, the Principle of Beauty,—and the Memory of great Men—When I am writing for myself for the mere sake of the Moment's enjoyment, perhaps nature has its course with me—but a Preface is written to the Public; a thing I cannot help looking upon as an Enemy, and which I cannot address without feelings of Hostility—If I write a Preface in a supple or subdued style, it will not be in character with me as a public speaker—I wo[d] be subdued before my friends, and thank them for subduing me—but among Multitudes of Men—I have no feel of stooping, I hate the idea of humility to them—

I never wrote one single Line of Poetry with the least Shadow of public thought.

Forgive me for vexing you and making a Trojan Horse of such a Trifle, both with respect to the matter in Question, and myself—but it eases me to tell you—I could not live without the love of my friends—I would jump down Ætna for any great Public good—but I hate a Mawkish Popularity.—I cannot be subdued before them—My glory would be to daunt and dazzle the thousand jabberers about Pictures and Books—I see swarms of Porcupines with their Quills erect 'like lime-twigs set to catch my Winged Book'° and I would fright 'em away with a torch—You will say my preface is not much of a Torch. It would have been too insulting 'to begin from Jove'° and I could not set a golden head upon a thing of clay—if there is any fault in the preface it is not affectation: but an undersong of disrespect to the Public.—if I write another preface. it must be done without a thought of those people—I will think about it. If it should not reach you in four—or five days—tell Taylor to publish it without a preface, and let the dedication simply stand 'inscribed to the memory of Thomas Chatterton.' I had resolved last night to write to you this morning—I wish it had been about something else—something to greet you towards the close of your long illness—I have had one or two intimations of your going to Hampstead for a space; and I regret to see your confounded Rheumatism keeps you in Little Brittain where I am sure the air is too confined—Devonshire continues rainy. As the drops beat against the window, they give me the same sensation as a quart of cold water offered to revive a half drowned devil—No feel of the clouds dropping fatness; but as if the roots of the Earth were rotten cold and drench'd—I have not been able to go to Kents' Cave at Babbicun°—however on one very beautiful day I had a fine clamber over the rocks all along as far as that place: I shall be in Town in about Ten days.—We go by way of Bath on purpose to call on Bailey. I hope soon to be writing to you about the things of the north, purposing to wayfare all over those parts. I have settled my accoutrements in my own mind, and will go to gorge wonders: However we'll have some days together before I set out—

I have many reasons for going wonder-ways: to make my winter chair free from spleen—to enlarge my vision—to escape disquisitions on Poetry and Kingston Criticism.—to promote digestion and economise shoe leather—I'll have leather buttons and belt; and if Brown holds his mind, over the Hills we go.—If my Books will help me to it,—thus will I take all Europe in turn, and see the Kingdoms of the Earth and the glory of them—Tom is getting better he hopes you may meet him at the top o' the hill—My Love to your nurses. I am ever

Your affectionate Friend,
JOHN KEATS.

Letter to John Taylor, 24 April 1818

Teignmouth Friday

My dear Taylor,

I think I Did very wrong to leave you to all the trouble of Endymion—but I could not help it then—another time I shall be more bent to all sort of troubles and disagreeables—Young Men for some time have an idea that such a thing as happiness is to be had and therefore are extremely impatient under any unpleasant restraining—in time however, of such stuff is the world about them, they know better and instead of striving from Uneasiness greet it as an habitual sensation, a pannier which is to weigh upon them through life.

And in proportion to my disgust at the task is my sense of your kindness & anxiety—the book pleased me much—it is very free from faults; and although there are one or two words I should wish replaced, I see in many places an improvement greatly to the purpose—

I think those speeches which are related—those parts where the speaker repeats a speech—such as Glaucus' repetition of Circe's words, should have inverted commas to every line—In this there is a little confusion. If we divide the speeches into *identical* and *related:* and to the former put merely one inverted comma at the beginning and another at the end; and to the latter inverted commas before every line, the book will be better understood at the first glance. Look at pages 126 and 127 you will find in the 3 line the beginning of a *related* speech marked thus "Ah! art awake°—while at the same time in the next page the continuation of the *identical speech* is mark'd in the same manner "Young Man of Latmos—You will find on the other side all the parts which should have inverted commas to every line—

I was purposing to travel over the north this Summer—there is but one thing to prevent me—I know nothing I have read nothing and I mean to follow Solomon's directions of 'get Wisdom—get understanding'—I find cavalier days are gone by. I find that I can have no enjoyment in the World but continual drinking of Knowledge—I find there is no worthy pursuit but the idea of doing some good for the world—some do it with their society—some with their wit—some with their benevolence—some with a sort of power of conferring pleasure and good humour on all they meet and in a thousand ways all equally dutiful to the command of Great Nature—there is but one way for me—the road lies through application study and thought. I will pursue it and to that end purpose retiring for some years. I have been hovering for some time between an exquisite sense of the luxurious and a love for Philosophy—were I calculated for the former I should be glad—but as I am not I shall turn all my soul to the latter. My Brother Tom is getting better and I hope I shall see both him and Reynolds well before I retire from the World. I shall see you soon and have some talk about what Books I shall take with me—

Your very sincere friend
JOHN KEATS

Remember me to Hessey—Woodhouse and Percy Street

I cannot discover any other error—the preface is well without those thing you have left out—Adieu—

Letter to J. H. Reynolds, 27 April 1818

Teignmouth Monday

My dear Reynolds.

It is an awful while since you have heard from me—I hope I may not be punished, when I see you well, and so anxious as you always are for me, with the remembrance of my so seldom writing when you were so horribly confined—the most unhappy hours in our lives are those in which we recollect times past to our own blushing—If we are immortal that must be the Hell—If I must be immortal, I hope it will be after having taken a little of 'that watery labyrinth'° in order to forget some of my schoolboy days & others since those.

I Have heard from George at different times how slowly you were recovering. it is a tedious thing—but all Medical Men will tell you how far a very gradual amendment is preferable; you will be strong after this, never fear.—We are here still enveloppd in clouds—I lay awake last night—listening to the Rain with a sense of being drown'd and rotted like a grain of wheat—There is a continual courtesy between the Heavens and the Earth.—the heavens rain down their unwelcomeness, and the Earth sends it up again to be returned to morrow. Tom has taken a fancy to a Physician here, Dr Turton, and I think is getting better—therefore I shall perhaps remain here some Months.—I have written to George for some Books—shall learn Greek, and very likely Italian—and in other ways prepare myself to ask Hazlitt in about a years time the best metaphysical road I can take.—For although I take poetry to be Chief, there is something else wanting to one who passes his life among Books and thoughts on Books—I long to feast upon old Homer as we have upon Shakespeare. and as I have lately upon Milton.—if you understand Greek, and would read me passages, now and then explaining their meaning, 't would be, from its mistiness, perhaps a greater luxury than reading the thing one's self.—I shall be happy when I can do the same for you.—I have written for my folio Shakespeare, in which there is the first few stanzas of my 'Pot of Basil': I have the rest here finish'd, and will copy the whole out fair shortly—and George will bring it to you—The Compliment is paid by us to Boccace, whether we publish or no:° so there is content in this world—mine is short—you must be deliberate about yours: you must not think of it till many months after you are quite well:—then put your passion to it,—and I shall be bound up with you in the shadows of mind, as we are in our matters of human life—Perhaps a Stanza or two will not be too foreign to your Sickness.

[*'Isabella' Stanzas 12, 13, 30 were given in the original letter and indicated here by the copyist*]

I heard from Rice this morning—very witty—and have just written to Bailey—Don't you think I am brushing up in the letter way? and being in for it,—you shall hear again from me very shortly:—if you will promise not to put hand to paper for me until you can do it with a tolerable ease of health—except it be a line or two—Give my Love to your Mother and Sisters Remember me to the Butlers —not forgetting Sarah

Your affectionate friend
JONN KEATS

Letter to J. H. Reynolds, 3 May 1818

Teignmouth May 3[d]

My dear Reynolds.

What I complain of is that I have been in so an uneasy a state of Mind as not to be fit to write to an invalid. I cannot write to any length under a dis-guised feeling. I should have loaded you with an addition of gloom, which I am sure you do not want. I am now thank God in a humour to give you a good groats worth—for Tom, after a Night without a Wink of sleep, and overburdened with fever, has got up after a refreshing day sleep and is better than he has been for a long time; and you I trust have been again round the Common° without any effect but refreshment.—As to the Matter I hope I can say with Sir Andrew 'I have matter enough in my head"° in your favor And now, in the second place, for I reckon that I have finished my Imprimis, I am glad you blow up the weather—all through your letter there is a leaning towards a climate-curse. and you know what a delicate satisfaction there is in having a vexation anathematized: one would think there has been growing up for these last four thousand years, a grandchild Scion of the old forbidden tree, and that some modern Eve had just violated it; and that there was come with double charge, 'Notus and Afer black with thunderous clouds from Sierra-leona"°—I shall breathe worsted stockings sooner than I thought for.° Tom wants to be in Town—we will have some such days upon the heath like that of last summer and why not with the same book: or what say you to a black Letter Chaucer printed in 1596: aye I've got one huzza! I shall have it bounden gothique a nice sombre binding—it will go a little way to unmodernize. And also I see no reason, because I have been away this last month, why I should not have a peep at your Spencerian°—notwithstanding you speak of your office, in my thought a little too early, for I do not see why a Mind like yours is not capable of harbouring and digesting the whole Mystery of Law as easily as Parson Hugh does Pepins°—which did not hinder him from his poetic Canary—Were I to study physic or rather Medicine again,—I feel it would not make the least difference in my Poetry; when the Mind is in its infancy a Bias is in reality a Bias, but when we have acquired more strength, a Bias becomes no Bias. Every department of knowledge we see excellent and calculated towards a great whole. I am so convinced of this, that I am glad at not having given away my medical Books, which I shall again look over to keep alive the little I know thitherwards; and moreover intend through you and Rice to become a sort of Pip-civilian. An extensive knowledge is needful to thinking people—it takes away the heat and fever; and helps, by widening speculation, to ease the Burden of the Mystery:° a thing I begin to understand a little, and which weighed upon you in the most gloomy and true sentence in your Letter. The difference of high Sensations with and without knowledge appears to me this—in the latter case we are falling continually ten thousand fathoms deep and being blown up again without wings and with all the horror of a bare shoulderd Creature—in the former case, our shoulders are fledge, and we go thro' the same air and space without fear. This is running one's rigs° on the score of abstracted benefit—when we come to human Life and the affections it is impossible how a parallel of breast and head can be drawn—(you will forgive me for thus privately treading out my depth and take it for treading as schoolboys tread the water)—it is impossible to know how far knowledge will console us for the death of a friend and the ill 'that flesh is heir to'—With respect to the affections and Poetry you must know by a sympathy my thoughts that way; and I dare say these few lines will be but a ratification: I wrote them on May-day—and intend to finish the ode all in good time.—

[*'Mother of Hermes! and still youthful Maia!' follows*]

You may be anxious to know for fact to what sentence in your Letter I allude. You say 'I fear there is little chance of any thing else in this life.' You seem by that to have been going through with a more painful and acute zest the same labyrinth that I have—I have come to the same conclusion thus far. My Branchings out therefrom have been numerous: one of them is the consideration of Wordsworth's genius and as a help, in the manner of gold being the meridian Line of worldly wealth,—how he differs from Milton.—And here I have nothing but surmises, from an uncertainty whether Miltons apparently less anxiety for Humanity proceeds from his seeing further or no than Wordsworth: And whether Wordsworth has in truth epic passion, and martyrs himself to the human heart, the main region of his song°—In regard to his genius alone—we find what he says true as far as we have experienced and we can judge no further but by larger experience—for axioms in philosophy are not axioms until they are proved upon our pulses: We read fine———things but never feel them to thee full until we have gone the same steps as the Author.—I know this is not plain; you will know exactly my meaning when I say, that now I shall relish Hamlet more than I ever have done—Or, better—You are sensible no man can set down Venery as a bestial or joyless thing until he is sick of it and therefore all philosophizing on it would be mere wording. Until we are sick, we understand not;—in fine, as Byron says, 'Knowledge is Sorrow';° and I go on to say that 'Sorrow is Wisdom'—and further for aught we can know for certainty! 'Wisdom is folly'—So you see how I have run away from Wordsworth, and Milton; and shall still run away from what was in my head, to observe, that some kind of letters are good squares others handsome ovals, and others some orbicular, others spheroid—and why should there not be another species with two rough edges like a Rat-trap? I hope you will find all my long letters of that species, and all will be well; for by merely touching the spring delicately and etherially, the rough edged will fly immediately into a proper compactness, and thus you may make a good wholesome loaf, with your own leven in it, of my fragments—If you cannot find this said Rat-trap sufficiently tractable—alas for me, it being an impossibility in grain for my ink to stain otherwise: If I scribble long letters I must play my vagaries. I must be too heavy, or too light, for whole pages—I must be quaint and free of Tropes and figures—I must play my draughts° as I please, and for my advantage and your erudition, crown a white with a black, or a black with a white, and move into black or white, far and near as I please—I must go from Hazlitt to Patmore,° and make Wordsworth and Coleman° play at leap-frog—or keep one of them down a whole half holiday at fly the garter—'From Gray to Gay, from Little° to Shakespeare'—Also as a long cause requires two or more sittings of the Court, so a long letter will require two or more sittings of the Breech wherefore I shall resume after dinner.—

Have you not seen a Gull, an orc, a sea Mew, or any thing to bring this Line to a proper length, and also fill up this clear part; that like the Gull I may *dip* —I hope, not out of sight—and also, like a Gull, I hope to be lucky in a good sized fish—This crossing a letter is not without its association—for chequer work leads us naturally to a Milkmaid, a Milkmaid to Hogarth Hogarth to Shakespeare Shakespear to Hazlitt—Hazlitt to Shakespeare and thus by merely pulling an apron string we set a pretty peal of Chimes at work—Let them chime on while, with your patience,—I will return to Wordsworth—whether or no he has an extended vision or a circumscribed grandeur—whether he is an eagle in his nest, or on the wing—And to be more explicit and to show you how tall I stand by the giant, I will put down a simile of human life as far as I now perceive it; that is, to the point to which I say we both have arrived at— Well—I compare human life to a large Mansion of Many Apartments, two of which I can only describe, the doors of the rest being as yet shut upon me—The first we step into we call the infant or thoughtless Chamber, in which we remain as long as we do not think—We remain there a long while, and notwithstanding the doors of the second Chamber remain wide open, showing a bright appearance, we care not to hasten to it; but are at length imperceptibly impelled by the awakening of the thinking principle—within us—we no sooner get into the second Chamber, which I shall call the Chamber of Maiden-Thought, than we become intoxicated with the light and the atmosphere, we see nothing but pleasant wonders, and think of delaying there for ever in delight: However among the effects this breathing is father of is that tremendous one of sharpening one's vision into the heart and nature of Man—of convincing ones nerves that the World is full of Misery and Heratbreak, Pain, Sickness and oppression—whereby This Chamber of Maiden Thought becomes gradually darken'd and at the same time on all sides of it many doors are set open—but all dark—all leading to dark passages—We see not the balance of good and evil. We are in a Mist—*We* are now in that state—We feel the 'burden of the Mystery,' To this point was Wordsworth come, as far as I can conceive when he wrote 'Tintern Abbey' and it seems to me that his Genius is explorative of those dark Passages. Now if we live, and go on thinking, we too shall explore them. he is a Genius and superior to us, in so far as he can, more than we, make discoveries, and shed a light in them—Here I must think Wordsworth is deeper than Milton—though I think it has depended more upon the general and gregarious advance of intellect, than individual greatness of Mind—From the Paradise Lost and the other Works of Milton, I hope it is not too presuming, even between ourselves to say, his Philosophy, human and divine, may be tolerably understood by one not much advanced in years, In his time englishmen were just emancipated from a great superstition—and Men had got hold of certain points and resting places in reasoning which were too newly born to be doubted, and too much opposed by the Mass of Europe not to be thought etherial and authentically divine—who could gainsay his ideas on virtue, vice, and Chastity in Comus, just at the time of the dismissal of Cod-pieces and a hundred other disgraces? who would not rest satisfied with his hintings at good and evil in the Paradise Lost, when just free from the inquisition and burrning in Smithfield? The Reformation produced such immediate and great benefits, that Protestantism was considered under the immediate eye of heaven, and its own remaining Dogmas and superstitions, then, as it were, regenerated, constituted those resting places and seeming sure points of Reasoning—from that I have mentioned, Milton, whatever he may have thought in the sequel, appears to have been content with these by his writings—He did not think into the human heart, as Wordsworth has done—Yet Milton as a Philosopher, had sure as great powers as Wordsworth—What is then to be inferr'd? O many things—It proves there is really a grand march of intellect—, It proves that a mighty providence subdues the mightiest Minds to the service of the time being, whether it be in human Knowledge or Religion—I have often pitied a Tutor who has to hear 'Nom[e]: Musa'°—so often dinn'd into his ears—I hope you may not have the same pain in this scribbling—I may have read these things before, but I never had even a thus dim perception of them: and moreover I like to say my lesson to one will endure my tediousness for my own sake—After all there is certainly something real in the World—Moore's present to Hazlitt is real—I like that Moore, and am glad I saw him at the Theatre just before I left Town. Tom has spit a leetle blood this afternoon, and that is rather a damper—but I know—the truth is there is something real in the World Your third Chamber of Life shall be a lucky and a gentle one—stored with the wine of love—and the Bread of Friendship—When you see George if he should not have recēd a letter from me tell him he will find one at home most likely—tell Bailey I hope soon to see him—Remember me to all The leaves have been out here, for MONY a day—I have written to George for the first stanzas of my Isabel—I shall have them soon and will copy the whole out for you.

Your affectionate friend
JOHN KEATS.

Letter to Benjamin Bailey, 10 June 1818

My dear Bailey, London—

I have been very much gratified and very much hurt by your Letters in the Oxford Paper:° because independant of that unlawful and mortal feeling of pleasure at praise, there is a glory in enthusiam; and because the world is malignant enough to chuckle at the most honorable Simplicity. Yes on my Soul my dear Bailey you are too simple for the World—and that Idea makes me sick of it—How is it that by extreme opposites we have as it were got disconted nerves—you have all your Life (I think so) believed every Body—I have suspected every Body—and although you have been so deceived you make a simple appeal—the world has something else to do, and I am glad of it—were it in my choice I would reject a petrarchal coronation—on accout of my dying day, and because women have Cancers. I should not by rights speak in this tone to you—for it is an incendiary spirit that would do so. Yet I am not old enough or magnanimous enough to anihilate self—and it would perhaps be paying you an ill compliment. I was in hopes some little time back to be able to releive your dullness by my spirits—to point out things in the world worth your enjoyment—and now I am never alone without rejoicing that there is such a thing as death—without placing my ultimate in the glory of dying for a great human purpose Perphaps if my affairs were in a different state I should not have written the above—you shall judge—I have two Brothers one is driven by the 'burden of Society' to America the other, with an exquisite love of Life, is in a lingering state—My Love for my Brothers from the early loss of our parents and even for earlier Misfortunes has grown into a affection 'passing the Love of Women'—I have been ill temper'd with them, I have vex'd them—but the thought of them has always stifled the impression that any woman might otherwise have made upon me—I have a sister too and may not follow them, either to America or to the Grave—Life must be undergone, and I certainly derive a consolation from the thought of writing one or two more Poems before it ceases—I have heard some hints of your retireing to scotland—I would like to know your feeling on it—it seems rather remote—perhaps Gleg will have a duty near you. I am not certain whether I shall be able to go my Journey on account of my Brother Tom and a little indisposition of my own—If I do not you shall see me soon—if no on my return—or I'll quarter myself upon you in Scotland next Winter. I had know my sister in Law some time before she was my Sister and was very fond of her. I like her better and better—she is the most disinterrested woman I ever knew—that is to say she goes beyond degree in it—To see an entirely disinterrested Girl quite happy is the most pleasant and extraordinary thing in the world—it depends upon a thousand Circumstances—on my word 'tis extraordinary. Women must want Imagination and they may thank God for it—and so may we that a delicate being can feel happy without any sense of crime. It puzzles me and I have no sort of Logic to comfort me—I shall think it over. I am not at home and your letter being there I cannot look it over to answer any particular—only I must say I felt that passage of Dante—if I take any book with me it shall be those minute volumes of carey° for they will go into the aptest corner. Reynolds is getting I may say robust—his illness has been of service to him—like eny one just recovered he is high-spirited. I hear also good accounts of Rice—With respects to domestic Literature—the Endinburgh Magasine in another blow up against Hunt calls me 'the amiable Mister Keats'° and I have more than a Laurel from the Quarterly Reviewers for they have *smothered* me in 'Foliage'° I want to read you my 'Pot of Basil' if you go to scotland I should much like to read it there to you among the Snows of next Winter. My Brothers' remembrances to you.

Your affectionate friend
JOHN KEATS—

Letter to Tom Keats, 25–27 June 1818

Here beginneth my journal, this Thursday, the 25th day of June, Anno Domini 1818. This morning we arose at 4, and set off in a Scotch mist; put up once under a tree, and in fine, have walked wet and dry to this place, called in the vulgar tongue Endmoor, 17 miles; we have not been incommoded by our knapsacks; they serve capitally, and we shall go on very well.

June 26—I merely put *pro forma*, for there is no such thing as time and space, which by the way came forcibly upon me on seeing for the first hour the Lake and Mountains of Winander—I cannot describe them—they surpass my expectation—beautiful water—shores and islands green to the marge—mountains all round up to the clouds. We set out from Endmoor this morning, breakfasted at Kendal with a soldier who had been in all the wars for the last seventeen years—then we have walked to Bowne's° to dinner—said Bowne's situated on the Lake where we have just dined, and I am writing at this present. I took an oar to one of the islands to take up some trout for dinner, which they keep in porous boxes. I enquired of the waiter for Wordsworth—he said he knew him, and that he had been here a few days ago, canvassing for the Lowthers. What think you of that—Wordsworth versus Brougham!!° Sad—sad—sad—and yet the family has been his friend always. What can we say? We are now about seven miles from Rydale, and expect to see him to-morrow. You shall hear all about our visit.

There are many disfigurements to this Lake—not in the way of land or water. No; the two views we have had of it are of the most noble tenderness—they can never fade away—they make one forget the divisions of life; age, youth, poverty and riches; and refine one's sensual vision into a sort of north star which can never cease to be open lidded and stedfast over the wonders of the great Power. The disfigurement I mean is the miasma of London. I do suppose it contaminated with bucks and soldiers, and women of fashion—and hat-band ignorance. The border inhabitants are quite out of keeping with the romance about them, from a continual intercourse with London rank and fashion. But why should I grumble? They let me have a prime glass of soda water—O they are as good as their neighbors. But Lord Wordsworth, instead of being in retirement, has himself and his house full in the thick of fashionable visitors quite convenient to be pointed at all the summer long. When we had gone about half this morning, we began to get among the hills and to see the mountains grow up before us—the other half brought us to Wynandermere, 14 miles to dinner. The weather is capital for the views, but is now rather misty, and we are in doubt whether to walk to Ambleside to tea—it is five miles along the borders of the Lake. Loughrigg will swell up before us all the way—I have an amazing partiality for mountains in the clouds. There is nothing in Devon like this, and Brown says there is nothing in Wales to be compared to it. I must tell you, that in going through Cheshire and Lancashire, I saw the Welsh mountains at a distance. We have passed the two castles, Lancaster and Kendal. 27th—We walked here to Ambleside yesterday along the border of Winandermere all beautiful with wooded shores and Islands—our road was a winding lane, wooded on each side, and green overhead, full of Foxgloves—every now and then a glimpse of the Lake, and all the while Kirkstone and other large hills nestled together in a sort of grey black mist. Ambleside is at the northern extremity of the Lake. We arose this morning at six, because we call it a day of rest, having to call on Wordsworth who lives only two miles hence—before breakfast we went to see the Ambleside water fall. The morning beautiful—the walk easy among the hills. We, I may say, fortunately, missed the direct path, and after wandering a little, found it out by the noise—for, mark you, it is buried in trees, in the bottom of the valley—the stream itself is interesting throughout with 'mazy error over pendant shades.'° Milton meant a smooth river—this is buffetting all the way on a rocky bed ever various—but the waterfall itself, which I came suddenly upon, gave me a pleasant twinge. First we stood a little below the head about half way down the first fall, buried deep in trees, and saw it streaming down two more descents to the depth of near fifty feet—then we went on a jut of rock nearly level with the second fall-head, where the first fall was above us, and the third below our feet still—at the same time we saw that the water was divided by a sort of cataract island on whose other side burst out a glorious stream—then the thunder and the freshness. At the same time the different falls have as different characters; the first darting down the slate-rock like an arrow; the second spreading out like a fan—the third dashed into a mist—and the one on the other side of the rock a sort of mixture of all these. We afterwards moved away a space, and saw nearly the whole more mild, streaming silverly through the trees. What astonishes me more than any thing is the tone, the coloring, the slate, the stone, the moss, the rock-weed; or, if I may so say, the intellect, the countenance of such places. The space, the magnitude of mountains and waterfalls are well imagined before one sees them; but this countenance or intellectual tone must surpass every imagination and defy any remembrance. I shall learn poetry here and shall henceforth write more than ever, for the abstract endeavor of being able to add a mite to that mass of beauty which is harvested from these grand materials, by the finest spirits, and put into etherial existence for the relish of one's fellows. I cannot think with Hazlitt that these scenes make man appear little. I never forgot my stature so completely—I live in the eye; and my imagination, surpassed, is at rest—We shall see another waterfall near Rydal to which we shall proceed after having put these letters in the post office. I long to be at Carlisle, as I expect there a letter from George and one from you. Let any of my friends see my letters—they may not be interested in descriptions—descriptions are bad at all times—I did not intend to give you any; but how can I help it? I am anxious you should taste a little of our pleasure; it may not be an unpleasant thing, as you have not the fatigue. I am well in health. Direct henceforth to Port Patrick till the 12th July. Content that probably three or four pair of eyes whose owners I am rather partial to will run over these lines I remain; and moreover that I am your affectionate brother JOHN.

Letter to Tom Keats, 29 June, 1, 2 July 1818

Keswick—June 29th 1818.

My dear Tom

I cannot make my Journal as distinct & actual as I could wish, from having been engaged in writing to George. & therefore I must tell you without circumstance that we proceeded from Ambleside to Rydal, saw the Waterfalls there, & called on Wordsworth, who was not at home. nor was any one of his family. I wrote a note & left it on the Mantlepiece. Thence on we came to the foot of Helvellyn, where we slept, but could not ascend it for the mist. I must mention that from Rydal we passed Thirlswater, & a fine pass in the Mountains from Helvellyn we came to Keswick on Derwent Water. The approach to Derwent Water surpassed Winandermere—it is richly wooded & shut in with rich-toned Mountains. From Helvellyn to Keswick was eight miles to Breakfast, After which we took a complete circuit of the Lake going about ten miles, & seeing on our way the Fall of Low-dore. I had an easy climb among the streams, about the fragments of Rocks & should have got I think to the summit, but unfortunately I was damped by slipping one leg into a squashy hole. There is no great body of water, but the accompaniment is delightful; for it ooses out from a cleft in perpendicular Rocks, all fledged with Ash & other beautiful trees. It is a strange thing how they got there. At the south end of the Lake, the Mountains of Burrowdale, are perhaps as fine as any thing we have seen—On our return from this circuit, we ordered dinner, & set forth about a mile & a half on the Penrith road, to see the Druid temple. We had a fag up hill, rather too near dinner time, which was rendered void, by the gratification of seeing those aged stones, on a gentle rise in the midst of Mountains, which at that time darkened all round, except at the fresh opening of the vale of St. John. We went to bed rather fatigued, but not so much so as to hinder us getting up this morning, to mount Skiddaw It promised all along to be fair, & we had fagged & tugged nearly to the top, when at halfpast six there came a mist upon us & shut out the view; we did not however lose anything by it, we were high enough without mist, to see the coast of Scotland; the Irish sea; the hills beyond Lancaster; & nearly all the large ones of Cumberland & Westmoreland, particularly Helvellyn & Scawfell: It grew colder & colder as we ascended, & we were glad at about three parts of the way to taste a little rum which the Guide brought with him, mixed, mind ye with mountain water, I took two glasses going & one returning—It is about six miles from where I am writing to the top. so we have walked ten miles before Breakfast today. We went up with two others, very good sort of fellows, All felt on arising into the cold air, that same elevation, which a cold bath gives one—I felt as if I were going to a Tournament. Wordsworth's house is situated just on the rise of the foot of mount Rydall, his parlor window looks directly down Winandermere; I do not think I told you how fine the vale of Grassmere is, & how I discovered 'the ancient woman seated on Helm Crag.'°—We shall proceed immediately to Carlisle, intending to enter Scotland on the 1st of July via —— July 1st—We are this morning at Carlisle—After Skiddaw, we walked to Ireby the oldest market town in Cumberland—where we were greatly amused by a country dancing school, holden at the Sun, it was indeed 'no new cotillon fresh from France.'° No they kickit & jumpit with mettle extraordinary, & whiskit, & fleckit, & toe'd it, & go'd it, & twirld it, & wheel'd it, & stampt it, & sweated it, tattooing the floor like mad; The difference between our country dances & these scotch figures, is about the same as leisurely stirring a cup o' Tea & beating up a batter pudding. I was extremely gratified to think, that if I had pleasures they knew nothing of. they had also some into which I could not possibly enter I hope I shall not return without having got the Highland fling, there was as fine a row of boys & girls as you ever saw, some beautiful faces, & one exquisite mouth. I never felt so near the glory of Patriotism, the glory of making by any means a country happier. This is what I like better than scenery. I fear our continued moving from place to place, will prevent our becoming learned in village affairs; we are mere creatures of Rivers, Lakes, & mountains. Our yesterday's journey was from Ireby to Wigton, & from Wigton to Carlisle—The Cathedral does not appear very fine; The Castle is very Ancient, & of Brick The City is very various, old white washed narrow streets; broad red brick ones more modern—I will tell you anon, whether the inside of the Cathedral is worth looking at. It is built of a sandy red stone or Brick. We have now walked 114 miles & are merely a little tired in the thighs, & a little blistered; We shall ride 38 miles to Dumfries, where we shall linger a while, about Nithsdale & Galloway, I have written two letters to Liverpool. I found a letter from sister George. very delightful indeed. I shall preserve it in the bottom of my knapsack for you.

[*'On visiting the Tomb of Burns' follows*]

You will see by this sonnet that I am at Dumfries, we have dined in Scotland. Burns' tomb is in the Churchyard corner, not very much to my taste, though on a scale, large enough to show they wanted to honour him—Mrs Burns lives in this place, most likely we shall see her tomorrow—This Sonnet I have written in a strange mood, half asleep. I know not how it is, the Clouds, the sky, the Houses, all seem anti Grecian & anti Charlemagnish—I will endeavour to get rid of my prejudices, & tell you fairly about the Scotch— July 2nd In Devonshire they say 'Well where be yee going.' Here it is, 'How is it all wi yoursel'—A man on the Coach said the horses took a Hellish heap o' drivin—the same fellow pointed out Burns' tomb with a deal of life, 'There de ye see it, amang the trees; white, wi a roond tap.' The first well dressed Scotchman we had any conversation with, to our surprise confessed himself a Deist. The careful manner of his delivering his opinions, not before he had received several encouraging hints from us, was very amusing—Yesterday was an immense Horse fair at Dumfries, so that we met numbers of men & women on the road, the women nearly all barefoot, with their shoes & clean stockings in hand, ready to put on & look smart in the Towns. There are plenty of wretched Cottages, where smoke has no outlet but by the door—We have now begun upon whiskey, called here *whuskey* very smart stuff it is—Mixed like our liquors with sugar & water tis called toddy, very pretty drink, & much praised by Burns.°

Letter to Fanny Keats, 2, 3, 5 July 1818

My dear Fanny, Dumfries July 2nd

I intended to have written to you from Kirkudbright the town I shall be in tomorrow—but I will write now becuse my Knapsack has worn my coat in the Seams, my coat has gone to the Taylors and I have but one Coat to my back in these parts. I must tell you how I went to Liverpool with George and our new Sister and the Gentleman my fellow traveller through the Summer and Autumn—We had a tolerable journey to Liverpool—which I left the next morning before George was up for Lancaster—Then we set off from Lancaster on foot with our Knapsacks on, and have walked a Little zig zag through the mountains and Lakes of Cumberland and Westmoreland—We came from Carlisle yesterday to this place—We are employed in going up Mountains, looking at Strange towns prying into old ruins and eating very hearty breakfasts. Here we are full in the Midst of broad Scotch 'How is it a' wi yoursel'—the Girls are walking about bare footed and in the worst cottages the Smoke finds its way out of the door—I shall come home full of news for you and for fear I should choak you by too great a dose at once I must make you used to it by a letter or two—We have been taken for travelling Jewellers, Razor sellers and Spectacle venders because friend Brown wears a pair—The first place we stopped at with our Knapsacks contained one Richard Bradshaw a notorious tippler—He stood in the shape of a ʒ° and ballanced himself as well as he could saying with his nose right in Mr Browns face 'Do— yo u sell Spect—ta—cles?' Mr Abbey says we are Don Quixotes—tell him we are more generally taken for Pedlars—All I hope is that we may not be taken for excisemen in this whiskey country—We are generally up about 5 walking before breakfast and we complete our 20 Miles before dinner—Yesterday we visited Burns's Tomb and this morning the fine Ruins of Lincluden—I had done thus far when my coat came back fortified at all points—so as we lose notime we set forth again through Galloway—all very pleasant and pretty with no fatigue when one is used to it—We are in the midst of Meg Merrilies' country of whom I suppose—you have heard—°

[*'Old Meg she was a Gipsey' follows*]

If you like these sort of Ballads I will now and then scribble one for you—if I send any to Tom I'll tell him to send them to you—I have so many interruptions that I cannot manage to fill a Letter in one day—since I scribbled the Song we have walked through a beautiful Country to Kirkudbright—at which place I will write you a song about myself—

[*'There was a naughty Boy' follows*]

My dear Fanny I am ashamed of writing you such stuff, nor would I if it were not for being tired after my days walking, and ready to tumble into bed so fatigued that when I am asleep you might sew my nose to my great toe and trundle me round the town like a Hoop without waking me—Then I get so hungry—a Ham goes but a very little way and fowls are like Larks to me—A Batch of Bread I make no more ado with than a sheet of parliament;° and I can eat a Bull's head as easily as I used to do Bull's eyes—I take a whole string of Pork Sausages down as easily as a Pen'orth of Lady's fingers°—Oh dear I must soon be contented with an acre or two of oaten cake a hogshead of Milk and a Cloaths basket of Eggs morning noon and night when I get among the Highlanders—Before we see them we shall pass into Ireland and have a chat with the Paddies, and look at the Giant's Cause-way which you must have heard of—I have not time to tell you particularly for I have to send a Journal to Tom of whom you shall hear all particulars or from me when I return—Since I began this we have walked sixty miles to newton stewart at which place I put in this Letter—tonight we sleep at Glenluce—tomorrow at Portpatrick and the next day we shall cross in the passage boat to Ireland—I hope Miss Abbey has quite recovered—Present my Respects to her and to M[r] And M[rs] Abbey—God bless you—

Your affectionate Brother JOHN—

Do write me a Letter directed to *Inverness*. Scotland—

Letter to Tom Keats, 3, 5, 7, 9 July 1818

My dear Tom, Auchencairn July 3rd

I have not been able to keep up my journal completely on accout of other letters to George and one which I am writing to Fanny from which I have turned to loose no time whilst Brown is coppying a song about Meg Merrilies which I have just written for her—We are now in Meg Merrilies county and have this morning passed through some parts exactly suited to her—Kirkudbright County is very beautiful, very wild with craggy hills somewhat in the westmoreland fashion—we have come down from Dumfries to the sea coast part of it—The song I mention you would have from Dilke: but perhaps you would like it here—

[*A fair copy of 'Old Meg she was a Gipsey' follows*]

Now I will return to Fanny—it rains. I may have time to go on here presently. July 5—You see I have missed a day from fanny's Letter. Yesterday was passed in Kircudbright—the Country is very rich—very fine—and with a little of Devon—I am now writing at Newton Stuart six Miles into Wigton—Our Landlady of yesterday said very few Southrens passed these ways—The children jabber away as in a foreign Language—The barefooted Girls look very much in keeping—I mean with the Scenery about them—Brown praises their cleanliness and appearance of comfort—the neatness of their cottages &c It may be—they are very squat among trees and fern and heaths and broom, on levels slopes and heights—They are very pleasant because they are very primitive—but I wish they were as snug as those up the Devonshire vallies—We are lodged and entertained in great varieties—we dined yesterday on dirty bacon dirtier eggs and dirtiest Potatoes with a slice of Salmon—we breakfast this morning in a nice carpeted Room with Sofa hair bottomed chairs and green-baized mehogany—A spring by the road side is always welcome—we drink water for dinner diluted with a Gill of wiskey. July 7th Yesterday Morning we set out from Glenluce going some distance round to see some Ruins—they were scarcely worth the while—we went on towards Stranrawier in a burning sun and had gone about six Miles when the Mail overtook us—we got up—were at Portpatrick in a jiffy, and I am writing now in little Ireland—The dialect on the neighbouring shores of Scotland and Ireland is much the same—yet I can perceive a great difference in the nations from the Chambermaid at this nate Inn kept by Mr Kelly—She is fair, kind and ready to laugh, because she is out of the horrible dominion of the Scotch kirk—A Scotch Girl stands in terrible awe of the Elders—poor little Susannas°—They will scarcely laugh—they are greatly to be pitied and the kirk is greatly to be damn'd. These kirkmen have done scotland good (Query?) they have made Men, Women, Old Men Young Men old Women, young women boys, girls and infants all careful—so that they are formed into regular Phalanges of savers and gainers—such a thrifty army cannot fail to enrich their Country and give it a greater apperance of comfort than that of their poor irish neighbours—These kirkmen have done Scotland harm—they have banished puns and laughing and kissing (except in cases where the very danger and crime must make it very fine and gustful. I shall make a full stop at kissing for after that there should be a better paren*t*-thesis: and go on to remind you of the fate of Burns. Poor unfortunate fellow—his disposition was southern—how sad it is when a luxurious imagination is obliged in self defence to deaden its delicacy in vulgarity, and riot in thing attainable that it may not have leisure to go mad after thing which are not. No Man in such matters will be content with the experience of others—It is true that out of suffrance there is no greatness, no dignity; that in the most abstracted Pleasure there is no lasting happiness: yet who would not like to discover over again that Cleopatra was a Gipsey, Helen a Rogue and Ruth a deep one? I have not sufficient reasoning faculty to settle the doctrine of thrift—as it is consistent with the dignity of human Society—with the happiness of Cottagers—All I can do is by plump contrasts—Were the fingers made to squeeze a guinea or a white hand? Were the Lips made to hold a pen or a kiss? And yet in Cities Man is shut out from his fellows if he is poor, the Cottager must be dirty and very wretched if she be not thrifty—The present state of society demands this and this convinces me that the world is very young and in a verry ignorant state—We live in a barbarous age. I would sooner be a wild deer than a Girl under the dominion of the kirk, and I would sooner be a wild hog than be the occasion of a Poor Creatures pennance before those execrable elders—It is not so far to the Giant's Cause way as we supposed—we thought it 70 and hear it is only 48 Miles—so we shall leave one of our knapsacks here at Donoghadee, take our immediate wants and be back in a week—when we shall proceed to the County of Ayr. In the Packet Yesterday we heard some Ballads from two old Men—one was a romance which seemed very poor—then there was the Battle of the Boyne—then Robin Huid as they call him—'Before the king you shall go, go, go, before the king you shall go.'° There were no Letters for me at Port Patrick so I am behind hand with you I dare say in news from George. Direct to Glasgow till the 17th of this month. 9th We stopped very little in Ireland and that you may not have leisere to marvel at our speedy return to Portpatrick I will tell you that is it as dear living in Ireland as at the Hummums°—thrice the expence of Scotland—it would have cost us £15 before our return—Moreover we found those 48 Miles to be irish ones which reach to 70 english—So having walked to Belfast one day and back to Donoghadee the next we left Ireland with a fair breeze—We slept last night at Port patrick where I was gratified by a letter from you. On our walk in Ireland we had too much opportunity to see the worse than nakedness, the rags, the dirt and misery of the poor common Irish—A Scotch cottage, though in that some times the Smoke has no exit but at the door, is a pallace to an irish one—We could observe that impetiosity in Man and boy and Woman—We had the pleasure of finding our way through a Peat-Bog—three miles long at least—dreary, black, dank, flat and spongy: here and there were poor dirty creatures and a few strong men cutting or carting peat. We heard on passing into Belfast through a most wretched suburb that most disgusting of all noises worse than the Bag pipe, the laugh of a Monkey, the chatter of women *solus* the scream of Macaw—I mean the sound of the Shuttle°—What a tremendous difficulty is the improvement of the condition of such people—I cannot conceive how a mind 'with child' of Philantrophy could grasp at possibility—with me it is absolute despair. At a miserable house of entertainment half way between Donaghadee and Bellfast were two Men Sitting at Whiskey one a Laborer and the other I took to be a drunken Weaver—The Laborer took me for a Frenchman and the other hinted at Bounty Money saying he was ready to take it—On calling for the Letters at Port patrick the man snapp'd out 'what Regiment'? On our return from Bellfast we met a Sadan—the Duchess of

Dunghill—It is no laughing matter tho—Imagine the worst dog kennel you ever saw placed upon two poles from a mouldy fencing—In such a wretched thing sat a squalid old Woman squat like an ape half starved from a scarcity of Buiscuit in its passage from Madagascar to the cape,—with a pipe in her mouth and looking out with a round-eyed skinny lidded, inanity—with a sort of horizontal idiotic movement of her head—squab and lean she sat and puff'd out the smoke while two ragged tattered Girls carried her along—What a thing would be a history of her Life and sensations. I shall endeavour when I know more and have thought a little more, to give you my ideas of the difference between the scotch and irish—The two Irishmen I mentioned were speaking of their treatment in England when the Weaver said—'Ah you were a civil Man but I was a drinker' Remember me to all—I intend writing to Haslam—but dont tell him for fear I should delay—We left a notice at Portpatrick that our Letters should be thence forwarded to Glasgow—Our quick return from Ireland will occasion our passing Glasgow sooner than we thought—so till further notice you must direct to Inverness

Your most affectionate Brother JOHN—

Remember me to the Bentleys

Letter to J. H. Reynolds, 11, 13 July 1818

Maybole July 11.

My Dear Reynolds.

I'll not run over the Ground we have passed. that would be merely as bad as telling a dream—unless perhaps I do it in the manner of the Laputan printing press—that is I put down Mountains, Rivers Lakes, dells, glens, Rocks, and Clouds, With beautiful enchanting, gothic picturesque fine, delightful, enchancting, Grand, sublime—a few Blisters &c—and now you have our journey thus far: where I begin a letter to you because I am approaching Burns's Cottage very fast—We have made continual enquiries from the time we saw his Tomb at Dumfries—his name of course is known all about—his great reputation among the plodding people is 'that he wrote a good MONY sensible things'—One of the pleasantest means of annulling self is approaching such a shrine as the Cottage of Burns—we need not think of his misery—that is all gone—bad luck to it—I shall look upon it hereafter with unmixed pleasure as I do upon my Stratford on Avon day with Bailey—I shall fill this sheet for you in the Bardies Country, going no further than this till I get into the Town of Ayr which will be a 9 miles' walk to Tea—We were talking on different and indifferent things, when on a sudden we turned a corner upon the immediate County of Air—the Sight was as rich as possible—I had no Conception that the native place of Burns was so beautiful—the Idea I had was more desolate, his rigs of Barley seemed always to me but a few strips of Green on a cold hill—O prejudice! it was rich as Devon—I endeavour'd to drink in the Prospect, that I might spin it out to you as the silkworm makes silk from Mulbery leaves—I cannot recollect it—Besides all the Beauty, there were the Mountains of Arran Isle, black and huge over the Sea—We came down upon every thing suddenly—there were in our way, the 'bonny Doon,' with the Brig that Tam O' Shanter cross'ed—Kirk Alloway, Burns's Cottage and then the Brigs of Ayr—First we stood upon the Bridge across the Doon; surrounded by every Phantasy of Green in tree, Meadow, and hill,—the Stream of the Doon, as a Farmer told us, is covered with trees from head to foot—you know those beautiful heaths so fresh against the weather of a summers evening—there was one stretching along behind the trees. I wish I knew always the humour my friends would be in at opening a letter of mine, to suit it to them nearly as possible I could always find an egg shell for Melancholy—and as for Merriment a Witty humour will turn any thing to Account—my head is sometimes in such a whirl in considering the million likings and antipathies of our Moments—that I can get into no settled strain in my Letters—My Wig! Burns and sentimentality coming across you and frank Floodgate in the office—O scenery that thou shouldst be crush'd between two Puns—As for them I venture the rascalliest in the Scotch Region—I hope Brown does not put them punctually in his journal—If he does I must sit on the cutty-stool° all next winter. We Went to Kirk allow'y 'a Prophet is no Prophet in his own Country'—We went to the Cottage and took some Whiskey—I wrote a sonnet for the mere sake of writing some lines under the roof—they are so bad I cannot transcribe them—The Man at the Cottage was a great Bore with his Anecdotes—I hate the rascal—his Life consists in fuz, fuzzy, fuzziest—He drinks glasses five for the Quarter and twelve for the hour,—he is a mahogany faced old Jackass who knew Burns—He ought to be kicked for having spoken to him. He calls himself 'a curious old Bitch'—but he is a flat old Dog—I sho[d] like to employ Caliph Vatheck° to kick him—O the flummery of a birth place! Cant! Cant! Cant! It is enough to give a spirit the guts-ache—Many a true word they say is spoken in jest—this may be because his gab hindered my sublimity.—The flat dog made me write a flat sonnet—My dear Reynolds—I cannot write about scenery and visitings—Fancy is indeed less than a present palpable reality, but it is greater than remembrance—you would lift your eyes from Homer only to see close before you the real Isle of Tenedos.—you would rather read Homer afterwards than remember yourself—One song of Burns's is of more worth to you than all I could think for a whole year in his native country—His Misery is a dead weight upon the nimbleness of one's quill—I tried to forget it—to drink Toddy without any Care—to write a merry Sonnet—it wont do—he talked with Bitches—he drank with Blackguards, he was miserable—We can see horribly clear in the works of such a man his whole life, as if we were God's spies.°—What were his addresses to Jean in the latter part of his life—I should not speak so to you—yet why not—you are not in the same case—you are in the right path, and you shall not be deceived—I have spoken to you against Marriage, but it was general—the Prospect in those matters has been to me so blank, that I have not been unwilling to die—I would not now, for I have inducements to Life—I must see my little Nephews in America, and I must see you marry your lovely Wife—My sensations are sometimes deadened for weeks together—but believe me I have more than once yearne'd for the time of your happiness to come, as much as I could for myself after the lips of Juliet.—From the tenor of my occasional rhodomontade in chitchat, you might have been deceived concerning me in these points—upon my soul, I have been getting more and more close to you every day, ever since I knew you, and now one of the first pleasures I look to is your happy Marriage—the more, since I have felt the pleasure of loving a sister in Law. I did not think it possible to become so much attached in so short a time—Things like these, and they are real, have made me resolve to have a care of my health—you must be as careful—The rain has stopped us to day at the end of a dozen Miles, yet we hope to see Loch-Lomond the day after to Morrow;—I will piddle out my information, as Rice says, next Winter at any time when a substitute is wanted for Vingt-un.° We bear the fatigue very well.—20 Miles a day in general—A cloud came over us in getting up Skiddaw—I hope to be more lucky in Ben Lomond—and more lucky still in Ben Nevis—what I think you wo[d] enjoy is poking about Ruins—sometimes Abbey, sometimes Castle. The short stay we made in Ireland has left few remembrances—but an old woman in a dog-kennel Sedan with a pipe in her Mouth, is what I can never forget—I wish I may be able to give you an idea of her—Remember me to your Mother and Sisters, and tell your Mother how I hope she will pardon me for having a scrap of paper pasted in the Book sent to her. I was driven on all sides and had not time to call on Taylor—So Bailey is coming to Cumberland°—well, if you'll let me know where at Inverness, I call on my return and pass a little time with him—I am glad 'tis not scotland—Tell my friends I do all I can for them, that is drink their healths in Toddy—Perhaps I may have some lines by and by to send you fresh on your own Letter—Tom has a few to shew you.

your affectionate friend
JOHN KEATS

Letter to Mrs. James Wylie, 6 August 1818

My dear Madam— Inverness 6th August 1818

It was a great regret to me that I should leave all my friends, just at the moment when I might have helped to soften away the time for them. I wanted not to leave my Brother Tom, but more especially, beleive me, I should like to have remained near you, were it but for an atom of consolation, after parting with so dear a daughter; My brother George has ever been more than a brother to me, he has been my greatest friend, & I can never forget the sacrifice you have made for his happiness. As I walk along the Mountains here, I am full of these things, & lay in wait, as it were, for the pleasure of seeing you, immediately on my return to town. I wish above all things, to say a word of Comfort to you, but I know not how. It is impossible to prove that black is white, It is impossible to make out, that sorrow is joy or joy is sorrow———Tom tells me that you called on Mr Haslam with a Newspaper giving an account of a Gentleman in a Fur cap, falling over a precipice in Kirkudbrightshire. If it was me, I did it in a dream, or in some magic interval between the first & second cup of tea; which is nothing extraordinary, when we hear that Mahomet, in getting out of Bed, upset a jug of water, & whilst it was falling, took a fortnight's trip as it seemed to Heaven: yet was back in time to save one drop of water being spilt.° As for Fur caps I do not remember one beside my own, except at Carlisle—this was a very good Fur cap, I met in the High Street, & I daresay was the unfortunate one. I daresay that the fates seeing but two Fur caps in the North, thought it too extraordinary, & so threw the Dies which of them should be drowned. The lot fell upon Jonas—I daresay his name was Jonas. All I hope is, that the gaunt

Ladies said not a word about hanging, if they did, I shall one day regret that I was not half drowned in Kirkudbright. Stop! let me see!—being half drowned by falling from a precipice is a very romantic affair—Why should I not take it to myself? Keep my secret & I will. How glorious to be introduced in a drawing room to a Lady who reads Novels, with—'M^{r} so & so—Miss so & so—Miss so & so. this is M^{r} so & so. who fell off a precipice, & was half drowned' Now I refer it to you whether I should loose so fine an opportunity of making my fortune—No romance lady could resist me—None—Being run under a Waggon; side lamed at a playhouse; Apopletic, through Brandy; & a thousand other tolerably decent things for badness would be nothing; but being tumbled over a precipice into the sea—Oh it would make my fortune—especially if you could continue to hint, from this bulletins authority, that I was not upset on my own account, but that I dashed into the waves after Jessy of Dumblane°—& pulled her out by the hair—But that, Alas! she was dead or she would have made me happy with her hand—however in this you may use your own discretion—But I must leave joking & seriously aver, that I have been *werry* romantic indeed, among these Mountains & Lakes. I have got wet through day after day, eaten oat cake, & drank whiskey, walked up to my knees in Bog, got a sore throat, gone to see Icolmkill & Staffa, met with wholesome food, just here & there as it happened; went up Ben Nevis, & N.B. came down again; Sometimes when I am rather tired, I lean rather languishingly on a Rock, & long for some famous Beauty to get down from her Palfrey in passing; approach me with—her saddle bags—& give me—a dozen or two capital roast beef sandwiches—When I come into a large town, you know there is no putting ones Knapsack into ones fob; so the people stare—We have been taken for Spectacle venders, Razor sellers, Jewellers, travelling linnen drapers, Spies, Excisemen, & many things else, I have no idea of—When I asked for letters at the Post Office, Port Patrick; the man asked what Regiment? I have had a peep also at little Ireland. Tell Henry I have not Camped quite on the bare Earth yet; but nearly as bad, in walking through Mull—for the Shepherds huts you can scarcely breathe in, for the smoke which they seem to endeavour to preserve for smoking on a large scale. Besides riding about 400, we have walked above 600 Miles, & may therefore reckon ourselves as set out.   I wish my dear Madam, that one of the greatest pleasures I shall have on my return, will be seeing you & that I shall ever be

Yours with the greatest Respect & sincerity

JOHN KEATS—

Letter to C. W. Dilke, 20, 21 September 1818

My dear Dilke,

According to the Wentworth place Bulletin you have left Brighton much improved: therefore now a few lines will be more of a pleasure than a bore. I have a few things to say to you and would fain begin upon them in this forth line: but I have a Mind too well regulated to proceed upon any thing without due preliminary remarks—you may perhaps have observed that in the simple process of eating radishes I never begin at the root but constantly dip the little green head in the salt—that in the Game of Whist if I have an ace I constantly play it first—So how can I with any face begin without a dissertation on letter writing—Yet when I consider that a sheet of paper contains room only for three pages, and a half° how can I do justice to such a pregnant subject? however as you have seen the historry of the world stamped as it were by a diminishing glass in the form of a chronological Map, so will I 'with retractile claws'° draw this in to the form of a table—whereby it will occupy merely the remainder of this first page—

Folio——	Parsons, Lawyers, Statesmen, Physians out of place—Ut—Eustace°—Thornton°out of practice or on their travels—
Fools cap—	1 superfine! rich or noble poets—ut Byron. 2 common ut egomet—
Quarto—	Projectors, Patentees, Presidents, Potatoe growers—
Bath°	Boarding schools, and suburbans in general
Gilt edge	Dandies in general, male female and literary—
Octavo or tears	All who make use of a lascivious seal—
Duodec—	May be found for the most part on Milliners and Dressmakers Parlour tables—
Strip	At the Playhouse doors, or any where—
Slip	Being but a variation—
Snip	So called from its size being disguised by a twist—

I suppose you will have heard that Hazlitt has on foot a prosecution against Blackwood°—I dined with him a few days since at Hessey's—there was not a word said about, though I understand he is excessively vexed—Reynolds by what I hear is almost over happy and Rice is in town. I have not seen him nor shall I for some time as my throat has become worse after getting well, and I am determined to stop at home till I am quite well—I was going to Town tomorrow with Mrs D.° but I thought it best, to ask her excuse this morning—I wish I could say Tom was any better. His identity presses upon me so all day that I am obliged to go out—and although I intended to have given some time to study alone I am obliged to write, and plunge into abstract images to ease myself of his countenance his voice and feebleness—so that I live now in a continual fever—it must be poisonous to life although I feel well. Imagine 'the hateful siege of contraries'°—if I think of fame of poetry it seems a crime to me, and yet I must do so or suffer—I am sorry to give you pain—I am almost resolv'd to burn this—but I really have not self possession and magninimity enough to manage the thing othewise—after all it may be a nervousness proceeding from the Mercury—

Bailey I hear is gaining his Spirits and he will yet be what I once thought impossible a cheerful Man—I think he is not quite so much spoken of in Little Brittain. I forgot to ask Mrs Dilke if she had any thing she wanted to say immediately to you—This morning look'd so unpromising that I did not think she would have gone—but I find she has on sending for some volumes of Gibbon—I was in a little funk yesterday, for I sent an unseal'd note of sham abuse, until I recollected from what I had heard Charles° say, that the servant could neither read nor write—not even to her Mother as Charles observed. I have just had a Letter from Reynolds—he is going on gloriously. The following is a translation of a Line of Ronsard—

'Love poured her Beauty into my warm veins'—.°

You have passed your Romance and I never gave into it or else I think this line a feast for one of your Lovers—How goes it with Brown?

Your sincere friend
JOHN KEATS—

Letter to J. H. Reynolds, ?22 September 1818

My dear Reynolds,

Believe me I have rather rejoiced in your happiness than fretted at your silence. Indeed I am grieved on your account that I am not at the same time happy—But I conjure you to think at Present of nothing but pleasure 'Gather the rose &c'° Gorge the honey of life. I pity you as much that it cannot last for ever, as I do myself now drinking bitters.—Give yourself up to it—you cannot help it—and I have a Consolation in thinking so—I never was in love—Yet the voice and the shape of a woman has haunted me these two days—at such a time when the relief, the feverous relief of Poetry seems a much less crime—This morning Poetry has conquered—I have relapsed into those abstractions which are my only life—I feel escaped from a new strange and threatening sorrow.—And I am thankful for it—There is an awful warmth about my heart like a load of Immortality.

Poor Tom—that woman°—and Poetry were ringing changes in my senses—now I am in comparison happy—I am sensible this will distress you—you must forgive me. Had I known you would have set out so soon I could have sent you the 'Pot of Basil' for I had copied it out ready.—Here is a free translation of a Sonnet of Ronsard, which I think will please you—I have the loan of his works—they have great Beauties.

[*A draft of 'Nature withheld Cassandra in the Skies' follows*]

I had not the original by me when I wrote it, and did not recollect the purport of the last lines—I should have seen Rice ere this—but I am confined by Sawrey's mandate in the house now, and have as yet only gone out in fear of the damp night—You know what an undangerous matter it is. I shall soon be quite recovered—Your offer I shall remember as though it had even now taken place in fact—I think it can not be—Tom is not up yet—I can not say he is better. I have not heard from George.

Y^r^ affect^te^ friend JOHN KEATS

Letter to J. A. Hessey, 8 October 1818

My dear Hessey.

You are very good in sending me the letter from the Chronicle°—and I am very bad in not acknowledging such a kindness sooner.—pray forgive me—It has so chanced that I have had that paper every day—I have seen today's. I cannot but feel indebted to those Gentlemen who have taken my part—As for the rest, I begin to get a little acquainted with my own strength and weakness.—Praise or blame has but a momentary effect on the man whose love of beauty in the abstract makes him a severe critic on his own Works. My own domestic criticism has given me pain without comparison beyond what Blackwood or the Quarterly could possibly inflict. and also when I feel I am right, no external praise can give me such a glow as my own solitary reperception & ratification of what is fine. J. S. is perfectly right in regard to the slipshod Endymion. That it is so is no fault of mine.—No!—though it may sound a little paradoxical. It is as good as I had power to make it—by myself—Had I been nervous about its being a perfect piece, & with that view asked advice, & trembled over every page, it would not have been written; for it is not in my nature to fumble—I will write independantly.—I have written independently *without Judgment*—I may write independently *& with judgment* hereafter.—The Genius of Poetry must work out its own salvation in a man: It cannot be matured by law & precept, but by sensation & watchfulness in itself—That which is creative must create itself—In Endymion, I leaped headlong into the Sea, and thereby have become better acquainted with the Soundings, the quicksands, & the rocks, than if I had stayed upon the green shore, and piped a silly pipe, and took tea & comfortable advice.—I was never afraid of failure; for I would sooner fail than not be among the greatest—But I am nigh getting into a rant. So, with remembrances to Taylor & Woodhouse &c I am

Yrs very sincerely
JOHN KEATS

Letter to Richard Woodhouse, 27 October 1818

My dear Woodhouse,

Your Letter gave me a great satisfaction; more on account of its friendliness, than any relish of that matter in it which is accounted so acceptable in the 'genus irritabile' The best answer I can give you is in a clerklike manner to make some observations on two principle points, which seem to point like indices into the midst of the whole pro and con, about genius, and views and atchievements and ambition and cœtera. 1st As to the poetical Character itself, (I mean that sort of which, if I am any thing, I am a Member; that sort distinguished from the wordsworthian or egotistical sublime; which is a thing per se and stands alone) it is not itself—it has no self—it is every thing and nothing—It has no character—it enjoys light and shade; it lives in gusto, be it foul or fair, high or low, rich or poor, mean or elevated—It has as much delight in conceiving an Iago as an Imogen. What shocks

the virtuous philosoper, delights the camelion Poet. It does no harm from its relish of the dark side of things any more than from its taste for the bright one; because they both end in speculation. A Poet is the most unpoetical of any thing in existence; because he has no Identity—he is continually in for—and filling some other Body—The Sun, the Moon, the Sea and Men and Women who are creatures of impulse are poetical and have about them an unchangeable attribute—the poet has none; no identity—he is certainly the most unpoetical of all God's Creatures. If then he has no self, and if I am a Poet, where is the Wonder that I should say I would write no more? Might I not at that very instant been cogitating on the Characters of saturn and Ops?° It is a wretched thing to confess; but is a very fact that not one word I ever utter can be taken for granted as an opinion growing out of my identical nature—how can it, when I have no nature? When I am in a room with People if I ever am free from speculating on creations of my own brain, then not myself goes home to myself: but the identity of every one in the room begins so to press upon me that, I am in a very little time anhilated—not only among Men; it would be the same in a Nursery of children: I know not whether I make myself wholly understood: I hope enough so to let you see that no dependence is to be placed on what I said that day.
In the second place I will speak of my views, and of the life I purpose to myself—I am ambitious of doing the world some good: if I should be spared that may be the work of maturer years—in the interval I will assay to reach to as high a summit in Poetry as the nerve bestowed upon me will suffer. The faint conceptions I have of Poems to come brings the blood frequently into my forehead—All I hope is that I may not lose all interest in human affairs—that the solitary indifference I feel for applause even from the finest Spirits, will not blunt any acuteness of vision I may have. I do not think it will—I feel assured I should write from the mere yearning and fondness I have for the Beautiful even if my night's labours should be burnt every morning and no eye ever shine upon them. But even now I am perhaps not speaking from myself; but from some character in whose soul I now live. I am sure however that this next sentence is from myself. I feel your anxiety, good opinion and friendliness in the highest degree, and am

Your's most sincerely
JOHN KEATS

Letter to George and Georgiana Keats, 14, 16, 21. 24, 31 October 1818

My dear George; There was a part in your Letter which gave me a great deal of pain, that where you lament not receiving Letters from England—I intended to have written immediately on my return from Scotland (which was two Months earlier than I had intended on account of my own as well as Tom's health) but then I was told by M[rs] W—— that you had said you would not wish any one to write till we had heard from you. This I thought odd and now I see that it could not have been so; yet at the time I suffered my unreflecting head to be satisfied and went on in that sort of abstract careless and restless Life with which you are well acquainted. This sentence should it give you any uneasiness do not let it last for before I finish it will be explained away to your satisfaction—

I am grieved to say that I am not sorry you had not Letters at Philadelphia; you could have had no good news of Tom and I have been withheld on his account from beginning these many days; I could not bring myself to say the truth, that he is no better, but much worse—However it must be told, and you must my dear Brother and Sister take example frome me and bear up against any Calamity for my sake as I do for your's. Our's are ties which independent of their own Sentiment are sent us by providence to prevent the deleterious effects of one great, solitary grief. I have Fanny and I have you—three people whose Happiness to me is sacred—and it does annul that selfish sorrow which I should otherwise fall into, living as I do with poor Tom who looks upon me as his only comfort—the tears will come into your Eyes—let them—and embrace each other—thank heaven for what happiness you have and after thinking a moment or two that you suffer in common with all Mankind hold it not a sin to regain your cheerfulness—I will relieve you of one uneasiness of overleaf: I returned I said on account of my health—I am now well from a bad sore throat which came of bog trotting in the Island of Mull—of which you shall hear by the coppies I shall make from my Scotch Letters—Your content in each other is a delight to me which I cannot express—the Moon is now shining full and brilliant—she is the same to me in Matter, what you are to me in Spirit—If you were here my dear Sister I could not pronounce the words which I can write to you from a distance: I have a tenderness for you, and an admiration which I feel to be as great and more chaste than I can have for any woman in the world. You will mention Fanny—her character is not formed; her identity does not press upon me as yours does. I hope from the bottom of my heart that I may one day feel as much for her as I do for you—I know not how it is, but I have never made any acquaintance of my own—nearly all through your medium my dear Brother—through you I know not only a Sister but a glorious human being—And now I am talking of those to whom you have made me known I cannot forbear mentioning Haslam as a most kind and obliging and constant friend—His behaviour to Tom during my absence and since my return has endeared him to me for ever—besides his anxiety about you. Tomorrow I shall call on your Mother° and exchange information with her—On Tom's account I have not been able to pass so much time with her as I would otherwise have done—I have seen her but twice—one I dined with her and Charles—She was well, in good Spirits and I kept her laughing at my bad jokes—We went to tea at Mrs Millar's and in going were particularly struck with the light and shade through the Gate way at the Horse Guards. I intend to write you such Volumes that it will be impossible for me to keep any order or method in what I write: that will come first which is uppermost in my Mind, not that which is uppermost in my heart—besides I should wish to give you a picture of our Lives here whenever by a touch I can do it; even as you must see by the last sentence our walk past Whitehall all in good health and spirits—this I am certain of, because I felt so much pleasure from the simple idea of your playing a game at Cricket—At Mrs Millars I saw Henry quite well—there was Miss Keasle—and the goodnatured Miss Waldegrave—Mrs Millar began a long story and you know it is her Daughter's way to help her on as though her tongue were ill of the gout—Mrs M. certainly tells a Story as though she had been taught her Alphabet in Crutched Friars.° Dilke has been very unwell; I found him very ailing on my return—he was under Medical care for some time, and then went to the Sea Side whence he has returned well—Poor little M^rs^ D—has had another gall-stone attack; she was well ere I returned—she is now at Brighton—Dilke was greatly pleased to hear from you and will write a Letter for me to enclose—He seems greatly desirous of hearing from you of the Settlement° itself—I came by ship from Inverness and was nine days at Sea without being sick—a little Qualm now and then put me in mind of you—however as soon as you touch the shore all the horrors of sickness are soon forgotten; as was the case with a Lady on board who could not hold her head up all the way. We had not been in the Thames an hour before her tongue began to some tune; paying off as it was fit she should all old scores. I was the only Englishman on board. There was a

downright Scotchman who hearing that there had been a bad crop of Potatoes in England had brought some triumphant Specimens from Scotland—these he exhibited with national pride to all the Lightermen,° and Watermen from the Nore to the Bridge. I fed upon beef all the way; not being able to eat the thick Porridge which the Ladies managed to manage with large awkward horn spoons into the bargain. Severn has had a narrow escape of his Life from a Typhous fever: he is now gaining strength—Reynolds has returned from a six weeks enjoyment in Devonshire, he is well and persuades me to publish my pot of Basil as an answer to the attacks made on me in Blackwood's Magazine and the Quarterly Review. There have been two Letters in my defence in the Chronicle and one in the Examiner, coppied from the Alfred Exeter paper,° and written by Reynolds—I do not know who wrote those in the Chronicle—This is a mere matter of the moment—I think I shall be among the English Poets after my death. Even as a Matter of present interest the attempt to crush me in the Quarterly has only brought me more into notice and it is a common expression among book men 'I wonder the Quarterly should cut its own throat.'
It does me not the least harm in Society to make me appear little and rediculous: I know when a Man is superior to me and give him all due respect—he will be the last to laugh at me and as for the rest I feel that I make an impression upon them which insures me personal respect while I am in sight whatever they may say when my back is turned—Poor Haydon's eyes will not suffer him to proceed with his picture—he has been in the Country—I have seen him but once since my return—I hurry matters together here because I do not know when the Mail sails—I shall enqure tomorrow and then shall know whether to be particular or general in my letter—you shall have at least two sheets a day till it does sail whether it be three days or a fortnight—and then I will begin a fresh one for the next Month. The Miss Reynoldses are very kind to me—but they have lately displeased me much and in this way—Now I am coming the Richardson.° On my return, the first day I called they were in a sort of taking or bustle about a Cousin of theirs° who having fallen out with her Grandpapa in a serious manner, was invited by M^rs R— to take Asylum in her house—She is an east indian and ought to be her Grandfather's Heir. At the time I called Mrs R. was in conference with her up stairs and the young Ladies were warm in her praises down stairs calling her genteel, interresting and a thousand other pretty things to which I gave no heed, not being partial to 9 days wonders—Now all is completely changed—they hate her; and from what I hear she is not without faults—of a real kind: but she has others

which are more apt to make women of inferior charms hate her. She is not a Cleopatra; but she is at least a Charmian. She has a rich eastern look; she has fine eyes and fine manners. When she comes into a room she makes an impression the same as the Beauty of a Leopardess. She is too fine and too concious of her Self to repulse any Man who may address her—from habit she thinks that nothing *particular*.° I always find myself more at ease with such a woman; the picture before me always gives me a life and animation which I cannot possibly feel with any thing inferiour—I am at such times too much occupied in admiring to be awkward or on a tremble. I forget myself entirely because I live in her. You will by this time think I am in love with her; so before I go any further I will tell you I am not—she kept me awake one Night as a tune of Mozart's might do—I speak of the thing as a passtime and an amuzement than which I can feel none deeper than a conversation with an imperial woman the very 'yes' and 'no' of whose Lips is to me a Banquet. I dont cry to take the moon home with me in my Pocket nor do I fret to leave her behind me. I like her and her like because one has no *sensations*—what we both are is taken for granted—You will suppose I have by this had much talk with her—no such thing—there are the Miss Reynoldses on the look out—They think I dont admire her because I did not stare at her—They call her a flirt to me—What a want of knowledge? she walks across a room in such a manner that a Man is drawn towards her with a magnetic Power. This they call flirting! they do not know things. They do not know what a Woman is. I believe tho' she has faults—the same as Charmian and Cleopatra might have had—Yet she is a fine thing speaking in a worldly way: for there are two distinct tempers of mind in which we judge of things—the worldly, theatrical and pantomimical; and the unearthly, spiritual and etherial—in the former Buonaparte, Lord Byron and this Charmian hold the first place in our Minds; in the latter John Howard, Bishop Hooker rocking his child's cradle° and you my dear Sister are the conquering feelings. As a Man in the world I love the rich talk of a Charmian; as an eternal Being I love the thought of you. I should like her to ruin me, and I should like you to save me. Do not think my dear Brother from this that my Passions are head long or likely to be ever of any pain to you—no

> 'I am free from Men of Pleasure's cares
> By dint of feelings far more deep than theirs'

This is Lord Byron,° and is one of the finest things he has said—I have no town talk for you, as I have not been much among people—as for Politics they are in my opinion only sleepy because they will soon be too wide awake—Perhaps not—for the long and continued Peace of England itself has given us notions of personal safety which are likely to prevent the reestablishment of our national Honesty—There is of a truth nothing manly or sterling in any part of the Government. There are many Madmen In the Country, I have no doubt, who would like to be beheaded on tower Hill merely for the sake of eclat, there are many Men like Hunt° who from a principle of taste would like to see things go on better, there are many like Sir F. Burdett° who like to sit at the head of political dinners—but there are none prepared to suffer in obscurity for their Country—the motives of our worst Men are interest and of our best Vanity—We have no Milton, no Algernon Sidney—Governers in these days loose the title of Man in exchange for that of Diplomat and Minister—We breathe in a sort of Officinal Atmosphere—All the departments of Government have strayed far from Spimpicity which is the greatest of Strength—there is as much difference in this respect between the present Government and oliver Cromwell's, as there is between the 12 Tables of Rome and the volumes of Civil Law which were digested by Justinian. A Man now entitlerd Chancellor has the same honour paid to him whether he be a Hog or a Lord Bacon. No sensation is created by Greatness but by the number of orders a Man has at his Button holes Notwithstand the part which the Liberals take in the Cause of Napoleon I cannot but think he has done more harm to the life of Liberty than any one else could have done: not that the divine right Gentlemen° have done or intend to do any good—no they have taken a Lesson of him and will do all the further harm he would have done without any of the good—The worst thing he has done is, that he has taught them how to organize their monstrous armies —The Emperor Alexander it is said intends to divide his Empire as did Diocletian—creating two Czars besides himself, and continuing the supreme Monarch of the whole—Should he do this and they for a series of Years keep peacable among themselves Russia may spread her conquest even to China—I think a very likely thing that China itself may fall Turkey certainly will—Meanwhile european north Russia will hold its horns against the rest of Europe, intrieguing constantly with France. Dilke, whom you know to be a Godwin perfectibily Man,° pleases himself with the idea that America will be the country to take up the human intellect where england leaves off—I differ there with him greatly—A country like the united states whose greatest Men are Franklins and Washingtons will never do that—They are great Men doubtless but how are they to be compared to those our countrey men Milton and the two Sidneys—The one is a philosophical Quaker full of mean and thrifty maxims the other sold the very Charger who had taken him through all his Battles—Those American's are great but they are not sublime Man—the humanity of the United States can never reach the sublime—Birkbeck's mind is too much in the American Stryle—you must endeavour to infuse a little Spirit of another sort into the Settlement, always with great caution, for thereby you may do your descendents more good than you may imagine. If I had a prayer to make for any great good, next to Tom's recovery, it should be that one of your Children should be the first American Poet. I have a great mind to make a prophecy and they say prophecies work out their own fullfillment.

[*''Tis "the witching time of night"' follows*]

This is friday, I know not what day of the Month°—I will enquire tomorrow for it is fit you should know the time I am writing. I went to Town yesterday, and calling at Mrs Millar's was told that your Mother would not be found at home—I met Henry as I turned the corner—I had no leisure to return, so I left the letters with him—He was looking very well—Poor Tom is no better tonight—I am affraid to ask him what Message I shall send from him—And here I could go on complaining of my Misery, but I will keep myself cheerful for your Sakes. With a great deal of trouble I have succeeded in getting Fanny to Hampstead—she has been several times—Mr Lewis has been very kind to Tom all the Summer there has scarce a day passed but he has visited him, and not one day without bringing or sending some fruit of the nicest kind. He has been very assiduous in his enquiries after you—It would give the old Gentleman a great pleasure if you would send him a Sheet enclosed in the next parcel to me, after you receive this—how long it will be first—Why did I not write to Philadelphia? Really I am sorry for that neglect—I wish to go on writing ad infinitum, to you—I wish for interresting matter, and a pen as swift as the wind—But the fact is I go so little into the Crowd now that I have nothing fresh and fresh every day to speculate upon, except my own Whims and Theroies—I have been but once to Haydon's, onece to Hunt's, once to Rices, once to Hessey's I have not seen Taylor, I have not been to the Theatre—Now if I had been many times to all these and was still in the habit of going I could on my return at night have each day something new to tell you of without any stop—But now I have such a dearth that when I get to the end of this sentence and to the bottom of this page I much wait till I can find something interesting to you before I begin another.—After all it is not much matter what it may be about; for the very words from such a distance penned by this hand will be grateful to you—even though I were to coppy out the tale of Mother Hubbard or Little Red Riding Hood—I have been over to Dilke's this evening—there with Brown we have been talking of different and indifferent Matters—of Euclid, of Metaphisics of the Bible, of Shakspeare of the horrid System and conseques of the fagging at great Schools—I know not yet how large a parcel I can send—I mean by way of Letters—I hope there can be no objection to my dowling° up a quire made into a small compass—That is the manner in which I shall write. I shall send you more than Letters—I mean a tale—which I must begin on account of the activity of my Mind; of its inability to remain at rest—It must be prose and not very exciting. I must do this because in the way I am at present situated I have too many interruptions to a train of feeling to be able to wite Poetry—So I shall write this Tale, and if I think it worth while get a duplicate made before I send it off to you—This is a fresh beginning the 21st October—Charles and Henry were with us on Sunday and they brought me your Letter to your Mother—we agreed to get a Packet off to you as soon as possible. I shall dine with your Mother tomorrow, when they have promised to have their Letters ready. I shall send as soon as possible without thinking of the little you may have from me in the first parcel, as I intend as I said before to begin another Letter of more regular information. Here I want to communicate so largely in a little time that I am puzzled where to direct my attention. Haslam has promised to let me know from Capper and Hazlewood.° For want of something better I shall proceed to give you some extracts from my Scotch Letters—Yet now I think on it why not send you the letters themselves—I have three of them at present—I belive Haydon has two which I will get in time. I dined with your Mother & Henry at Mrs Millar's on thursday when they gave me their Letters Charles's I have not yet he has promised to send it. The thought of sending my scotch Letters has determined me to enclose a few more which I have received and which will give you the best cue to how I am going on better than you could otherwise know—Your Mother was well and was sorry I could not stop later. I called on Hunt yesterday—it has been always my fate to meet Ollier there—On thursday I walked with Hazlitt as far as covent Garden: he was going to play Rackets°—I think Tom has been rather better these few last days—he has been less nervous. I expect Reynolds tomorrow Since I wrote thus far I have met with that same Lady° again, whom I saw at Hastings and whom I met when we were going to the English Opera. It was in a street which goes from Bedford Row to Lamb's Conduit

Street—I passed her and turned back—she seemed glad of it; glad to see me and not offended at my passing her before We walked on towards Islington where we called on a friend of her's who keeps a Boarding School. She has always been an enigma to me—she has been in a Room with you and with Reynolds and wishes we should be acquainted without any of our common acquaintance knowing it. As we went along, some times through shabby, sometimes through decent Street I had my guessing at work, not knowing what it would be and prepared to meet any surprise—First it ended at this House at Islington: on parting from which I pressed to attend her home. She consented and then again my thoughts were at work what it might lead to, tho' now they had received a sort of genteel hint from the Boarding School. Our Walk ended in 34 Gloucester Street Queen Square—not exactly so for we went up stairs into her sitting room—a very tasty sort of place with Books, Pictures a bronze statue of Buonaparte, Music, æolian Harp; a Parrot a Linnet—A Case of choice Liquers &c &c &. she behaved in the kindest manner—made me take home a Grouse for Tom's dinner—Asked for my address for the purpose of sending more game—As I had warmed with her before and kissed her—I thought it would be living backwards not to do so again—she had a better taste: she perceived how much a thing of course it was and shrunk from it—not in a prudish way but in as I say a good taste—She contived to disappoint me in a way which made me feel more pleasure than a simple kiss could do—she said I should please her much more if I would only press her hand and go away. Whether she was in a different disposition when I saw her before—or whether I have in fancy wrong'd her I cannot tell—I expect to pass some pleasant hours with her now and then: in which I feel I shall be of service to her in matters of knowledge and taste: if I can I will—I have no libidinous thought about her—she and your George are the only women à peu près de mon age whom I would be content to know for their mind and friendship alone—I shall in a short time write you as far as I know how I intend to pass my Life—I cannot think of those things now Tom is so unwell and weak. Notwithstand your Happiness and your recommendation I hope I shall never marry. Though the most beautiful Creature were waiting for me at the end of a Journey or a Walk; though the carpet were of Silk, the Curtains of the morning Clouds; the chairs and Sofa stuffed with Cygnet's down; the food Manna, the Wine beyond Claret, the Window opening on Winander mere, I should not feel—or rather my Happiness would not be so fine, as my Solitude is sublime. Then instead of what I have described, there is a Sublimity to welcome me home—The roaring of the wind is my wife and the Stars through the window pane are my Children. The mighty abstract Idea I have of Beauty in all things stifles the more divided and minute domestic happiness—an amiable wife and sweet Children I contemplate as a part of that Beauty. but I must have a thousand of those beautiful particles to fill up my heart. I feel more and more every day, as my imagination strengthens, that I do not live in this world alone but in a thousand worlds—No sooner am I alone than shapes of epic greatness are stationed around me, and serve my Spirit the office which is equivalent to a king's body guard—then 'Tragedy, with scepter'd pall, comes sweeping by' According to my state of mind I am with Achilles shouting in the Trenches or with Theocritus in the Vales of Sicily. Or I throw my whole being into Triolus and repeating those lines, 'I wander, like a lost soul upon the stygian Banks staying for waftage,"° I melt into the air with a voluptuousness so delicate that I am content to be alone—These things combined with the opinion I have of the generallity of women—who appear to me as children to whom I would rather give a Sugar Plum than my time, form a barrier against Matrimony which I rejoice in. I have written this that you might see I have my share of the highest pleasures and that though I may choose to pass my days alone I shall be no Solitary. You see therre is nothing spleenical in all this. The only thing that can ever affect me personally for more than one short passing day, is any doubt about my powers for poetry—I seldom have any, and I look with hope to the nighing time when I shall have none. I am as happy as a Man can be—that is in myself I should be happy if Tom was well, and I knew you were passing pleasant days—Then I should be most enviable—with the yearning Passion I have for the beautiful, connected and made one with the ambition of my intellect. Thnk of my Pleasure in Solitude, in comparison of my commerce with the world—there I am a child—there they do not know me not even my most intimate acquaintance—I give into their feelings as though I were refraining from irritating a little child—Some think me middling, others silly, others foolish—every one thinks he sees my weak side against my will; when in truth it is with my will—I am content to be thought all this because I have in my own breast so great a resource. This is one great reason why they like me so; because they can all show to advantage in a room, and eclipese from a certain tact one who is reckoned to be a good Poet—I hope I am not here playing tricks 'to make the angels weep':° I think not: for I have not the least contempt for my species; and though it may sound paradoxical: my greatest elevations of soul leaves me every time more humbled—Enough of this—

though in your Love for me you will not think it enough. Haslam has been here this morning, and has taken all the Letter's except this sheet, which I shall send him by the Twopenny, as he will put the Parcel in the Boston post Bag by the advice of Capper and Hazlewood, who assure him of the safety and expedition that way—the Parcel will be forwarded to Warder and thence to you all the same. There will not be a Philadelphia Ship for these six weeks—by that time I shall have another Letter to you. Mind you I mark this Letter A. By the time you will receive this you will have I trust passed through the greatest of your fatigues. As it was with your Sea sickness I shall not hear of them till they are past. Do not set to your occupation with too great an axiety—take it calmly—and let your health be the prime consideration. I hope you will have a Son, and it is one of my first wishes to have him in my Arms—which I will do please God before he cuts one double tooth. Tom is rather more easy than he has been: but is still so nervous that I can not speak to him of these Matters—indeed it is the care I have had to keep his Mind aloof from feelings too acute that has made this Letter so short a one—I did not like to write before him a Letter he knew was to reach your hands—I cannot even now ask him for any Message—his heart speaks to you—Be as happy as you can. Think of me and for my sake be cheerful. Believe me my dear Brother and sister

Your anxious and affectionate Brother

JOHN—

This day is my Birth day—
All our friends have been anxious in their enquiries and all send their rembrances

Letter to B. R. Haydon, 22 December 1818

My dear Haydon, Tuesday Wentworth Place—

Upon my Soul I never felt your going out of the room at all—and believe me I never rhodomontade any where but in your Company—my general Life in Society is silence. I feel in myself all the vices of a Poet, irritability, love of effect and admiration—and influenced by such devils I may at times say more rediculous things than I am aware of—but I will put a stop to that in a manner I have long resolved upon—I will buy a gold ring and put it on my finger—and from that time a Man of superior head shall never have occasion to pity me, or one of inferior

Nunskull to chuckle at me—I am certainly more for greatness in a Shade than in the open day—I am speaking as a mortal—I should say I value more the Priviledge of seeing great things in loneliness—than the fame of a Prophet—Yet here I am sinning—so I will turn to a thing I have thought on more—I mean your means till your Picture be finished: not only now but for this year and half have I thought of it. Believe me Haydon I have that sort of fire in my Heart that would sacrifice every thing I have to your service—I speak without any reserve—I know you would do so for me—I open my heart to you in a few words—I will do this sooner than you shall be distressed: but let me be the last stay—ask the rich lovers of art first—I'll tell you why—I have a little money which may enable me to study and to travel three or four years—I never expect to get any thing by my Books: and moreover I wish to avoid publishing—I admire Human Nature but I do not like *Men*—I should like to compose things honourable to Man—but not fingerable over by *Men*. So I am anxious to exist with° troubling the printer's devil or drawing upon Men's and Women's admiration—in which great solitude I hope God will give me strength to rejoice Try the long purses—but do not sell your drawing or I shall consider it a breach of friendship. I am sorry I was not at home when Salmon° called—Do write and let me know all you present whys and wherefores—

Your's most faithfully
JOHN KEATS

Letter to George and Georgiana Keats, 16–18, 22, ?29, 31 December 1818, 2–4 January 1819

B° My dear Brother and Sister,

You will have been prepared, before this reaches you for the worst news you could have, nay if Haslam's letter arrives in proper time, I have a consolation in thinking the first shock will be past before you receive this. The last days of poor Tom were of the most distressing nature; but his last moments were not so painful, and his very last was without a pang—I will not enter into any parsonic comments on death—yet the common observations of the commonest people on death are as true as their proverbs. I have scarce a doubt of immortality of some nature of other—neither had Tom. My friends have been exceedingly kind to me every one of them—Brown detained me at his

House. I suppose no one could have had their time made smoother than mine has been. During poor Tom's illness I was not able to write and since his death the task of beginning has been a hindrance to me. Within this last Week I have been every where—and I will tell you as nearly as possible how all go on—With Dilke and Brown I am quite thick—with Brown indeed I am going to domesticate—that is we shall keep house together—I Shall have the front parlour and he the back one—by which I shall avoid the noise of Bentley's Children—and be the better able to go on with my Studies—which ave been greatly interrupted lately, so that I have not the Shadow of an idea of a book in my head, and my pen seems to have grown too goutty for verse. How are you going on now? The going on of the world make me dizzy—there you are with Birkbeck—here I am with brown—sometimes I fancy an immense separation, and sometimes, as at present, a direct communication of spirit with you. That will be one of the grandeurs of immortality—there will be no space and consequently the only commerce between spirits will be by their intelligence of each other—when they will completely understand each other—while we in this world merely compehend each other in different degrees—the higher the degree of good so higher is our Love and friendship—I have been so little used to writing lately that I am affraid you will not smoke my meaning so I will give an example—Suppose Brown or Haslam or any one whom I understand in the nether degree to what I do you, were in America, they would be so much the farther from me in proportion as their identity was less impressed upon me. Now the reason why I do not feel at the present moment so far from you is that I remember your Ways and Manners and actions; I know you manner of thinking, you manner of feeling: I know what shape your joy or your sorrow would take, I know the manner of you walking, standing, sauntering, sitting down, laughing, punning, and evey action so truly that you seem near to me. You will remember me in the same manner—and the more when I tell you that I shall read a passage of Shakspeare every Sunday at ten o Clock—you read one at the same time and we shall be as near each other as blind bodies can be in the same room—I saw your Mother the day before yesterday, and intend now frequently to pass half a day with her—she sceem'd tolerably well. I called in Henrietta Street and so was speaking with you Mother about Miss Millar—we had a chat about Heiresses—she told me I think of 7 or eight dying Swains. Charles was not at home. I think I have heard a little more talk about Miss Keasle°—all I know of her is, she had a new sort of shoe on of bright leather like our Knapsacks—Miss Millar gave me one of her confounded pinches. N.B. did not like it. M[rs] Dilke went with me to see Fanny last week, and Haslam went with me last Sunday—she was well—she gets a little plumper and had a little Colour. On Sunday I brought from her a present of facescreens and a work bag for M[rs] D. they were really very pretty—From walthamstow we walked to Bethnal green°—were I fell so tired from my long walk that I was obliged to go to Bed at ten—M[r] and M[rs] ...° were there—Haslam has been excessively kind—and his anxiety about you is great. I never meet him but we have some chat thereon. He is always doing me some good turn—he gave me this thin paper° for the purpose of writing to you. I have been passing an hour this morning with M[r] Lewis—he wants news of you very much. Haydon was here yesterday—he amused us much by speaking of young Hopner who went with Capt[n] Ross on a voyage of discovery to the Poles°—The Ship was sometimes entirely surrounded with vast mountains and crags of ice and in a few Minutes not a particle was to be seen all round the Horizon. Once they met with with so vast a Mass that they gave themselves over for lost; their last recourse was in meeting it with the Bowspit, which they did, and split it asunder and glided through it as it parted for a great distance—one Mile ane more Their eyes were so fatigued with the eternal dazzle and whiteness that they lay down on their backs upon deck to relieve their sight on the blue Sky. Hopner describes his dreadful weriness at the continual day—the sun ever moving in a circle round above their heads—so pressing upon him that he could not rid himself of the sensation even in the dark Hold of the Ship—The Esquimaux are described as the most wretched of Beings—they float from the Summer to their winter residences and back again like white Bears on the ice floats—They seem never to have washed, and so when their features move, the red skin shows beneath the cracking peal of dirt. They had no notion of any inhabitants in the World but themselves. The sailors who had not seen a Star for some time, when they came again southwards, on the hailing of the first revision, of one all ran upon deck with feelings of the most joyful nature. Haydon's eyes will not suffer him to proceed with his Picture—his Physician tells him he must remain two months more, inactive. Hunt keeps on his old way—I am completely tired of it all—He has lately publish'd a Pocket-Book call'd the litrerary Pocket-Book—full of the most sickening stuff you can imagine.° Reynolds is well—he has become an edinburgh Reviewer°—I have not heard from Bailey Rice I have seen very little of lately—and I am very sorry for it. The Miss R's are all as usual—Archer° above all people called on me one day—he wanted some information, by my means, from Hunt and Haydon, concerning some Man they knew. I got him what he wanted, but know none of the whys and wherefores. Poor Kirkman left wentworth place one evening about half past eight and was stopped, beaten and robbed of his Watch in Pond Street. I saw him a few days since, he had not recovered from his bruize I called on Hazlitt the day I went to Romney Street—I gave John Hunt extracts from your Letters—he has taken no notice. I have seen Lamb lately—Brown and I were taken by Hunt to Novello's—there we were devastated and excruciated with bad and repeated puns—Brown dont want to go again. We went the other evening to see Brutus a new Trageday by Howard Payne, an American—Kean was excellent—the play was very bad—It is the first time I have been since I went with you to the Lyceum—

Mrs Brawne who took Brown's house for the Summer, still resides in Hampstead—she is a very nice woman—and her daughter senior° is I think beautiful and elegant, graceful, silly, fashionable and strange we have a little tiff now and then—and she behaves a little better, or I must have sheered off—I find by a sidelong report from your Mother that I am to be invited to Miss Millar's birthday dance—Shall I dance with Miss Waldegrave? Eh! I shall be obliged to shirk a good many there—I shall be the only Dandy there—and indeed I merely comply with the invitation that the party may no be entirely destitute of a specimen of that Race. I shall appear in a complete dress of purple Hat and all—with a list of the beauties I have conquered embroidered round my Calves.

hurs-day This morning° is so very fine, I should have walked over to Walthamstow if I had thought of it yesterday—What are you doing this morning? Have you a clear hard frost as we have? How do you come on with the gun? Have you shot a Buffalo? Have you met with any Pheasants? My Thoughts are very frequently in a foreign Country—I live more out of England than in it—The Mountains of Tartary are a favourite lounge, if I happen to miss the Allegany ridge,° or have no whim for Savoy.° There must be great pleasure in pursuing game—pointing your gun—no, it wont do—now no—rabbit it—now bang—smoke and feathers—where is it? Shall you be able to get a good pointer or so? Have you seen Mr Trimmer—He is an acquaintance of Peachey's. Now I am not addressing miself to G. minor, and yet I am—for you are one—Have you some warm furs? By your next Letters I shall expect to hear exactly how you go on—smother nothing—let us have all—fair and foul all plain—Will the little bairn have made his entrance before you have this? Kiss it for me, and when it can first know a cheese from a Caterpillar show it my picture twice a Week—You will be glad to hear that Gifford's° attack upon me has done me service—it has got my Book among several *Sets*—Nor must I forget to mention once more, what I suppose Haslam has told you, the present of a £25 note I had anonymously sent me—I have many things to tell you—the best way will be to make coppies of my correspondence; and I must not forget the Sonnet I received with the Note—Last Week I received the following from Woodhouse, whom you must recollect—'My dear Keats,—I send enclosed a Letter which, when read take the trouble to return to me. The History of its reaching me is this. My Cousin, Miss Frogley of Hounslow borrowed my copy of Endymion for a specified time—Before she had time to look into it; she and my friend M[r] H[y] Neville of Esher, who was house Surgeon to the late Princess Charlotte, insisted upon having it to read for a day or two, and undertook to make my Cousin's peace with me on account of the extra delay—Neville told me that one of the Misses Porter° (of romance Celebrity) had seen it on his table, dipped into it, and expressed a wish to read it—I desired he would keep it as long, and lend it to as many, as he pleased, provided it was not allowed to slumber on any one's shelf. I learned subsequently from Miss Frogley that these Ladies had requested of M[r] Neville, if he was acquainted with the Author the Pleasure of an introduction—About a week back the enclosed was transmitted by M[r] Neville to my Cousin, as a species of apology for keeping her so long without the Book—And she sent it to me, knowing it would give me Pleasure—I forward it to you for somewhat the same reason, but principally because it gives me the opportunity of naming to you (which It would have been fruitless to do before) the opening there is for an introduction to a class of society, from which you may possibly derive advantage as well as gratification, if you think proper to avail yourself of it. In such case I should be very happy to further your Wishes. But do just as you please. The whole is entirely entre nous—Your's &c—R.W.' Well—now this is Miss Porter's Letter to Neville—'Dear Sir, As my Mother is sending a Messenger to Esher, I cannot but make the same the bearer of my regrets for not having had the pleasure of seeing you, the morning you called at the gate—I had given orders to be denied: I was so very unwell with my still adhæsive cold; but had I known it was you I should have taken off the interdict for a few minutes, to say, how very much I am delighted with Endymion—I had just finished the Poem, and have done as you permitted lent it to Miss Fitzgerald. I regret you are not personally acquainted with the Author: for I should have been happy to have acknowledged to him, through the advantage of your Communication the very rare delight my Sister and myself have enjoyed from this first fruits of Genius. I hope the ill-natured Review will not have damaged (or damped) such true Parnassian fire—it ought not for when Life is granted &c' and so she goes on—Now I feel more obliged than flattered by this—so obliged that I will not at present give you an extravaganza of a Lady Romancer. I will be introduced to them if it be merely for the pleasure of writing to you about it—I shall certainly see a new race of People—I shall more certainly have no time for them—Hunt has asked me to meet Tom Moore some day—so you shall hear of him. The night we went to Novello's there was a complete set to of Mozart and punning—I was so completely tired of it that if I were to follow my own inclinations I should never meet any one of that set again, not even Hunt—who is certainly a pleasant fellow in the main when you are with him—but in reallity he is vain, egotistical and disgusting in matters of taste and in morals—He understands many a beautiful thing; but then, instead of giving other minds credit for the same degree of perception as he himself possesses—he begins an explanation in such a curious manner that our taste and self-love is offended continually. Hunt does one harm by making fine things petty and beautiful things hateful—Through him I am indifferent to Mozart, I care not for white Busts—and many a glorious thing when associated with him becames a nothing—This distorts one's mind—makes one's thoughts bizarre—perplexes one in the standard of Beauty—Martin is very much irritated against Blackwood for printing some Letters in his Magazine which were Martin's property—he always found excuses for Blackwood till he himself was injured and now he is enraged—I have been several times thinking whether or not I should send you the examiners as Birkbeck no doubt has all the good periodical Publications—I will save them at all events.—I must not forget to mention how attentive and useful Mrs Bentley has been—I am sorry to leave her—but I must, and I hope she will not be much a looser by it—Bentley is very well—he has just brought me a cloathes' basket of Books. Brown has gone to town to day to take his Nephews who are on a visit here to see the Lions°—I am passing a Quiet day—which I have not done a long while—and if I do continue so—I feel I must again begin with my poetry—for if I am not in action mind or Body I am in pain—and from that I suffer greatly by going into parties where from the rules of society and a natural pride I am obliged to smother my Spirit and look like an Idiot—because I feel my impulses given way to would too much amaze them—I live under an everlasting restraint—Never relieved except when I am composing—so I will write away. Friday. I think you knew before you left England that my next subject would be 'the fall of

Hyperion' I went on a little with it last night—but it will take some time to get into the vein again. I will not give you any extracts because I wish the whole to make an impression—I have however a few Poems which you will like and I will copy out on the next sheet—I shall dine with Haydon on Sunday and go over to Walthamstow on Monday if the frost hold—I think also of going into Hampshire this Christmas to Mr Snooks —they say I shall be very much amused—But I dont know—I think I am in too huge a Mind for study—I must do it—I must wait at home, and let those who wish come to see me. I cannot always be (how do you spell it?) trapsing—Here I must tell you that I have not been able to keep the journal or write the Tale I promised—now I shall be able to do so—I will write to Haslam this morning to know when the Packet sails and till it does I will write somethng evey day—after that my journal shall go on like clockwork—and you must not complain of its dullness—for what I wish is to write a quantity to you—knowing well that dullness itself will from me be interesting to you—You may conceive how this not having been done has weighed upon me—I shall be able to judge from your next what sort of information will be of most service or amusement to you. Perhaps as you were fond of giving me sketches of character you may like a little pic nic of scandal even across the Atlantic—But now I must speak particularly to you my dear Sister—for I know you love a little quizzing, better than a great bit of apple dumpling—Do you know Uncle Redall?° He is a little Man with an innocent, powdered, upright head; he lisps with a protruded under lip—he has two Neices each one would weigh three of him—one for height and the other for breadth—he knew Barttolozzi°—he gave a supper and ranged his bottles of wine all up the kitchen and cellar stairs—quite ignorant of what might be drank—it might have been a good joke to pour on the sly bottle after bottle into a washing tub and roar for more—If you were to trip him up it would discompose a Pigtail and bring his under lip nearer to his nose. He never had the good luck to loose a silk Handkerchef in a Crowd and therefore has only one topic of conversation—Bartolotzzi—Shall I give you Miss Brawn? She is about my height—with a fine style of countenance of the lengthen'd sort—she wants sentiment in every feature—she manages to make her hair look well—her nostrills are fine—though a little painful—he mouth is bad and good—he Profil is better than her full-face which indeed is not full but pale and thin without showing any bone—Her shape is very graceful and so are her movements—her Arms are good her hands badish—her feet tolerable—she is not seventeen°—but she is ignorant—monstrous in her behaviour flying out in all directions, calling people such names—that I was forced lately to make use of the term *Minx*—this is I think no from any innate vice but from a penchant she has for acting stylishly. I am however tired of such style and shall decline any more of it—She had a friend to visit her lately—you have known plenty such—Her face is raw as if she was standing out in a frost—her lips raw and seem always ready for a Pullet—she plays the Music without one sensation but the feel of the ivory at her fingers—she is a downright Miss without one set off—we hated her and smoked her and baited her, and I think drove her away—Miss B— thinks her a Paragon of fashion, and says she is the only woman she would change persons with—What a Stupe—She is superior as a Rose to a Dandelion—When we went to bed Brown observed as he put out the Taper what an ugly old woman that Miss Robinson would make—at which I must have groan'd aloud for I'm sure ten minutes. I have not seen the thing Kingston again—George will describe him to you—I shall insinuate some of these Creatures into a Comedy some day—and perhaps have Hunt among them—Scene, a little Parlour—Enter Hunt—Gattie°—Hazlitt—Mrs Novello—Ollier—Gattie) Ha! Hunt! got into you new house? Ha! Mrs Novello seen Altam and his Wife?° Mrs N. Yes (with a grin) *its* Mr Hunts is'nt it? Gattie. Hunts' no ha! Mr Olier I congratulate you upon the highest compliment I ever heard paid to the Book. Mr Haslit, I hope you are well (Hazlitt—yes Sir, no Sir—Mr Hunt (at the Music) La Biondina° &c Hazlitt did you ever hear this—La Biondina &c—Hazlitt—O no Sir—I never—Olier—Do Hunt give it us over again—divino—Gattie/divino—Hunt when does your Pocket Book come out—/Hunt/What is this absorbs me quite?° O we are spinning on a little, we shall floridize soon I hope—Such a thing was very much wanting—people think of nothing but money-getting—now for me I am rather inclined to the liberal side of things—but I am reckoned lax in my christian principles—& & & &c—It is some days since I wrote the last page—and what have I been about since I have no Idea—I dined at Haslam's on sunday—with Haydon yesterday and saw Fanny in the morning—she was well—Just now I took out my poem to go on with it—but the thought of my writing so little to you came upon me and I could not get on—so I have began at random—and I have not a word to say—and yet my thoughts are so full of you that I can do nothing else. I shall be confined at Hampstead a few days on account of a sore throat—the first thing I do will be to visit your Mother again—The last time I saw Henry he show'd me his first engraving which I thought capital—Mr Lewis called this morning and brought some american Papers. I have not look'd into them—I think we ought to have heard of you before this—I am in daily expectation of Letters—Nil desperandum—Mrs Abbey wishes to take Fanny from School—I shall strive all I can against that—There has happened great Misfortune in the Drewe Family°—old Drewe has been dead some time; and lately George Drewe expired in a fit—on which account Reynolds has gone into Devonshire—He dined a few days since at Horace Twisse's° with Liston° and Charles Kemble°—I see very little of him now, as I seldom go to little Britain because the Ennui always seizes me there, and John Reynolds is very dull at home—Nor have I seen Rice—How you are now going on is a Mystery to me—I hope a few days will clear it up. I never know the day of the Month—It is very fine here to-day though I expect a Thundercloud or rather a snow cloud in less than an hour—I am at present alone at Wentworth place—Brown being at Chichester and Mr & Mrs Dilke making a little stay in Town. I know not what I should do without a Sunshiny morning now and then—it clears up one's spirits—Dilke and I frequently have some chat about you—I have now and then some doubts but he seems to have a great confidence—I think there will soon be perceptible a change in the fashionable slang literature of the day—it seems to me that Reviews have had their day—that the public have been surfeited—there will soon be some new folly to keep the Parlours in talk—What it is I care not—We have seen three literary kings in our Time—Scott—Byron—and then the scotch novels.° All now appears to be dead—or I may mistake—literary Bodies may still keep up the Bustle which I do not hear—Haydon show'd me a letter he had received from Tripoli—Ritchey was well and in good Spirits, among Camels, Turbans, Palm Trees and sands—You may remember I promised to send him an Endymion which I did not—howeever he has one—you have one—One is in the Wilds of america—the other is on a Camel's back in the plains of Egypt. I am looking into a Book of Dubois's°—he has written directions to the Players—one of them is very good. 'In singing never mind the music—observe what time you please. It would be a pretty degradation indeed if you were obliged to confine your genius to the dull regularity of a fiddler—horse hair and cat's guts—no, let him keep *your* time and play *your* tune—*dodge him*'—I will now copy out the Letter and Sonnet I have spoken of—The outside cover was thus directed 'Messrs Taylor and Hessey (Booksellers) No 93 Fleet Street London' and it contained this 'Messrs Taylor and Hessey are requested to forward the enclosed letter by some *safe* mode of conveyance to the Author of Endymion, who is not known at Teignmouth: or if they have not his address, they will return the letter by post, directed as below, within *a fortnight* (Mr P. Fenbank

P.O. Teignmouth' 9th Nov[r] 1818—In this sheet was enclosed the following—with a superscription 'M[r] John Keats Teignmouth'—Then came Sonnet to John Keats—which I would not copy for any in the world but you—who know that I scout 'mild light and loveliness' or any such nonsense in myself

Star of high promise!—not to this dark age
  Do thy mild light and loveliness belong;—
  For it is blind intolerant and wrong;
Dead to empyreal soarings, and the rage
Of scoffing spirits bitter war doth wage
  With all that hold integrity of song.
  Yet thy clear beam shall shine through ages strong
To ripest times a light—and heritage.
And there breathe now who dote upon thy fame,
  Whom thy wild numbers wrap beyond their being,
Who love the freedom of thy Lays—their aim
  Above the scope of a dull tribe unseeing—
And there is one whose hand will never scant
From his poor store of fruits all *thou* can'st want.

November, 1818 turn over

I tun'd over and found a £25-note—Now this appears to me all very proper—if I had refused it—I should have behaved in a very bragadochio dunderheaded manner—and yet the present galls me a little. and I do not know whether I shall not return it if I ever meet with the donor—after whom to no purpose I have written—I have your Minature on the Table George the great—its very like—though not quite about the upper lip—I wish we had a better of you little George—I must not forget to tell you that a few days since I went with Dilke a shooting on the heath and shot a Tom-tit—There were as many guns abroad as Birds—I intended to have been at Chichester this Wednesday—but on account of this sore throat I wrote him (Brown) my excuse yesterday—

Thursday (I will date when I finish—I received a Note from Haslam yesterday—asking if my letter is ready—now this is only the second sheet—notwithstanding all my promises—But you must reflect what hindrances I have had—However on sealing this I shall have nothing to prevent my proceeding in a gradual journal—which will increase in a Month to a considerable size. I will insert any little pieces I may write—though I will not give any extracts from my large poem° which is scarce began—I what° to hear very much whether Poetry and literature in general has gained or lost interest with you—and what sort of writing is of the highest gust with you now. With what sensation do you read Fielding?—and do not Hogarth's pictures seem an old thing to you? Yet you are very little more removed from general association than I am—recollect that no Man can live but in one society at a time—his enjoyment in the different states of human society must depend upon the Powers of his Mind—that is you can imagine a roman triumph, or an olympic game as well as I can. We with our bodily eyes see but the fashion and Manners of one country for one age—and then we die—Now to me manners and customs long since passed whether among the Babylonians or the Bactrians° are as real, or eveven more real than those among which I now live—My thoughts have turned lately this way—The more we know the more inadequacy we discover in the world to satisfy us—this is an old observation; but I have made up my Mind never to take any thing for granted—but even to examine the truth of the commonest proverbs—This however is true—Mrs Tighe° and Beattie° once delighted me—now I see through them and can find nothing in them—or weakness—and yet how many they still delight! Perhaps a superior being may look upon Shakspeare in the same light—is it possible? No—This same inadequacy is discovered (forgive me little George you know I don't mean to put you in the mess) in Women with few exceptions—the Dress Maker, the blue Stocking and the most charming sentimentalist differ but in a Slight degree, and are equally smokeable—But I'll go no further—I may be speaking sacrilegiously—and on my word I have thought so little that I have not one opinion upon any thing except in matters of taste—I never can feel certain of any truth but from a clear perception of its Beauty—and I find myself very young minded even in that perceptive power—which I hope will encrease—A year ago I could not understand in the slightest degree Raphael's cartoons—now I begin to read them a little—and how did I learn to do so? By seeing something done in quite an opposite spirit—I mean a picture of Guido's in which all the Saints, instead of that heroic simplicity and unaffected grandeur which they inherit from Raphael, had each of them both in countenance and gesture all the canting, solemn melo dramatic mawkishness of Mackenzie's father Nicholas°—When I was last at Haydon's I look over a Book of Prints° taken from the fresco of the Church of Milan the name of which I forget—in it are comprised Specimens of the first and second age of art in Italy—I do not think I ever had a greater treat out of Shakspeare—Full of Romance and the most tender feeling—magnificence of draperies beyond any I ever saw not excepting Raphael's—But

Grotesque to a curious pitch—yet still making up a fine whole—even finer to me than more accomplish'd works—as there was left so much room for Imagination. I have not heard one of this last course of Hazlitt's lecture's—They were upon 'Wit and Humour,' the english comic writers.' Saturday Jany 2nd Yesterday M^r^ M^rs^ D and myself dined at Mrs Brawne's—nothing particular passed. I never intend here after to spend any time with Ladies unless they are handsome—you lose time to no purpose—For that reason I shall beg leave to decline going again to Redall's or Butlers or any Squad where a fine feature cannot be mustered among them all—and where all the evening's amusement consists in saying 'your good health' *your* good health, and YOUR good health—and (o I beg you pardon) your's Miss——. and such thing not even dull enough to keep one awake—with respect to amiable speaking I can read—let my eyes be fed or I'll never go out to dinner any where—Perhaps you may have heard of the dinner given to Thos Moore in Dublin, because I have the account here by me in the Philadelphia democratic paper°—The most pleasant thing that accured was the speech Mr Tom made on his Farthers health being drank—I am affraid a great part of my Letters are filled up with promises and what I will do rather than any great deal written—but here I say once for all—that circumstances prevented me from keeping my promise in my last, but now I affirm that as there will be nothing to hinder me I will keep a journal for you. That I have not yet done so you would forgive if you knew how many hours I have been repenting of my neglect—For I have no thought pervading me so constantly and frequently as that of you—my Poeem cannot frequently drive it away—you will retard it much more that You could by taking up my time if you were in England—I never forget you except after seeing now and then some beautiful woman—but that is a fever—the thought of you both is a passion with me but for the most part a calm one—I asked Dilke for a few lines for you—he has promised them—I shall send what I have written to Haslam on Monday Morning. what I can get into another sheet tomorrow I will—there are one or two little poems you might like—I have given up snuff very nearly quite—Dilke has promised to sit with me this evening, I wish he would come this minute for I want a pinch of snuff very much just now—I have none though in my own snuff box—My sore throat is much better to day—I think I might venture on a crust—Here are the Poems—they will explain themselves—as all poeems should do without any comment

[*A draft of 'Fancy' follows*]

I did not think this had been so long a Poem—I have another not so long—but as it will more conveniently be coppied on the other side I will just put down here some observations on Caleb Williams by Hazlitt—I mean to say St Leon for although he has mentioned all the Novels of Godwin very finely I do not quote them, but this only on account of its being a specimen of his usual abrupt manner, and fiery laconiscism—He says of St Leon 'He is a limb torn off from Society. In possession of eternal youth and beauty, he can feel no love; surrounded, tantalized and tormented with riches, he can do no good. The faces of Men pass before him as in a speculum; but he is attached to them by no common tie of sympathy or suffering. He is thrown back into himself and his own thoughts. He lives in the solitude of his own breast,—without wife or child or friend or Enemy in the world. *His is the solitude of the Soul, not of woods, or trees or mountains*—but the desert of society—the waste and oblivion of the heart. He is himself alone. His existence is purely intellectual, and is therefore intolerable to one who has felt the rapture of affection, or the anguish of woe . . .'° As I am about it I might as well give you his caracter of Godwin as a Romancer 'Whoever else is, it is pretty clear that the author of Caleb Williams is not the Author of waverly. Nothing can be more distinct or excellent in their several ways than these two writers. If the one owes almost every thing to external observation and traditional character, the other owes every thing to internal conception and contemplation of the possible workings of the human Mind. There is little knowledge of the world, little variety, neither an eye for the picturesque, nor a talent for the humourous in Caleb Williams, for instance, but you can not doubt for a moment of the originality of the work and the force of the conception. The impression made upon the reader is the exact measure of the strength of the authors genius. For the effect in Caleb Williams and St Leon, is entirely made out, not by facts nor dates, by blackletter or magazine learning, by transcript nor record, but by intense and patient study of the human heart, and by an imagination projecting itself into certain situations, and capable of working up its imaginary feelings to the height of reality.' This appears to me quite correct—now I will copy the other Poem—it is on the double immortality of Poets—

[*A draft of 'Bards of Passion and of Mirth' follows*]

These are specimens of a sort of rondeau which I think I shall become partial to—because you have one idea amplified with greater ease and more delight and freedom than in the sonnet—It is my intention to wait a few years before I publish any minor poems—and then I hope to have a volume of some worth—and which those people will realish who cannot bear the burthen of a long poem—In my journal I intend to copy the poems I write the days they are written—there is just room I see in this page to copy a little thing I wrote off to some Music as it was playing—

[*A draft of 'I had a dove and the sweet dove died' follows*]

Sunday.°

I have been dining with Dilke to day—He is up to his Ears in Walpole's letters Mr Manker° is there; I have come round to see if I can conjure up any thing for you—Kirkman° came down to see me this morning—his family has been very badly off lately—He told me of a villainous trick of his Uncle William in Newgate Street who became sole Creditor to his father under pretence of serving him, and put an execution on his own Sister's goods—He went in to the family at Portsmouth; conversed with them, went out and sent in the Sherif's officer—He tells me too of abominable behaviour of Archer to Caroline Mathew—Archer has lived nearly at the Mathews these two years; he has been amusing Caroline all this time—and now he has written a Letter to M^rs^ M—declining on pretence of inability to support a wife as he would wish, all thoughts of marriage. What is the worst is, Caroline is 27 years old —It is an abominable matter—He has called upon me twice lately—I was out both times—What can it be for—There is a letter° to day in the Examiner to the Electors of westminster on Mr Hobhouse's account—In it there is a good Character of Cobbet—I have not the paper by me or I would copy it—I do not think I have mentioned the Discovery of an african kingdom—the account is much the same as the first accounts of Mexico—all magnificence—there is a Book being written about it°—I will read it and give you the cream in my next. The ramance we have heard upon it runs thus: they have window frames of gold—100,000 infantry—human sacrifices—The Gentleman who is the adventurer has his wife with him—she I am told is a beautiful little sylphid woman—her husband was to have been sacrificed to their Gods and was led through a Chamber filled with different instruments of torture with priveledge to choose what death he would die, without their having a thought of his aversion to such a death they considering it a supreme distinction—However he was let off and became a favorite with the King, who at last openly patronised him; though at first on account of the Jealousy of his Ministers he was wont to hold conversations with his Majesty in the dark middle of the night—All this sounds a little Blue-beardish—but I hope it is true—There is another thing I must mention of the momentous kind;—but I must mind my periods in it—M[rs] Dilke has two Cats—A Mother and a Daughter—now the Mother is a tabby and the daughter a black and white like the spotted child—Now it appears ominous to me for the doors of both houses are opened frequently—so that there is a complete thorough fare for both Cats (there being no board up to the contrary) they may one and several of them come into my room ad libitum. But no—the Tabby only comes—whether from sympathy from ann the maid or me I can not tell—or whether Brown has left behind him any atmospheric sprit of Maidenhood I can not tell. The Cat is not an old Maid herself—her daughter is a proof of it—I have questioned her—I have look'd at the lines of her paw—I have felt her pulse—to no purpose—Why should the *old* Cat come to me? I ask myself— and myself has not a word to answer. It may come to light some day; if it does you shall hear of it. Kirkman this morning promised to write a few lines for you and send them to Haslam. I do not think I have any thing to say in the Business way—You will let me know what you would wish done with your property in England—What things you would wish sent out—but I am quite in the dark about what you are doing—if I do not hear soon I shall put on my Wings and be after you—I will in my next, and after I have seen your next letter—tell you my own particular idea of America. Your next letter will be the key by which I shall open your hearts and see what spaces want filling, with any particular information—Whether the affairs of Europe are more or less interesting to you—whether you would like to hear of the Theatre's—of the bear Garden—of the Boxers—the Painters—The Lecturers—the Dress—The Progress of Dandyism—The Progress of Courtship—or the fate of Mary Millar—being a full true and très particular account of Miss M's ten Suitors—How the first tried the effect of swearing; the second of stammering; the thid of whispering;—the fourth of sonnets—the fifth of spanish leather boots the sixth of flattering her body—the seventh of flattering her mind—the eighth of flattering himself—the ninth stuck to the Mother—the tenth kissed the Chambermaid and told her to tell her Mistress—But he was soon discharged his reading lead him into an error—he could not sport the Sir Lucius° to any advantage—And now for this time I bid you good by—I have been thing of these sheets so long that I appear in closing them to take my leave of you—but that is not it—I shall immediately as I send this off begin my journal—when some days I shall write no more than 10 lines and others 10 times as much. M[rs] Dilke is knocking at the wall for Tea is ready—I will tell you what sort of a tea it is and then bid you—Good bye—This is monday morning—no thing particular happened yesterday evening, except that just when the tray came up M[rs] Dilke and I had a battle with celery stalks—she sends her love to you—I shall close this and send it immediately to Haslam—remaining ever

My dearest brother and sister
Your most affectionate Brother
JOHN—

Letter to B. R. Haydon, 8 March 1819

My dear Haydon,

You must be wondering where I am and what I am about! I am mostly at Hampstead, and about nothing; being in a sort of qui bono temper, not exactly on the road to an epic poem. Nor must you think I have forgotten you. No, I have about every three days been to Abbey's and to the Lawers. Do let me know how you have been getting on, and in what spirits you are.

You got out gloriously in yesterday's Examiner.° What a set of little people we live amongst. I went the other day into an ironmonger's shop, without any change in my sensations—men and tin kettles are much the same in these days. They do not study like children at five and thirty, but they talk like men at twenty. Conversation is not a search after knowledge, but an endeavour at effect. In this respect two most opposite men, Wordsworth and Hunt, are the same. A friend of mine observed the other day that if Lord Bacon were to make any remark in a party of the present day, the conversation would stop on the sudden. I am convinced of this, and from this I have come to the resolution never to write for the sake of writing, or making a poem, but from running over with any little knowledge and experience which many years of reflection may perhaps give me—otherwise I will be dumb. What Imagination I have I shall enjoy, and greatly, for I have experienced the satisfaction of having great conceptions without the toil of sonnetteering. I will not spoil my love of gloom by writing an ode to darkness; and with respect to my livelihood I will not write for it, for I will not mix with that most vulgar of all crowds the literary. Such things I ratify by looking upon myself, and trying myself at lifting mental weights, as it were. I am three and twenty with little knowledge and middling intellect. It is true that in the height of enthusiasm I have been cheated into some fine passages, but that is nothing.

I have not been to see you because all my going out has been to town, and that has been a great deal. Write soon.

Yours constantly,
JOHN KEATS

Letter to Joseph Severn, 29 March 1819

My dear Severn, Wentworth Place—
Monday—af[t]

Your note gave me some pain, not on my own account, but on yours—Of course I should mev suffer any petty vanity of mine to hinder you in any wise; and therefore I should say, 'put the miniature in the exhibition' if only myself was to be hurt.° But, will it not hurt you? What good can it do to any future picture—Even a large picture is lost in that canting place—what a drop of water in the ocean is a Miniature. Those who might chance to see it for the most part if they had ever heard of either of us—and know what we were and of what years would laugh at the puff of the one and the vanity of the other I am however in these matters a very bad judge—and would advise you to act in a way that appears to yourself the best for your interest. As your Hemia and Helena is finished send that without the prologue of a Miniature. I shall see you soon, if you do not pay me a visit sooner—there's a Bull for you.

Yours ever sincerely
JOHN KEATS—

To Fanny Keats, 12 April 1819

My dear Fanny, Wentworth Place

I have been expecting a Letter from you about what the Parson said to your answers°—I have thought also of writing to you often, and I am sorry to confess that my neglect of it has been but a small instance of my idleness of late—which has been growing upon me, so that it will require a great shake to get rid of it. I have written nothing, and almost read nothing—but I must turn over a new leaf—One most discouraging thing hinders me—we have no news yet from George—so that I cannot with any confidence continue the Letter I have been preparing for him.

Many are in the same state with us and many have heard from the Settlement—They must be well however: and we must consider this silence as good news—I ordered some bulbous roots for you at the Gardeners, and they sent me some, but they were all in bud—and could not be sent, so I put them in our Garden There are some beautiful heaths now in bloom in Pots—either heaths or some seasonable plants I will send you instead—perhaps some that are not yet in bloom that you may see them come out—Tomorrow night I am going to a rout°—a thing I am not at all in love with—Mr Dilke and his Family have left Hampstead—I shall dine with them to day in Westminster where I think I told you they were going to reside for the sake of sending their Son Charles to the Westminster School. I think I mentioned the Death of Mr Haslam's Father—Yesterday week the two Mr Wylies dined with me. I hope you have good store of double violets—I think they are the Princesses of flowers and in a shower of rain, almost as fine as barley sugar drops are to a schoolboy's tongue. I suppose this fine weather the lambs tails give a frisk or two extraordinary—when a boy would cry huzza and a Girl O my! a little Lamb frisks its tail. I have not been lately through Leicester Square—the first time I do I will remember your Seals°—I have thought it best to live in Town this Summer, chiefly for the sake of books, which cannot be had with any comfort in the Country—besides my Scotch jouney gave me a doze of the Picturesque with which I ought to be contented for some time. Westminster is the place I have pitched upon—the City or any place very confined would soon turn me pale and thin—which is to be avoided. You must make up your mind to get Stout this summer—indeed I have an idea we shall both be corpulent old folkes with tripple chins and stumpy thumbs—

Your affectionate Brother
JOHN

Letter to B. R. Haydon, 13 April 1819

My dear Haydon, Tuesday—

When I offered you assistance I thought I had it in my hand; I thought I had nothing to do, but to do. The difficulties I met with arose from the alertness and suspicion of Abbey; and especially from the affairs being still in a Lawer's hand—who has been drainng our Property for the last 6 years of evey charge he could make—I cannot do two things at once, and thus this affair has stopped my pursuits in every way—from the first prospect I had of difficulty. I assure you I have harrassed myself 10 times more than if I alone had been concernned in so much gain or loss. I have also ever told you the exact particulars as well as and as literally as my hopes or fear could translate them—for it was only by parcels that I found all those petty obstacles which for my own sake should not exist a moment—and yet why not—for from my own imprudence and neglect all my accounts are entirely in my Guardians Power—This has taught me a Lesson. hereafter I will be more correct. I find myself possessed of much less than I thought for and now if I had all on the table all I could do would be to take from it a moderate two years subsistence and lend you the rest; but I cannot say how soon I could become possessed of it. This would be no sacrifice nor any matter worth thinking of—much less than parting as I have more than once done with little sums which might have gradually formed a library to my taste—These sums amount to gether to nearly 200, which I have but a chance of ever being repaid or paid at a very distant period. I am humble enough to put this in writing from the sense I have of your struggling situation and the great desire that you should do me the justice to credit the unostentatious and willing state of my nerves on all such occasions. It has not been my fault—I am doubly hurt at the slightly repoachful tone of your note and at the occasion of it,—for it must be some other disappointment; you seem'd so sure of some important help when I last saw you—now you have maimed me again; I was whole I had began reading again—when your note came I was engaged in a Book—I dread as much as a Plague the idle fever of two months more without any fruit. I will walk over the first fine day: then see what aspect you affairs have taken, and if they should continue gloomy walk into the City to Abbey and get his consent for I am persuaded that to me alone he will not concede a jot°

Letter to Fanny Keats, 1 May 1819

My dear Fanny, Wentworth Place Saturday—

If it were but six o Clock in the morning I would set off to see you to day: if I should do so now I could not stop long enough for a how d'ye do—it is so long a walk through Hornsey and Tottenham—and as for Stage Coaching it besides that it is very expensive it is like going into the Boxes by way of the pit—I cannot go out on Sunday—but if on

Monday it should promise as fair as to day I will put on a pair of loose easy palatable boots and me rendre chez vous—I continue increasing my letter to George to send it by one of Birkbeck's sons who is going out soon—so if you will let me have a few more lines, they will be in time—I am glad you got on so well with Monsr le Curè—is he a nice Clergyman—a great deal depends upon a cock'd hat and powder—not gun powder, lord love us, but lady-meal, violet-smooth, dainty-scented lilly-white, feather-soft, wigsby-dressing, coat-collar-spoiling whisker-reaching, pig-tail loving, swans down-puffing, parson-sweetening powder—I shall call in passing at the tottenham nursery and see if I can find some seasonable plants for you. That is the nearest place—or by our la' kin or lady kin, that is by the virgin Mary's kindred, is there not a twig-manufacturer in Walthamstow? M^r^ & Mrs Dilke are coming to dine with us to day—they will enjoy the country after Westminster—O there is nothing like fine weather, and health, and Books, and a fine country, and a contented Mind, and Diligent-habit of reading and thinking, and an amulet against the ennui—and, please heaven, a little claret-wine cool out of a cellar a mile deep—with a few or a good many ratafia cakes—a rocky basin to bathe in, a strawberry bed to say your prayers to Flora in, a pad nag to go you ten miles or so; two or three sensible people to chat with; two or thee spiteful folkes to spar with; two or three odd fishes to laugh at and two or three numskuls to argue with—instead of using dumb bells on a rainy day—

[*'Two or three Posies' follows*]

Good bye I've an appoantment—can't stop pon word—good bye—now dont get up—open the door myself—go-o-o d bye—see ye Monday

J—K—

Letter to George and Georgiana Keats, 14, 19 February, ?3, 12, 13, 17, 19 March, 15, 16, 21, 30 April, 3 May 1819

Letter C°— sunday Morn Feby 14—

My dear Brother & Sister—How is it we have not heard from you from the Settlement yet? The Letters must surely have miscarried—I am in expectation every day—Peachey wrote me a few days ago saying some more acquaintances of his were preparing to set out for Birkbeck—therefore I shall take the opportunity of sending you what I can muster in a sheet or two—I am still at Wentworth Place—indeed I have kept in doors lately. resolved if possible to rid myself of my sore throat—consequently i have not been to see your Mother since my return from Chichester—but my absence from her has been a great weight upon me—I say since my return from Chichester—I believe I told you I was going thither—I was nearly a fortnight at Mr John Snook's and a few days at old Mr Dilke's—Nothing worth speaking of happened at either place—I took down some of the thin paper and wrote on it a little Poem call'd 'St Agnes Eve'—which you shall have as it is when I have finished the blank part of the rest for you—I went out twice at Chichester to old Dowager card parties—I see very little now, and very few Persons—being almost tired of Men and things—Brown and Dilke are very kind and considerate towards me—The Miss Reynoldses have been stopping next door lately—but all very dull—Miss Brawne and I have every now and then a chat and a tiff—Brown and Dilke are walking round their Garden hands in Pockets making observations. The Literary world I know nothing about—There is a Poem from Rogers dead born°—and another satire is expected from Byron call'd Don Giovanni—Yesterday I went to town for the first time for these three weeks—I met people from all parts and of all sets—Mr Towers —one of the Holts—Mr Domine Williams—Mr Woodhouse Mrs Hazlitt and Son—Mrs Webb—Mrs Septimus Brown—Mr Woodhouse was looking up at a Book-window in newgate street and being short-sighted twisted his Muscles into so queer a stupe° that I stood by in doubt whether it was him or his brother, if he has one and turning round saw Mrs Hazlitt with that little Nero her son—Woodhouse on his features subsiding proved to be Woodhouse and not his Brother—I have had a little business with Mr Abbey—From time to time he has behaved to me with a little Brusquerie—this hurt me a little especially wheen I knew him to be the only Man in England who dared to say a thing to me I did not approve of without its being resented or at least noticed—so I wrote him about it and have made an alteration in my favor—I expect from this to see more of Fanny—who has been quite shut out from me. I see Cobbet has been attacking the Settlement°—but I cannot tell what to believe—and shall be all out at elbous till I hear from you. I am invited to Miss Millar's Birthday dance on the 19th I am nearly sure I shall not be able to go—a Dance would injure my throat very much. I see very little of Reynolds. Hunt I hear is going on very badly—I mean in money Matters I shall not be surprised to hear of the worst—Haydon too in consequence of his eyes is out at elbows. I live as prudently as it is possible for me to do. I have not seen Haslam lately—I have not seen

Richards for this half year—Rice for three Months or C C. C. for god knows when—When I last called in Henrietta Street—Mrs Millar was verry unwell—Miss Waldegrave as staid and self possessed as usual—Miss Millar was well—Henry was well—There are two new tragedies—one by the Apostate Man,° and one by Miss Jane Porter—Next week I am going to stop at Taylor's for a few days when I will see them bothe and tell you what they are—M^{rs} and Mr Bentley are well and all the young Carrots. I said nothing of consequence passed at Snook's—no more than this that I like the family very much M^{r} and Mrs Snook were very kind—we used to have over a little Religion and politicts together almost every evening—and sometimes about you—He proposed writing out for me all the best part of his experience in farming to send to you if I should have an opportunity of talking to him about it I will get all I can at all events—but you may say in your answer to this what value you place upon such information. I have not seen Mr Lewis lately for I have shrunk from going up the hill—Mr Lewis went a few morning ago to town with Mrs Brawne they talked about me—and I heard that M^{r} L Said a thing I am not at all contented with—Says he 'O, he is quite the little Poet' now this is abominable—you might as well say Buonaparte is quite the little Soldier—You see what it is to be under six foot and not a lord—There is a long fuzz to day in the examiner about a young Man who delighted a young woman with a Valentine—I think it must be Ollier's.° Brown and I are thinking of passing the summer at Brussels if we do we shall go about the first of May—We i e Brown and I sit opposite one another all day authorizing (N.B. an s instead of a z would give a different meaning) He is at present writing a Story of an old Woman who lived in a forest and to whom the Devil or one his Aid de feus came one night very late and in disguise—The old Dame sets before him pudding after pudding—mess after mess—which he devours and moreover casts his eyes up at a side of Bacon hanging over his head and at the same time asks whither her Cat is a Rabbit—On going he leaves her three pips of eve's apple—and some how—she, having liv'd a virgin all her life, begins to repent of it and wishes herself beautiful enough to make all the world and even the other world fall in love with her—So it happens—she sets out from her smoaky Cottage in magnificent apparel; the first city She enters evey one falls in love with her—from the Prince to the Blacksmith. A young gentleman on his way to the church to be married leaves his unfortunate Bride and follows this nonsuch—A whole regiment of soldiers are smitten at once and follow her—A whole convent of Monks in corpus christi procession join the Soldiers—The Mayor and Corporation follow the same road—Old and young, deaf and dumb—all but the blind are smitten and form an immense concourse of people who—what Brown will do with them I know not—The devil himself falls in love with her flies away with her to a desert place—in consequence of which she lays an infinite number of Eggs—The Eggs being hatched from time to time fill the world with many nuisances such as John Knox—George Fox—Johanna Southcote—Gifford—There have been within a fortnight eight failures of the highest consequence in London—Brown went a few evenings since to Davenport's° and on his coming in he talk'd about bad news in the City with such a face, I began to think of a national Bankruptcy—I did not feel much surprised—and was rather disappointed. Carlisle,° a Bookseller on the *Hone* principle has been issuing Pamphlets from his shop in fleet Street Called the Deist—he was conveyed to Newgate last Thursday—he intends making his own defence. I was surprised to hear from Taylor the amount of Murray the Booksellers last sale—what think you of £25,000? He sold 4000 coppies of Lord Byron. I am sitting opposite the Shakspeare I brought from the Isle of wight—and I never look at it but the silk tassels on it give me as much pleasure as the face of the Poet itself—except that I do not know how you are going on—In my next Packet as this is one by the way, I shall send you the Pot of Basil, St Agnes eve, and if I should have finished it a little thing call'd the 'eve of St Mark' you see what fine mother Radcliff° names I have—it is not my fault—I did not search for them—I have not gone on with Hyperion—for to tell the truth I have not been in great cue for writing lately—I must wait for the spring to rouse me up a little—The only time I went out from Bedhampton was to see a Chapel consecrated—Brown I and John Snook the boy, went in a chaise behind a leaden horse Brown drove, but the horse did not mind him—This Chapel is built by a Mr Way° a great Jew converter—who in that line has spent one hundred thousand Pounds—He maintains a great number of poor Jews—Of course his communion plate was stolen—he spoke to the Clerk about it—The Clerk said he was very sorry adding—'I dare shay your honour its among ush' The Chapel is built in Mr Way's park—The Consecration was—not amusing—there were numbers of carriages, and his house crammed with Clergy—they sanctified the Chapel—and it being a wet day consecrated the burial ground through the vestry window. I begin to hate Parsons—they did not make me love them that day—when I saw them in their proper colours—A Parson is a Lamb in a drawing room and a lion in a Vestry—The notions of Society will not permit a Parson to give way to his temper in any shape—so he festers in himself—his features get a peculiar diabolical self sufficient iron stupid expession—He is continually acting—His mind is against every Man and every Mans mind is against him—He is an Hippocrite to the Believer and a Coward to the unbeliever—He must be either a Knave or an Ideot—And there is no Man so much to be pitied as an ideot parson. The soldier who is cheated into an esprit du corps—by a red coat, a Band and Colours for the purpose of nothing—is not half so pitiable as the Parson who is led by the nose by the Bench of Bishops—and is smothered in absurdities—a poor necessary subaltern of the Church—

Friday Feby 18°—The day before yesterday I went to Romney Street—your Mother was not at home—but I have just written her that I shall see her on wednesday. I call'd on Mr Lewis this morning—he is very well—and tells me not to be uneasy about Letters the chances being so arbitary—He is going on as usual among his favorite democrat papers—We had a chat as usual about Cobbett: and the westminster electors. Dilke has lately been verry much harrassed about the manner of educating his Son—he at length decided for a public school—and then he did not know what school—he at last has decided for Westminster; and as Charley is to be a day boy, Dilke will remove to Westminster. We lead verry quiet lives here—Dilke is at present in greek histories and antiquities—and talks of nothing but the electors of Westminster and the retreat of the ten-thousand—I never drink now above three glasses of wine—and never any spirits and water. Though by the bye the other day—Woodhouse took me to his coffee house—and ordered a Bottle of Claret—now I like Claret whenever I can have Claret I must drink it.—'t is the only palate affair that I am at all sensual in—Would it not be a good Speck° to send you some vine roots—could I be done? I'll enquire—If you could make some wine like Claret to dink on summer evenings in an arbour! For really 't is so fine—it fills the mouth one's mouth with a gushing freshness—then goes down cool and feverless—then you do not feel it quarelling with your liver—no it is rather a Peace maker and lies as quiet as it did in the grape—then it is as fragrant as the Queen Bee; and the more ethereal Part of it mounts into the brain, not assaulting the cerebral apartments like a bully in a bad house looking for his trul and hurrying from door to door bouncing against the waist-coat; but rather walks like Aladin about his own enchanted palace so gently that you do not feel his step—Other wines of a heavy and spirituous nature transform a Man to a Silenus; this makes him a Hermes—and gives a Woman the soul and imortality of Ariadne for whom Bacchus always kept a good cellar of claret—and even of that he could never persuade her to take above two cups—I said this same Claret is the only palate-passion I have I forgot game I must plead guilty to the breast of a Partridge, the back of a hare, the back-bone of a grouse, the wing and side of a Pheasant and a Woodcock *passim* Talking of game (I wish I could make it) the Lady whom I met at Hastings and of whom I said something in my last I think, has lately made me many presents of game, and enabled me to make as many—She made me take home a Pheasant the other day which I gave to Mrs Dilke; on which, tomorrow, Rice, Reynolds and the Wentworthians will dine next door—The next I intend for your Mother. These moderate sheets of paper are much more pleasant to write upon than those large thin sheets which I hope you by this time have received—though that cant be now I think of it—I have not said in any Letter yet a word about my affairs—in a word I am in no despair about them—my poem° has not at all succeeded—in the course of a year or so I think I shall try the public again—in a selfish point of view I should suffer my pride and my contempt of public opinion to hold me silent—but for your's and fanny's sake I will pluck up a spirit, and try again—I have no doubt of success in a course of years if I persevere—but it must be patience—for the Reviews have enervated and made indolent mens minds—few think for themselves—These Reviews too are getting more and more powerful and especially the Quarterly—They are like a superstition which the more it prostrates the Crowd and the longer it continues the more powerful it becomes just in proportion to their increasing weakness—I was in hopes that when people saw, as they must do now, all the trickery and iniquity of these Plagues they would scout them, but no they are like the spectators at the Westminster cock-pit—they like the battle and do not care who wins or who looses—Brown is going on this morning with the story of his old woman and the Devil—He makes but slow progreess—the fact is it is a Libel on the Devil and as that person is Brown's Muse, look ye, if he libels his own Muse how can he expect to write—Either Brown or his muse must turn tale—Yesterday was Charley Dilkes birth day—Brown and I were invited to tea—During the evening nothing passed worth notice but a little conversation between Mrs Dilke and Mrs Brawne—The subject was the watchman—It was ten o'Clock and Mrs Brawne, who lived during the summer in Brown's house and now lives in the Road, recognized her old Watchman's voice and said that he came as far as her now: 'indeed'; said 'Mrs D. 'does he turn the Corner?' There have been some Letters pass between me and Haslam: but I have not seen him lately—the day before yesterday—which I made a day of Business, I call'd upon him—he was out as usual—Brown has been walking up and down the room a breeding—now at this moment he is being delivered of a couplet—and I dare say will be as well as can be expected—Gracious—he has twins! I have a long Story to tell you about Bailey—I will say first the circumstances as plainly and as well as I can remember, and then I will make my comment. You know that Bailey was very much cut up about a little Jilt in the country somewhere; I thought he was in a dying state about it when at Oxford with him: little supposing as I have since heard, that he was at that very time making impatient Love to Marian Reynolds— and guess my astonishment at hearing after this that he had been trying at Miss Martin—So matters have been. So Matters stood—when he got ordained and went to a Curacy near Carlisle where the family of the Gleigs reside—There his susceptible heart was conquered by Miss Gleig—and thereby all his connections in town have been annulled—both male and female—I do not now remember clearly the facts—These however I know—He showed his correspondence with Marian to Gleig—returnd all her Letters and asked for his own—he also wrote very abrubt Letters to Mrs Reynolds—I do not know any more of the Martin affair than I have written above—No doubt his conduct has been verry bad. The great thing to be considered is—whether it is want of delicacy and principle or want of Knowledge and polite experience—And again Weakness—yes that is it—and the want of a Wife—yes that is it—and then Marian made great Bones of him, although her Mother and sister have teased her very much about it. Her conduct has been very upright throughout the whole affair—She liked Bailey as a Brother—but not as a Husband—especially as he used to woo her with the Bible and Jeremy Taylor under his arm—they walked in no grove but Jeremy Taylors°—Marians obstinacy is some excuse—but his so quickly taking to miss Gleig can have no excuse—except that of a Ploughmans who wants a wife—The thing which sways me more against him than any thing else is Rice's conduct on the occasion; Rice would not make an immature resolve: he was ardent in his friendship for Bailey; he examined the whole for and against minutely; and he has abandoned Bailey entirely. All this I am not supposed by the Reynoldses to have any hint of—It will be a good Lesson to the Mother and Daughters—nothing would serve but Bailey—If you mentioned the word Tea pot—some one of them came out with an a propos about Bailey—noble fellow—fine fellow! was always in their mouths—this may teach them that the man who redicules romance is the most romantic of Men—that he who abuses women and slights them—loves them the most—that he who talks of roasting a Man alive would not do it when it came to the push—and above all that they are very shallow people who take every thing literal A Man's life of any worth is a continual allegory—and very few eyes can see the Mystery of his life—a life like the scriptures, figurative—which such people can no more make out than they can the hebrew Bible. Lord Byron cuts a figure—but he is not figurative—Shakspeare led a life of Allegory; his works are the comments on it—

On Monday° we had to dinner Severn & Cawthorn the Bookseller & print virtuoso; in the evening Severn went home to paint & we other three went to the play to see Sheild's new tragedy ycleped Evadné—In the morning Severn & I took a turn round the Museum, There is a Sphinx there of a giant size, & most voluptuous Egyptian expression, I had not seen it before—The play was bad even in comparison with *1818* the Augustan age of the Drama, 'Comme on sait' as Voltaire says.—the whole was made up of a virtuous young woman, an indignant brother, a suspecting lover, a libertine prince, a gratuitous villain, a street in Naples, a Cypress grove, lillies & roses, virtue & vice, a bloody sword, a spangled jacket, One Lady Olivia, One Miss ONeil alias Evadné, alias Bellamira, alias—Alias—Yea & I say unto you a greater than Elias—there was Abbot, & talking of Abbot his name puts me in mind of a Spelling book lesson, descriptive of the whole Dramatis personae—Abbot—Abbess—Actor—Actress—The play is a fine amusement as a friend of mine once said to me—'Do what you will' says he 'A poor gentleman who wants a guinea, cannot spend his two shillings better than at the playhouse'— The pantomime was excellent, I had seen it before & enjoyed it again— Your Mother & I had some talk about Miss H.—says I will Henry have that Miss H. a lath with a boddice, she who has been fine drawn—fit for nothing but to cut up into Cribbage pins, to the tune of 15.2;° One who is all muslin; all feathers & bone; Once in travelling she was made use of as a lynch pin; I hope he will not have her, though it is no uncommon thing to be *smitten with a staff;* though she might be very useful as his walking stick, his fishing rod, his toothpic—his hat stick (she runs so much in his head) let him turn farmer, she would cut into hurdles; let him write poetry she would be his turnstyle; Her gown is like a flag on a pole; she would do for him if he turn freemason; I hope she will prove a flag of truce; When she sits languishing with her one foot, on a stool, & one elbow on the table, & her head inclined, she looks like the sign of the crooked billet—or the frontispeice to Cinderella or a teapaper wood cut of Mother Shipton at her studies; she is a make-believe— She is bon a side a thin young—'Oman—But this is mere talk of a fellow creature; yet pardie I would not that Henry have her—Non volo ut eam possideat, nam, for it would be a bam,° for it would be a sham—Don't think I am writing a petition to the Governors of St Lukes;° no, that would be in another style. May it please your worships; forasmuch as the undersigned has committed, transferred, given up, made over, consigned, and aberrated himself to the art & mystery of poetry; for as much as he hath cut, rebuffed, affronted, huffed, & shirked, and taken stint, at all other employments, arts, mysteries, & occupations honest, middling & dishonest; for as much as he hath at sundry times, & in diverse places, told truth unto the men of this generation, & eke to the women, moreover; for as much as he hath kept a pair of boots that did not fit, & doth not admire Sheild's play, Leigh Hunt, Tom Moore, Bob Southey & Mr Rogers; & does admire Wm Hazlitt: more over for as more, as he liketh half of Wordsworth, & none of Crabbe; more overest for for as most; as he hath written this page of penmanship—he prayeth your Worships to give him a lodging—witnessed by R^d^ Abbey & Co. cum familiaribus & Consanguiniis (signed) Count de Cockaigne°— The nothing of the day is a machine called the Velocepede°—It is a wheel-carriage to ride cock horse upon, sitting astride & pushing it along with the toes, a rudder wheel in hand. they will go seven miles an hour, A handsome gelding will come to eight guineas, however they will soon be cheaper, unless the army takes to them I look back upon the last month, & find nothing to write about, indeed I do not recollect one thing particular in it—It's all alike, we keep on breathing. The only amusement is a little scandal of however fine a shape, a laugh at a pun—& then after all we wonder how we could enjoy the scandal, or laugh at the pun,

I have been at different times turning it in my head whether I should go to Edinburgh & study for a physician; I am afraid I should not take kindly to it, I am sure I could not take fees—& yet I should like to do so; it is not worse than writing poems, & hanging them up to be flyblown on the Reviewshambles—Every body is in his own mess— Here is the parson at Hampstead quarreling with all the world, he is in the wrong by this same token; when the black Cloth was put up in the Church for the Queen's mourning, he asked the workmen to hang it the wrong side outwards, that it might be better when taken down, it being his perquisite—Parsons will always keep up their Character, but as it is said there are some animals, the Ancients knew, which we do not; let us hope our posterity will miss the black badger with tri-cornered hat; Who knows but some Revisor of Buffon or Pliny, may put an account of the parson in the Appendix; No one will then believe it any more than we beleive in the Phoenix. I think we may class the lawyer in the same natural history of Monsters; a green bag will hold as much as a lawn sleeve— The only difference is that the one is fustian, & the other flimsy; I am not unwilling to read Church history at present & have Milnes° in my eye his is reckoned a very good one—March 12 Friday—I went to town yesterday chiefly for the purpose of seeing some young Men who were to take some Letters for us to you—through the medium of Peachey. I was surprised and disappointed at hearing they had changed their minds and did not purpose going so far as Birkbeck's—I was much disappointed; for I had counted upon seeing some persons who were to see you—and upon your seeing some who had seen me—I have not only lost this opportunity—but the sail of the Post-Packet to new york or Philadelphia—by which last, your Brothers have sent some Letters—The weather in town yesterday was so stifling that I could not remain there though I wanted much to see Kean in Hotspur—I have by me at present Hazlitt's Letter to Gifford°—perhaps you would like an extract or two from the high seasond parts—It begins thus. 'Sir, You have an ugly trick of saying what is not true of any one you do not like; and it will be the object of this Letter to cure you of it. You say what you please of others; it is time you were told what you are. In doing this give me leave to borrow the familiarity of your style:—for the fidelity of the picture I shall be answerable. You are a little person, but a considerable cat's paw; and so far worthy of notice. Your clandestine connection with persons high in office constantly influences your opinions, and alone gives importence to them. You are the government critic, a character nicely differing from that of a government spy—the invisible link, that connects literature with the Police.' Again—'Your employers Mr Gifford, do not pay their hirelings for nothing—for condescending to notice weak and wicked sophistry; for pointing out to contempt what excites no admiration; for cautiously selecting a few specimens of bad taste and bad grammar where nothing else is to be found They want your invincible pertness, your mercenary malice, your impenetrable dullness, your barefaced impudence, your pragmatical self sufficiency, your hypocritical zeal, your pious frauds to stand in the gap of their Prejudices and pretensions, to fly blow and taint public opinion, to defeat independent efforts, to apply not the touch of the scorpion but the touch of the Torpedo to youthful hopes, to crawl and leave the slimy track of sophistry and lies over every work that does not "dedicate its sweet leaves"° to some Luminary of the tresury bench, or is not fostered in the hot bed of corruption—This is your office; "this is what is look'd for at your hands and this you do not baulk"—to sacrifice what little honesty and prostitute what little intellect you possess to any dirty job you are commission'd to execute. "They keep you as an ape does an apple in the corner of his jaw, first mouth'd to be at last swallow'd"°—You are by appointment literary toad eater to greatness and taster to the court—You have a natural aversion to whatever differs from your own pretensions, and an acquired one for what gives offence to your superiors. Your vanity panders to your interest, and your malice truckles only to your love of Power. If your instinctive or premeditated abuse of your enviable trust were found wanting in a single instance; if you were to make a single slip in getting up your select committee of enquiry and greenbag report of the state of Letters, your occupation would be gone. You would never after obtain a squeeze of the hand from a great man, or a smile from a Punk of Quality. The great and powerful (whom you call wise and good) do not like to have the privacy of their self love startled by the obtrusive and unmanageable claims of Literature and Philosophy, except through the intervention of people like you, whom; if they have common penetration, they soon find out to be without any superiority of intellect; or if they do not whom they can despise for their meanness of soul. You "have the office opposite to saint Peter"° You "keep a corner in the public mind, for foul prejudice and corrupt power to knot and gender in";° you volunteer your services to people of quality to ease scruples of mind and qualmes of conscience; you lay the flattering unction of venal prose and laurell'd verse to their souls°—You persuade them that there is neither purity of morals, not depth of understanding, except in themselves and their hangers on; and would prevent the unhallow'd names of Liberty and humanity from ever being whispered in years polite! You, sir, do you not all this? I cry you mercy then: I took you for° the Editor of the Quarterly Review!' This is the sort of feu de joie he keeps up—there is another extract or two—one especially which I will copy tomorrow—for the candles are burnt down and I am using the wax taper—which has a long snuff on it—the fire is at its last click—I am sitting with my back to it with one foot rather askew upon the rug and the other with the heel a little elevated from the carpet—I am writing this on the Maid's tragedy which I have read since tea with Great pleasure—Besides this volume of Beaumont & Fletcher—there are on the table two volumes of chaucer and a new work of Tom Moores call'd 'Tom Cribb's memorial to Congress'°—nothing in it—These are trifles—but I require nothing so much of you as that you will give me a like description of yourselves, however it may be when you are writing to me—Could I see the same thing done of any great Man long since dead it would be a great delight: as to know in what position Shakspeare sat when he began 'To be or not to be'—such thing become interesting from distance of time or place. I hope you are both now in that sweet sleep which no two beings deserve more than you do—I must fancy you so—and please myself in the fancy of speaking a prayer and a blessing over you and your lives—God bless you—I whisper good night in your ears and you will dream of me—

Saturday 13 March. I have written to Fanny this morning; and received a note from Haslam—I was to have dined with him to morrow: he give me a bad account of his Father, who has not been in Town for 5 weeks—and is not well enough for company—Haslam is well—and from the prosperous state of some love affair he does not mind the double tides he has to work—I have been a Walk past westend—and was going to call at Mr Monkhouse's—but I did not, not being in the humour—I know not why Poetry and I have been so distant lately I must make some advances soon or she will cut me entirely. Hazlitt has this fine Passage in his Letter Gifford, in his Review of Hazlitt's characters of Shakspeare's plays, attacks the Coriolanus critique—He says that Hazlitt has slandered Shakspeare in saying that he had a leaning to the arbitary side of the question. Hazlitt thus defends himself 'My words are "Coriolanus is a storehouse of political commonplaces. The Arguments for and against aristocracy and democracy, on the Preveleges of the few and the claims of the many, on Liberty and slavery, power and the abuse of it, peace and war, are here very ably handled, with the spirit of a poet and the acuteness of a Philosopher. Shakspeare himself seems to have had a leaning to the arbitary side of the question, perhaps from some feeling of contempt for his own origin, and to have spared no occasion of baiting the rabble. *What he says of them is very true; what he says of their betters is also very true, though he dwells less upon it.*" I then proceed to account for this by shewing how it is that "the cause of the people is but little calculated for a subject for Poetry; or that the language of Poetry naturally falls in with the language of power." I affirm, Sir, that Poetry, that the imagination, generally speaking, delights in power, in strong excitement, as well as in truth, in good, in right, whereas pure reason and the moral sense approve only of the true and good. I proceed to show that this general love or tendency to immediate excitement or theatrical effect, no matter how produced, gives a Bias to the imagination often consistent° with the greatest good, that in Poetry it triumphs over Principle, and bribes the passions to make a sacrifice of common humanity. You say that it does not, that there is no such original Sin in Poetry, that it makes no such sacrifice or unworthy compromise between poetical effect and the still small voice of reason—And how do you prove that there is no such principle giving a bias to the imagination, and a false colouring to poetry? Why by asking in reply to the instances where this principle operates, and where no other can with much modesty and simplicity—"But are these the only topics that afford delight in Poetry &c" No; but these objects do afford delight in poetry, and they afford it in proportion to their strong and often tragical effect, and not in proportion to their strong and often tragical effect,° and not in proportion to the good produced, or their desireableness in a moral point of view? "Do we read with more pleasure of the ravages of a beast of prey than of the Shepherds pipe upon the Mountain?" No but we do read with pleasure of the ravages of a beast of prey, and we do so on the principle I have stated, namely from the sense of power abstracted from the sense of good; and it is the same principle that makes us read with admiration and reconciles us in fact to the triumphant progress of the conquerers and mighty Hunters of mankind, who come to stop the shepherd's Pipe upon the Mountains and sweep away his listening flock. Do you mean to deny that there is any thing imposing to the imagination in power, in grandeur, in outward shew, in the accumulation of individual wealth and luxury, at the expense of equal justice and the common weal? Do you deny that there is any thing in the "Pride, Pomp and Circumstance of glorious war, that makes ambition virtue"?° in the eyes of admiring multitudes? Is this a new theory of the Pleasures of the imagination which says that the pleasures of the imagination do not take rise soly in the calculations of the understanding? Is it a paradox of my creating that "one murder makes a villain millions a Hero!" or is it not true that here, as in other cases, the enormity of the evil overpowers and makes a convert of the imagination by its very magnitude? You contradict my reasoning, because you know nothing of the question, and you think that no one has a right to understand what you do not. My offence against purity in the passage alluded to, "which contains the concentrated venom of my malignity," is, that I have admitted that there are tyrants and slaves abroad in the world; and you would hush the matter up, and pretend that there is no such thing in order that there may be nothing else. Farther I have explained the cause, the subtle sophistry of the human mind, that tolerates and pampers the evil in order to guard against its approaches; you would conceal the cause in order to prevent the cure, and to leave the proud flesh about the heart to harden and ossify into one impenetrable mass of selfishness and hypocrisy, that we may not "sympathise in the distresses of suffering virtue" in any case in which they come in competition with the fictitious° wants and "imputed weaknesses of the great." You ask "are we gratified by the cruelties of Domitian or Nero?" No, not we—they were too petty and cowardly to strike the imagination at a distance; but the Roman senate tolerated them, addressed their perpetrators, exalted them into gods, the fathers of the people; they had pimps and scribblers of all sorts in their pay, their Senecas, &c till a turbulent rabble thinking that there were no injuries to Society greater than the endurance of unlimited and wanton oppression, put an end to the farce and abated the nuisance as well as they could. Had you and I lived in those times we should have been what we are now, I "a sour mal content," and you "a sweet courtier."' The manner in which this is managed: the force and innate power with which it yeasts and works up itself—the feeling for the costume of society; is in a style of genius—He hath a demon as he himself says of Lord Byron—We are to have a party this evening—The Davenports from Church row—I dont think you know any thing of them—they have paid me a good deal of attention—I like Davenport himself—The names of the rest are Miss Barnes Miss Winter with the Children—

*March 17*th—Wednesday—On sunday I went to Davenports' were I dined—and had a nap. I cannot bare a day anhilated in that manner—there is a great difference between an easy and an uneasy indolence—An indolent day—fill'd with speculations even of an unpleasant colour—is bearable and even pleasant alone—when one's thoughts cannot find out any thing better in the world; and experience has told us that locomotion is no change: but to have nothing to do, and to be surrounded with unpleasant human identities; who press upon one just enough to prevent one getting into a lazy position; and not enough to interest or rouse one; is a capital punishment of a capital crime: for is not giving up, through goodnature, one's time to people who have no light and shade a capital crime? Yet what can I do?—they have been very kind and attentive to me. I do not know what I did on monday—nothing—nothing—nothing—I wish this was any thing extraordinary—Yesterday I went to town: I called on Mr Abbey; he began again (he has don it frequently lately) about that hat-making concern—saying he wish you had hearkened to it: he wants to make me a Hat-maker—I really believe 't is all interested: for from the manner he spoke withal and the card he gave me I think he is concerned in Hat-making himself. He speaks well of Fanny's health—Hodgkinson is married—From this I think he takes a little Latitude—Mr A was waiting very impatiently for his return to the counting house—and mean while observed how strange it was that Hodgkinson should have been not able to walk two months ago and that now he should be married.—'I do not,' says he 'think it will do him any good: I should not be surprised if he should die of a consumption in a year or two.' I called at Taylor's and found that he and Hilton had set out to dine with me: so I followed them immediately back—I walk'd with them townwards again as far as Cambden Town and smoak'd home a Segar°—This morning I have been reading the '*False one*'° I have been up to Mrs Bentley's—shameful to say I was in bed at ten—I mean this morning—The Blackwood's review has committed themselves in a scandalous heresy—they have been putting up Hogg the ettrick shepherd° against Burns—The senseless villains. I have not seen Reynolds Rice or any of our set lately—. Reynolds is completely limed in the law: he is not only reconcil'd to it but hobbyhorses upon it—Blackwood wanted very much to see him—the scotch cannot manage by themselves at all—they want imagination—and that is why they are so fond of Hogg, who has a little of it—

Friday 19th Yesterday I got a black eye—the first time I took a Cricket bat—Brown who is always one's friend in a disaster applied a leech to the eyelid, and there is no inflammation this morning though the ball hit me directly on the sight—'t was a white ball—I am glad it was not a clout—This is the second black eye I have had since leaving school—during all my school days I never had one at all—we must eat a peck before we die—This morning I am in a sort of temper indolent and supremely careless: I long after a stanza or two of Thompson's Castle of indolence—My passions are all alseep from my having slumbered till nearly eleven and weakened the animal fibre all over me to a delightful sensation about three degrees on this side of faintness—if I had teeth of pearl and the breath of lillies I should call it langour—but as I am + I must call it Laziness—In this state of effeminacy the fibres of the brain are relaxed in common with the rest of the body, and to such a happy degree that pleasure has no show of enticement and pain no unbearable frown. Neither Poetry, nor Ambition, nor Love have any alertness of countenance as they pass by me: they seem rather like three figures on a greek vase—a Man and two women—whom no one but myself could distinguish in their disguisement.° This is the only happiness; and is a rare instance of advantage in the body overpowering the Mind. I have this moment received a note from Haslam in which he expects the death of his Father who has been for some time in a state of insen-

+ especially as I have a black eye

sibility--his mother bears up he says very well—I shall go to town tommorrow to see him. This is the world—thus we cannot expect to give way many hours to pleasure—Circumstances are like Clouds continually gathering and bursting—While we are laughing the seed of some trouble is put into the wide arable land of events—while we are laughing it sprouts it grows and suddenly bears a poison fruit which we must pluck —Even so we have leisure to reason on the misfortunes of our friends; our own touch us too nearly for words. Very few men have ever arrived at a complete disinterestedness of Mind: very few have been influenced by a pure desire of the benefit of others—in the greater part of the Benefactors & to Humanity some meretricious motive has sullied their greatness—some melodramatic scenery has facinated them—From the manner in which I feel Haslam's misfortune I perceive how far I am from any humble standard of disinterestedness—Yet this feeling ought to be carried to its highest pitch, as there is no fear of its ever injuring society—which it would do I fear pushed to an extremity—For in wild nature the Hawk would loose his Breakfast of Robins and the Robin his of Worms° The Lion must starve as well as the swallow—The greater part of Men make their way with the same instinctiveness, the same unwandering eye from their purposes, the same animal eagerness as the Hawk—The Hawk wants a Mate, so does the Man—look at them both they set about it and procure one in the same manner—They want both a nest and they both set about one in the same manner—they get their food in the same manner—The noble animal Man for his amusement smokes his pipe—the Hawk balances about the Clouds—that is the only difference of their leisures. This it is that makes the Amusement of Life—to a speculative Mind. I go among the Feilds and catch a glimpse of a stoat or a fieldmouse peeping out of the withered grass—the creature hath a purpose and its eyes are bright with it—I go amongst the buildings of a city and I see a Man hurrying along—to what? The Creature has a purpose and his eyes are bright with it. But then as Wordsworth says, 'We have all one human heart"°—there is an ellectric fire in human nature tending to purify—so that among these human creatures there is continully some birth of new heroism—The pity is that we must wonder at it: as we should at finding a pearl in rubbish—I have no doubt that thousands of people never heard of have had hearts completely disinterested: I can remember but two—Socrates and Jesus—their Histories evince it—What I heard a little time ago, Taylor observe with respect to Socrates, may be said of Jesus—That he was so great a man that though he transmitted no writing of his own to posterity, we have his Mind and his sayings and his greatness handed to us by others. It is to be lamented that the history of the latter was written and revised by Men interested in the pious frauds of Religion. Yet through all this I see his splendour. Even here though I myself am pursueing the same instinctive course as the veriest human animal you can think of—I am however young writing at random—straining at particles of light in the midst of a great darkness—without knowing the bearing of any one assertion of any one opinion. Yet may I not in this be free from sin? May there not be superior beings amused with any graceful, though instinctive attitude my mind may fall into, as I am entertained with the alertness of a Stoat or the anxiety of a Deer? Though a quarrel in the streets is a thing to be hated, the energies displayed in it are fine; the commonest Man shows a grace in his quarrel—By a superior being our reasoning may take the same tone—though erroneous they may be fine—This is the very thing in which consists poetry; and if so it is not so fine a thing as philosophy—For the same reason that an eagle is not so fine a thing as a truth—Give me this credit—Do you not think I strive—to know myself? Give me this credit—and you will not think that on my own accout I repeat Milton's lines

> 'How charming is divine Philosophy
> Not harsh and crabbed as dull fools suppose
> But musical as is Apollo's lute'—°

No—no for myself—feeling grateful as I do to have got into a state of mind to relish them properly—Nothing ever becomes real till it is experienced—Even a Proverb is no proverb to you till your Life has illustrated it—I am ever affraid that your anxiety for me will lead you to fear for the violence of my temperament continually smothered down: for that reason I did not intend to have sent you the following sonnet—but look over the two last pages and ask yourselves whether I have not that in me which will well bear the buffets of the world. It will be the best comment on my sonnet; it will show you that it was written with no Agony but that of ignorance; with no thirst of any thing but knowledge when pushed to the point though the first steps to it were throug my human passions—they went away, and I wrote with my Mind—and perhaps I must confess a little bit of my heart—

[*'Why did I laugh tonight? No voice will tell' follows*]

I went to bed, and enjoyed an uninterrupted sleep—Sane I went to bed and sane I arose. || This is the 15th of April—you see what a time it is since I wrote—all that time I have been day by day expecting Letters from you—I write quite in the dark—In the hopes of a Letter daily I have deferred that I might write in the light—I was in town yesterday and at Taylors heard that young Brikbeck had been in Town and was set to forward in six or seven days—so I shall dedicate that time to making up this parcel ready for him—I wish I could hear from you to make me 'whole and general as the casing air'° A few days after the 19th of april I received a note from Haslam containng the news of his father's death—The Family has all been well—Haslam has his father's situation. The Framptons° have behaved well to him—The day before yesterday I went to a rout at Sawrey's—it was made pleasant by Reynolds being there, and our getting into conversation with one of the most beautiful Girls I ever saw—She gave a remarkable prettiness to all those commonplaces which most women who talk must utter—I liked Mrs Sawrey very well. The Sunday before last your Brothers were to come by a long invitation—so long that for the time I forgot it when I promised Mrs Brawne to dine with her on the same day—On recollecting my engagement with your Brothers I immediately excused myself with Mrs Brawn but she would not hear of it and insisted on my bringing my friends with me. so we all dined at Mrs Brawne's. I have been to Mrs Bentley's this morning and put all the Letters two and from you and poor Tom and me—I have found some of the correspondence between him and that degraded Wells and Amena°—It is a wretched business. I do not know the rights of it—but what I do know would I am sure affect you so much that I am in two Minds whether I will tell you any thing about it—And yet I do not see why—for any thing tho' it be unpleasant, that calls to mind those we still love, has a compensation in itself for the pain it occasions—so very likely tomorrow I may set about coppying thee whole of what I have about it: with no sort of a Richardson self satisfaction—I hate it to a sickness—and I am affraid more from indolence of mind than any thing else I wonder how people exist with all their worries. I have not been to Westminster but once lately and that was to see Dilke in his new Lodgings—I think of living somewhere in the neighbourhood myself—Your mothers was well by your Brothers' account. I shall see her perhaps tomorrow—yes I shall—We have had the Boys° here lately—they make a bit of a racket—I shall not be sorry when they go. I found also this morning in a note from George to you my dear sister a lock of your hair which I shall this moment put in the miniature case. A few days ago Hunt dined here and Brown invited Davenport to meet him. Davenport from a sense of weakness thought it incumbent on him to show off—and pursuant to that never ceased talking and boaring all day, till I was completely fagged out—Brown grew melancholy—but Hunt perceiving what a complimentary tendency all this had bore it remarkably well—Brown grumbled about it for two or three days—I went with Hunt to Sir John Leicester's gallery° there I saw Northcote°—Hilton—Bewick and many more of great and Little note. Haydons picture is of very little progress this last year—He talk about finishing it next year—Wordsworth is going to publish a Poem called Peter Bell—what a perverse fellow it is! Why wilt he talk about Peter Bells—I was told not to tell—but to you it will not be tellings—Reynolds hearing that said Peter Bell was coming out, took it into his head to write a skit upon it call'd Peter Bell. He did it as soon as thought on it is to be published this morning,° and comes out before the real Peter Bell, with this admirable motto from the 'Bold stroke for a Wife'. 'I am the real Simon Pure'° It would be just as well to trounce Lord Byron in the same manner. I am still at a stand in versifying—I cannot do it yet with any pleasure—I mean however to look round at my resources and means—and see what I can do without poetry—To that end I shall live in Westminster—I have no doubt of making by some means a little to help on or I shall be left in the Lurch—with the burden of a little Pride—However I look in time—The Dilkes like their lodging in Westminster tolerably well. I cannot help thinking what a shame it is that poor Dilke should give up his comfortable house & garden for his Son, whom he will certainly ruin with too much care—The boy has nothing in his ears all day but himself and the importance of his education—Dilke has continually in his mouth 'My Boy' This is what spoils princes: it may have the same effect with Commoners. Mrs Dilke has been very well lately—But what a shameful thing it is that for that obstinate Boy Dilke should stifle himself in Town Lodgings and wear out his Life by his continual apprehension of his Boys fate in Westminsterschool, with the rest of the Boys and the Masters—Evey one has some wear and tear—One would think Dilke ought to be quiet and happy—but no—this one Boy—makes his face pale, his society silent and his vigilance jealous—He would I have no doubt quarrel with any one who snubb'd his Boy—With all this he has no notion how to manage him O what a farce is our greatest cares! Yet one must be in the pother for the sake of Clothes food and Lodging. There has been a squabble between Kean and one Mr Bucke —There are faults on both sides—on Bucks the faults are positive to the Question: Keans fault is a want of genteel knowledge and high Policy—The formor writes knavishly foolish and the other silly bombast. It was about a Tragedy° written by said Mr Bucke; which it appears Mr Kean kick'd at—it was so bad—. After a little struggle of Mr Bucke's against Kean—drury

Lane had the Policy to bring it on and Kean the impolicy not to appear in it—It was damn'd—The people in the Pit had a favouite call on the night of 'Buck Buck rise up' and 'Buck Buck how many horns do I hold up'. Kotzebue the German Dramatist and traitor to his country was murdered lately by a young student whose name I forget—he stabbed himself immediately after crying out Germany! Germany!° I was unfortunate to miss Richards the only time I have been for many months to see him. Shall I treat you with a little extempore.

[*'When they were come unto the Faery's Court' follows*]

Brown is gone to bed—and I am tired of rhyming—there is a north wind blowing playing young gooseberry° with the trees—I dont care so it heps even with a side wind a Letter to me—for I cannot put faith in any reports I hear of the Settlement some are good some bad—Last Sunday I took a Walk towards highgate and in the lane that winds by the side of Lord Mansfield's park I met Mr Green our Demonstrator at Guy's in conversation with Coleridge—I joined them, after enquiring by a look whether it would be agreeable—I walked with him at his alderman-after dinner pace for near two miles I suppose In those two Miles he broached a thousand things—let me see if I can give you a list—Nightingales, Poetry—on Poetical sensation—Metaphysics—Different genera and species of Dreams—Nightmare—a dream accompanied by a sense of touch—single and double touch—A dream related—First and second consciousness—the difference explained between will and Volition—so my metaphysicians from a want of smoking the second consiousness—Monsters—the Kraken—Mermaids—southey believes in them—southeys belief too much diluted—A Ghost story—Good morning—I heard his voice as he came towards me—I heard it as he moved away—I had heard it all the interval—if it may be called so. He was civil enough to ask me to call on him at Highgate Good Night! It looks so much like rain I shall not go to town to day; but put it off till tomorrow—Brown this morning is writing some spenserian stanzas against Mrs Miss Brawne and me; so I shall amuse myself with him a little: in the manner of Spenser—

[*'He is to weet a melancholy Carle' follows*]

This character would ensure him a situation in the establishment of patient Griselda—The servant has come for the little Browns this morning—they have been a toothache to me which I shall enjoy the riddance of—Their little voices are like wasps stings—'Some times am I all wound with Browns."° We had a claret feast some little while ago There were Dilke, Reynolds, Skinner,° Mancur, John Brown,° Martin, Brown and I—We all got a little tipsy—but pleasantly so—I enjoy Claret to a degree—I have been looking over the correspondence of the pretended Amena and Wells this evening—I now see the whole cruel deception—I think Wells must have had an accomplice in it—Amena's Letters are in a Man's language, and in a Man's hand imitating a woman's—The instigations to this diabolical scheme were vanity, and the love of intrigue. It was no thoughtless hoax—but a cruel deception on a sanguine Temperament, with every show of friendship. I do not think death too bad for the villain—The world would look upon it in a different light should I expose it—they would call it a frolic—so I must be wary—but I consider it my duty to be prudently revengeful. I will hang over his head like a sword by a hair. I will be opium to his vanity—if I cannot injure his interests—He is a rat and he shall have ratsbane to his vanity—I will harm him all I possibly can—I have no doubt I shall be able to do so—Let us leave him to his misery alone except when we can throw in a little more—The fifth canto of Dante pleases me more and more—it is that one in which he meets with Paulo and Francesca—I had passed many days in rather a low state of mind and in the midst of them I dreamt of being in that region of Hell. The dream was one of the most delightful enjoyments I ever had in my life—I floated about the whirling atmosphere as it is described with a beautiful figure to whose lips mine were joined at it seem'd for an age—and in the midst of all this cold and darkness I was warm—even flowery tree tops sprung up and we rested on them sometimes with the lightness of a cloud till the wind blew us away again—I tried a Sonnet upon it—there are fourteen lines but nothing of what I felt in it—o that I could dream it every night—

[*'A dream, after reading Dante's Episode of Paolo and Francesca' follows*]

I want very very much a little of your wit my dear sister—a Letter or two of yours just to bandy back a pun or two across the Atlantic and send a quibble over the Floridas—Now you have by this time crumpled up your large Bonnet, what do you wear—a cap! do you put your hair in papers of a night? do you pay the Miss Birkbeck's a morning visit—have you any tea? or to you milk and water with them—What place of Worship do you go to—the Quakers the Moravians, the Unitarians or the Methodists—Are there any flowers in bloom you like—any beautiful heaths—Any Streets full of Corset Makers. What sort of shoes have you to fit those pretty feet of yours? Do you desire Comp[ts] to one another? Do you ride on Horseback? What do you have for breakfast, dinner and supper? without mentioning lunch and bever and wet and snack°—and a bit to stay one's stomach—Do you get any spirits—now you might easily distill some whiskey—and going into the woods set up a whiskey shop for the Monkeys—Do you and the miss Birkbecks get groggy on any thing—a little so so ish so as to be obliged to be seen home with a Lantern—You may perhaps have a game at puss in the corner—Ladies are warranted to play at this game though they have not whiskers. Have you a fiddle in the Settlement—or at any rate a jew's harp—which will play in spite of ones teeth—When you have nothing else to do for a whole day I tell you how you may employ it—First get up and when you are dress'd, as it would be pretty early, with a high wind in the woods give George a cold Pig° with my Complements. Then you may saunter into the nearest coffee-house and after taking a dram and a look at the chronicle—go and frighten the wild boars upon the strength—you may as well bring one home for breakfast serving up the hoofs garnished with bristles and a grunt or two to accompany the singing of the kettle—then if George is not up give him a colder Pig always with my Compliments—When you are both set down to breakfast I advise you to eat your full share—but leave off immediately on feeling yourself inclined to any thing on the other side of the puffy—avoid that for it does not become young women—After you have eaten your breakfast—keep your eye upon dinner—it is the safest way—You should keep a Hawk's eye over your dinner and keep hovering over it till due time then pounce taking care not to break any plates—While you are hovering with your dinner in prospect you may do a thousand things—put a hedgehog into George's hat—pour a little water into his rifle—soak his boots in a pail of water—cut his jacket round into shreds like a roman kilt or the back of my grandmothers stays—sow *off* his buttons.

Yesterday I could not write a line I was so fatigued for the day before, I went to town in the morning called on your Mother, and returned in time for a few friends we had to dinner. There were Taylor Woodhouse, Reynolds—we began cards at about 9 o'Clock, and the night coming on and continuing dark and rainy they could not think of returning to town—so we played at Cards till very daylight—and yesterday I was not worth a sixpence—Your mother was very well but anxious for a Letter. We had half an hours talk and no more for I was obliged to be home. M[rs] and Miss Millar were well—and so was Miss

Waldegrave—I have asked your Brothers here for next Sunday—When Reynolds was here on Monday—he asked me to give Hunt a hint to take notice of his Peter Bell in the Examiner—the best thing I can do is to write a little notice of it myself which I will do here and copy it out if it should suit my Purpose—*Peter-Bell* There have been lately advertized two Books both Peter Bell by name; what stuff the one was made of might be seen by the motto, 'I am the real Simon Pure'. This false florimel has hurried from the press and obtruded herself into public notice while for ought we know the real one may be still wandering about the woods and mountains. Let us hope she may soon appear and make good her right to the magic girdle—The Pamphleteering Arch-image we can perceive has rather a splenetic love than a downright hatred to real florimels°—if indeed they had been so christened—or had even a pretention to play at bob cherry with Barbara Lewthwaite:° but he has a fixed aversion to those three rhyming Graces Alice Fell, Susan Gale and Betty Foy;° and now at lenght especially to Peter Bell—fit Apollo. It may be seen from one or two passages in this little skit, that the writer of it has felt the finer parts of M[r] Wordsworth and perhaps expatiated with his more remote and sublimer muse;° This as far as it relates to Peter Bell is unlucky. The more he may love the sad embroidery of the Excursion; the more he will hate the coarse Samplers of Betty Foy and Alice Fell; and as they come from the same hand, the better will be able to imitate that which can be imitated. to wit Peter Bell—as far as can be imagined from the obstinate Name—We repeat, it is very unlucky— this real Simon Pure is in parts the very Man—There is a pernicious likeness in the scenery a 'pestilent humour' in the rhymes and an inveterate cadence in some of the Stanzas that must be lamented—If we are one part amused at this we are three parts sorry that an appreciator of Wordsworth should show so much temper at this really provoking name of Peter Bell—! This will do well enough—I have coppied it and enclosed it to Hunt—You will call it a little politic—seeing I keep clear of all parties. I say something for and against both parties—and suit it to the tune of the examiner—I mean to say I do not unsuit it—and I believe I think what I say nay I am sure I do—I and my conscience are in luck to day—which is an excellent thing—The other night I went to the Play with Rice, Reynolds and Martin—we saw a new dull and half damnd opera call'd 'the heart of Mid Lothian'° that was on Saturday—I stopt at Taylors on sunday with Woodhouse—and passed a quiet sort of pleasant day. I have been very much pleased with the Panorama of the ships at the north Pole°—with the icebergs, the Mountains, the Bears the Walrus—the seals the

Penguins—and a large whale floating back above water—it is impossible to describe the place—Wednesday Evening—

[*A draft of 'La belle dame sans merci' follows*]

Why four kisses—you will say—why four because I wish to restrain the headlong impetuosity of my Muse—she would have fain said 'score' without hurting the rhyme—but we must temper the Imagination as the Critics say with Judgment. I was obliged to choose an even number that both eyes might have fair play: and to speak truly I think two a piece quite sufficient—Suppose I had said seven; there would have been three and a half a piece—a very awkward affair—and well got out of on my side—

[*'Song of Four Fairies' follows*]

I have been reading lately two very different books Robertson's America° and Voltaire's Siecle De Louis xiv It is like walking arm and arm between Pizarro and the great-little Monarch. In How lementabl a case do we see the great body of the people in both instances: in the first, where Men might seem to inherit quiet of Mind from unsophisticated senses; from uncontamination of civilisation; and especially from their being as it were estranged from the mutual helps of Society and its mutual injuries—and thereby more immediately under the Protection of Providence—even there they had mortal pains to bear as bad; or even worse than Baliffs, Debts and Poverties of civilised Life—The whole appears to resolve into this—that Man is originally 'a poor forked creature'° subject to the same mischances as the beasts of the forest, destined to hardships and disquietude of some kind or other. If he improves by degrees his bodily accomodations and comforts—at each stage, at each accent there are waiting for him a fresh set of annoyances—he is mortal and there is still a heaven with its Stars abov his head. The most interesting question that can come before us is, How far by the persevering endeavours of a seldom appearing Socrates Mankind may be made happy—I can imagine such happiness carried to an extreme—but what must it end in?—Death—and who could in such a case bear with death—the whole troubles of life which are now frittered away in a series of years, would then be accumulated for the last days of a being who instead of hailing its approach, would leave this world as Eve left Paradise—But in truth I do not at all believe in this sort of perfectibility—the nature of the world will not admit of it—the inhabitants of the world will correspond to itself—Let the fish philosophise the ice away from the Rivers in winter time and they shall be at continual play in the tepid delight of summer. Look at the Poles and at the sands of Africa, Whirlpools and volcanoes—Let men exterminate them and I will say that they may arrive at earthly Happiness—The point at which Man may arrive is as far as the paralel state in inanimate nature and no further—For instance suppose a rose to have sensation, it blooms on a beautiful morning it enjoys itself—but there comes a cold wind, a hot sun—it cannot escape it, it cannot destroy its annoyances—they are as native to the world as itself: no more can man be happy in spite, the worldy elements will prey upon his nature—The common cognomen of this world among the misguided and superstitious is 'a vale of tears' from which we are to be redeemed by a certain arbitary interposition of God and taken to Heaven—What a little circumscribed straightened notion! Call the world if you Please 'The vale of Soul-making' Then you will find out the use of the world (I am speaking now in the highest terms for human nature admitting it to be immortal which I will here take for granted for the purpose of showing a thought which has struck me concerning it) I say '*Soul making*' Soul as distinguished from an Intelligence—There may be intelligences or sparks of the divinity in millions—but they are not Souls till they acquire identities, till each one is personally itself. Intelligences are atoms of perception—they know and they see and they are pure, in short they are God—how then are Souls to be made? How then are these sparks which are God to have identity given them—so as ever to possess a bliss peculiar to each ones individual existence? How, but by the medium of a world like this? This point I sincerely wish to consider because I think it a grander system of salvation than the chrystain religion—or rather it is a system of Spirit-creation—This is effected by three grand materials acting the one upon the other for a series of years—These three Materials are the *Intelligence*—the *human heart* (as distinguished from intelligence or Mind) and the *World* or *Elemental space* suited for the proper action of *Mind and Heart* on each other for the purpose of forming the *Soul* or *Intelligence destined to possess the sense of Identity*. I can scarcely express what I but dimly perceive—and yet I think I perceive it—that you may judge the more clearly I will put it in the most homely form possible—I will call the *world* a School instituted for the purpose of teaching little children to read—I will call the *human heart* the *horn Book* used in that School—and I will call the *Child able to read, the Soul* made from that *school* and its *hornbook*. Do you not see how necessary a World of Pains and troubles is to school an Intelligence and make it a soul? A Place where the heart must feel and suffer in a thousand diverse ways! Not merely is the Heart a Hornbook, It is the Minds Bible, it is the Minds experience, it is the teat from which the Mind or intelligence sucks its identity—As various as the Lives of Men are—so various become their souls, and thus does God make individual beings, Souls, Identical Souls of the sparks of his own essence—This appears to me a faint sketch of a system of Salvation which does not affront our reason and humanity—I am convinced that many difficulties which christians labour under would vanish before it—There is one whch even now Strikes me—the Salvation of Children—In them the Spark or intelligence returns to God without any identity—it having had no time to learn of, and be altered by, the heart—or seat of the human Passions—It is pretty generally suspected that the christian scheme has been coppied from the ancient persian and greek Philosophers. Why may they not have made this simple thing even more simple for common apprehension by introducing Mediators and Personages in the same manner as in the hethen mythology abstractions are personified—Seriously I think it probable that this System of Soul-making—may have been the Parent of all the more palpable and personal Schemes of Redemption, among the Zoroastrians the Christians and the Hindoos. For as one part of the human species must have their carved Jupiter; so another part must have the palpable and named Mediator and saviour, their Christ their Oromanes and their Vishnu—If what I have said should not be plain enough, as I fear it may not be, I will put you in the place where I began in this series of thoughts—I mean, I began by seeing how man was formed by circumstances—and what are circumstances?—but touchstones of his heart—? and what are touch stones?—but proovings of his hearrt?—and what are proovings of his heart but fortifiers or alterers of his nature? and what is his altered nature but his soul?—and what was his soul before it came into the world and had These provings and alterations and perfectionings?—An intelligence—without Identity—and how is this Identity to be made? Through the medium of the Heart? And how is the heart to become this Medium but in a world of Circumstances?—There now I think what with Poetry and Theology you may thank your Stars that my pen is not very long winded—Yesterday I received two Letters from your Mother and Henry which I shall send by young Birkbeck with this—

Friday—April 30—Brown has been rummaging up some of my old sins—that is to say sonnets I do not think you remember them, so I will copy them out as well as two or three lately written—I have just written one on Fame—which Brown is transcribing and he has his book and mine I must employ myself perhaps in a sonnet on the same subject—

[*The two sonnets 'On Fame' and the 'Sonnet To Sleep' follow*]

The following Poem—the last I have written is the first and the only one with which I have taken even moderate pains—I have for the most part dash'd of my lines in a hurry—This I have done leisurely—I think it reads the more richly for it and will I hope encourage me to write other thing in even a more peacable and healthy spirit. You must recollect that Psyche was not embodied as a goddess before the time of Apulieus the Platonist who lived afteir the Agustan age, and consequently the Goddess was never worshipped or sacrificed to with any of the ancient fervour—and perhaps never thought of in the old religion—I am more orthodox than to let a hethen Goddess be so neglected—

[*'Ode to Psyche' follows*]

Here endethe ye Old to Psyche

Incipit altera Sonneta.

I have been endeavouring to discover a better sonnet stanza than we have. The legitimate does not suit the language over-well from the pouncing rhymes—the other kind appears too elegiac—and the couplet at the end of it has seldom a pleasing effect—I do not pretend to have succeeded—It will explain itself—

[*'If by dull rhymes our English must be chain'd' follows*]

Here endeth the other Sonnet—this is the 3^{d} of May & every thing is in delightful forwardness; the violets are not withered, before the peeping of the first rose; You must let me know every thing, how parcels go & come, what papers you have, & what Newspapers you want, & other things—God bless you my dear Brother & Sister

Your ever Affectionate Brother
JOHN KEATS

Letter to Mary-Ann Jeffery, 31 May 1819

C. Brown Esq$^{re's}$
My dear Lady, Wentworth Place—Hampstead—

I was making a day or two ago a general conflagration of all old Letters and Memorandums, which had become of no interest to me—I made however, like the Barber-inquisitor in Don Quixote some reservations—among the rest your and your Sister's Letters. I assure you you had not entirely vanished from my Mind, or even become shadows in my remembrance: it only needed such a memento as your Letters to bring you back to me—Why have I not written before? Why did I not answer your Honiton Letter? I had no good news for you—every concern of ours, (ours I wish I could say) and still I must say *ours*—though George is in America and I have no Brother left—Though in the midst of my troubles I had no relation except my young sister I have had excellent friends. Mr B. at whose house I now am, invited me,—I have been with him ever since. I could not make up my mind to let you know these things. Nor should I now—but see what a little interest will do—I want you to do me a Favor; which I will first ask and then tell you the reasons. Enquire in the Villages round Teignmouth if there is any Lodging commodious for its cheapness; and let me know where it is and what price. I have the choice as it were of two Poisons (yet I ought not to call this a Poison) the one is voyaging to and from India for a few years;° the other is leading a fevrous life alone with Poetry—This latter will suit me best—for I cannot resolve to give up my Studies It strikes me it would not be quite so proper for you to make such inquiries—so give my love to your Mother and ask her to do it. Yes, I would rather conquer my indolence and strain my nerves at some grand Poem—than be in a dunderheaded indiaman—Pray let no one in Teignmouth know any thing of this—Fanny° must by this time have altered her name—perhaps you have also—are you all alive? Give my Comp^ts^ to M^rs^—your Sister. I have had good news, (tho' 'tis a queerish world in which such things are call'd good) from George—he and his wife are well—I will tell you more soon—Especially dont let the Newfoundland fisherman know it—and especially no one else—I have been always till now almost as careless of the world as a fly—my troubles were all of the Imagination—My Brother George always stood between me and any dealings with the world—Now I find I must buffet it—I must take my stand upon some vantage ground and begin to fight—I must choose between despair & Energy—I choose the latter—though the world has taken on a quakerish look with me, which I once thought was impossible—

> 'Nothing can bring back the hour
> Of splendour in the grass and glory in the flower"°

I once thought this a Melancholist's dream—

But why do I speak to you in this manner? No believe me I do not write for a mere selfish purpose—the manner in which I have written of myself will convince you. I do not do so to Strangers. I have not quite made up my mind—Write me on the receipt of this—and again at your Leisure; between whiles you shall hear from me again—

Your sincere friend
JOHN KEATS

Letter to Mary-Ann Jeffery, 9 June 1819

Wentworth Place

My Dear young Lady,—I am exceedingly obliged by your two letters—Why I did not answer your first immediately was that I have had a little aversion to the South of Devon from the continual remembrance of my Brother Tom. On that account I do not return to my old Lodgins in Hampstead though the people of the house have become friends of mine—This however I could think nothing of, it can do no more than keep one's thoughts employed for a day or two. I like your description of Bradley very much and I dare say shall be there in the course of the summer; it would be immediately but that a friend with ill health and to whom I am greatly attached° call'd on me yesterday and proposed my spending a Month with him at the back of the Isle of Wight. This is just the thing at present—the morrow will take care of itself—I do not like the name of Bishop's Teigntown—I hope the road from Teignmouth to Bradley does not lie that way—Your advice about the Indiaman is a very wise advice, because it just suits me, though you are a little in the wrong concerning its destroying the energies of Mind: on the contrary it would be the finest thing in the world to strengthen them—To be thrown among people who care not for you, with whom you have no sympathies forces the Mind upon its own resources, and leaves it free to make its speculations of the differences of human character and to class them with the calmness of a Botanist. An Indiaman is a little world. One of the great reasons that the english have produced the finest writers in the world; is, that the English world has ill-treated them during their lives and foster'd them after their deaths. They have in general been trampled aside into the bye paths of life and seen the festerings of Society. They have not been treated like the Raphaels of Italy. And where is the Englishman and Poet who has given a magnifacent Entertainment at the christening of one of his Hero's Horses as Boyardo did?° He had a Castle in the Appenine. He was a noble Poet of

Romance; not a miserable and mighty Poet of the human Heart. The middle age of Shakspeare was all clouded over; his days were not more happy than Hamlet's who is perhaps more like Shakspeare himself in his common every day Life than any other of his Characters—Ben Johnson was a common Soldier and in the Low countries, in the face of two armies, fought a single combat with a french Trooper and slew him—For all this I will not go on board an Indiaman, nor for examples sake run my head into dark alleys: I dare say my discipline is to come, and plenty of it too. I have been very idle lately, very averse to writing; both from the overpowering idea of our dead poets and from abatement of my love of fame. I hope I am a little more of a Philosopher than I was, consequently a little less of a versifying Pet-lamb. I have put no more in Print or you should have had it. You will judge of my 1819 temper when I tell you that the thing I have most enjoyed this year has been writing an ode to Indolence. Why did you not make your long-haired sister put her great brown hard fist to paper and cross your Letter? Tell her when you write again that I expect chequer-work°—My friend Mr Brown is sitting opposite me employed in writing a Life of David. He reads me passages as he writes them stuffing my infidel mouth as though I were a young rook—Infidel Rooks do not provender with Elisha's Ravens. If he goes on as he has begun your new Church had better not proceed, for parsons will be superseeded—and of course the Clerks must follow. Give my love to your Mother with the assurance that I can never forget her anxiety for my Brother Tom. Believe also that I shall ever remember our leave-taking with *you.*

Ever sincerely yours'
JOHN KEATS

Letter to Fanny Brawne, 1 July 1819

Shanklin,
Isle of Wight, Thursday.

My dearest Lady,

I am glad I had not an opportunity of sending off a Letter which I wrote for you on Tuesday night—'twas too much like one out of Rousseau's Heloise. I am more reasonable this morning. The morning is the only proper time for me to write to a beautiful Girl whom I love so much: for at night, when the lonely day has closed, and the lonely, silent, unmusical Chamber is waiting to receive me as into a Sepulchre, then believe me my passion gets entirely the sway, then I would not have you see those Rapsodies which I once thought it impossible I should ever give way to, and which I have often laughed at in another, for fear you should° either too unhappy or perhaps a little mad. I am now at a very pleasant Cottage window, looking onto a beautiful hilly country, with a glimpse of the sea; the morning is very fine. I do not know how elastic my spirit might be, what pleasure I might have in living here and breathing and wandering as free as a stag about this beautiful Coast if the remembrance of you did not weigh so upon me. I have never known any unalloy'd Happiness for many days together: the death or sickness of some one has always spoilt my hours—and now when none such troubles oppress me, it is you must confess very hard that another sort of pain should haunt me. Ask yourself my love whether you are not very cruel to have so entrammelled me, so destroyed my freedom. Will you confess this in the Letter you must write immediately and do all you can to console me in it—make it rich as a draught of poppies to intoxicate me—write the softest words and kiss them that I may at least touch my lips where yours have been. For myself I know not how to express my devotion to so fair a form: I want a brighter word than bright, a fairer word than fair. I almost wish we were butterflies and liv'd but three summer days—three such days with you I could fill with more delight than fifty common years could ever contain. But however selfish I may feel, I am sure I could never act selfishly: as I told you a day or two before I left Hampstead, I will never return to London if my Fate does not turn up Pam° or at least a Court-card. Though I could centre my Happiness in you, I cannot expect to engross your heart so entirely—indeed if I thought you felt as much for me as I do for you at this moment I do not think I could restrain myself from seeing you again tomorrow for the delight of one embrace. But no—I must live upon hope and Chance. In case of the worst that can happen, I shall still love you—but what hatred shall I have for another! Some lines I read the other day are continually ringing a peal in my ears:

> To see those eyes I prize above mine own
> Dart favors on another—
> And those sweet lips (yielding immortal nectar)
> Be gently press'd by any but myself—
> Think, think Francesca, what a cursed thing
> It were beyond expression!°

J.

Do write immediately. There is no Post from this Place, so you must address Post Office, Newport, Isle of Wight. I know before night I shall curse myself for having sent you so cold a Letter; yet it is better to do it as much as in my senses as possible. Be as kind as the distance will permit to your

J. KEATS.

Present my Compliments to your mother, my love to Margaret° and best remembrances to your Brother—if you please so.

Letter to Fanny Brawne, 8 July 1819

July 8th

My sweet Girl,

Your Letter gave me more delight, than any thing in the world but yourself could do; indeed I am almost astonished that any absent one should have that luxurious power over my senses which I feel. Even when I am not thinking of you I receive your influence and a tenderer nature steeling upon me. All my thoughts, my unhappiest days and nights have I find not at all cured me of my love of Beauty, but made it so intense that I am miserable that you are not with me: or rather breathe in that dull sort of patience that cannot be called Life. I never knew before, what such a love as you have made me feel, was; I did not believe in it; my Fancy was affraid of it, lest it should burn me up. But if you will fully love me, though there may be some fire, 't will not be more than we can bear when moistened and bedewed with Pleasures. You mention 'horrid people' and ask me whether it depend upon them, whether I see you again—Do understand me, my love, in this—I have so much of you in my heart that I must turn Mentor when I see a chance of harm beffaling you. I would never see any thing but Pleasure in your eyes, love on your lips, and Happiness in your steps. I would wish to see you among those amusements suitable to your inclinations and spirits; so that our loves might be a delight in the midst of Pleasures agreeable enough, rather than a resource from vexations and cares—But I doubt much, in case of the worst, whether I shall be philosopher enough to follow my own Lessons: if I saw my resolution give you a pain I could not. Why may I not speak of your Beauty, since without that I could never have lov'd you—I cannot conceive any beginning of such love as I have for you but Beauty. There may be a sort of love for which, without the least sneer at it, I have the highest respect and can admire it in others: but it has not the richness, the bloom, the full form, the enchantment of love after my own heart. So let me speak of you Beauty, though to my own endangering; if you could be so cruel to me as to try elsewhere its Power. You say you are affraid I shall think you do not love me—in saying this you make me ache the more to be near you. I am at the diligent use of my faculties here, I do not pass a day without sprawling some blank verse or tagging some rhymes; and here I must confess, that, (since I am on that subject,) I love you the more in that I believe you have liked me for my own sake and for nothing else—I have met with women whom I really think would like to be married to a Poem and to be given away by a Novel. I have seen your Comet,° and only wish it was a sign that poor Rice would get well whose illness makes him rather a melancholy companion: and the more so as so to conquer his feelings and hide them from me, with a forc'd Pun. I kiss'd your writing over in the hope you had indulg'd me by leaving a trace of honey—What was your dream? Tell it me and I will tell you the interpretation thereof.

Ever yours my love!
JOHN KEATS—

Do not accuse me of delay—we have not here an opportunity of sending letters every day—Write speedily—

Letter to J. H. Reynolds, 11 July 1819

My dear Reynolds,

. . .

You will be glad to hear under my own hand (tho' Rice says we are like sauntering Jack & Idle Joe)° how diligent I have been, & am being. I have finish'd the Act° and in the interval of beginning the 2^{d} have proceeded pretty well with Lamia, finishing the 1st part which consists of about 400 lines. I have great hopes of success, because I make use of my Judgment more deliberately than I yet have done; but in Case of failure with the world, I shall find my content. And here (as I know you have my good at heart as much as a Brother,) I can only repeat to you what I have said to George—that however I sho^{d} like to enjoy what the competences of life procure, I am in no wise dashed at a different prospect. I have spent too many thoughtful days & moralized thro' too many nights for that, and fruitless wo^{d} they be indeed, if they did not by degrees make me look upon the affairs of the world with a healthy deliberation. I have of late been moulting: not for fresh feathers & wings: they are gone, and in their stead I hope to have a pair of patient sublunary legs. I have altered, not from a Chrysalis into a butterfly, but the Contrary. having two little loopholes, whence I may look out into the stage of the world: and that world on our coming here I almost forgot. The first time I sat down to write, I co[d] scarcely believe in the necessity of so doing. It struck me as a great oddity—Yet the very corn which is now so beautiful, as if it had only took to ripening yesterday, is for the market: So, why sho[d] I be delicate.—

Letter to Fanny Brawne, ?15 July 1819

Shanklin
Thursday Evening

My love,

I have been in so irritable a state of health these two or three last days, that I did not think I should be able to write this week. Not that I was so ill, but so much so as only to be capable of an unhealthy teasing letter. To night I am greatly recovered only to feel the languor I have felt after you touched with ardency. You say you perhaps might have made me better: you would then have made me worse: now you could quite effect a cure: What fee my sweet Physician would I not give you to do so. Do not call it folly, when I tell you I took your letter last night to bed with me. In the morning I found your name on the sealing wax obliterated. I was startled at the bad omen till I recollected that it must have happened in my dreams, and they you know fall out by contraries. You must have found out by this time I am a little given to bode ill like the raven; it is my misfortune not my fault; it has proceeded from the general tenor of the circumstances of my life, and rendered every event suspicious. However I will no more trouble either you or myself with sad Prophecies; though so far I am pleased at it as it has given me opportunity to love your disinterestedness towards me. I can be a raven no more; you and pleasure take possession of me at the same moment. I am afraid you have been unwell. If through me illness have touched you (but it must be with a very gentle hand) I must be selfish enough to feel a little glad at it. Will you forgive me this? I have been reading lately an oriental tale of a very beautiful color—° It is of a city of melancholy men, all made so by this circumstance. Through a series of adventures each one of them by turns reach some gardens of Paradise where they meet with a most enchanting Lady; and just as they are going to embrace her, she bids them shut their eyes—they shut them—and on opening their eyes again find themselves descending to the earth in a magic basket. The remembrance of this Lady and their delights lost beyond all recovery render them melancholy ever after. How I applied this to you, my dear; how I palpitated at it; how the certainty that you were in the same world with myself, and though as beautiful, not so talismanic as that Lady; how I could not bear you should be so you must believe because I swear it by yourself. I cannot say when I shall get a volume ready. I have three or four stories° half done, but as I cannot write for the mere sake of the press, I am obliged to let them progress or lie still as my fancy chooses. By Christmas perhaps they may appear, but I am not yet sure they ever will. 'Twill be no matter, for Poems are as common as newspapers and I do not see why it is a greater crime in me than in another to let the verses of an half-fledged brain tumble into the reading-rooms and drawing room windows. Rice has been better lately than usual: he is not suffering from any neglect of his parents who have for some years been able to appreciate him better than they did in his first youth, and are now devoted to his comfort. Tomorrow I shall, if my health continues to improve during the night, take a look father about the country, and spy at the parties about here who come hunting after the picturesque like beagles. It is astonishing how they raven down scenery like children do sweetmeats. The wondrous Chine here is a very great Lion: I wish I had as many guineas as there have been spy-glasses in it. I have been, I cannot tell why, in capital spirits this last hour. What reason? When I have to take my candle and retire to a lonely room, without the thought as I fall asleep, of seeing you tomorrow morning? or the next day, or the next—it takes on the appearance of impossibility and eternity—I will say a month—I will say I will see you in a month at most, though no one but yourself should see me; if it be but for an hour. I should not like to be so near you as London without being continually with you: after having once more kissed you Sweet I would rather be here alone at my task than in the bustle and hateful literary chit-chat. Meantime you must write to me—as I will every week—for your letters keep me alive. My sweet Girl I cannot speak my love for you. Good night! and

Ever yours
JOHN KEATS.

Letter to Fanny Brawne, 25 July 1819

Sunday Night.

My sweet Girl,

I hope you did not blame me much for not obeying your request of a Letter on Saturday: we have had four in our small room playing at cards night and morning leaving me no undisturb'd opportunity to write. Now Rice and Martin are gone, I am at liberty. Brown to my sorrow confirms the account you give of your ill health. You cannot conceive how I ache to be with you: how I would die for one hour—for what is in the world? I say you cannot conceive; it is impossible you should look with such eyes upon me as I have upon you: it cannot be. Forgive me if I wander a little this evening, for I have been all day employ'd in a very abstract Poem° and I am in deep love with you—two things which must excuse me. I have, believe me, not been an age in letting you take possession of me; the very first week I knew you I wrote myself your vassal; but burnt the Letter as the very next time I saw you I thought you manifested some dislike to me. If you should ever feel for Man at the first sight what I did for you, I am lost. Yet I should not quarrel with you, but hate myself if such a thing were to happen—only I should burst if the thing were not as fine as a Man as you are as a Woman. Perhaps I am too vehement, then fancy me on my knees, especially when I mention a part of you Letter which hurt me; you say speaking of Mr. Severn 'but you must be satisfied in knowing that I admired you much more than your friend.' My dear love, I cannot believe there ever was or ever could be any thing to admire in me especially as far as sight goes—I cannot be admired, I am not a thing to be admired. You are, I love you; all I can bring you is a swooning admiration of your Beauty. I hold that place among Men which snub-nos'd brunettes with meeting eyebrows do among women—they are trash to me—unless I should find one among them with a fire in her heart like the one that burns in mine. You absorb me in spite of myself—you alone: for I look not forward with any pleasure to what is call'd being settled in the world; I tremble at domestic cares—yet for you I would meet them, though if it would leave you the happier I would rather die than do so. I have two luxuries to brood over in my walks, your Loveliness and the hour of my death. O that I could have possession of them both in the same minute. I hate the world: it batters too much the wings of my self-will, and would I could take a sweet poison from your lips to send me out of it. From no others would I take it. I am indeed astonish'd to find myself so careless of all charms but yours—remembring as I do the time when even a bit of ribband was a matter of interest with me. What softer words can I find for you after this—what it is I will not read. Nor will I say more here, but in a Postscript answer any thing else you may have mentioned in your Letter in so many words—for I am distracted with a thousand thoughts. I will imagine you Venus to night and pray, pray, pray to your star like a Hethen.

Your's ever, fair Star,
JOHN KEATS.

My seal is mark'd like a family table cloth with my mother's initial F for Fanny: put between my Father's initials. You will soon hear from me again. My respectful Compts to your Mother. Tell Margaret I'll send her a reef of best rocks and tell Sam° I will give him my light bay hunter if he will tie the Bishop hand and foot and pack him in a hamper and send him down for me to bathe him for his health with a Necklace of good snubby stones about his Neck.

Letter to Fanny Brawne, 5, 6 August 1819

My dear Girl, Shanklin Thursday Night—

You say you must not have any more such Letters as the last: I'll try that you shall not by running obstinate the other way—Indeed I have not fair play—I am not idle enough for proper downright love-letters—I leave this minute a scene in our Tragedy° and see you (think it not blasphemy) through the mist of Plots speeches, counterplots and counterspeeches—The Lover is madder than I am—I am nothing to him—he has a figure like the Statue of Maleager° and double distilled fire in his heart. Thank God for my diligence! were it not for that I should be miserable. I encourage it, and strive not to think of you—but when I have succeeded in doing so all day and as far as midnight, you return as soon as this artificial excitement goes off more severely from the fever I am left in—Upon my soul I cannot say what you could like me for. I do not think myself a fright any more than I do M^{r} A Mr B. and M^{r} C—yet if I were a woman I should not like A—B. C. But enough of this—So you intend to hold me to my promise of seeing you in a short time. I shall keep it with as much sorrow as gladness: for I am not one of the Paladins of old who livd upon water grass and smiles for years together—What though would I not give to night for the gratification of my eyes alone? This day week we shall move to Winchester; for I feel the want of a Library. Brown will leave me there to pay a visit to Mr Snook at Bedhampton: in his absence I will flit to you and back. I will stay very little while; for as I am in a train of writing now I fear to disturb it—let it have its course bad or good—in it I shall try my own strength and the public pulse. At Winchester I shall get your Letters more readily; and it being a cathedral City I shall have a pleasure always a great one to me when near a Cathedral, of reading them during the service up and down the Aisle—

Friday Morning Just as I had written thus far last night, Brown came down in his morning coat and nightcap, saying he had been refresh'd by a good sleep and was very hungry—I left him eating and went to bed being too tired to enter into any discussions. You would delight very greatly in the walks about here; the Cliffs, woods, hills, sands, rocks &c about here. They are however not so fine but I shall give them a hearty good bye to exchange them for my Cathedrall—Yet again I am not so tired of Scenery as to hate Switzerland—We might spend a pleasant Year at Berne or Zurich—if it should please Venus to hear my 'Beseech thee to hear us O Goddess' And if she should hear god forbid we should what people call, *settle*—turn into a pond, a stagnant Lethe—a vile crescent, row or buildings. Better be imprudent moveables than prudent fixtures—Open my Mouth at the Street door like the Lion's head at Venice to receive hateful cards Letters messages. Go out and wither at tea parties; freeze at dinners; bake at dances, simmer at routs. No my love, trust yourself to me and I will find you nobler amusements; fortune favouring. I fear you will not receive this till Sunday or Monday; as the irishman would write do not in the mean while hate me—I long to be off for Winchester for I begin to dislike the very door posts here—the names, the pebbles. You ask after my health, not telling me whether you are better. I am quite well. You going out is no proof that you are—how is it? Late hours will do you great harm—What fairing is it? I was alone for a couple of days while Brown went gadding over the country with his ancient Knapsack. Now I like his society as wells as any Man's, yet regretted his return—it broke in upon me like a Thunderbolt—I had got in a dream among my Books—really luxuriating in a solitude and silence you alone should have disturb'd—

Your ever affectionate
JOHN KEATS—

Letter to Benjamin Bailey, 14 August 1819

. . .

We removed to Winchester for the convenience of a Library and find it an exceeding pleasant Town, enriched with a beautiful Cathedrall and surrounded by a fresh-looking country. We are in tolerably good and cheap Lodgings. Within these two Months I have written 1500 Lines, most of which besides many more of prior composition you will probably see by next Winter. I have written two Tales, one from Boccacio call'd the Pot of Basil; and another call'd St Agnes' Eve on a popular superstition; and a third call'd Lamia—(half finished—I have also been writing parts of my Hyperion and completed 4 Acts of a Tragedy. It was the opinion of most of my friends that I should never be able to write a scene—I will endeavour to wipe away the prejudice—I sincerely hope you will be pleased when my Labours since we last saw each other shall reach you—One of my Ambitions is to make as great a revolution in modern dramatic writing as Kean has done in acting—another to upset the drawling of the blue stocking literary world—if in the course of a few years I do these two things I ought to die content—and my friends should drink a dozen of Claret on my Tomb—I am convinced more and more every day that (excepting the human friend Philosopher) a fine writer is the most genuine Being in the World—Shakspeare and the paradise Lost every day become greater wonders to me—I look upon fine Phrases like a Lover—I was glad to see, by a Passage in one of Brown's Letters some time ago from the north that you were in such good Spirits—Since that you have been married and in congralating you I wish you every continuance of them—Present my Respects to Mrs Bailey. This sounds oddly to me, and I dare say I do it awkwardly enough: but I suppose by this time it is nothing new to you—Brown's remembrances to you—As far as I know we shall remain at Winchester for a goodish while—

Ever your sincere friend
JOHN KEATS.

Letter to Fanny Brawne, 16 August 1819

Winchester August 17th°

My dear Girl—what shall I say for myself? I have been here four days and not yet written you—'t is true I have had many teasing letters of business to dismiss—and I have been in the Claws, like a Serpent in an Eagle's, of the last act of our Tragedy—This is no excuse; I know it; I do not presume to offer it—I have no right either to ask a speedy answer to let me know how lenient you are—I must remain some days in a Mist—I see you through a Mist: as I dare say you do me by this time—Believe in the first Letters I wrote you: I assure you I felt as I wrote—I could not write so now—The thousand images I have had pass through my brain—my uneasy spirits—my unguess'd fate—all spead as a veil between me and you—Remember I have had no idle leisure to brood over you—'t is well perhaps I have not—I could not have endured the throng of Jealousies that used to haunt me before I had plunged so deeply into imaginary interests. I would feign, as my sails are set, sail on without an interruption for a Brace of Months longer—I am in complete cue—in the fever; and shall in these four Months do an immense deal—This Page as my eye skims over it I see is excessively unloverlike and ungallant—I cannot help it—I am no officer in yawning quarters; no Parson-romeo—My Mind is heap'd to the full; stuff'd like a cricket ball—if I strive to fill it more it would burst—I know the generallity of women would hate me for this; that I should have so unsoften'd so hard a Mind as to forget them; forget the brightest realities for the dull imaginations of my own Brain—But I conjure you to give it a fair thinking; and ask yourself whether 't is not better to explain my feelings to you, than write artificial Passion—Besides you would see through it—It would be vain to strive to deceive you—'T is harsh, harsh, I know it—My heart seems now made of iron—I could not write a proper answer to an invitation to Idalia°—You are my Judge: my forehead is on the ground—You seem offended at a little simple innocent childish playfulness in my last—I did not seriously mean to say that you were endeavouring to make me keep my promise—I beg your pardon for it—'T is but *just* you Pride should take the alarm—*seriously*—You say I may do as I please—I do not think with any conscience I can; my cash-recourses are for the present stopp'd; I fear for some time—I spend no money but it increases my debts—I have all my life thought very little of these matters—they seem not to belong to me—It may be a proud sentence; but, by heaven, I am as entirely above all matters of interest as the Sun is above the Earth—And though of my own money I should be careless; of my Friends I must be spare. You see how I go on—like so many strokes of a Hammer—I cannot help it—I am impell'd, driven to it. I am not happy enough for silken Phrases, and silver sentences—I can no more use soothing words to you than if I were at this moment engaged in a charge of Cavalry—Then you will say I should not write at all—Should I not? This Winchester is a fine place; a beautiful Cathedral and many other ancient building in the Environs. The little coffin of a room at Shanklin, is changed for a large room—where I can promenade at my pleasure—looks out onto a beautiful—blank side of a house—It is strange I should like it better than the view of the sea from our window at Shanklin—I began to hate the very posts there—the voice of the old Lady over the way was getting a great Plague—The Fisherman's face never altered any more than our black tea-pot—the nob however was knock'd off to my little relief. I am getting a great dislike of the picturesque; and can only relish it over again by seeing you enjoy it—One of the pleasantest things I have seen lately was at Cowes—The Regent in his Yatch (I think they spell it)° was anchored oppoisite—a beautiful vessel—and all the Yatchs and boats on the coast, were passing and repassing it; and curcuiting and tacking about it in every direction—I never beheld any thing so, silent, light, and graceful. As we pass'd over to Southampton, there was nearly an accident—There came by a Boat well mann'd; with to naval officers at the stern—Our Bow-lines took the top of their little mast and snapped it off close by the bord—Had the mast been a little stouter they would have been upset—In so trifling an event I could not help admiring our seamen—Neither Officer nor man in the whole Boat moved a Muscle—they scarcely notic'd it even with words—Forgive me for this flint-worded Letter—and believe and see that I cannot think of you without some sort of energy—though mal a propos—Even as I leave off—it seems to me that a few more moments thought of you would uncrystallize and dissolve me°—I must not give way to it—but turn to my writing again—If I fail I shall die hard—O my love, your lips are growing sweet again to my fancy—I must forget them—Ever your affectionate

KEATS—

Letter to John Taylor, 23 August 1819

My dear Taylor— Winchester Monday morn.

24 Aug[st]

You will perceive that I do not write you till I am forced by necessity: that I am sorry for. You must forgive me for entering abruptly on the subject, merely pefixing an intreaty that you will not consider my business manner of wording and proceeding any distrust of, or stirrup standing against you; but put it to the account of a desire of order and regularity—I have been rather unfortunate lately in money concerns—from a threatened chancery suit°—I was deprived at once of all recourse to my Guardian I relied a little on some of my debts being paid—which are of a tolerable amount—but I have not had one pound refunded—For these three Months Brown has advanced me money: he is not at all flush and I am anxious to get some elsewhere—We have together been engaged (this I should wish to remain secret) in a Tragedy which I have just finish'd; and from which we hope to share moderate Profits—Being thus far connected, Brown proposed to me, to stand with me responsible for any money you may advance to me to drive through the summer—I must observe again that it is not from want of reliance on you readiness to assist me that I offer a Bill; but as a relief to myself from a too lax sensation of Life—which ought to be responsible which requires chains for its own sake—duties to fulfil with the more earnestness the less strictly they are imposed Were I completely without hope—it might be different—but am I not right to rejoice in the idea of not being Burthensome to my friends? I feel every confidence that if I choose I may be a popular writer; that I will never be; but for all that I will get a livelihood—I equally dislike the favour of the public with the love of a woman—they are both a cloying treacle to the wings of independence. I shall ever consider them (People) as debtors to me for verses, not myself to them for admiration—which I can do without. I have of late been indulging my spleen by composing a preface *at* them: after all resolving never to write a preface at all. 'There are so many verses,' would I have said to them, 'give me so much means to buy pleasure with as a relief to my hours of labour'—You will observe at the end of this if you put down the Letter 'How a solitarry life engenders pride and egotism!' True: I know it does but this Pride and egotism will enable me to write finer things than any thing else could—so I will indulge it—Just so much as I am humbled by the genius above my grasp, am I exalted and look with hate and contempt upon the literary world—A Drummer boy who holds out his hand familiarly to a field marshall—that Drummer boy with me is the good word and favour of the public—Who would wish to be among the commonplace crowd of the little-famous—who are each individually lost in a throng made up of themselfes? is this worth louting° or playing the hypocrite for? To beg suffrages for a seat on the benches of a myriad aristocracy in Letters? This is not wise—I am not a wise man—T is Pride—I will give you a definition of a proud Man—He is a Man who has neither vanity nor wisdom—one fill'd with hatreds cannot be vain—

neither can he be wise—Pardon me for hammering instead of writing —Remember me to Woodhouse, Hessey and all in Percey street—

Ever yours sincerely
JOHN KEATS

P.S. I have read what Brown has said on the other side°—He agrees with me that this manner of proceeding might appear to harsh, distant and indelicate with you. This however will place all in a clear light. Had I to borrow money of Brown and were in your house, I should request the use of your name in the same manner—

Letter to J. H. Reynolds, 24 August 1819

My dear Reynolds, Winchest^{r} August 25th—°

By this Post I write to Rice who will tell you why we have left Shanklin; and how we like this Place—I have indeed scarely any thing else to say, leading so monotonous a life except I was to give you a history of sensations, and day-nightmares—You would not find me at all unhappy in it; as all my thoughts and feelings which are of the selfish nature, home speculations every day continue to make me more Iron—I am convinced more and more day by day that fine writing is next to fine doing the top thing in the world; the Paradise Lost becomes a greater wonder—The more I know what my diligence may in time probably effect; the more does my heart distend with Pride and Obstinacy—I feel it in my power to become a popular writer—I feel it in my strength to refuse the poisonous suffrage of a public—My own being which I know to be becomes of more consequence to me than the crowds of Shadows in the Shape of Man and women that inhabit a Kingdom. The Soul is a world of itself and has enough to do in its own home—Those whom I know already and who have grown as it were a part of myself I could not do without: but for the rest of Mankind they are as much a dream to me as Miltons Hierarchies. I think if I had a free and healthy and lasting organisation of heart and Lungs—as strong as an ox's—so as to be able° unhurt the shock of extreme thought and sensation without weariness, I could pass my Life very nearly alone though it should last eighty years. But I feel my Body too weak to support me to the height; I am obliged continually to check myself and strive to be nothing. It would be vain for me to endeavour after a more reasonable manner of writing to you: I have nothing to speak of but myself—and what can I say but what I feel? If you should have any reason to regret this state of excitement in me, I will turn the tide of your feelings in the right channel by mentioning that it is the only state for the best sort of Poetry—that is all I care for, all I live for. Forgive me for not filling up the whole sheet; Letters become so irksome to me that the next time I leave London I shall petition them all to be spar'd me. To give me credit for constancy and at the same time wave letter writing will be the highest indulgence I can think of.

Ever your affectionate friend
JOHN KEATS

Letter to J. H. Reynolds, 21 September 1819

Winchester. Tuesday

My dear Reynolds,

I was very glad to hear from Woodhouse that you would meet in the Country. I hope you will pass some pleasant time together. Which I wish to make pleasanter by a brace of letters, very highly to be estimated, as really I have had very bad luck with this sort of game this season. I 'kepen in solitarinesse,"° for Brown has gone a visiting. I am surprized myself at the pleasure I live alone in. I can give you no news of the place here, or any other idea of it but what I have to this effect written to George. Yesterday I say to him was a grand day for Winchester. They elected a Mayor—It was indeed high time the place should receive some sort of excitement. There was nothing going on: all asleep: not an old maid's sedan returning from a card party: and if any old woman got tipsy at Christenings they did not expose it in the streets. The first night tho' of our arrival here, there was a slight uproar took place at about 10 o' the Clock. We heard distinctly a noise patting down the high Street as of a walking cane of the good old Dowager breed; and a little minute after we heard a less voice observe 'What a noise the ferril° made—it must be loose'—Brown wanted to call the Constables, but I observed 'twas only a little breeze, and would soon pass over.—The side streets here are excessively maiden-lady like: the door steps always fresh from the flannel. The knockers have a staid serious, nay almost awful quietness about them.—I never saw so quiet a collection of Lions' & Rams' heads—The doors most part black, with a little brass handle just above the keyhole, so that in Winchester a man may very quietly shut himself out of his own house. How beautiful the season is now—How fine the air. A temperate sharpness about it. Really, without joking, chaste weather—Dian skies°—I never lik'd stubble fields so much as now—Aye better than the chilly green of the spring. Somehow a stubble plain looks warm—in the same way that some pictures look warm—this struck me so much in my sunday's walk that I composed upon it.° I hope you are better employed than in gaping after weather. I have been at different times so happy as not to know what weather it was—No I will not copy a parcel of verses. I always somehow associate Chatterton with autumn. He is the purest writer in the English Language. He has no French idiom, or particles like Chaucer—'tis genuine English Idiom in English words. I have given up Hyperion—there were too many Miltonic inversions in it—Miltonic verse cannot be written but in an artful or rather artist's humour. I wish to give myself up to other sensations. English ought to be kept up. It may be interesting to you to pick out some lines from Hyperion and put a mark × to the false beauty proceeding from art, and one || to the true voice of feeling. Upon my soul 'twas imagination I cannot make the distinction—Every now & then there is a Miltonic intonation—But I cannot make the division properly. The fact is I must take a walk; for I am writing so long a letter to George; and have been employed at it all the morning. You will ask, have I heard from George. I am sorry to say not the best news—I hope for better—This is the reason among others that if I write to you it must be in such a scraplike way. I have no meridian to date Interests from, or measure circumstances—To night I am all in a mist; I scarcely know what's what—But you knowing my unsteady & vagarish disposition, will guess that all this turmoil will be settled by tomorrow morning. It strikes me to night that I have led a very odd sort of life for the two or three last years—Here & there—No anchor—I am glad of it.—If you can get a peep at Babbicomb before you leave the country,° do.—I think it is the finest place I have seen, or—is to be seen in the South. There is a Cottage there I took warm water at, that made up for the tea. I have lately skirk'd some friends of ours, and I advise you to do the same, I mean the blue-devils—I am never at home to them. You need not fear them while you remain in Devonshire. there will be some of the family waiting for you at the Coach office—but go by another Coach.—I shall beg leave to have a third opinion in the first discussion you have with Woodhouse—just half way—between both. You know I will not give up my argument—In my walk to day I stoop'd under a rail way that lay across my path, and ask'd myself 'Why I did not get over' Because, answered I, 'no one wanted to force you under'—I would give a guinea to be a reasonable man—good sound sense—a says what he thinks, and does what he says man—and did not take snuff—They say men near death however mad they may have been, come to their senses—I hope I shall here in this letter—there is a decent space to be very sensible in—many a good proverb has been in less—Nay I have heard of the statutes at large being chang'd into the Statutes at Small and printed for a watch paper. Your sisters by this time must have got the Devonshire ees—short ees—you know 'em—they are the prettiest ees in the Language. O how I admire the middle siz'd delicate Devonshire girls of about 15. There was one at an Inn door holding a quartern of brandy—the very thought of her kept me warm a whole stage°—and a 16 miler too—'You'll pardon me for being jocular.'°

Ever your affectionate friend
JOHN KEATS—

Letter to Richard Woodhouse, 21, 22 September 1819

Dear Woodhouse, Tuesday—

If you see what I have said to Reynolds before you come to your own dose you will put it between the bars unread; provided they have begun fires in Bath—I should like a bit of fire to night—one likes a bit of fire—How glorious the Blacksmiths' shops look now—I stood to night before one till I was verry near listing for one. Yes I should like a bit of fire—at a distance about 4 feet 'not quite hob nob'—as wordsworth says°—The fact was I left Town on Wednesday—determined to be in a hurry—You don't eat travelling—you're wrong—beef—beef—I like the look of a sign—The Coachman's face says eat eat, eat—I never feel more contemptible than when I am sitting by a good looking coachman—One is nothing—Perhaps I eat to persuade myself I am somebody. You must be when slice after slice—but it wont do—the Coachman nibbles a bit of bread—he's favour'd—he's had a Call—a Hercules Methodist—Does he live by bread alone? O that I were a Stage Manager—perhaps that's as old as 'doubling the Cape'—'How are ye old 'un? hey! why dont'e speak?' O that I had so sweet a Breast to sing as the Coachman hath! I'd give a penny for his Whistle—and bow to the Girls on the road—Bow—nonsense—'t is a nameless graceful slang action—Its effect on the women suited to it must be delightful. It touches 'em in the ribs—en passant—very off hand—very fine—Sed thongum formosa vale vale inquit Heigh ho la!° You like Poetry better—so you shall have some I was going to give Reynolds—

[*'Ode to Autumn' follows*]

I will give you a few lines from Hyperion on account of a word in the last line of a fine sound—

[*'The Fall of Hyperion' ii. 1–4, 6 follow*]

I think you will like the following description of the Temple of Saturn—

[*'The Fall of Hyperion' i. 61–86 follow*]

I see I have completely lost my direction—So I e'n make you pay double postage°. I had begun a sonnet in french of Ronsard—on my word 't is verry capable of poetry—I was stop'd by a circumstance not worth mentioning—I intended to call it La Platonique Chevalresque—I like the second line—

Non ne suis si audace a languire
De m'empresser au cœur vos tendres mains. &c

Here is what I had written for a sort of induction—

[*'The Fall of Hyperion' i. 1–11 follow*]

My Poetry will never be fit for any thing it does n't cover its ground well—You see he she is off her guard and does n't move a peg though Prose is coming up in an awkward style enough—Now a blow in the spondee will finish her—But let it get over this line of circumvallation° if it can. These are unpleasant Phrases. Now for all this you two must write me a letter apiece—for as I know you will interread one another—I am still writing to Reynolds as well as yourself—As I say to George I am writing to you but *at* your Wife—And dont forget to tell Reynold's of the fairy tale Undine°—Ask him if he has read any of the American Brown's novels that Hazlitt speaks so much of°—I have read one call'd Wieland—very powerful—something like Godwin—Between Schiller and Godwin—A Domestic prototype of Shiller's Armenian°—More clever in plot and incident than Godwin—A strange american scion of the German trunk. Powerful genius—accomplish'd horrors—I shall proceed tomorrow—Wednesday—I am all in a Mess here—embowell'd in Winchester. I wrote two Letters to Brown one from said Place, and one from London, and neither of them has reach'd him—I have written him a long one this morning and am so perplex'd as to be an object of Curiosity to you quiet People. I hire myself a show waggan and trumpetour. Here's the wonderful Man whose Letters wont go!—All the infernal imaginarry thunderstorms from the Post-office are beating upon me—so that 'unpoeted I write'° Some curious body has detained my Letters—I am sure of it. They know not what to make of me—not an acquaintance in the Place—what can I be about? so they open my Letters—Being in a lodging house, and not so self will'd, but I am a little cowardly I dare not spout my rage against the Ceiling—Besides I should be run through the Body by the major in the next room—I don't think his wife would attempt such a thing—Now I am going to be serious—After revolving certain circumstances in my Mind; chiefly connected with a late american letter—I have determined to take up my abode in a cheap Lodging in Town and get employment in some of our elegant Periodical Works—I will no longer live upon hopes—I shall carry my plan into execution speedily—I shall live in Westminster—from which a walk to the British Museum will be noisy and muddy—but otherwise pleasant enough—I shall enquire of Hazlitt how the figures of the market stand. O that I could somthing agrestrural,° pleasant, fountain-voic'd—not plague you with unconnected nonsense—But things won't leave me *alone*. I shall be in Town as soon as either of you—I only wait for an answer from Brown: if he receives mine which is now a very moot point—I will give you a few reasons why I shall persist in not publishing The Pot of Basil—It is too smokeable°—I can get it smoak'd at the Carpenters shaving chimney much more cheaply—There is too much inexperience of live, and simplicity of knowledge in it—which might do very well after one's death—but not while one is alive. There are very few would look to the reality. I intend to use more finesse with the Public. It is possible to write fine things which cannot be laugh'd at in any way. Isabella is what I should call were I a reviewer 'A weak-sided Poem' with an amusing sober-sadness about it. Not that I do not think Reynolds and you are quite right about it—it is enough for me. But this will not do to be public—If I may so say, in my dramatic capacity I enter fully into the feeling: but in Propria Persona I should be apt to quiz it myself—There is no objection of this kind to Lamia—A good deal to S[t] Agnes Eve—only not so glaring—Would a I say I could write you something sylvestran. But I have no time to think: I am an otiosus-peroccupatus° Man—I thnk upon crutches, like the folks in your Pump room—Have you seen old Bramble° yet—they say he's on his last legs—The gout did not treat the old Man well so the Physician superseded it, and put the dropsy in office, who gets very fat upon his new employment, and behaves worse than the other to the old Man—But he'll have his house about his ears soon—We shall have another fall of Siege-arms—I suppose M[rs] Humphrey persists in a big-belley—poor thing she little thinks how she is spoiling the corners of her mouth—and making her nose quite a piminy.° M[r] Humphrey I hear was giving a Lecture in the gaming-room—When some one call'd out Spousey! I hear too he has received a challenge from a gentleman who lost that evening—The fact is M[r] H. is a mere nothing out of his Bed-room.—Old Tabitha died in being bolstered up for a whist-party. They had to cut again—Chowder died long ago—M[rs] H. laments that the last time they *put him* (i.e. to breed) he didn't take—They say he was a direct descendent of Cupid and Veney in the Spectator°—This may be eisily known by the Parish Books—If you do not write in the course of a day or two: direct to me at Rice's—Let me know how you pass your times and how you are—

Your sincere friend
JOHN KEATS—

Hav'nt heard from Taylor—

Letter to Charles Brown, 22 September 1819

. . .

Now I am going to enter on the subject of self. It is quite time I should set myself doing something, and live no longer upon hopes. I have never yet exerted myself. I am getting into an idle minded, vicious way of life, almost content to live upon others. In no period of my life have I acted with any self will, but in throwing up the apothecary-profession. That I do not repent of. Look at xxxxxx: if he was not in the law he would be acquiring, by his abilities, something towards his support. My occupation is entirely literary; I will do so too. I will write, on the liberal side of the question, for whoever will pay me. I have not known yet what it is to be diligent. I purpose living in town in a cheap lodging, and endeavouring, for a beginning, to get the theatricals of some paper. When I can afford to compose deliberate poems I will. I shall be in expectation of an answer to this. Look on my side of the question. I am convinced I am right. Suppose the Tragedy should succeed,—there will be no harm done. And here I will take an opportunity of making a remark or two on our friendship, and all your good offices to me. I have a natural timidity of mind in these matters: liking better to take the feeling between us for granted, than to speak of it. But, good God! what a short while you have known me! I feel it a sort of duty thus to recapitulate, however unpleasant it may be to you. You have been living for others more than any man I know. This is a vexation to me; because it has been depriving you, in the very prime of your life, of pleasures which it was your duty to procure. As I am speaking in general terms this may appear nonsense; you perhaps will not understand it: but if you can go over, day by day, any month of the last year,—you will know what I mean. On the whole, however, this is a subject that I cannot express myself upon. I speculate upon it frequently; and, believe me, the end of my speculations is always an anxiety for your happiness. This anxiety will not be one of the least incitements to the plan I purpose pursuing. I had got into a habit of mind of looking towards you as a help in all difficulties. This very habit would be the parent of idleness and difficulties. You will see it is a duty I owe myself to break the neck of it. I do nothing for my subsistence—make no exertion. At the end of another year, you shall applaud me,—not for verses, but for conduct. If you live at Hampstead next winter—I like xxxxxxxxx and I cannot help it. On that account I had better not live there. While I have some immediate cash, I had better settle myself quietly, and fag on as others do. I shall apply to Hazlitt, who knows the market as well as any one, for something to bring me in a few pounds as soon as possible. I shall not suffer my pride to hinder me. The whisper may go round; I shall not hear it. If I can get an article in the 'Edinburg', I will. One must not be delicate. Nor let this disturb you longer than a moment. I look forward, with a good hope, that we shall one day be passing free, untrammelled, unanxious time together. That can never be if I continue a dead lump. xxxxxxxxxxxxxx I shall be expecting anxiously an answer from you. If it does not arrive in a few days, this will have miscarried, and I shall come straight to xxxx before I go to town, which you, I am sure, will agree had better be done while I still have some ready cash. By the middle of October I shall expect you in London. We will then set at the Theatres. If you have any thing to gainsay, I shall be even as the deaf adder which stoppeth her ears.

. . .

Letter to C. W. Dilke, 22 September 1819

My dear Dilke, Winchester Wednesday Eve—

Whatever I take too for the time I cannot leave off in a hury; letter writing is the go now; I have consumed a Quire at least. You must give me credit, now, for a free Letter when it is in reality an interested one, on two points, the one requestive, the other verging to the pros and cons—As I expect they will lead me to seeing and conferring with you in a short time, I shall not enter at all upon a letter I have lately received from george of not the most comfortable intelligence: but proceed to these two points, which if you can theme out in sexions and subsexions for my edification, you will oblige me. The first I shall begin upon, the other will follow like a tail to a Comet. I have written to Brown on the subject, and can but go over the same Ground with you in a very short time, it not being more in length than the ordinary paces between the Wickets. It concerns a resolution I have taken to endeavour to acquire something by temporary writing in periodical works. You must agree with me how unwise it is to keep feeding upon hopes, which depending so much on the state of temper and imagination, appear gloomy or bright, near or afar off just as it happens—Now an act has three parts—to act, to do, and to perform—I mean I should *do* something for my immediate welfare—Even if I am swept away like a Spider from a drawing room I am determined to spin—home spun any thing for sale. Yea I will trafic. Any thing but Mortgage my Brain to Blackwood. I am determined not to lie like a dead lump. If Reynolds had not taken to the law, would he not be earning something? Why cannot I—You may say I want tact—that is easily acquired. You may be up to the slang of a cock pit in three battles. It is fortunate I have not before this been tempted to venture on the common.° I should a year or two ago have spoken my mind on every subject with the utmost simplicity. I hope I have learnt a little better and am confident I shall be able to cheat as well as any literary Jew of the Market and shine up an article on any thing without much knowledge of the subject, aye like an orange. I would willingly have recourse to other means. I cannot; I am fit for nothing but literature. Wait for the issue of this Tragedy? No—there cannot be greater uncertainties east west, north, and south than concerning dramatic composition. How many months must I wait! Had I not better begin to look about me now? If better events supersede this necessity what harm will be done? I have no trust whatever on Poetry—I dont wonder at it—the marvel is to me how people read so much of it. I think you will see the reasonableness of my plan. To forward it I purpose living in cheap Lodging in Town, that I may be in the reach of books and information, of which there is here a plentiful lack. If I can any place tolerably comfitable I will settle myself and fag till I can afford to buy Pleasure—which if I never can afford I must go Without—Talking of Pleasure, this moment I was writing with one hand, and with the other holding to my Mouth a Nectarine—good god how fine—It went down soft pulpy, slushy, oozy—all its delicious embonpoint melted down my throat like a large beatified Strawberry. I shall certainly breed.° Now I come to my request. Should you like me for a neighbour again? Come, plump it out, I wont blush. I should also be in the neighbourhood of Mrs Wylie, which I shoud be glad of, though that of course does not influence me. Therefore will you look about Marsham, or rodney street for a couple of rooms for me. Rooms like the gallants legs in massingers time 'as good as the times allow, Sir.'° I have written to day to Reynolds, and to Woodhouse. Do you know him? He is a Friend of Taylors at whom Brown has taken one of his funny odd dislikes. I'm sure he's wrong, because Woodhouse likes my Poetry—conclusive. I ask your opinion and yet I must say to you as to him, Brown that if you have any thing to say against it I shall be as obstinate & heady as a Radical. By the Examiner coming in your hand writing you must be in Town. They have put me into spirits: Notwithstand my aristocratic temper I cannot help being verry much pleas'd with the present public proceedings.° I hope sincerely I shall be able to put a Mite of help to the Liberal side of the Question before I die. If you should have left Town again (for your Holidays cannot be up yet) let me know—when this is forwarded to you—A most extraordinary mischance has befallen two Letters I wrote Brown—one from London whither I was obliged to go on business for George; the other from this place since my return. I cant make it out. I am excessively sorry for it. I shall hear from Brown and from you almost together for I have sent him a Letter to day: you must positively agree with me or by the delicate toe nails of the virgin I will not open your Letters. If they are as David says 'suspicious looking letters'° I wont open them—If St John had been half as cunning he might have seen the revelations comfortably in his own room, without giving Angels the trouble of breaking open Seals. Remember me to Mrs D.—and the Westmonisteranian° and believe me

Ever your sincere friend
JOHN KEATS—

Letter to George and Georgiana Keats, 17, 18, 20, 21, 24, 25, 27 September 1819

My dear George, Winchester Sep^{tr} Friday—

I was closely employed in reading and composition, in this place, whither I had come from Shanklin, for the convenience of a library, when I received your last, dated July 24th You will have seen by the short Letter I wrote from Shanklin, how matters stand beetween us and Mrs Jennings.° They had not at all mov'd and I knew no way of overcoming the inveterate obstinacy of our affairs. On receiving your last I immediately took a place in the same night's coach for London—Mr Abbey behaved extremely well to me, appointed Monday evening at 7 to meet me and observed that he should drink tea at that hour. I gave him the inclosed note and showed him the last leaf of yours to me. He really appeared anxious about it; promised he would forward your money as quickly as possible—I think I mention'd that Walton was dead—He will apply to Mr Gliddon the partner;° endeavour to get rid of Mrs Jennings's claim and be expeditious. He has received an answer from my Letter to Fry°—that is something. We are certainly in a very low estate: I say we, for I am in such a situation that were it not for the assistance of Brown & Taylor, I must be as badly off as a Man can be. I could not raise any sum by the promise of any Poem—no, not by the mortgage of my intellect. We must wait a little while. I really have hopes of success. I have finish'd a Tragedy which if it succeeds will enable me to sell what I may have in manuscript to a good avantage. I have pass'd my time in reading, writing and fretting—the last I intend to give up and stick to the other two. They are the only chances of benefit to us. Your wants will be a fresh spur to me. I assure you you shall more than share what I can get, whilst I am still young—the time may come when age will make me more selfish. I have not been well treated by the world—and yet I have capitally well—I do not know a Person to whom so many purse strings would fly open as to me—if I could possibly take advantage of them—which I cannot do for none of the owners of these purses are rich—Your present situation I will not suffer myself to dwell upon—when misfortunes are so real we are glad enough to escape them, and the thought of them. I cannot help thinking Mr Audubon° a dishonest man—Why did he make you believe that he was a Man of Property? How is it his circumstances have altered so suddenly? In truth I do not believe you fit to deal with the world; or at least the american worrld—But good God—who can avoid these chances—You have done your best—Take matters as coolly as you can and confidently expecting help from England, act as if no help was nigh. Mine I am sure is a tolerable tragedy—it would have been a bank to me, if just as I had finish'd it I had not heard of Kean's resolution to go to America. That was the worst news I could have had. There is no actor can do the principal character besides Kean. At Covent Garden there is a great chance of its being damn'd. Were it to succeed even there it would lift me out of the mire. I mean the mire of a bad reputation which is continually rising against me. My name with the literary fashionables is vulgar—I am a weaver boy° to them—a Tragedy would lift me out of this mess. And mess it is as far as it regards our Pockets—But be not cast down any more than I am. I feel I can bear real ills better than imaginary ones. Whenever I find myself growing vapourish, I rouse myself, wash and put on a clean shirt brush my hair and clothes, tie my shoestrings neatly and in fact adonize as I were going out—then all clean and comfortable I sit down to write. This I find the greatest relief—Besides I am becoming accustom'd to the privations of the pleasures of sense. In the midst of the world I live like a Hermit. I have forgot how to lay plans for enjoyment of any Pleasure. I feel I can bear any thing, any misery, even impisonment—so long as I have neither wife nor child. Perpaps you will say yours are your only comfort—they must be. I return'd to Winchester the day before yesterday and am now here alone, for Brown some days before I left, went to Bedhampton and there he will be for the next fortnight. The term of his house will be up in the middle of next month when we shall return to Hampstead. On Sunday I dined with your Mother and Henry and Charles in Henrietta Street—M[rs] and Miss Millar were in the Country—Charles had been but a few days returned from Paris. I dare say you will have letters expessing the motives of his journey. M[rs] Wylie and Miss Waldegrave seem as quiet as two Mice there alone. I did not show your last—I thought it better not. For better times will certainly come and why should they be unhappy in the main time. On Monday Morning I went to Walthamstow—Fanny look'd better than I had seen her for some time. She complains of not hearing from you appealing to me as if it was half my fault—I had been so long in retirement that London appeared a very odd place I could not make out I had so many acquaintance, and it was a whole day before I could feel among Men—I had another strange sensation there was not one house I felt any pleasure to call at. Reynolds was in the Country and saving himself I am prejudiced against all that family. Dilke and his wife and child were in the Country. Taylor was at Nottingham—I was out and every body was out. I walk'd about the Streets as in a strange land—Rice was the only one at home—I pass'd some time with him. I know him better since we have liv'd a month together in the isle of Wight. He is the most sensible, and even wise Man I know—he has a few John Bull prejudices; but they improve him. His illness is at times alarming. We are great friends, and there is no one I like to pass a day with better. Martin call'd in to bid him good bye before he set out for Dublin. If you would like to hear one of his jokes here is one which at the time we laugh'd at a good deal. A Miss—with three young Ladies, one of them Martin's sister had come a gadding in the Isle of wight and took for a few days a Cottage opposite ours—we dined with them one day, and as I was saying they had fish—Miss—said she thought *they tasted of the boat*—No says Martin very seriously they haven't been kept long enough. I saw Haslam he is very much occupied with love and business being one of Mr Saunders executors and Lover to a young woman He show'd me her Picture by Severn—I think she is, though not very cunning, too cunning for him. Nothing strikes me so forcibly with a sense of the rediculous as love—A Man in love I do think cuts the sorryest figure in the world—Even when I know a poor fool to be really in pain about it, I could burst out laughing in his face—His pathetic visage becomes irrisistable. Not that I take Haslam as a pattern for Lovers—he is a very worthy man and a good friend. His love is very amusing. Somewhere in the Spectator° is related an account of a Man inviting a party of stutterers and squinters to his table. 't would please me more to scrape together a party of Lovers, not to dinner—no to tea. There would be no fighting as among Knights of old—

[*'Pensive they sit, and roll their languid eyes' follows*]

You see I cannot get on without writing as boys do at school a few nonsense verses—I begin them and before I have written six the whim has pass'd—if there is any thing deserving so respectable a name in them. I shall put in a bit of information any where just as it strikes me. Mr Abbey is to write to me as soon as he can bring matters to bear, and then I am to go to Town to tell him the means of forwarding to you through Capper and Hazlewood—I wonder I did not put this before—I shall go on tomorrow—it is so fine now I must take a bit of a walk—

Saturday—

With my inconstant disposition it is no wonder that this morning, amid all our bad times and misfortunes, I should feel so alert and well spirited. At this moment you are perhaps in a very different state of

Mind. It is because my hopes are very paramount to my despair. I have been reading over a part of a short poem I have composed lately call'd 'Lamia'—and I am certain there is that sort of fire in it which must take hold of people in some way—give them either pleasant or unpleasant sensation. What they want is a sensation of some sort. I wish I could pitch the key of your spirits as high as mine is—but your organ loft is beyond the reach of my voice—I admire the exact admeasurement of my niece in your Mother's letter—O the little span long elf°—I am not in the least judge of the proper weight and size of an infant. Never trouble yourselves about that: she is sure to be a fine woman—Let her have only delicate nails both on hands and feet and teeth as small as a May-fly's. who will live you his life on a square inch of oak-leaf. And nails she must have quite different from the market women here who plough into the butter and make a quatter pound taste of it. I intend to wite a letter to you Wifie and there I may say more on this little plump subject—I hope she's plump—'Still harping on my daughter"°—This Winchester is a place tolerably well suited to me; there is a fine Cathedral, a College, a Roman-Catholic Chapel, a Methodist do, an independent do,—and there is not one loom or any thing like manufacturing beyond bread & butter in the whole City. There are a number of rich Catholics in the place. It is a respectable, ancient aristocratical place—and moreover it contains a nunnery—Our set are by no means so hail fellow, well met, on literary subjects as we were wont to be. Reynolds has turn'd to the law. Bye the bye, he brought out a little piece at the Lyceum call'd *one, two thee, four, by advertisement*. It met with complete success. The meaning of this odd title is explained when I tell you the principal actor is a mimic who takes off four of our best performers in the course of the farce—Our stage is loaded with mimics. I did not see the Piece being out of Town the whole time it was in progress. Dilke is entirely swallowed up in his boy: 't is really lamentable to what a pitch he carries a sort of parental mania—I had a Letter from him at Shanklin—He went on a word or two about the isle of Wight which is a bit of hobby horse of his; but he soon deviated to his boy. 'I am sitting' says he 'at the window expecting my Boy from School.' I suppose I told you some where that he lives in Westminster, and his boy goes to the School there, where he gets beaten, and every bruise he has and I dare say deserves is very bitter to Dilke. The Place I am speaking of, puts me in mind of a circumstace occured lately at Dilkes—I think it very rich and dramatic and quite illustrative of the little quiet fun that he will enjoy sometimes. First I must tell you their house is at the corner of Great Smith Street, so that some of the windows look into one Street, and the back windows into another round the corner—Dilke had some old people to dinner, I know not who—but there were two old ladies among them—Brown was there—they had known him from a Child. Brown is very pleasant with old women, and on that day, it seems, behaved himself so winningly they they became hand and glove together and a little complimentary. Brown was obliged to depart early. He bid them good bye and pass'd into the passage—no sooner was his back turn'd than the old women began lauding him. When Brown had reach'd the Street door and was just going, Dilke threw up the Window and call'd 'Brown! Brown! They say you look younger than ever you did!' Brown went on and had just turn'd the corner into the other street when Dilke appeared at the back window crying 'Brown! Brown! By God, they say you're handsome!' You see what a many words it requires to give any identity to a thing I could have told you in half a minute. I have been reading lately Burton's Anatomy of Melancholy; and I think you will be very much amused with a page I here coppy for you.° I call it a Feu de joie round the batteries of Fort S[t] Hyphen-de-Phrase on the birthday of the Digamma. The whole alphabet was drawn up in a Phalanx on the cover of an old Dictionary. Band playing 'Amo, Amas &c' 'Every Lover admires his Mistress, though she be very deformed of herself, ill-favored, wrinkled, pimpled, pale, red, yellow, tann'd, tallow-fac'd, have a swoln juglers platter face, or a thin, lean, chitty face, have clouds in her face, be crooked, dry, bald, goggle-eyed, blear-eyed or with staring eyes, she looks like a squis'd cat, hold her head still awry, heavy, dull, hollow-eyed, black or yellow about the eyes, or squint-eyed, sparrow-mouth'd, Persean-hook-nosed, have a sharp fox nose, a red nose, China flat, great nose, nare simo patuloque,° a nose like a promontory, gubber-tush'd, rotten teeth, black, uneven, brown teeth, beetle-brow'd, a witches beard, her breath stink all over the room, her nose drip winter and summer, with a Bavarian poke under her chin, a sharp chin, lave-eared, with a long crane's neck, which stands awry too, pendulis mammis° her dugs like two double jugs, or else no dugs in the other extream, bloody-falln° fingers, she have filthy, long, unpaired, nails, scabbed hands or wrists, a tan'd skin, a rotten carcass, crooked back, she stoops, is lame, splea footed, as slender in the middle as a cow in the wast, gowty legs, her ankles hang over her shooes, her feet stink, she breed lice, a meer changeling, a very monster, an aufe° imperfect, her whole complexion savors, an harsh voice, incondite gesture, vile gate, a vast virago, or an ugly tit, a slug, a fat fustilugs, a trusse, a long lean rawbone, a Skeleton, a Sneaker, (si qua patent meliora puta)° and to thy Judgement looks like a mard in a Lanthorn, whom thou couldst not fancy for a world, but hatest, loathest, and wouldst have spit in her face, or blow thy nose in her bosom, remedium amoris° to another man, a dowdy, a Slut, a scold, a nasty rank, rammy, filthy, beastly quean, dishonest peradventure, obscene, base, beggarly, rude, foolish, untaught—peevish, Irus' daughter, Thersite's sister,° Grobian's Scholler;° if he love her once, he admires her for all this, he takes no notice of any such errors or imperfections of boddy or mind—' There's a dose for you—fine!! I would give my favourite leg to have written this as a speech in a Play: with what effect could Mathews° pop-gun it at the pit! This I thnk will amuse you more than so much Poetry. Of that I do not like to copy any as I am affraid it is too mal apropo for you at present—and yet I will send you some—for by the time you receive it things in England may have taken a different turn. When I left Mr Abbey on monday evening I walk'd up Cheapside but returned to put some letters in the Post and met him again in Bucklersbury: we walk'd together through the Poultry as far as the hatter's shop he has some concern in—He spoke of it in such a way to me, I thought he wanted me to make an offer to assist him in it. I do believe if I could be a hatter I might be one. He seems anxious about me. He began blowing up Lord Byron while I was sitting with him, however Says he the fellow says true things now & then; at which he took up a Magasine and read me some extracts from Don Juan, (Lord Byron's last flash poem) and particularly one against literary ambition.° I do think I must be well spoken of among sets, for Hodgkinson is more than polite, and the coffee-german° endeavour'd to be very close to me the other night at covent garden where I went at half-price before I tumbled into bed—Every one however distant an acquaintance behaves in the most conciliating manner to me—You will see I speak of this as a matter of interest. On the next Sheet I will give you a little politics. In every age there has been in England for some two or thee centuries subjects of great popular interest on the carpet: so that however great the uproar one can scarcely prophesy any material change in the government; for as loud disturbances have agitated this country many times. All civiled countries become gradually more enlighten'd and there should be a continual change for the better. Look at this Country at present and remember it when it was even thought impious to doubt the justice of a trial by Combat—From that time there has been a gradual change—Three great changes have been in progress—First for the better, next for the worse, and a third time for the better once more. The first was the gradual annihilation of the tyranny of the nobles. when kings found it their interest to conciliate the common people, elevate them and be just to them. Just when baronial Power ceased and before standing armies were so dangerous, Taxes were few. kings were lifted by the people over the heads of their nobles, and those people held a rod over kings. The change for the worse in Europe was again this. The obligation of kings to the Multitude began to be forgotten—Custom had made noblemen the humble servants of kings—Then kings turned to the Nobles as the adorners of their power, the slaves of it, and from the people as creatures continually endeavouring to check them. Then in every kingdom therre was a long struggle of kings to destroy all popular privileges. The english were the only people in europe who made a grand kick at this. They were slaves to Henry 8th but were freemen under william 3rd at the time the french were abject slaves under Lewis 14th The example of England, and the liberal writers of france and england sowed the seed of opposition to this Tyranny—and it was swelling in the ground till it burst out in the french revolution—That has had an unlucky termination. It put a stop to the rapid progress of free sentiments in England; and gave our Court hopes of turning back to the despotism of the 16 century. They have made a handle of this event in every way to undermine our freedom. They spread a horrid superstition against all inovation and improvement—The present struggle in England of the people is to destroy this superstition. What has rous'd them to do it is their distresses—Perpaps on this account the pres'ent distresses of this nation are a fortunate thing—tho so horrid in their experience. You will see I mean that the french Revolution put a temporry stop to this third change, the change for the better—Now it is in progress again and I thing in an effectual one. This is no contest beetween whig and tory—but between right and wrong. There is scarcely a grain of party spirit now in England—Right and Wrong considered by each man abstractedly is the fashion. I know very little of these things. I am convinced however that apparently small causes make great alterations. There are little signs wherby we many know how matters are going on—This makes the business about Carlisle the Bookseller of great moment in my mind. He has been selling deistical pamphlets, republished Tom Payne and many other works held in superstitious horror. He even has been selling for some time immense numbers of a work call 'The Deist' which comes out in weekly numbers—For this Conduct he I think has had above a dozen inditements issued against him; for which he has found Bail to the amount of many thousand Pounds—After all they are affraid to prosecute: they are affraid of his defence: it would be published in all the papers all over the Empire: they shudder at this: the Trials would light a flame they could not extinguish. Do you not think this of great import? You will hear by the papers of the proceedings at Manchester and Hunt's triumphal entry into London°—It would take me a whole day and a quire of paper to give you any thing like detail—I will merely mention that it is calculated that 30.000 people were in the streets waiting for him—The whole distance from the Angel Islington to the Crown and anchor° was lined with Multitudes. As I pass'd Colnaghi's° window I saw a profil Portraict of Sands the destroyer of Kotzebue. His very look must interest every one in his favour—I suppose they have represented him in his college dress—He seems to me like a young Abelard—A fine Mouth, cheek bones (and this is no joke) full of sentiment; a fine unvulgar nose and plump temples. On looking over some Letters I found the one I wrote intended for you from the foot of Helvellyn to Liverpool—but you had sail'd and therefore It was returned to me. It contained among other nonsense an Acrostic of my Sister's name—and a pretty long name it is. I wrote it in a great hurry which you will see. Indeed I would not copy it if I thought it would ever be seen by any but yourselves— . . .° I ought to make a large Q° here: but I had better take the opportunity of telling you I have got rid of my haunting sore throat—and conduct myself in a manner not to catch another You speak of Lord Byron and me—There is this great difference between us. He describes what he sees—I describe what I imagine—Mine is the hardest task. You see the immense difference—The Edinburgh review are affraid to touch upom my Poem°—They do not know what to make of it—they do not like to condemn it and they will not praise it for fear—They are as shy of it as I should be of wearing a Quaker's hat—The fact is they have no real taste—they dare not compromise their Judgements on so puzzling a Question—If on my next Publication they should praise me and so lug in Endymion—I will address in a manner they will not at all relish—The Cowardliness of the Edinburgh is worse than the abuse of the Quarterly. Monday°—This day is a grand day for winchester—they elect the Mayor. It was indeed high time the place should have some sort of excitement. There was nothing going on—all asleep—Not an old Maids Sedan returning from a card party—and if any old women have got tipsy at christenings they have not exposed themselves in the Street—The first night tho' of our arrival here there was a slight uproar took place at about ten of the clock—We heard distinctly a noise patting down the high street as of a walking Cane of the good old dowager breed; and a little minute after we heard a less voice observe 'what a noise the ferr*i*l made.—it must be loose.' Brown wanted to call the Constables, but I observed 't was only a little breeze and would soon pass over. The side-streets here are excessively maiden lady like—The door steps always fresh from the flannel. The knockers have a very staid serious, nay almost awful quietness about them—I never saw so quiet a collection of Lions, and rams heads—The doors most part black with a little brass handle just above the key hole—so that you may easily shut yourself out of your own house—he! he! There is none of your Lady Bellaston° rapping and ringing here—no thundering-Jupiter footmen no opera-trebble-tattoos—but a modest lifting up of the knocker by a set of little wee old fingers that peep through the grey mittens, and a dying fall thereof—The great beauty of Poetry is, that it makes every thing every place interesting—The palatine venice and the abbotine Winchester are equally interesting—Some time since I began a Poem call'd 'the Eve of St Mark' quite in the spirit of Town quietude. I thnk it will give you the sensation of walking about an old county Town in a coolish evening. I know not yet whether I shall ever finish it—I will give it far as I have gone. *Ut tibi placent!*°

[*'The Eve of St. Mark' follows*]

What follows is an imitation of the Authors in Chaucer's time—'t is more ancient than Chaucer himself and perhaps between him and Gower

[*lines 99–114 of 'The Eve of St. Mark' follow*]

I hope you will like this for all its Carelessness—I must take an opportunity here to observe that though I am writing *to* you I am all the while writing *at* your Wife—This explanation will account for my speaking sometimes *hoity-toityishly*. Whereas if you were alone I should sport a little more sober sadness. I am like a squinting gentleman who saying soft things to one Lady ogles another—or what is as bad in arguing with a person on his left hand appeals with his eyes to one on the right. His Vision is elastic he bends it to a certain object but having a patent spring it flies off. Writing has this disadvantage of speaking. one cannot write a wink, or a nod, or a grin, or a purse of the Lips, or a *smile—O law!* One can-not put ones pinger to one's nose, or yerk ye in the ribs, or lay hold of your button in writing—but in all the most lively and titterly parts of my Letter you must not fail to imagine me as the epic poets say—now here, now there, now with one foot pointed at the ceiling, now with another—now with my pen on my ear, now with my elbow in my mouth—O my friends you loose the action—and attitude is every thing as Fusili° said when he took up his leg like a Musket to shoot a Swallow just darting behind his shoulder. And yet does not the word mum! go for ones finger beside the nose—I hope it does. I have to make use of the word Mum! before I tell you that Severn has got a little Baby—all his own let us hope—He told Brown he had given up painting and had turn'd modeller. I hope sincerely tis not a party concern; that no M[r] — or **** is the real *Pinxit* and Severn the poor *Sculpsit*° to this work of art—You know he has long studied in the Life-Academy.° Haydon—yes your wife will say, 'here is a sum total account of Haydon again I wonder your Brother don't put a monthly bulleteen in the Philadelphia Papers about him—I wont hear—no—skip down to the bottom—aye and there are some more of his verses, skip (lullaby-by) them too' 'No, lets go regularly through' 'I wont hear a word about Haydon—bless the child, how rioty she is!—there go on there' Now pray go on here for I have a few words to say about Haydon—Before this Chancery threat had cut of every legitimate supply of Cash from me I had a little at my disposal: Haydon being very much in want I lent him 30£ of it. Now in this se-saw game of Life I got nearest to the ground and this chancery business rivetted me there so that I was sitting in that uneasy position where the seat slants so abominably. I applied to him for payment—he could not—that was no wonder. but goodman Delver,° where was the wonder then, why marry, in this, he did not seem to care much about it—and let me go without my money with almost nonchalance when he aught to have sold his drawings to supply me. I shall perhaps still be acquainted with him, but for friendship that is at an end. Brown has been my friend in this he got him to sign a Bond payable at three Months—Haslam has assisted me with the return of part of the money you lent him. Hunt—'there,' says your wife, 'there's another of those dull folkes—not a syllable about my friends—well—Hunt—what about Hunt pray—you little thing see how she bites my finger—my! is not this a tooth'—Well, when you have done with the tooth, read on—Not a syllable about your friends Here are some syllables. As far as I could smoke things on the Sunday before last, thus matters stood in Henrietta street—Henry was a greater blade than ever I remember to have seen him. He had on a very nice coat, a becoming waistcoat and buff trowsers—I think his face has lost a little of the spanish-brown, but no flesh. He carv'd some beef exactly to suit my appetite, as if I had been measured for it. As I stood looking out of the window with Charles after dinner, quizzing the Passengers, at which, I am sorry to say he is too apt, I observed that his young, son of a gun's whiskers had begun to curl and curl—little twists and twists; all down the sides of his face getting properly thickish on the angles of the the visage, He certainly will have a notable pair of Whiskers. 'How shiny your gown is in front' says Charles 'Why, dont you see 't is an apron says Henrry' Whereat I scrutiniz'd and behold your mother had a purple stuff gown on, and over it an apron of the same colour, being the same cloth that was used for the lining—and furthermore to account for the shining it was the first day of wearing. I guess'd as much of the Gown—but that is entre-nous. Charles likes england better than france. They've got a fat, smiling, fair Cook as ever you saw—she is a little lame, but that improves her. it makes her go more swimmingly. When I ask'd 'Is Mrs Wylie within' she gave such a large, five-and-thirty-year-old smile, it made me look round upon the forth stair—it might have been the fifth—but that's a puzzle. I shall never be able if I were to set myself a recollecting for a year, to recollect that—I think I remember two or three specks in her teeth but I really cant say exactly. Your mother said something about Miss Keasle—what that was is quite a riddle to me now—Whether she had got fatter or thinner, or broader or longer—straiter, or had taken to the zig zags—Whether she had taken to, or left off, asses Milk—that by the by she ought never to touch—how much better it would be to put her out to nurse with the Wise woman of Brentford.° I can say no more on so spare a subject. Miss Millar now is a different morsell if one knew how to divide and subdivide, theme her out into sections and subsections—Say a little on every part of her body as it is divided in common with all her fellow creatures, in Moor's Almanac. But Alas! I have not heard a word about her. no cue to begin upon. There was indeed a buzz about her and her mother's being at Old M^rs^ So and So's *who was like to die*—as the jews say—but I dare say, keeping up their dialect, *she was not like to die*. I must tell you a good thing Reynolds *did*: 't was the best thing he ever *said*. You know at taking leave of a party at a door way, sometimes a Man dallies and foolishes and gets awkward, and does not know how to make off to advantage—Good bye—well—good-bye—and yet he does not—go—good bye and so on—well—good bless you—You know what I mean. Now Reynolds was in this predicament and got out of it in a very witty way. He was leaving us at Hampstead. He delay'd, and we were joking at him and even said, 'be off'—at which he put the tails of his coat between his legs, and sneak'd off as nigh like a spanial as could be. He went with flying colours: this is very clever—I must, being upon the subject, tell you another good thing of him; He began, for the service it might be of to him in the law, to learn french. He had Lessons at the cheap rate of 2.6 per fag.° and observed to Brown 'Gad says he, the man sells his Lessons so cheap he must have stolen 'em.' You have heard of Hook the farce writer.° Horace Smith said to one who ask'd him if he knew Hook 'Oh yes Hook and I are very intimate.' Theres a page of Wit for you—to put John Bunyan's emblems° out of countenance.

Tuesday—You see I keep adding a sheet daily till I send the packet off—which I shall not do for a few days as I am inclined to write a good deal: for there can be nothing so remembrancing and enchaining as a good long letter be it composed of what it may—From the time you left me, our friends say I have altered completely—am not the same person—perhaps in this letter I am for in a letter one takes up one's existence from the time we last met—I dare say you have altered also—evey man does—Our bodies every seven years are completely fresh-materiald—seven years ago it was not this hand that clench'd itself against Hammond°—We are like the relict garments of a Saint: the same and not the same: for the careful Monks patch it and patch it for St Anthony's shirt. This is the reason why men who had been bosom friends, on being separated for any number of years, afterwards meet coldly, neither of them knowing why—The fact is they are both altered—Men who live together have a silent moulding and influencing power over each other—They interassimulate. 'T is an uneasy thought that in seven years the same hands cannot greet each other again. All this may be obviated by a willful and dramatic exercise of our Minds towards each other. Some think I have lost that poetic ardour and fire 't is said I once had—the fact is perhaps I have: but instead of that I hope I shall substitute a more thoughtful and quiet power. I am more frequently, now, contented to read and think—but now & then, haunted with ambitious thoughts. Quieter in my pulse, improved in my digestion; exerting myself against vexing speculations—scarcely content to write the best verses for the fever they leave behind. I want to compose without this fever. I hope I one day shall. You would scarcely imagine I could live alone so comfortably 'Kepen in solitarinesse' I told Anne, the servent here, the other day, to say I was not at home if any one should call. I am not certain how I should endure loneliness and bad weather together. Now the time is beautiful. I take a walk every day for an hour before dinner and this is generally my walk—I go out at the back gate across one street, into the Cathedral yard, which is always interesting; then I pass under the trees along a paved path, pass the beautiful front of the Cathedral, turn to the left under a stone door way—then I am on the other side of the building—which leaving behind me I pass on through two college-like squares seemingly built for the dwelling place of Deans and Prebendaries—garnished with grass and shaded with trees. Then I pass through one of the old city gates and then you are in one College-Street through which I pass and at the end thereof crossing some meadows and at last a country alley of gardens I arrive, that is, my worship arrives at the foundation of Saint Cross, which is a very interesting old place, both for its gothic tower and alms-square and for the appropriation of its rich rents to a relation of the Bishop of Winchester—Then I pass across St Cross meadows till you come to the most beautifully clear river—now this is only one mile of my walk I will spare you the other two till after supper when they would do you more good—You must avoid going the first mile just after dinner. I could almost advise you to put by all this nonsense until you are lifted out of your difficulties—but when you come to this part feel with confidence what I now feel that though there can be no stop put to troubles we are inheritors of there can be and must be and end to immediate difficulties. Rest in the confidence that I will not omit any exertion to benefit you by some means or other. If I cannot remit you hundreds, I will tens and if not that ones. Let the next year be managed by you as well as possible—the next month I mean for I trust you will soon receive Abbey's remittance. What he can send you will not be a sufficient capital to ensure you any command in America. What he has of mine I nearly have anticipated by debts. So I would advise you not to sink it, but to live upon it in hopes of my being able to encrease it—To this end I will devote whatever I may gain for a few years to come—at which period I must begin to think of a security of my own comforts when quiet will become more pleasant to me than the World—Still I would have you doubt my success—'T is at present the cast of a die with me. You say 'these things will be a great torment to me.' I shall not suffer them to be so. I shall only exert myself the more—while the seriousness of their nature will prevent me from missing up imaginary griefs. I have not had the blue devils once since I received your last—I am advised not to publish till it is seen whether the Tragedy will or not succeed—Should it, a few months may see me in the way of acquiring property; should it not it will be a drawback and I shall have to perform a longer literary Pilgrimage—You will perceive that it is quite out of my interest to come to America—What could I do there? How could I employ myself? Out of the reach of Libraries. You do not mention the name of the gentleman who assists you. 'T is an extraordinary thing. How could you do without that assistance? I will not trust myself with brooding over this. The following is an extract from a Letter of Reynolds to me 'I am glad to hear you are getting on so well with your writings. I hope you are not neglecting the revision of your Poems for the press: from which I expect more than you do'—the first thought that struck me on reading your last, was to mortgage a Poem to Murray: but on more consideration I made up my mind not to do so: my reputation is very low: he would perhaps not have negociated my bill of intellect or given me a very small sum. I should have bound myself down for some time. 'T is best to meet present misfortunes; not for a momentary good to sacrifice great benefits which one's own untramell'd and free industry may bring one in the end. In all this do never think of me as in any way unhappy: I shall not be so. I have a great pleasure in thinking of my responsibility to you and shall do myself the greatest luxury if I can succeed in any way so as to be of assistance to you. We shall look back upon these times—even before our eyes are at all dim—I am convinced of it. But be careful of those Americans—I could almost advise you to come whenever you have the sum of 500£ to England—Those Americans will I am affraid still fleece you—If ever you should think of such a thing you must bear in mind the very different state of society here—The immense difficulties of the times—The great sum required per annum to maintain yourself in any decency. In fact the whole is with Providence. I know not how to advise you but by advising you to advise with yourself. In your next tell me at large your thoughts, about america; what chance there is of succeeding there: for it appears to me you have as yet been somehow deceived. I cannot help thinking Mr Audubon has deceived you. I shall not like the sight of him—I shall endeavour to avoid seeing him—You see how puzzled I am—I have no meridian to fix you to—being the Slave of what is to happen. I think I may bid you finally remain in good hopes: and not teise yourself with my changes and variations of Mind—If I say nothing decisive in any one particular part of my Letter. you may glean the truth from the whole pretty correctly—You may wonder why I had not put your affairs with Abbey in train on receiving your Letter before last, to which there will reach you a short answer dated from shanklin. I did write and speak to Abbey but to no purpose. You last, with the enclosed note has appealed home to him—He will not see the necessity of a thing till he is hit in the mouth. 'T will be effectual—I am sorry to mix up foolish and serious things together—but in writing so much I am obliged to do so—and I hope sincerely the tenor of your mind will maintain itself better. In the course of a few months I shall be as good an Italian Scholar as I am a french one—I am reading Ariosto at present: not manageing more than six or eight stanzas at a time. When I have done this language so as to be able to read it tolerably well—I shall set myself to get complete in latin and there my learning must stop. I do not think of venturing upon Greek. I would not go even so far if I were not persuaded of the power the knowlege of any language gives one. the fact is I like to be acquainted with foreign languages. It is besides a nice way of filling up intervals &c Also the reading of Dante is well worth the while. And in latin there is a fund of curious literature of the middle ages—The Works of many great Men Aretine and Sanazarius and Machievel—I shall never become attach'd to a foreign idiom so as to put it into my writings. The Paradise lost though so fine in itself is a curruption of our Language—it should be kept as it is unique—a curiosity. a beautiful and grand Curiosity. The most remarkable Production of the world—A northern dialect accommodating itself to greek and latin inversions and intonations. The purest english I think—or what ought to be the purest—is Chatterton's—The Language had existed long enough to be entirely uncorrupted of Chaucer's gallicisms and still the old words are used—Chatterton's language is entirely northern—I prefer the native music of it to Milton's cut by feet I have but lately stood on my guard against Milton. Life to him would be death to me. Miltonic verse cannot be written but in the vein of art—I wish to devote myself to another sensation—

I have been obliged to intermiten your Letter for two days (this being Friday morn) from having had to attend to other correspondence. Brown who was at Bedhampton, went thence to Chichester, and I still directing my letters Bedhampton—there arose a misunderstand about them—I began to suspect my Letters had been stopped from curiosity. However yesterday Brown had four Letters from me all in a Lump—and the matter is clear'd up—Brown complained very much in his Letter to me of yesterday of the great alteration the Disposition of Dilke has undergone—He thinks of nothing but 'Political Justice'° and his Boy—Now the first political duty a Man ought to have a Mind to is the happiness of his friends. I wrote Brown a comment on the subject, wherein I explained what I thought of Dilke's Character. Which resolved itself into this conclusion. That Dilke was a Man who cannot feel he has a personal identity unless he has made up his Mind about every thing. The only means of strengthening one's intellect is to make up one's mind about nothing—to let the mind be a thoroughfare for all thoughts. Not a select party. The genus is not scarce in population. All the stubborn arguers you meet with are of the same brood—They never begin upon a subject they have not preresolved on. They want to hammer their nail into you and if you turn the point, still they think you wrong. Dilke will never come at a truth as long as he lives; because he is always trying at it. He is a Godwin-methodist. I must not forget to mention that your mother show'd me the lock of hair—'t is of a very dark colour for so young a creature. When it is two feet in length I shall not stand a barley corn higher. That's not fair—one ought to go on growing as well as others—At the end of this sheet I shall stop for the present—and sent it off. you may expect another Letter immediately after it. As I never know the day of the month but by chance I put here that this is *the 24th September.* I would wish you here to stop your ears, for I have a word or two to say to your Wife—My dear sister, In the first place I must quarrel with you for sending me such a shabby sheet of paper—though that is in some degree made up for by the beautiful impresson of the seal. You should like to know what I was doing—The first of May—let me see—I cannot recollect. I have all the Examiners ready to send—They will be a great treat to you when they reach you—I shall pack them up when my Business with Abbey has come to a good conclusion and the remittance is on the road to you—I have dealt round your best wishes to our friends, like a pack of cards but being always given to cheat, myself, I have turned up ace. You see I am making game of you. I see you are not all all happy in that America. England however would not be over happy for us if you were here. Perpaps 'twould be better to be teased herre than there. I must preach patience to you both. No step hasty or injurious to you must be taken. Your observation on the moschetos gives me great pleasure T is excessively poetical and humane. You say let one large sheet be all to me: You will find more than that in diffrent parts of this packet for you. Certainly I have been caught in rains. A Catch in the rain occasioned my last sore throat—but As for red-hair'd girls upon my word I do not recollect ever having seen one—Are you quizzing me or Miss Waldegrave when you talk of promenading. As for Pun-making I wish it was as good a trade as pin-making—there is very little business of that sort going on now. We struck for wages like the manchester weavers°—but to no purpose—so we are all out of employ—I am more lucky than some you see by having an oportunity of exporting a few—getting into a little foreign trade—which is a comfortable thing. I wish one could get change for a pun in silver currency. I would give three and a half any night to get into Drury-pit—But they wont ring at all. No more will notes you will say—but notes are differing things—though they make together a Pun mote°—as the term goes. If I were your Son I shouldn't mind you, though you rapt me with the Scissars—But lord! I should be out of favor sin the little un be comm'd. You have made an Uncle of me, you have, and I don't know what to make of myself. I suppose next there'll be a Nevey. You say—in may last, write directly. I have not received your Letter above 10 days. The thought of you little girl puts me in mind of a thing I heard a M[r] Lamb say. A child in arms was passing by his chair toward the mother, in the nurses arms—Lamb took hold of the long clothes saying 'Where, god bless me, Where does it leave off?' *Saturday*. If you would prefer a joke or two to any thing else I have too for you fresh hatchd. just ris as the Baker's wives say by the rolls. The first I play'd off at Brown—the second I play'd *on* on myself. Brown when he left me 'Keats!' says he 'my good fellow' (staggering upon his left heel, and fetching an irregular pirouette with his right) 'Keats' says he (depressing his left eyebrow and elevating his right one ((tho by the way, at the moment, I did not know which was the right one)) 'Keats' says he (still in the same posture but forthermore both his hands in his waistcoat pockets and jutting out his stomach) 'Keats—my—go-o-ood fell o-o-o-ooh!' says he (interlarding his exclamation with certain ventriloquial parentheses)—no this is all a lie—He was as sober as a Judge when a judge happens to be sober; and said 'Keats, if any Letters come for me—Do not forward them, but open them and give me the marrow of them in few words.' At the time when I wrote my first to him no Letters had arrived—I thought I would invent one, and as I had not time to manufacture a long one I dabbed off as short one—and that was the reason of the joke succeeding beyond my expectations. Brown let his house to a M[r] Benjamin a Jew. Now the water which furnishes the house is in a tank sided with a composition of lime and the lime impregnates the water unpleasantly—Taking advantage of this circumstance I pretended that M[r] Benjamin had written the following short note—'Sir. By drinking your damn'd tank water I have got the gravel—what reparation can you make to me and my family? Nathan Benjamin' By a fortunate hit, I hit upon his right hethen name)—his right Pronomen. Brown in consequence it appears wrote to the surprised M[r] Benjamin the following 'Sir, I cannot offer you any remuneration until your gravel shall have formed itself into a Stone when I will cut you with Pleasure. C. Brown' This of Browns M[r] Benjamin has answered insisting on an explatinon of this singular circumstance. B. says 'when I read your Letter and his following I roared, and in came Mr Snook who on reading them seem'd likely to burst the hoops of his fat sides—so the Joke has told well—Now for the one I played on myself—I must first give you the scene and the dramatis Personæ—There are an old Major and his youngish wife live in the next apartments to me—His bed room door opens at an angle with my sitting room door. Yesterday I was reading as demurely as a

Parish Clerk when I heard a rap at the door—I got up and opened it—no one was to be seen—I listened and heard some one in the Major's room—Not content with this I went up stairs and down look'd in the cubboards—and watch'd—At last I set myself to read again not quite so demurely—when there came a louder rap—I arose determin'd to find out who it was—I look out the Stair cases were all silent—'This must be the Major's wife said I—at all events I will see the truth' so I rapt me at the Major's door and went in to the utter surprise and confusion of the Lady who was in reality there—after a little explanation, which I can no more describe than fly, I made my retreat from her convinced of my mistake. She is to all appearance a silly body and is really surprised about it—She must have been—for I have discovered that a little girl in the house was the Rappee—I assure you she has nearly make me sneeze.° If the Lady tells tits I shall put a very grave and moral face on the matter with the old Gentleman, and make his little Boy a present of a humming top—My Dear George—This Monday morning the 27th I have received your last dated July 12th You say you have not heard from England these three months—Then my Letter from Shanklin wrtten I think at the end of July cannot have reach'd you. You shall not have cause to think I neglect you. I have kept this back a little time in expectation of hearing from Mr Abbey—You will say I might have remained in Town to be Abbey's messenger in these affairs. That I offer'd him—but he in his answer convinced me he was anxious to bring the Business to an issue—He observed that by being himself the agent in the whole, people might be more expeditious. You say you have not heard for three months and yet you letters have the tone of knowing how our affairs are situated by which I conjecture I acquainted you with them in a Letter previous to the Shanklin one. That I may not have done. To be certain I will here state that it is in consequence of Mrs Jennings threatning a Chancery suit that you have been kept from the receipt of monies and myself deprived of any help from Abbey—I am glad you say you keep up your Spirits—I hope you make a true statement on that score—Still keep them up—for we are all young—I can only repeat here that you shall hear from me again immediately—Notwithstanding their bad intelligence I have experienced some pleasure in receiving so correctly two Letters from you, as it gives me if I may so say a distant Idea of Proximity. This last improves upon my little niece—Kiss her for me. Do not fret yourself about the delay of money on account of any immediate opportunity being lost: for in a new country whoever has money must have opportunity of employing it in many ways. The report runs now more in favor of Kean stopping in England. If he should I have confident hopes of our Tragedy—If he smokes the hotblooded character of Ludolph—and he is the only actor that can do it—He will add to his own fame, and improve my fortune—I will give you a half dozen lines of it before I part as a specimen—

> 'Not as a Swordsman would I pardon crave,
> But as a Son: the bronz'd Centurion
> Long-toil'd in foreign wars, *and whose high deeds*
> *Are shaded in a forest of tall spears,*
> *Known only to his troop*, hath greater plea
> Of favour with my Sire than I can have—'°

Believe me my dear brother and Sister—

Your affectionate and anxious Brother
JOHN KEATS

Letter to Fanny Brawne, 13 October 1819

25 College Street.

My dearest Girl,

This moment I have set myself to copy some verses out fair. I cannot proceed with any degree of content. I must write you a line or two and see if that will assist in dismissing you from my Mind for ever so short a time. Upon my Soul I can think of nothing else—The time is passed when I had power to advise and warn you against the unpromising morning of my Life—My love has made me selfish. I cannot exist without you—I am forgetful of every thing but seeing you again—my Life seems to stop there—I see no further You have absorb'd me. I have a sensation at the present moment as though I was dissolving—I should be exquisitely miserable without the hope of soon seeing you I should be affraid to separate myself far from you. My sweet Fanny, will your heart never change? My love, will it? I have no limit now to my love—You note came in just here—I cannot be happier away from you—'T is richer than an Argosy of Pearles. Do not threat me even in jest. I have been astonished that Men could die Martyrs for religion—I have shudder'd at it—I shudder no more—I could be martyr'd for my Religion—Love is my religion—I could die for that—I could die for you. My Creed is Love and you are its only tenet—You have ravish'd me away by a Power I cannot resist; and yet I could resist till I saw you;

and even since I have seen you I have endeavoured often 'to reason against the reasons of my Love.'° I can do that no more—the pain would be too great—My Love is selfish—I cannot breathe without you.

Yours for ever
JOHN KEATS

Letter to John Taylor, 17 November 1819

Wentworth Place
Wednesday,

My dear Taylor,

I have come to a determination not to publish any thing I have now ready written; but for all that to publish a Poem before long and that I hope to make a fine one. As the marvellous is the most enticing and the surest guarantee of harmonious numbers I have been endeavouring to persuade myself to untether Fancy and let her manage for herself—I and myself cannot agree about this at all. Wonders are no wonders to me. I am more at Home amongst Men and women. I would rather read Chaucer than Ariosto—The little dramatic skill I may as yet have however badly it might show in a Drama would I think be sufficient for a Poem—I wish to diffuse the colouring of St Agnes eve throughout a Poem in which Character and Sentiment would be the figures to such drapery—Two or three such Poems, if God should spare me, written in the course of the next six years, would be a famous gradus ad Parnassum altissimum°—I mean they would nerve me up to the writing of a few fine Plays—my greatest ambition—when I do feel ambitious. I am sorry to say that is very seldom. The subject we have once or twice talked of appears a promising one, The Earl of Leicester's historry. I am this morning reading Holingshed's Elisabeth,° You had some Books awhile ago, you promised to lend me, illustrative of my Subject. If you can lay hold of them or any others which may be serviceable to me I know you will encourage my low-spirited Muse by sending them—or rather by letting me know when our Errand cart Man shall call with my little Box. I will endeavour to set my self selfishly at work on this Poem that is to be—

Your sincere friend
JOHN KEATS—

Letter to Fanny Keats, 8 February 1820

Wentworth Place
Tuesday morn.

My dear Fanny—

I had a slight return of fever last night, which terminated favourably, and I am now tolerably well, though weak from small quantity of food to which I am obliged to confine myself: I am sure a mouse would starv upon it. Mrs Wylie came yesterday. I have a very pleasant room for a sick person. A Sopha bed is made up for me in the front Parlour which looks on to the grass plot as you remember Mrs Dilkes does. How much more comfortable than a dull room up stairs, where one gets tired of the pattern of the bed curtains. Besides I see all that passes—for instanc now, this morning, if I had been in my own room I should not have seen the coals brought in. On sunday between the hours of twelve and one I descried a Pot boy. I conjectured it might be the one o'Clock beer—Old women with bobbins and red cloaks and unpresuming bonnets I see creeping about the heath. Gipseys after hare skins and silver spoons. Then goes by a fellow with a wooden clock under his arm that strikes a hundred and more. Then comes the old french emigrant (who has been very well to do in france) whith his hands joined behind on his hips, and his face full of political schemes. Then passes Mr David Lewis a very goodnatured, goodlooking old gentleman whas has been very kind to Tom and George and me. As for those fellows the Brickmakers they are always passing to and fro. I mus'n't forget the two old maiden Ladies in well walk who have a Lap dog between them, that they are very anxious about. It is a corpulent Little Beast whom it is necessary to coax along with an ivory-tipp'd cane. Carlo our Neighbour Mrs Brawne's dog and it meet sometimes. Lappy thinks Carlo a devil of a fellow and so do his Mistresses. Well they may—he would sweep 'em all down at a run; all for the Joke of it. I shall desire him to peruse the fable of the Boys and the frogs: though he prefers the tongues and the Bones.° You shall hear from me again the day after tomorrow—

Your affectionate Brother
JOHN KEATS

Letter to James Rice, 14, 16 February 1820

Wentworth Place
My dear Rice, Monday Morn.

I have not been well enough to make any tolerable rejoinder to your kind Letter. I will as you advise be very chary of my health and spirits. I am sorry to hear of your relapse and hypochondriac symptoms attending it. Let us hope for the best as you say. I shall follow your example in looking to the future good rather than brooding upon present ill. I have not been so worn with lengthen'd illnesses as you have therefore cannot answer you on your own ground with respect to those haunting and deformed thoughts and feelings you speak of. When I have been or supposed myself in health I have had my share of them, especially within this last year. I may say that for 6 Months before I was taken ill I had not passed a tranquil day—Either that gloom overspred me or I was suffering under some passionate feeling, or if I turn'd to versify that acerbated the poison of either sensation. The Beauties of Nature had lost their power over me. How astonishingly (here I must premise that illness as far as I can judge in so short a time has relieved my Mind of a load of deceptive thoughts and images and makes me perceive things in a truer light)—How astonishingly does the chance of leaving the world impress a sense of its natural beauties on us. Like poor Falstaff, though I do not babble, I think of green fields.° I muse with the greatest affection on every flower I have known from my infancy—their shapes and coulours as are new to me as if I had just created them with a superhuman fancy—It is because they are connected with the most thoughtless and happiest moments of our Lives—I have seen foreign flowers in hothouses of the most beautiful nature, but I do not care a straw for them. The simple flowers of our spring are what I want to see again.

Brown has left the inventive and taken to the imitative art—he is doing his forte which is copying Hogarth's heads.°

He has just made a purchace of the methodist meeting Picture, which gave me a horrid dream a few nights ago. I hope I shall sit under the trees with you again in some such place as the isle of Wight—I do not mind a game at cards in a saw pit or waggon; but if ever you catch me on a stage coach in the winter full against the wind bring me down with a brace of bullets and I promise not to 'peach. Remberme to Reynolds and say how much I should like to hear from him: that Brown returned immediately after he went on Sunday, and that I was vex'd at forgetting to ask him to lunch for as he went towards the gate I saw he was fatigued and hungry.

I am
my dear Rice
ever most sincerely yours
JOHN KEATS

I have broken this open to let you know I was surprised at seeing it on the table this morning; thinking it had gone long ago.

Letter to Fanny Brawne, ?February 1820

My dearest Girl,

According to all appearances I am to be separated from you as much as possible. How I shall be able to bear it, or whether it will not be worse than your presence now and then, I cannot tell. I must be patient, and in the meantime you must think of it as little as possible. Let me not longer detain you from going to Town—there may be no end to this emprisoning of you. Perpaps you had better not come before tomorrow evening: send me however without fail a good night You know our situation—what hope is there if I should be recovered ever so soon—my very health will not suffer me to make any great exertion. I am reccommended not even to read poetry much less write it. I wish I had even a little hope. I cannot say forget me—but I would mention that there are impossibilities in the world. No more of this—I am not strong enough to be weaned—take no notice of it in your good night. Happen what may I shall ever be my dearest Love

Your affectionate
J—K—

Letter to Fanny Brawne, ?February 1820

My dearest Fanny,

I read your note in bed last night, and that might be the reason of my sleeping so much better. I thnk Mr Brown is right in supposing you may stop too long with me, so very nervous as I am. Send me every evening a written Good night. If you come for a few minutes about six it may be the best time. Should you ever fancy me too low-spirited I must warn you to ascbribe it to the medicine I am at present taking which is of a nerve-shaking nature—I shall impute any depression I may experience to this cause. I have been writing with a vile old pen the whole week, which is excessively ungallant. The fault is in the Quill: I have mended it and still it is very much inclin'd to make blind es. However these last lines are in a much better style of penmanship thof a little disfigured by the smear of black currant jelly; which has made a little mark on one of the Pages of Brown's Ben Jonson, the very best book he has. I have lick'd it but it remains very purple°—I did not know whether to say purple or blue, so in the mixture of the thought wrote purplue which may be an excellent name for a colour made up of those two, and would suit well to start next spring. Be very careful of open doors and windows and going without your duffle grey—God bless you Love!—

J. KEATS—

P.S. I am sitting in the back room—Remember me to your Mother—

Letter to Fanny Brawne, ?27 February 1820

My dearest Fanny,

I had a better night last night than I have had since my attack, and this morning I am the same as when you saw me. I have been turning over two volumes of Letters written between Rosseau and two Ladies° in the perplexed strain of mingled finesse and sentiment in which the Ladies and gentlemen of those days were so clever, and which is still prevalent among Ladies of this Country who live in a state of resoning romance. The Likeness however only extends to the mannerism not to the dexterity. What would Rousseau have said at seeing our little correspondence! What would his Ladies have said! I don't care much—I would sooner have Shakspeare's opinion about the matter. The common gossiping of washerwomen must be less disgusting than the continual and eternal fence and attack of Rousseau and these sublime Petticoats. One calls herself Clara and her friend Julia two of Rosseau's Heroines—they all the same time christen poor Jean Jacques St Preux—who is the pure cavalier of his famous novel.° Thank God I am born in England with our own great Men before my eyes—Thank god that you are fair and can love me without being Letter-written and sentimentaliz'd into it—M[r] Barry Cornwall has sent me another Book, his first, with a polite note—I must do what I can to make him sensible of the esteem I have for his kindness. If this north east would take a turn it would be so much the better for me. Good bye, my love, my dear love, my beauty—

love me for ever—

J—K—

Letter to C. W. Dilke, 4 March 1820

My dear Dilke,

Since I saw you I have been gradually, too gradually perhaps, improving; and though under an interdict with respect to animal food living upon pseudo victuals, Brown says I have pick'd up a little flesh lately. If I can keep off inflammation for the next six weeks I trust I shall do very well. You certainly should have been at Martin's dinner for making an index is surely as dull work as engraving. Have you heard that the Bookseller is going to tie himself to the manager eat or not as he pleases? He says Rice shall have his foot on the fender notwithstanding. Reynolds is going to sail on the salt seas. Brown has been mightily progressing with his Hogarth. A damn'd melancholy picture it is, and during the first week of my illness it gave me a psalm singing nightmare, that made me almost faint away in my sleep. I know I am better, for I can bear the Picture. I have experienced a specimen of great politeness from M[r] Barry Cornwall. He has sent me his books. Some time ago he had given his first publish'd book to Hunt for me; Hunt forgot to give it and Barry Cornwall thinking I had received it must have thought me very neglectful fellow. Notwithstanding he sent me his second book and on my explaining that I had not received his first he sent me that also.° I am sorry to see by M[rs] D's note that she has been so unwell with the spasms. Does she continue the Medicines that benefited her so much? I am affraid not. Remember me to her and say I shall not expect her at Hampstead next week unless the Weather changes for the warmer. It is better to run no chance of a supernumery cold in March. As for you you must come. You must improve in your penmanship; your writing is like the speaking of a child of three years old, very understandable to its father but to no one else. The worst is it looks well—no that is not the worst—the worst is, it is worse than Bailey's. Bailey's looks illegible and may perchance be read: your's looks very legible and may perchance not be read—I would endeavour to give you a facsimile of your word Thistlewood if I were not minded on the instant that Lord chesterfield has done some such thing to his Son.° Now I would not bathe in the same River with lord C. though I had the upper hand of the stream. I am grieved that in writing and speaking it is necessary to make use of the same particles as he did. Cobbet is expected to come in. O that I had two double plumpers° for him. The ministry are not so inimical to him but it would like to put him out of Coventry. Casting my eye on the other side I see a long word written in a most vile manner, unbecoming a Critic. You must recollect I have served no apprenticeship to old plays. If the only copies of the greek and Latin Authors had been made by you, Bailey and Haydon they Were as good as lost. It has been said that the Character of a Man may be known by his hand writing—if the Character of the age may be known by the average goodness of said, what a slovenly age we live in. Look at Queen Elizabeth's Latin exercises and blush. Look at Milton's hand—I cant say a word for shakespeare—

Your sincere friend
JOHN KEATS

Letter to Fanny Brawne, ?March 1820

Sweetest Fanny,

You fear, sometimes, I do not love you so much as you wish? My dear Girl I love you ever and ever and without reserve. The more I have known you the more have I lov'd. In every way—even my jealousies have been agonies of Love, in the hottest fit I ever had I would have died for you. I have vex'd you too much. But for Love! Can I help it? You are always new. The last of your kisses was ever the sweetest; the last smile the brightest; the last movement the gracefullest. When you pass'd my window home yesterday, I was fill'd with as much admiration as if I had then seen you for the first time. You uttered a half complaint once that I only lov'd your Beauty. Have I nothing else then to love in you but that? Do not I see a heart naturally furnish'd with wings imprison itself with me? No ill prospect has been able to turn your thoughts a moment from me. This perhaps should be as much a subject of sorrow as joy—but I will not talk of that. Even if you did not love me I could not help an entire devotion to you: how much more deeply then must I feel for you knowing you love me. My Mind has been the most discontented and restless one that ever was put into a body too small for it. I never felt my Mind repose upon anything with complete and undistracted enjoyment—upon no person but you. When you are in the room my thoughts never fly out of window: you always concentrate my whole senses. The anxiety shown about our Loves in your last note is an immense pleasure to me: however you must not suffer such speculations to molest you any more: nor will I any more believe you can have the least pique against me. Brown is gone out—but here is Mrs. Wylie—when she is gone I shall be awake for you.—Remembrances to your Mother.

Your affectionate
J. KEATS.

Letter to Fanny Brawne, ?March 1820

My dearest Fanny, I slept well last night and am no worse this morning for it. Day by day if I am not deceived I get a more unrestrain'd use of my Chest. The nearer a racer gets to the Goal the more his anxiety becomes so I lingering upon the borders of health feel my impatience increase. Perhaps on your account I have imagined my illness more serious than it is: how horrid was the chance of slipping into the ground instead of into your arms—the difference is amazing Love—Death must come at last; Man must die, as Shallow says;° but before that is my fate I feign would try what more pleasures than you have given so sweet a creature as you can give. Let me have another oportunity of years before me and I will not die without being remember'd. Take care of yourself dear that we may both be well in the Summer. I do not at all fatigue myself with writing, having merely to put a line or two here and there, a Task which would worry a stout state of the body and mind, but which just suits me as I can do no more.

Your affectionate
J.K—

Letter to Fanny Brawne, ?March 1820

My dearest Fanny, whever you know me to be alone, come, no matter what day. Why will you go out this weather? I shall not fatigue myself with writing too much I promise you. Brown says I am getting stouter. I rest well and from last night do not remember any thing horrid in my dream, which is a capital symptom, for any organic derangement always occasions a Phantasmagoria. It will be a nice idle amusement to hunt after a motto for my Book which I will have if lucky enough to hit upon a fit one—not intending to write a preface.° I fear I am too late with my note—you are gone out—you will be as cold as a topsail in a north latitude—I advise you to furl yourself and come in a doors.

Good bye Love.
J.K.

Letter to Fanny Brawne, ?March 1820

My dearest Girl,

In consequence of our company I suppose I shall not see you before tomorrow. I am much better today—indeed all I have to complain of is want of strength and a little tightness in the Chest. I envied Sam's walk with you to day; which I will not do again as I may get very tired of envying. I imagine you now sitting in your new black dress which I like so much and if I were a little less selfish and more enthousiastic I should run round and surprise you with a knock at the door. I fear I am too prudent for a dying kind of Lover. Yet, there is a great difference between going off in warm blood like Romeo, and making one's exit like a frog in a frost. I had nothing particular to say to day, but not intending that there shall be any interruption to our correspondence (which at some future time I propose offering to Murray) I write something! God bless you my sweet Love! Illness is a long lane, but I see you at the end of it, and shall mend my pace as well as possible

J—K

Letter to Fanny Brawne, ?May 1820

Tuesday Morn—

My dearest Girl,

I wrote a Letter for you yesterday expecting to have seen your mother. I shall be selfish enough to send it though I know it may give you a little pain, because I wish you to see how unhappy I am for love of you, and endeavour as much as I can to entice you to give up your whole heart to me whose whole existence hangs upon you. You could not step or move an eyelid but it would shoot to my heart—I am greedy of you—Do not think of any thing but me. Do not live as if I was not existing—Do not forget me—But have I any right to say you forget me? Perhaps you think of me all day. Have I any right to wish you to be unhappy for me? You would forgive me for wishing it, if you knew the extreme passion I have that you should love me—and for you to love me as I do you, you must think of no one but me, much less write that sentence. Yesterday and this morning I have been haunted with a sweet vision—I have seen you the whole time in your shepherdess dress. How my senses have ached at it!° How my heart has been devoted to it! How my eyes have been full of Tears at it! Indeed I think a real Love is enough to occupy the widest heart—Your going to town alone, when I heard of it was a shock to me—yet I expected it—*promise me you will not for some time, till I get better*. Promise me this and fill the paper full of the most endearing mames.° If you cannot do so with good will, do my Love tell me—say what you think—confess if your heart is too much fasten'd on the world. Perhaps then I may see you at a greater distance, I may not be able to appropriate you so closely to myself. Were you to loose a favorite bird from the cage, how would your eyes ache after it as long as it was in sight; when out of sight you would recover a little. Perphaps if you would, if so it is, confess to me how many things are necessary to you besides me, I might be happier, by being less tantaliz'd. Well may you exclaim, how selfish, how cruel, not to let me enjoy my youth! to wish me to be unhappy! You must be so if you love me—upon my Soul I can be contented with nothing else. If you could really what is call'd enjoy yourself at a Party—if you can smile in peoples faces, and wish them to admire you *now*, you never have nor ever will love me—I see *life* in nothing but the cerrtainty of your Love—convince me of it my sweetest. If I am not somehow convinc'd I shall die of agony. If we love we must not live as other men and women do—I cannot brook the wolfsbane of fashion and foppery and tattle. You must be mine to die upon the rack if I want you. I do not pretend to say I have more feeling than my fellows—but I wish you seriously to look over my letters kind and unkind and consider whether the Person who wrote them can be able to endure much longer the agonies and uncertainties which you are so peculiarly made to create—My recovery of bodily health will be of no benefit to me if you are not all mine when I am well. For god's sake save me—or tell me my passion is of too awful a nature for you. Again God bless you

J.K.

No—my sweet Fanny—I am wrong. I do not want you to be unhappy—and yet I do, I must while there is so sweet a Beauty—my loveliest my darling! Good bye! I kiss you—O the torments!

Letter to Fanny Brawne, ?June 1820

My dearest Fanny,

My head is puzzled this morning, and I scarce know what I shall say though I am full of a hundred things. 'T is certain I would rather be writing to you this morning, notwithstanding the alloy of grief in such an occupation, than enjoy any other pleasure, with health to boot, unconnected with you. Upon my soul I have loved you to the extreme. I wish you could know the Tenderness with which I continually brood over your different aspects of countenance, action and dress. I see you come down in the morning: I see you meet me at the Window—I see every thing over again eternally that I ever have seen. If I get on the pleasant clue I live in a sort of happy misery, if on the unpleasant 'tis miserable misery. You complain of my illtreating you in word thought and deed—I am sorry,—at times I feel bitterly sorry that I ever made you unhappy—my excuse is that those words have been wrung from me by the sharpness of my feelings. At all events and in any case I have been wrong; could I believe that I did it without any cause, I should be the most sincere of Penitents. I could give way to my repentant feelings now, I could recant all my suspicions, I could mingle with you heart and Soul though absent, were it not for some parts of your Letters. Do you suppose it possible I could ever leave you? You know what I think of myself and what of you. You know that I should feel how much it was my loss and how little yours—My friends laugh at you! I know some of them—when I know them all I shall never think of them again as friends or even acquaintance. My friends have behaved well to me in every instance but one, and there they have bcome tattlers, and inquisitors into my conduct: spying upon a secret I would rather die than share it with any body's confidence. For this I cannot wish them well, I care not to see any of them again. If I am the Theme, I will not be the Friend of idle Gossips. Good gods what a shame it is our Loves should be so put into the microscope of a Coterie. Their laughs should not affect you (I may perhaps give you reasons some day for these laughs, for I suspect a few people to hate me well enough, *for reasons I know of*, who have pretended a great friendship for me) when in competition with one, who if he never should see you again would make you the saint of his memorry—These Laughters, who do not like you, who envy you for your Beauty, who would have God-bless'd-me from you for ever: who were plying me with disencouragements with respect to you eternally. People are revengeful—do not mind them—do nothing but love me—if I knew that for certain life and health will in such event be a heaven, and death itself will be less painful. I long to believe in immortality I shall never be able to bid you an entire farewell. If I am destined to be happy with you here—how short is the longest Life—I wish to believe in immortality—I wish to live with you for ever. Do not let my name ever pass between you and those laughers, if I have no other merit than the great Love for you, that were sufficient to keep me sacred and unmentioned in such society. If I had been cruel and injust I swear my love has ever been greater than my cruelty which last but a minute whereas my Love come what will shall last for ever If concessions to me has hurt your Pride, god knows I have had little pride in my heart when thinking of you. Your name never passes my Lips—do not let mine pass yours—Those People do not like me. After reading° my Letter you even then wish to see me. I am strong enough to walk over—but I dare not. I shall feel so much pain in parting with you again. My dearest love, I am affraid to see you, I am strong but not strong enough to see you. Will my arm be ever round you again. And if so shall I be obliged to leave you again. My sweet Love! I am happy whilst I believe your first Letter. Let me be but certain that you are mine heart and soul, and I could die more happily than I could otherwise live. If you think me cruel—if you think I have sleighted you—do muse it over again and see into my heart—My Love to you is 'true as truth's simplicity and simpler than the infancy of truth"° as I think I once said before How could I slight you? How threaten to leave you? not in the spirit of a Threat to you—no—but in the spirit of Wretchedness in myself. My fairest, my delicious, my angel Fanny! do not believe me such a vulgar fellow. I will be as patient in illness and as believing in Love as I am able—

Yours for ever my dearest
JOHN KEATS—

Letter to Fanny Brawne, ?5 July 1820

Wednesday Morng.

My dearest Girl,

I have been a walk this morning with a book in my hand, but as usual I have been occupied with nothing but you: I wish I could say in an agreeable manner. I am tormented day and night. They talk of my going to Italy. 'Tis certain I shall never recover if I am to be so long separate from you: yet with all this devotion to you I cannot persuade myself into any confidence of you. Past experience connected with the fact of my long separation from you gives me agonies which are scarcely to be talked of. When your mother comes I shall be very sudden and expert in asking her whether you have been to Mrs. Dilke's, for she might say no to make me easy. I am literally worn to death, which seems my only recourse. I cannot forget what has pass'd. What? nothing with a man of the world, but to me deathful. I will get rid of this as much as possible. When you were in the habit of flirting with Brown you would have left off, could your own heart have felt one half of one pang mine did. Brown is a good sort of Man—he did not know he was doing me to death by inches. I feel the effect of every one of those hours in my side now; and for that cause, though he has done me many services, though I know his love and friendship for me, though at this moment I should be without pence were it not for his assistance, I will never see or speak to him until we are both old men, if we are to be. I *will* resent my heart having been made a football. You will call this madness. I have heard you say that it was not unpleasant to wait a few years—you have amusements—your mind is away—you have not brooded over one idea as I have, and how should you? You are to me an object intensely desireable—the air I breathe in a room empty of you is unhealthy. I am not the same to you—no—you can wait—you have a thousand activities—you can be happy without me. Any party, any thing to fill up the day has been enough. How have you pass'd this month? Who have you smil'd with? All this may seem savage in me. You do not feel as I do—you do not know what it is to love—one day you may—your time is not come. Ask yourself how many unhappy hours Keats has caused you in Loneliness. For myself I have been a Martyr the whole time, and for this reason I speak; the confession is forc'd from me by the torture. I appeal to you by the blood of that Christ you believe in: Do not write to me if you have done anything this month which it would have pained me to have seen. You may have altered—if you have not—if you still behave in dancing rooms and other societies as I have seen you—I do not want to live—if you have done so I wish this coming night may be my last. I cannot live without you, and not only you but *chaste you; virtuous you.* The Sun rises and sets, the day passes, and you follow the bent of your inclination to a certain extent—you have no conception of the quantity of miserable feeling that passes through me in a day.—Be serious! Love is not a plaything—and again do not write unless you can do it with a crystal conscience. I would sooner die for want of you than—

Yours for ever
J. KEATS.

Letter to Fanny Brawne, ?August 1820

I do not write this till the last, that no eye may catch it°

My dearest Girl,
I wish you could invent some means to make me at all happy without you. Every hour I am more and more concentrated in you; every thing else tastes like chaff in my Mouth. I feel it almost impossible to go to Italy—the fact is I cannot leave you, and shall never taste one minute's content until it pleases chance to let me live with you for good. But I will not go on at this rate. A person in health as you are can have no conception of the horrors that nerves and a temper like mine go through. What Island do your friends propose retiring to? I should be happy to go with you there alone, but in company I should object to it; the backbitings and jealousies of new colonists who have nothing else to amuse them selves, is unbearable. M[r] Dilke came to see me yesterday, and gave me a very great deal more pain than pleasure. I shall never be able any more to endure to society of any of those who used to meet at Elm Cottage° and Wentworth Place. The last two years taste like brass

upon my Palate. If I cannot live with you I will live alone. I do not think my health will improve much while I am separated from you. For all this I am averse to seeing you—I cannot bear flashes of light and return into my glooms again. I am not so unhappy now as I should be if I had seen you yesterday. To be happy with you seems such an impossibility! it requires a luckier Star than mine! it will never be. I enclose a passage from one of your Letters which I want you to alter a little—I want (if you will have it so) the matter express'd less coldly to me. If my health would bear it, I could write a Poem which I have in my head, which would be a consolation for people in such a situation as mine. I would show some one in Love as I am, with a person living in such Liberty as you do. Shakspeare always sums up matters in the most sovereign manner. Hamlet's heart was full of such Misery as mine is when he said to Ophelia 'Go to a Nunnery, go, go!'° Indeed I should like to give up the matter at once—I should like to die. I am sickened at the brute world which you are smiling with. I hate men and women more. I see nothing but thorns for the future—wherever I may be next winter in Italy or nowhere Brown will be living near you with his indecencies°—I see no prospect of any rest. Suppose me in Rome—well, I should there see you as in a magic glass going to and from town at all hours,—I wish you could infuse a little confidence in human nature into my heart. I cannot muster any—the world is too brutal for me—I am glad there is such a thing as the grave—I am sure I shall never have any rest till I get there At any rate I will indulge myself by never seeing any more Dilke or Brown or any of their Friends. I wish I was either in your arms full of faith or that a Thunder bolt would strike me.

God bless you—J.K—

Letter to John Taylor, 13 August 1820

Wentworth Place
Sat[y] Morn.

My dear Taylor,

My Chest is in so nervous a State, that any thing extra such as speaking to an unaccostomed Person or writing a Note half suffocates me. This Journey to Italy wakes me at daylight every morning and haunts me horribly. I shall endeavour to go though it be with the sensation of marching up against a Batterry. The first step towards it is to know the expense of a Journey and a years residence: which if you will ascertain for me and let me know early you will greatly serve me. I have more to say but must desist for every line I write encreases the tightness of the Chest, and I have many more to do. I am convinced that this sort of thing does not continue for nothing—If you can come with any of our friends do.

Your sincere friend
JOHN KEATS—

Letter to Percy Bysshe Shelley, 16 August 1820

Hampstead August 16th

My dear Shelley,

I am very much gratified that you, in a foreign country, and with a mind almost over occupied, should write to me in the strain of the Letter beside me. If I do not take advantage of your invitation it will be prevented by a circumstance I have very much at heart to prophesy—There is no doubt that an english winter would put an end to me, and do so in a lingering hateful manner, therefore I must either voyage or journey to Italy as a soldier marches up to a battery. My nerves at present are the worst part of me, yet they feel soothed when I think that come what extreme may, I shall not be destined to remain in one spot long enough to take a hatred of any four particular bed-posts. I am glad you take any pleasure in my poor Poem;—which I would willingly take the trouble to unwrite, if possible, did I care so much as I have done about Reputation. I received a copy of the Cenci, as from yourself from Hunt. There is only one part of it I am judge of; the Poetry, and dramatic effect, which by many spirits now a days is considered the mammon. A modern work it is said must have a purpose, which may be the God—*an artist* must serve Mammon—he must have 'self concentration' selfishness perhaps. You I am sure will forgive me for sincerely remarking that you might curb your magnanimity and be more of an artist, and 'load every rift' of your subject with ore° The thought of such discipline must fall like cold chains upon you, who perhaps never sat with your wings furl'd for six Months together. And is not this extraordinay talk for the writer of Endymion? whose mind was like a pack of scattered cards—I am pick'd up and sorted to a pip. My Imagination is a Monastry and I am its Monk—you must explain my metap^cs° to yourself. I am in expectation of Prometheus every day. Could I have my own wish for its interest effected you would have it still in manuscript—or be but now putting an end to the second act. I remember you advising me not to publish my first-blights, on Hampstead heath—I am returning advice upon your hands. Most of the Poems in the volume I send you have been written above two years, and would never have been publish'd but from a hope of gain; so you see I am inclined enough to take your advice now. I must express once more my deep sense of your kindness, adding my sincere thanks and respects for Mrs Shelley. In the hope of soon seeing you I remain

most sincerely yours,
JOHN KEATS—

Letter to Charles Brown, ?August 1820

My dear Brown,

xxxxxxx I ought to be off at the end of this week, as the cold winds begin to blow towards evening;—but I will wait till I have your answer to this. I am to be introduced, before I set out, to a Dr Clarke,° a physician settled at Rome, who promises to befriend me in every way at Rome. The sale of my book is very slow, though it has been very highly rated. One of the causes, I understand from different quarters, of the unpopularity of this new book, and the others also, is the offence the ladies take at me. On thinking that matter over, I am certain that I have said nothing in a spirit to displease any woman I would care to please: but still there is a tendency to class women in my books with roses and sweetmeats,—they never see themselves dominant. If ever I come to publish 'Lucy Vaughan Lloyd',° there will be some delicate picking for squeamish stomachs. I will say no more, but, waiting in anxiety for your answer, doff my hat, and make a purse as long as I can.

Your affectionate friend,
JOHN KEATS.

Letter to Fanny Keats, 23 August 1820

Wentworth Place
Wednesday Morning

My dear Fanny,

It will give me great Pleasure to see you here, if you can contrive it; though I confess I should have written instead of calling upon you before I set out on my journey, from the wish of avoiding unpleasant partings. Meantime I will just notice some parts of your Letter. The Seal-breaking business° is over blown—I think no more of it. A few days ago I wrote to Mr Brown, asking him to befriend me with his company to Rome. His answer is not yet come, and I do not know when it will, not being certain how far he may be from the Post Office to which my communication is addressed. Let us hope he will go with me. George certainly ought to have written to you: his troubles, anxieties and fatigues are not quite a sufficient excuse. In the course of time you will be sure to find that this neglect, is not forgetfulness. I am sorry to hear you have been so ill and in such low spirits. Now you are better, keep so. Do not suffer Your Mind to dwell on unpleasant reflections—that sort of thing has been the destruction of my health—Nothing is so bad as want of health—it makes one envy Scavengers and Cinder-sifters. There are enough real distresses and evils in wait for every one to try the most vigorous health. Not that I would say yours are not real—but they are such as to tempt you to employ your imagination on them, rather than endeavour to dismiss them entirely. Do not diet your mind with grief, it destroys the constitution; but let your chief care be of your health, and with that you will meet with your share of Pleasure in the world—do not doubt it. If I return well from Italy I will turn over a new leaf for you. I have been improving lately, and have very good hopes of 'turning a Neuk'° and cheating the Consumption. I am not well enough to write to George myself—Mr Haslam will do it for me, to whom I shall write to day, desiring him to mention as gently as possible your complaint—I am my dear Fanny

Your affectionate Brother
JOHN.

Letter to Charles Brown, 30 September 1820

Saturday Sept[r] 28°
Maria Crowther
off Yarmouth isle
of wight—

My dear Brown,

The time has not yet come for a pleasant Letter from me. I have delayed writing to you from time to time because I felt how impossible it was to enliven you with one heartening hope of my recovery; this morning in bed the matter struck me in a different manner; I thought I would write 'while I was in some liking'° or I might become too ill to write at all and then if the desire to have written should become strong it would be a great affliction to me. I have many more Letters to write and I bless my stars that I have begun, for time seems to press,—this may be my best opportunity. We are in a calm and I am easy enough this morning. If my spirits seem too low you may in some degree impute it to our having been at sea a fortmight without making any way. I was very disappointed at not meeting you at bedhamption, and am very provoked at the thought of you being at Chichester to day. I should have delighted in setting off for London for the sensation merely—for what should I do there? I could not leave my lungs or stomach or other worse things behind me. I wish to write on subjects that will not agitate me much—there is one I must mention and have done with it. Even if my body would recover of itself, this would prevent it—The very thing which I want to live most for will be a great occasion of my death. I cannot help it. Who can help it? Were I in health it would make me ill, and how can I bear it in my state? I dare say you will be able to guess on what subject I am harping—you know what was my greatest pain during the first part of my illness at your house. I wish for death every day and night to deliver me from these pains, and then I wish death away, for death would destroy even those pains which are better than nothing. Land and Sea, weakness and decline are great seperators, but death is the great divorcer for ever. When the pang of this thought has passed through my mind, I may say the bitterness of death is passed. I often wish for you that you might flatter me with the best. I think without my mentioning it for my sake you would be a friend to Miss Brawne when I am dead. You think she has many faults—but, for my sake, think she has not one— —if there is any thing you can do for her by word or deed I know you will do it. I am in a state at present in which woman merely as woman can have no more power over me than stocks and stones, and yet the difference of my sensations with respect to Miss Brawne and my Sister is amazing. The one seems to absorb the other to a degree incredible. I seldom think of my Brother and Sister in america. The thought of leaving Miss Brawne is beyond every thing horrible—the sense of darkness coming over me—I eternally see her figure eternally vanishing. Some of the phrases she was in the habit of using during my last nursing at Wentworth place ring in my ears—Is there another Life? Shall I awake and find all this a dream? There must be we cannot be created for this sort of suffering. The receiving of this letter is to be one of yours—I will say nothing about our friendship or rather yours to me more than that as you deserve to escape you will never be so unhappy as I am. I should think of—you in my last moments. I shall endeavour to write to Miss Brawne if possible to day. A sudden stop to my life in the middle of one of these Letters would be no bad thing for it keeps one in a sort of fever awhile. Though fatigued with a Letter longer than any I have written for a long while it would be better to go on for ever than awake to a sense of contrary winds. We expect to put into Portland roads to night. The Captn the Crew and the Passengers are all ill-temper'd and weary. I shall write to dilke. I feel as if I was closing my last letter to you—My dear Brown

Your affectionate friend
JOHN KEATS

Letter to Mrs Samuel Brawne, ?24 October 1820

Octr 24 Naples Harbour
care Giovanni

My dear Mrs Brawne,

A few words will tell you what sort of a Passage we had, and what situation we are in, and few they must be on account of the Quarantine, our Letters being liable to be opened for the purpose of fumigation at the Health Office.° We have to remain in the vessel ten days and are, at present shut in a tier of ships. The sea air has been beneficial to me about to as great an extent as squally weather and bad accommodations and provisions has done harm—So I am about as I was—Give my Love to Fanny and tell her, if I were well there is enough in this Port of Naples to fill a quire of Paper—but it looks like a dream—every man who can row his boat and walk and talk seems a different being from myself—I do not feel in the world—It has been unfortunate for me that one of the Passengers is a young Lady in a Consumption—her imprudence has vexed me very much—the knowledge of her complaint—the flushings in her face, all her bad symptoms have preyed upon me—they would have done so had I been in good health. Severn now is a very good fellow but his nerves are too strong to be hurt by other peoples illnesses—I remember poor Rice wore me in the same way in the isle of wight—I shall feel a load off me when the Lady vanishes out of my sight. It is impossible to describe exactly in what state of health I am—at this moment I am suffering from indigestion very much, which makes such stuff of this Letter. I would always wish you to think me a little worse than I really am; not being of a sanguine disposition I am likely to succeed. If I do not recover your regret will be softened if I do your pleasure will be doubled—I dare not fix my Mind upon Fanny, I have not dared to think of her. The only comfort I have had that way has been in thinking for hours together of having the knife she gave me put in a silver-case—the hair in a Locket—and the Pocket Book in a gold net—Show her this. I dare say no more—Yet you must not believe I am so ill as this Letter may look for if ever there was a person born without the faculty of hoping I am he. Severn is writing to Haslam, and I have just asked him to request Haslam to send you his account of my health. O what an account I could give you of the Bay of Naples if I could once more feel myself a Citizen of this world—I feel a Spirit in my Brain would lay it forth pleasantly—O what a misery it is to have an intellect in splints! My Love again to Fanny—tell Tootts° I wish I could pitch her a basket of grapes—and tell Sam the fellows catch here with a line a little fish much like an anchovy, pull them up fast   Remember me to M^{rs} and Mr Dilke—mention to Brown that I wrote him a letter at Portmouth which I did not send and am in doubt if he ever will see it.

my dear Mrs Brawne
yours sincerely and affectionate
JOHN KEATS—

Good bye Fanny! god bless you

Letter to Charles Brown, 1 November 1820

Naples. Wednesday first in November.

My dear Brown,

Yesterday we were let out of Quarantine, during which my health suffered more from bad air and a stifled cabin than it had done the whole voyage. The fresh air revived me a little, and I hope I am well enough this morning to write to you a short calm letter;—if that can be called one, in which I am afraid to speak of what I would the fainest dwell upon. As I have gone thus far into it, I must go on a little;—perhaps it may relieve the load of WRETCHEDNESS which presses upon me. The persuasion that I shall see her no more will kill me. I cannot q°—— My dear Brown, I should have had her when I was in health, and I should have remained well. I can bear to die—I cannot bear to leave her. Oh, God! God! God! Every thing I have in my trunks that reminds me of her goes through me like a spear. The silk lining she put in my travelling cap scalds my head. My imagination is horribly vivid about her—I see her—I hear her. There is nothing in the world of sufficient interest to divert me from her a moment. This was the case when I was in England; I cannot recollect, without shuddering, the time that I was prisoner at Hunt's, and used to keep my eyes fixed on Hampstead all day. Then there was a good hope of seeing her again—Now!—O that I could be buried near where she lives! I am afraid to write to her—to receive a letter from her—to see her hand writing would break my heart—even to hear of her any how, to see her name written would be more than I can bear. My dear Brown, what am I to do? Where can I look for consolation or ease? If I had any chance of recovery, this passion would kill me. Indeed through the whole of my illness, both at your house and at Kentish Town, this fever has never ceased wearing me out. When you write to me, which you will do immediately, write to Rome (poste restante)—if she is well and happy, put a mark thus +,—if—Remember me to all. I will endeavour to bear my miseries patiently. A person in my state of health should not have such miseries to bear. Write a short note to my sister, saying you have heard from me. Severn is very well. If I were in better health I should urge your coming to Rome. I fear there is no one can give me any comfort. Is there any news of George? O, that something fortunate had ever happened to me or my brothers!—then I might hope,—but despair is forced upon me as a habit. My dear Brown, for my sake, be her advocate for ever. I cannot say a word about Naples; I do not feel at all concerned in the thousand novelties around me. I am afraid to write to her. I should like her to know that I do not forget her. Oh, Brown, I have coals of fire in my breast. It surprised me that the human heart is capable of containing and bearing so much misery. Was I born for this end? God bless her, and her mother, and my sister, and George, and his wife, and you, and all!

Your ever affectionate friend,
JOHN KEATS.

Letter to Charles Brown, 30 November 1820

Rome. 30 November 1820.

My dear Brown,

'Tis the most difficult thing in the world to me to write a letter. My stomach continues so bad, that I feel it worse on opening any book,—yet I am much better than I was in Quarantine. Then I am afraid to encounter the proing and conning of any thing interesting to me in England. I have an habitual feeling of my real life having past, and that I am leading a posthumous existence. God knows how it would have been—but it appears to me—however, I will not speak of that subject. I must have been at Bedhampton nearly at the time you were writing to me from Chichester—how unfortunate—and to pass on the river too! There was my star predominant! I cannot answer any thing in your letter, which followed me from Naples to Rome, because I am afraid to look it over again. I am so weak (in mind) that I cannot bear the sight of any hand writing of a friend I love so much as I do you. Yet I ride the little horse,°—and, at my worst, even in Quarantine, summoned up more puns, in a sort of desperation, in one week than in any year of my life. There is one thought enough to kill me—I have been well, healthy, alert &c, walking with her—and now—the knowledge of contrast, feeling for light and shade, all that information (primitive sense) necessary for a poem are great enemies to the recovery of the stomach. There, you rogue, I put you to the torture,—but you must bring your philosophy to bear—as I do mine, really—or how should I be able to live? Dr Clarke is very attentive to me; he says, there is very little the matter with my lungs, but my stomach, he says, is very bad. I am well disappointed in hearing goods news from George,—for it runs in my head we shall all die young.° I have not written to xxxxx yet, which he must think very neglectful; being anxious to send him a good account of my health, I have delayed it from week to week. If I recover, I will do all in my power to correct the mistakes made during sickness; and if I should not, all my faults will be forgiven. I shall write to xxx to-morrow, or next day. I will write to xxxxx in the middle of next week. Servern is very well, though he leads so dull a life with me. Remember me to all friends, and tell xxxx I should not have left London without taking leave of him, but from being so low in body and mind. Write to George as soon as you receive this, and tell him how I am, as far as you can guess;—and also a note to my sister—who walks about my imagination like a ghost—she is so like Tom. I can scarcely bid you good bye even in a letter. I always made an awkward bow.

God bless you!
JOHN KEATS.

APPENDIX I

St. Agnes' Eve

(from George Keats's transcript)

St. Agnes' Eve—Ah, bitter chill it was!
The owl, for all his Feathers, was a-cold;
The hare limp'd trembling thro' the frozen grass,
And silent was the flock in woolly fold:
Numb were the Beadsman's fingers, while he told
His rosary, and while his frosted breath
Like incense from a censer old,
Seem'd taking flight for heaven, without a death
Past the sweet Virgin's Picture, as his prayer he saith.

His prayer he saith, this patient, holy Man, 10
Then takes his lamp and riseth from his Knees
And back returneth, meagre, barefoot, wan,
Along the chapel aisle by slow degrees:
The sculptur'd dead, on each side, seem to freeze,
Emprison'd in black, purgatorial rails:
Knights, Ladies, praying in dumb Orat'ries,
He passeth by; and his weak spirit fails
To think how they may ache in icy hoods and mails.

Northward he turneth through a little door,
And scarce three steps, ere Music's golden tongue 20
Flatter'd to tears this aged Man and Poor;
But no—already had his deathbell rung;
The joys of all his life were said and sung:
His was harsh penance on St. Agnes' Eve:
Another way he went, and soon among
Rough ashes sat he for his Soul's reprieve,
And all night kept awake, for sinners' sake to grieve.

That ancient Beadsman heard the prelude soft;
And so it chanc'd, for many a door was wide,
From hurry to and fro. Soon, up aloft, 30
The silver, snarling trumpets 'gan to chide:
High-lamped chambers, ready with their pride
Were glowing to receive a thousand guests:
The carved angels, ever eager-eyed,
Star'd where upon their heads the cornice rests,
With hair blown back, and wings put cross-wise on their breasts.

At length burst in the argent revelry,
With tiara and plume and rich array,
Numerous as shadows haunting fairyly
The brain, new stuff'd, in youth, with triumphs gay 40
Of old romance. These let us wish away,
And turn sole-thoughted to one Lady there
Whose heart had brooded, all that wintry day,
On love and wing'd St. Agnes' saintly care,
As she had heard old Dames full many time declare.

They told her how upon St. Agnes' Eve
Young virgins might have visions of delight,
And soft adorings from their loves receive
Upon the honied middle of the night,
If ceremonies due they did aright; 50
As, supperless to bed they must retire,
And couch supine their beauties lily white;
Nor look behind nor sideways, but enquire
Of heaven with upward eyes for all that they desire.

'Twas said her future lord would there appear
Offering as sacrifice—all in the dream—
Delicious food even to her lips brought near;
Viands and wine and fruit and sugar'd cream,
To touch her palate with the fine extreme
Of relish: then soft music heard; and then 60
More pleasures followed in a dizzy stream
Palpable almost: then to wake again
Warm in the virgin morn, no weeping Magdalen.

Full of this whim was thoughtful Madeline:
The music, yearning like a god in pain
She scarcely heard: her maiden eyes divine,
Fix'd on the floor, saw many a sweeping train
Pass by—She heeded not at all—in vain
Came many a tiptoe, amorous Cavalier,
And back retir'd; not cool'd by high disdain, 70
But she saw not: her heart was otherwhere:
She sigh'd for Agnes' dreams the sweetest of the year.

She danc'd along with vague, uneager eyes,
Anxious her lips, her breathing quick and short.
The hallow'd hour was near at hand: she sighs
Amid the timbrels, and the throng'd resort
Of whisperers in anger and in sport;
Mid looks of love, defiance, hate, and scorn,
Hoodwink'd with faery fancy; a la mort,
Save to St. Agnes and her lambs unshorn, 80
And all the bliss to be before to-morrow morn.

So, purposing each moment to retire,
She linger'd still. Meantime, across the moors,
Had come young Porphyro, with heart afire
For Madeline. Beside the portal doors,
Buttress'd from moonlight, stands he, and implores
All saints to give him sight of Madeline,
But for one moment in the tedious hours,
That he might gaze and worship all unseen;
Perchance speak, kneel, touch, kiss—in sooth such things have
been. 90

He ventures in: let no buzz'd whisper tell:
All eyes be muffled, or a hundred swords
Will storm his heart—love's fev'rous citadel:
For him, those chambers held barbarian hordes,
Hyena foemen, and hot-blooded Lords,
Whose very dogs would execrations howl
Against his lineage: not one breast affords
Him any mercy, in that mansion foul,
Save one old Beldame, weak in body and in Soul.

Ah! happy chance! the aged creature came, 100
Shuffling along with ivory headed wand,
To where he stood, hid from the torches' flame,
Behind a broad hall pillar, far beyond
The sound of Merriment and chorus bland:
He startled her; but soon she knew his face,
And grasped his fingers in her palsied hand,
Saying, 'Mercy Jesu! hie thee from this place;
They are all here to-night, the whole blood thirsty race!

'Get hence! Get hence! there's dwarfish Hildebrand:
He had a fever late, and in the fit 110
He cursed thee and thine—both house and land:
Then there's that old Lord Maurice, not a whit
More tame for his gray hairs—Alas me! flit!
Flit like a Ghost away.'—'Ah Gossip dear,
We're safe enough; here in this arm chair sit,
And tell me how'—'Good Saints! not here, not here;
Follow me child or else these stones will be thy bier.'

He follow'd through a lowly arched way,
Brushing the cobwebs with his lofty plume,
And as she utter'd 'Wella—well-a-day!' 120
He found him in a little moonlight room,
Pale, lattic'd, chill, and silent as a tomb.
'Now tell me where is Madeline,' said he,
'O tell me Goody by the holy loom
Which none but secret sisterhood may see
When they St. Agnes' wool are weaving piously.'

'St. Agnes! Ah! it is St. Agnes' Eve—
Yet men will murder upon holy days:
Thou must hold water in a witch's sieve,
And be liege lord of all the Elves and Fays, 130
To venture so about these thorny ways
A tempting Be'lzebub—St. Agnes' Eve!
God's help! my Lady fair the conjuror plays
This very night: good Angels her deceive!
But let me laugh awhile, I've mickle time to grieve.'

Feebly she laugheth in the languid Moon,
While Porphyro upon her face doth look,
Like puzzled Urchin on an aged Crone,
Who keepeth clos'd a wond'rous riddle-book
As spectacled she sits in chimney nook. 140
Sudden his eyes grew brilliant, when she told
His Lady's purpose; and he scarce could brook
Sighs, at the thought of those enchantments cold:
Sweet Madeline asleep in lap of Legends old.

Sudden a thought came full blown like a rose,
Heated his brow, and in his pained heart
Made purple riot: then doth he propose
A stratagem, that makes the beldame start:
'A cruel Man and impious thou art;
Sweet Lady, let her pray, and sleep, and dream 150
Alone with her good angels, far apart
From wicked Men like thee. O Christ, I deem
Thou canst not surely be the same that thou didst seem.'

'I will not harm her: by the great St. Paul,'
Sweareth Porphyro; 'O may I ne'er find grace
When my weak voice shall unto heaven call,
If one of her soft ringlets I displace,
Or look with ruffian passion in her face:
Good Angela believe me by these tears:
Or I will, even in a Moment's space, 160
Awake, with horrid shout my foeman's ears,
And beard them, though they be more fang'd than wolves and bears.'

'How canst thou terrify a feeble Soul?
A poor, weak, palsy-stricken, churchyard thing,
Whose passing bell may ere the midnight toll;
Whose prayers for thee, each morn and evening,
Were never miss'd.'—Thus plaining doth she bring
A gentler speech from burning Porphyro;
So woeful, and of such deep sorrowing,
That Angela gives promise she will do 170
Whatever he shall wish, betide her weal or woe.

Which was, to lead him in close secrecy,
Even to Madeline's chamber, and there hide
Him in a closet, if such one there be;
That he might see her beauty unespied,
Or win perhaps that night a peerless bride,
While legion'd fairies pac'd the Coverlet,
And pale enchantment held her sleepy eyed.
Never on such a night have lovers met
Since Merlin pay'd his demon all the monstrous debt. 180

'It shall be as thou wishest,' said the Dame:
'All cates and dainties shall be stored there
Quickly on this feast-night:—by the tambour frame
Her own lute thou wilt see: no time to spare,
For I am slow and feeble, and scarce dare
On such a catering trust my dizzy head:
Wait here my child with patience: Kneel in prayer
The while: sooth thou must needs the Lady wed,
Or may I never leave my grave among the dead.'

So saying, she hobbled off with busy fear. 190
The lover's endless minutes quickly pass'd;
The dame return'd, and whisper'd in his ear
To follow her; with aged eyes agast
From fright of dim espial. Safe at last,
Through many a dusky gallery, they gain
The Maiden's chamber—silken, hush'd and chaste;
Where Porphyro took covert, pleas'd amain:
His poor guide hurried back with agues in her brain.

Her falt'ring hand upon the Ballustrade,
Old Angela was feeling for the stair, 200
When Madeline, St. Agnes' charmed Maid,
Rose like a spirit to her, unaware:
With silver taper light, and pious care
She turn'd, and down the aged gossip led
To a safe level matting. Now prepare,
Young Porphyro; a gazing on that Bed;
She comes, she comes again, like ring dove fray'd and fled.

Out went the taper as she floated in;
Its little smoke in pallid moonshine, died:
She clos'd the door, she panted, all a kin 210
To spirits of the air, and visions wide:
No uttered syllable, or, woe betide:
But to her heart, her heart was voluble,
Paining with eloquence her balmy side;
As though a tongueless nightingale should swell
Her throat in vain, and die, heart-stifled, in her dell.

A casement high and tripple-arch'd, there was,
All garlanded with carven imag'ries
Of fruits, and flowers, and bunches of knot-grass,
And diamonded with panes of quaint device; 220
Inumerable of stains and splendid dyes
As are the tiger moth's deep-damask'd wings;
And in the midst, 'mong thousand heraldries,
And twilight saints, and dim emblazonings
A shielded scutcheon blush'd with blood of Queens and Kings.

Full on this casement shone the wintry moon,
And threw warm gules on Madeline's fair breast,
As down she knelt for heaven's grace and boon:
Rose bloom fell on her hands together prest,
And on her silver cross soft amethyst, 230
And on her hair a glory, like a Saint:
She seem'd a splendid angel, newly drest,
Save wings, for heaven:—Porphyro grew faint:
She pray'd, too pure a thing, too free from mortal taint.

Anon his heart revives: her vespers done,
Of all its wreathed pearls her hair she frees;
Unclasps her warmed jewels one by one;
Loosens her fragrant boddice; by degrees
Her rich attire creeps rustling to her knees:
Half hidden, like a Mermaid in sea-weed,
Pensive awhile she dreams awake and sees
In fancy fair St. Agnes in her bed—
But dares not look behind or all the charm is fled.

Soon trembling in her soft and chilly nest,
In sort of wakeful swoon, perplex'd she lay,
Until the poppied warmth of sleep oppress'd
Her soothed limbs, and soul fatigued away:
Flown, like a thought, until the morrow day;
Blissfully haven'd both from joy and pain;
Clasp'd like a Missal where swart Paynims pray; 250
Blinded alike from sunshine and from rain,
As though a rose should shut and be a bud again.

Stol'n to this Paradise, and so entranc'd,
Porphyro gazed upon her empty dress,
And listen'd to her breathing if it chanced
To wake into a slumberous tenderness;
Which when he heard that minute did he bless
And breathed himself: then from the closet crept,
Noiseless as fear in a wide wilderness,
And over the hush'd carpet, silent, stept, 260
And 'tween the curtains peep'd, where, lo!—how fast she slept.

Then by the bed side, where the faded Moon
Made a dim silver twilight, soft he set
A table, and, half-anguished, threw thereon
A cloth of woven crimson, gold, and jet:—
O for some drowsy morphean amulet!
The boisterous, braying, festive clarion,
The kettle drum, and far-heard clarinet,
Affray his ears, though but in dying tone:—
The Hall door shuts again, and all the noise is gone. 270

And still she slept an azure-lidded sleep,
In blanched linen, smooth, and lavender'd,
While he, brought from the cabinet a heap
Of candied apple, quince, and plum, and gourd;
With jellies soother than the creamy curd,
And lucent syrops, tinct with cinnamon;
Manna and dates, in Argosy transferr'd
From Fez; and spiced dainties, every one,
From silken Samarcand to cedar'd Lebanon.

These delicates he heap'd with glowing hand 280
On golden dishes and in baskets bright
Of wreathed silver: sumptuous they stand
In the retired quiet of the night,
Filling the chilly room with perfume light.—
'And now my love my seraph fair awake!
Thou art my heaven, and I thine Eremite:
Open thine eyes for meek St. Agnes' sake,
Or I shall drowse beside thee, so my soul doth ache.'

Thus whispering, his warm, unnerved arm
Sunk in her pillow. Shaded was her dream 290
By the dusk curtains: 'twas a midnight charm
Impossible to melt as iced stream:
The lustrous salvers in the moonlight gleam;
Broad golden fringe upon the carpet lies:
It seem'd he never, never could redeem
From such a stedfast spell his Lady's eyes;
So mus'd awhile, entoil'd in woofed Phantasies.

Awakening up, he took her hollow lute,—
Tumultuous,—and, in chords that tenderest be,
He play'd an ancient ditty long since mute, 300
In provence call'd, 'La belle dame sans merci':
Close to her ear touching the melody;—
Wherewith disturbed, she uttered a soft moan:
He ceased—she panted quick,—and suddenly
Her blue affrayed eyes wide open shone:
Upon his knees he sunk, pale as fair-sculptur'd stone.

Her eyes were open, but she still beheld,
Now wide awake, the vision of her sleep:
There was a painful change, that nigh expell'd
The blisses of her dream so pure and deep: 310
At which fair Madeline began to weep,
And mourn forth witless words with many a sigh;
While still her gaze on Porphyro would keep;
Who knelt, with joined hands and piteous eyes,
Fearing to move or speak, she look'd so dreamingly.

'Ah, Porphyro!' said she, 'but even now
Thy voice was at sweet tremble in mine ear,
And tim'd, devout, with every softest vow;
And those sad eyes were spiritual and clear:
How chang'd thou art! how pallid, cold and drear! 320
Give me that voice again, my Porphyro,
Those looks immortal, those complainings dear!'
So while she speaks his arms encroaching slow
Have zon'd her, heart to heart—loud, loud the dark winds blow:

For on the midnight came a tempest fell.
More sooth for that his close rejoinder flows
Into her burning ear: and still the spell
Unbroken guards her in serene repose.
With her wild dream he mingled as a rose
Marryeth its odour to a violet. 330
Still, still she dreams, louder the frost wind blows,
Like love's alarum pattering the sharp sleet
Against the window panes: St. Agnes' moon hath set.

'Tis dark: still pattereth the flaw blown sleet:
'This is no dream my Bride, my Madeline!'
'Tis dark: the iced gusts still rave and beat:
'No dream—alas! alas! and woe is mine!
Porphyro will leave me here to fade and pine.—
Cruel—what traitor could thee hither bring?
I curse not, for my heart is lost in thine, 340
Though thou forsakest a deceived thing;—
A Dove forlorn and lost with sick unpruned wing.'

'My Madeline! sweet dreamer! lovely Bride!
Say, may I be, for aye thy vassal blest?
Thy beauty's shield, heart-shap'd and vermeil dyed?
Ah, silver shrine, here will I take my rest
After so many hours of toil and quest,
A famish'd Pilgrim,—saved by miracle.
Though I have found, I will not rob thy nest
Saving of thy sweet self; if thou think'st well 350
To trust, fair Madeline, to no rude Infidel.

'Hark! 'tis an elfin storm from faery land
Of haggard seeming, but a boon indeed:
Arise arise! the morning is at hand:—
The bloated wassaillers will never heed:—
Let us away, my love, with happy speed;
There are no ears to hear, or eyes to see,—
Drown'd all in Rhenish and the sleepy mead:
Awake! Arise! my love and fearless be
For o'er the southern Moors I have a home for thee.' 360

She hurried at his words, beset with fears,
For there were sleeping dragons all around,
At glaring watch, perhaps, with ready spears—
Down the wide stairs a darkling way they found.
In all the house was found no human sound.
A chain droop'd lamp was flickering by each door;
The Arras rich with horseman hawk and hound,
Fluttered in the besieging wind's uproar;
And the long carpets rose along the gusty floor.

They glide, like Phantoms, into the wide Hall; 370
Like Phantoms, to the iron porch, they glide;
Where lay the Porter, in uneasy sprawl,
With a huge empty beaker by his side:
The wakeful Bloodhound rose, and shook his hide
But his sagacious eye an inmate owns:
By one and one the bolts full easy slide:—
The chains lay silent on the footworn stones;—
The key turns, and the door upon its hinges groans—

And they are gone: aye ages long ago
These lovers fled away into the storm. 380
That night the Baron dreamt of many a woe,
And all his warrior-guests, with shade and form
Of witch, and demon, and large coffin worm,
Were long benightmar'd. Angela went off
Twitch'd with the Palsy; and with face deform
The beadsman stiffen'd, twixt a sigh and laugh
Ta'en sudden from his beads by one weak little cough.

APPENDIX II

La Belle Dame sans Mercy

(from the *Indicator*, 10 May 1820)

Ah, what can ail thee, wretched wight,
  Alone and palely loitering;
The sedge is wither'd from the lake,
  And no birds sing.

Ah, what can ail thee, wretched wight,
  So haggard and so woe-begone?
The squirrel's granary is full,
  And the harvest's done.

I see a lily on thy brow,
  With anguish moist and fever dew; 10
And on thy cheek a fading rose
  Fast withereth too.

I met a Lady in the mead
  Full beautiful, a fairy's child;
Her hair was long, her foot was light,
  And her eyes were wild.

I set her on my pacing steed,
  And nothing else saw all day long;
For sideways would she lean, and sing
  A fairy's song. 20

I made a garland for her head,
  And bracelets too, and fragrant zone:
She look'd at me as she did love,
  And made sweet moan.

She found me roots of relish sweet,
  And honey wild, and manna dew;
And sure in language strange she said,
  I love thee true.

She took me to her elfin grot,
  And there she gaz'd and sighed deep, 30
And there I shut her wild sad eyes—
  So kiss'd to sleep.

And there we slumber'd on the moss,
  And there I dream'd, ah woe betide,
The latest dream I ever dream'd
  On the cold hill side.

I saw pale kings, and princes too,
  Pale warriors, death-pale were they all;
Who cried, 'La belle dame sans mercy
  Hath thee in thrall!' 40

I saw their starv'd lips in the gloom
  With horrid warning gaped wide,
And I awoke, and found me here
  On the cold hill side.

And this is why I sojourn here
  Alone and palely loitering,
Though the sedge is wither'd from the lake,
  And no birds sing.

NOTES

POETRY

A list of Abbreviations will be found at the beginning of the book.

Imitation of Spenser

1 Composed probably early in 1814; published in *1817*. Brown calls this Keats's 'earliest attempt': 'It was the "Faery Queen" that awakened his genius. In Spenser's fairy land he was enchanted, breathed in a new world, and became another being; till, enamoured of the stanza, he attempted to imitate it, and succeeded' (*KC* ii. 56, 55).

Song: 'Stay, ruby breasted warbler, stay'

2 Composed '1814' (according to Georgiana Wylie's transcript on which this text is based); published in R. M. Milnes (ed.), *The Poetical Works of John Keats* (London 1876). According to Woodhouse 'This song was written at the request of some young ladies [the Mathew sisters] who were tired of singing the words printed with the air and desired fresh words to the same tune' (W3). The tune was by Reginald Spofforth.

'Fill for me a brimming Bowl'

3 Composed after 'obtaining a casual sight' of a woman at Vauxhall (W3) during August 1814; published by Ernest de Selincourt in *N&Q* 4 Feb. 1905, p. 81. Text from holograph fair copy (Morgan Library).

l. 26 *Arno*. The river that runs through Florence, where it is warmer than Lapland.

To Lord Byron

4 Composed Dec. 1814; published in *1848*; present text from W2.
For Keats's later thought on Byron see above, pp. 456 and 508.

Written on the Day That Mr. Leigh Hunt Left Prison

Composed 2 Feb. 1815; published in *1817*. Hunt (whom Keats had not met at this point) had been imprisoned for two years for libelling the Prince Regent in the *Examiner* (22 Mar. 1812, p. 179). Keats handed this poem to Cowden Clarke, who had introduced him to the *Examiner*. This was the first Cowden Clarke knew of Keats's poetry (see Ward, p. 41).

To Hope

5 Composed Feb. 1815; published in *1817*.

ll. 19–20. The Keats family had been broken up on the death of their grandmother in December 1814.

Ode to Apollo

6 Composed Feb. 1815; published in *1848*; present text from W3.

l. 12. Homer was traditionally blind.

7 l. 14. *Maro*: Virgil (Publius Virgilius Maro).

l. 17. *Aeneid* vi. 212–35.

l. 33. *FQ* iii tells the legend of Chastity and the whole work is dedicated to Elizabeth, the Virgin Queen.

l. 34. *Æolian lyre.* Wind-harp.

l. 36. *Tasso.* Keats owned and read Tasso's *Gerusamlemme Liberata* ('Jerusalem Delivered') in Edward Fairfax's translation of 1600, *Godfrey of Bouloigne.*

'Oh Chatterton! how very sad thy fate'

8 Composed 1815; published in *1848*; present text from W3. Keats dedicated *Endymion* to Chatterton who, by his suicide at the age of seventeen in 1770, came to represent all the abuse and neglect that poets may suffer. See Keats's remarks on pp. 493 and 515 above; also Coleridge's 'Monody on the Death of Chatterton' and Wordsworth's 'Resolution and Independence' 43–4. For a sterner view see Hazlitt, *Works*, v. 122.

l. 8. *amate.* The copy text notes 'affright—Spenser'. It is the kind of archaism used by Chatterton.

Lines Written on 29 May, the Anniversary of Charles's Restoration, on Hearing the Bells Ringing

Composed 29 May 1815; published 1925 (Lowell); present text from W3.

l. 5. Algernon Sidney (1622–83), Lord William Russell (1639–83), and Sir Henry Vane (1613–62) were executed for treason against Charles II and so were icons of republicanism to 19th-c. liberals. Cf. 'Oh! how I love, on a fair summer's eve' 9–10 and Keats's remarks on p. 424 above.

On receiving a curious Shell, and a Copy of Verses, from [*some*] *Ladies*

9 Composed summer 1815; published in *1817*. The poem is addressed to George Felton Mathew (his sisters being the 'Ladies' of the title). George Keats's transcript records that this was 'written on receiving a Copy of Tom Moore's "Golden Chain", and a most beautiful Dome shaped shell from a Lady'.

l. 1. *Golconda.* Hyderabad, known for diamonds.

l. 8. *Armida . . . Rinaldo.* Heroine and hero of Tasso's *Gerusalemme Liberata.*

l. 12. *Britomartis.* Heroine of *FQ* iii.

ll. 25–30. Keats knew William Sotheby's translation of Weiland's *Oberon* (1798) which dwells on the separation between Oberon and Titania more than does *A Midsummer Night's Dream.*

10 l. 41. *Eric.* Nickname for George Felton Mathew.

To George Felton Mathew

Composed Nov. 1815; published in *1817*. Sent as a verse letter to Mathew, perhaps in response to his 'To a Poetical Friend' (published in the *European Magazine*, Oct. 1816, p. 365).

l. 5. *brother Poets.* Beaumont and Fletcher.

11 l. 18. Milton, *L'Allegro* 136.

l. 43. *cassia.* Poeticism for fragrant shrub; Keats seems to mean honeysuckle (cf. 'Calidore' 96).

12 ll. 67–71. *Alfred . . . Tell . . . Wallace . . . Burns.* All patriot heroes (see above, Introduction, p. xxiii). Burns, as both poet and political activist, brings together the two classes of heroes already mentioned. Keats, in his enthusiasm for his new friend, ignores the fact that Mathew did not share his political views. Mathew later described Keats as 'of the sceptical and republican school. An advocate for the innovations which were making progress in his time. A faultfinder with everything established. I, on the other hand, hated controversy and dispute—dreaded discord and disorder—loved the institutions of my country' (*KC* ii. 186–7).

l. 75. *FQ* I. iii. 4.

l. 77. *the source.* The springs of Mount Helicon, sacred to the Muses.

ll. 84–7. Keats suggests a history for Mathew like those given in Ovid's *Metamorphoses.*

l. 93. *Kissing thy daily food.* As a fish.

'O Solitude! if I must with thee dwell'

13 Composed late 1815 or early 1816; published in the *Examiner* 5 May 1816, and in *1817*; present text from *1817*. This was Keats's first published poem and was written shortly after his move from Edmonton to London to study medicine.

Woman! when I behold thee flippant, vain

Date of composition uncertain—1815 or 1816 (see Bate, p. 40 n.); published in *1817*. Each stanza comprises a sonnet (leading some editors to print it as 3 distinct poems). It is an instance of Keats's testing the sonnet's—and the stanza's—capabilities.

ll. 12–13. *Calidore . . . Red Cross Knight . . . Leander.* 3 ardent lovers, the first two from *FQ* (Calidore's name suggests heat and picks up from 'I hotly burn').

*To ******: 'Had I a man's fair form, then might my sighs'*

14 Composed 1815 or 1816 (perhaps on 14 Feb. 1816 as a valentine for Mary Frogley; see 'Hadst thou liv'd in days of old'); published in *1817*. Woodhouse in his copy of *1817* notes that Keats is playing with the 'idea that the

diminutiveness of his size makes him contemptible and that no woman can like a man of small stature'. See Keats's comments on his height on p. 451 above. J. Burke Severs suggests that the speaker is a fairy (*KSJ* 6 (1957), 109–13).

'Give me women wine and snuff'

15 Composed late 1815 or early 1816; published in H. B. Forman (ed.), *The Poetical Works of John Keats* (London, 1884); present text from holograph in Trinity College, Cambridge. Keats's fellow medical student Henry Stephens recalls that:

> Whilst attending lectures [Keats] would sit & instead of Copying out the lecture, would often scribble some doggrell rhymes, among the Notes of Lecture . . . In my Syllabus of Chemical Lectures he scribbled many lines on the paper cover. This cover has been long torn off, except one small piece on which is the following fragment of Doggrel rhyme . . . (*KC* ii. 210)

l. 1. Perhaps suggested by Christopher Marlowe's reputed opinion that 'all they that loved not tobacco and boys were fools'.

l. 4. *resurrection*. This may be a medical student's joke since the men who dug up corpses for anatomies were known as 'resurrection men'.

'I am as brisk'

Composed probably Feb. 1816; published H. W. Garrod (ed.) *The Poetical works of John Keats* (Oxford, 1939); present text from holograph (Wisbech and Fenland Museum). It is written on the same sheet as 'Hadst thou liv'd in days of old'.

'O grant that like to Peter I'

Date of composition unknown; published in J. M. Murry, *The Poems and Verses of John Keats* (London, 1930); present text from holograph (Harvard).

*To ****: 'Hadst thou liv'd in days of old'*

16 Composed 14 Feb. 1816 'at his brother George's request and sent as a Valentine to a lady (Miss Frogley)' (W2). Published, with alterations, in *1817*.

ll. 3–34. Keats employs the medieval and Renaissance device of the 'blazon', in which a woman's beauties are itemized in sequence.

l. 6. *fane*. Temple.

l. 29. *little loves*. Putti; cf. Spenser's 'Epithalamion' 357–9.

l. 36. Cf. Shakespeare, *Sonnets* xxxviii. 9–10.

17 l. 40. *Graces four*. Instead of the usual three.

ll. 44–58. Mary Frogley is presented as Britomart, Spenser's woman knight of Chastity.

ll. 46, 57. *ivory . . . alabaster*. In Renaissance fashion Keats uses these terms to suggest colour and value but not substance.

l. 60. *northern lights*. The Aurora Borealis.

Specimen of an Induction to a Poem

Composed 1816, probably during spring; published in *1817*. The diction and metre recall Hunt's *The Story of Rimini* published in 1816.

18 l. 6. *Archimago*. Fashioner of false and deceptive images in *FQ*.

l. 38. *banneral*. Pennon.

l. 40. A heraldic device; the 'bloody field' is the red background.

19 l. 61. *Libertas*. 'Liberty'; Keats so names Leigh Hunt who had worn 'the chain for freedom's sake' ('Addressed to [Haydon]' (above, p. 44) 6).

Calidore: A Fragment

Composed 1816, probably during spring; published in *1817*. Calidore is the Knight of Courtesy and hero of *FQ* vi.

20 l. 19. *little boat*. Much used for imaginative exploration in the poetry of this time. Cf. Shelley's 'Alastor' (published Mar. 1816) and Wordsworth's 'Peter Bell' (composed 1798; published 1819).

l. 26. Cf. Chaucer, *Troilus and Criseyde* v. 1814–15 'And down from thennes faste he gan avyse | This litel spot of erthe'.

l. 28. *light blue mountains*. As in the backgrounds of many 15th- and 16th-c. paintings, both Italian and northern.

21 l. 49. *spiral*. Rising in a spire.

l. 67. *shallop*. Light boat.

22 l. 107. Cf. *Hamlet* II. ii. 299.

l. 119. *weed*. Clothing.

l. 122. *Gondibert*. The name comes from Sir William Davenant's epic *Gondibert* (1651); 'Clerimond' is a name invented by Keats.

23 l. 155. *incense*. Fragrance; cf. 'The Fall of Hyperion' 27.

'To one who has been long in city pent'

24 Composed June 1816; published in *1817*. 'Written in the fields' (G. Wylie's transcript) at a time when Keats was largely confined to London on account of his medical studies. There is an echo of *PL* ix. 444–59 ('Much he the place admired . . .| As one who long in populous city pent,| Where houses and thick sewers annoy the air' etc.)—a passage which Keats marked in his copy (though probably after he composed this poem). Cf. also Coleridge, 'This Lime-Tree Bower my Prison' 28–30 and 'To the Nightingale' 2.

'Oh! how I love, on a fair summer's eve'

Composed 1816, probably during summer; published in *1848*; present text from W2.

l. 10. Keats is thinking of Milton's republicanism and his combination of poetry and a political career; 'Sydney' is probably Algernon Sidney (see 'Lines written on the 29th May'), great-nephew of Sir Philip and staunch republican. He lived in poverty and exile after the Restoration until his execution. This poem may however conflate Algernon with Philip Sidney who shares with Milton the status of statesman-poet. See Keats's letter of 14 Oct. 1818 in which he compares modern statesmen with 'Milton and the two Sidneys' (above, p. 424).

l. 14. *spells*. Casts spells upon.

To My Brother George (sonnet)

25 Composed Aug. 1816; published in *1817*; Keats wrote this while on holiday at Margate after his July examinations.

ll. 1–3. Cf. Shakespeare, *Sonnets* xxxiii. 1–2.

ll. 10–11. Cf. Milton, 'On the Morning of Christ's Nativity' 229–31 'So when the sun in bed,| Curtained with cloudy red,| Pillows his chin upon an orient wave'.

ll. 13–14. Cf. Blake, *Marriage of Heaven and Hell*, i. 10 ('Where man is not, nature is barren'), and Keats's remark to Bailey (letter of 13 Mar. 1818; above, p. 382) 'Scenery is fine—but human nature is finer'.

To My Brother George (epistle)

Composed Aug. 1816 as a verse letter from Margate; published in *1817*.

l. 1. This echoes the sound of Shakespeare, *Sonnets* xxxiii.

l. 4. *sphery strains*. Music of the spheres.

26 l. 19. *the bay*. Poetry (rewarded by a laurel crown).

27 l. 66. *spell*. Enchant.

l. 78. Miltonic word order.

28 l. 130. Red-jacketed soldiers; see Keats's remarks about barracks in his letter to Reynolds of 17 Apr. 1817 (above, p. 350).

29 l. 141. *westward* towards London (where George was).

To Charles Cowden Clarke

Composed Sept. 1816 as a verse letter from Margate; published in *1817*. For Cowden Clarke see Index of Correspondents; the poem is a grateful acknowledgement of the mental worlds which he had opened for Keats.

l. 6. 'The "Naiad Zephyr" is a composite figure invented by Keats, not a synonym for the swan' (Barnard), 544.

29–30 ll. 29–31. *Baiæ's shore.* The Bay of Naples, home of Torquato Tasso whose work Cowden Clarke had introduced to Keats. *Armida*. The heroine of *Gerusalemme Liberata*.

30 ll. 33–7. Allusions to Spenser: *Mulla's stream* is the river near Spenser's home in Kilcolman, Ireland; 34 cf. 'Epithalamion' 175. *Belphoebe . . . Una . . . Archimago*. Figures from *FQ* i and ii.

l. 44. *Libertas*. Leigh Hunt whose *The Story of Rimini* is suggested in the next lines.

l. 56. *Spenserian vowels*. Cf. 'Lamia' 200; Bailey records that Keats had his own theory about 'the management of open & close vowels. . . . that the vowels should be so managed as not to clash with one another so as to mar the melody,—& yet that they should be interchanged, like differing notes of music to prevent monotony' (*KC* ii. 277).

ll. 69–71. Cf. 'To George Felton Mathew' 67–9.

31 ll. 110–12. Cowden Clarke was a good pianist and introduced Keats to much of the music that he loved. *Mozart*. See Keats's remark to George and Georgiana (letter of 14 Oct. 1818; above, p. 423) 'She kept me awake one Night as a tune of Mozart's might do'. Thomas *Arne* (1710–1778), an English composer who set many of Shakespeare's songs to music. *song of Erin*. Probably a reference to Thomas Moore's *Irish Melodies* (published in London and Dublin between 1808 and 1834).

32 l. 122. *bland*. Smooth (without pejorative sense).

On First Looking into Chapman's Homer

Composed Oct. 1816; published in the *Examiner*, 1 Dec. 1816 and in *1817* (the present text). Cowden Clarke recalls the 'memorable night' on which he introduced Keats to Chapman's translation of Homer and was rewarded by 'one of [Keats's] delighted stares' at the line 'The sea had soak'd his heart through' from the description of Ulysses' shipwreck (*Odyssey* v). Keats composed the sonnet immediately after this all-night session and delivered it to Cowden Clarke before breakfast: 'when I came down to breakfast the next morning, I found upon my table a letter with no other enclosure than his famous sonnet, "On First Looking into Chapman's Homer". We had parted . . . at day-spring, yet he contrived that I should receive the poem from a distance of, may be, two miles by ten o'clock' (*Recollections*, p. 130); Hunt's view was that this sonnet 'which terminates with so energetic a calmness . . . completely announced the new poet taking possession' (*Lord Byron and Some of his Contemporaries* (London, 1828), p. 248). George Chapman's *The Whole Works of Homer* was published in 1614. It is significant that Keats and most of his contemporaries (though not Byron) should have favoured this translation and not the more recent one by Pope.

l. 1. *of gold*. Rich for the imagination (cf. 'Isabella' 126).

l. 7. *serene*. A substantive, as in the Latin *serenum* ('a clear, bright, or serene sky').

l. 11. *Cortez.* In fact Balboa was the first European to sight the Pacific (Keats had read William Robertson's description of this in his *History of America* (Dublin 1777), i. 204: this was one of the school library books at Enfield).

'Keen, fitful gusts are whisp'ring here and there'

33 Composed Oct. or Nov. 1816 'very shortly after [Keats's] installation at [Leigh Hunt's] cottage' (*Recollections*, p. 134). Keats had first met Hunt in October. Published in *1817*.

l. 12. *drown'd.* Refers to 'gentle Lycid', Milton's friend Edward King, drowned 10 Aug. 1637 and the subject of 'Lycidas'.

ll. 13–14. There was a portrait of Petrarch and Laura (the woman to whom he dedicated his poetry but also, punningly, his laurel crown) at Hunt's cottage.

Sleep and Poetry

Composed between Oct. and Nov. 1816 'in the library at Hunt's cottage, where an extemporary bed had been made up for him on the sofa' (*Recollections*, pp. 133–4); published in *1817* as the final poem in the volume.

Motto 'The Floure and the Leafe' 17–21; this poem is no longer attributed to Chaucer (see 'This pleasant Tale is like a little Copse' (above, p. 55).

34 l. 33. *limning.* Painting, drawing; an orthographically good rhyme.

35 l. 74. *Meander.* A river in Asia Minor notorious for its windings (hence the verb).

35–6 ll. 85–95. This string of verbal emblems shows the influence of Keats's reading in Elizabethan and Jacobean literature.

l. 89. *Montmorenci.* A river in Quebec with a sheer waterfall.

36 ll. 96–154. Keats's programme for poetic and spiritual development owes much to Wordsworth's 'Lines written a few miles above Tintern Abbey'. See also Keats's letter to Reynolds of 3 May 1818 (above, pp. 395–6).

ll. 102–21. The visual details of this passage may owe something to Nicholas Poussin's *L'Empire de Flore.* This is the world of pagan pastoral.

l. 126. *car.* Chariot.

ll. 127–8. 'Personification of the Epic poet when the enthusiasm of inspiration is upon him' (Woodhouse in *1817*).

37 l. 157. *real.* A key word for Keats; see John Bayley, 'Keats and Reality', *Proceedings of the British Academy*, 48 (1962).

l. 158. The 'muddy stream' may owe something to the mire of effluent in which the dullards of Pope's *Dunciad* live (ll. 162–229). In Keats's account of English poetry the fulness of Elizabethan and Jacobean writing (ll. 171–180) was wilfully betrayed by the neoclassicists (ll. 181–206).

38 l. 168. *ether.* The clear sky (*OED*).

l. 172. *paragon*. A verb (so used by Shakespeare and Milton).

ll. 186–7. *rocking horse . . . Pegasus*. Keats is attacking the mechanical evenness of the heroic couplet. Hazlitt had used the image in the *Examiner* (20 Aug. 1815, p. 542): 'Dr Johnson and Pope would have converted [Milton's] vaulting Pegasus into a rocking-horse' (*Works*, iv. 40). Cf. Blake's anti-heroic-couplet (written in Chapmanesque fourteeners) 'Her whole life is an epigram, smack-smooth and neatly penned,| Plainted quite neat to catch applause, with a sliding noose at the end.'

l. 198. *certain wands of Jacob's wit*. Gen. 30: 37–42 (Jacob uses wands of poplar, hazel, and chestnut to control the breeding of his cattle at the expense of Laban).

l. 202. *the bright Lyrist*. Phoebus Apollo.

l. 206. *Boileau*. (1636–1711), author of the neoclassicist *Art Poétique*.

39 l. 209. *boundly*. A coinage (on analogy with (e.g.) 'goodly').

l. 218. *lone spirits*. 'alluding to H[enry] Kirke White [1785–1806], Chatterton—& other poets of great promise neglected by the age, who died young' (Woodhouse in *1817*).

ll. 221–2. Cf. Wordsworth, 'Ode: Intimations of Immortality' 136–7. This Wordsworthian echo prepares for the allusion to Wordsworth's Lakeland home in 226.

ll. 226–8. 'Leigh Hunt's poetry is here alluded to, in terms too favourable' (Woodhouse in *1817*).

ll. 230–5. Woodhouse takes this to be a reference to Coleridge; De Selincourt suggests Byron.

ll. 241–2. The attraction of power, irrespective of application, was much discussed at this time. Hazlitt writes: 'The sense of power is as strong a principle in the mind as the love of pleasure. Objects of terror and pity exercise the same despotic control over it as those of love or beauty' (*Works*, v. 7). Keats quotes Hazlitt on this subject in his letter of 14 Feb.–3 May 1819 (above, pp. 460–2).

ll. 242–5. Gothic romances and ballads were much in vogue.

40 l. 245. Cf. 'Ode on a Grecian Urn' 48 (above, p. 289).

l. 248. 'Allusion to the coming age of poetry under the type of a myrtle' (Woodhouse in *1817*).

l. 257. *Yeaned*. Brought forth, born (used of lambs and kids).

41 l. 303. *Dedalian wings* see Glossary q.v. *Daedalus*.

42 l. 322. *rout*. Usually a disorderly crowd, but here in the (not necessarily contradictory) sense of 'fashionable gathering'.

ll. 333–6. This image recalls Titian's *Bacchus and Ariadne*, a painting which Keats admired when it was exhibited at the British Institution in 1816 (see Jack, pp. 130–1).

l. 338. Hunt introduced Keats to many examples of visual art through his portfolio collections of engravings (the 19th-c. alternative to photographic plates). Keats describes some of the contents of Hunt's cottage in ll. 354–91. For suggestions about the works alluded to see Jack.

43 l. 379. *unshent*. Not disfigured.

l. 381. A bust of Sappho (b. Lesbos *c*.620 BC), woman poet and first great lyrist.

l. 387. *Kosciusko*. See note to 'To Kosciusko' (below, note to p. 45).

To My Brothers

44 Composed 18 Nov. 1816 (Tom Keats's seventeenth birthday); published in *1817*.

l. 8. *condoles*. Used transitively ('lore' is the subject, 'care' the object).

Addressed to [*Haydon*]

Composed 19 or 20 Nov. 1816 and enclosed in a letter of 20 Nov. to Haydon as apropos the previous evening ('Last Evening wrought me up and I cannot forbear sending you the following' (*L* i. 117); published in *1817*. Haydon promised to send a copy to Wordsworth (Keats wrote that 'the idea . . . put me out of breath' (*L* i. 118)) who thought it 'assuredly vigorously conceived and well expressed . . . and . . . very agreeably concluded' (20 Jan. 1817 in De Selincourt (ed.), *The Letters of William and Dorothy Wordsworth: The Later Years*, rev. Mary Moorman and Alan Hill (Oxford, 1970), iii. 361).

ll. 2–8. 2–4 refer to Wordsworth (Helvellyn is a mountain near his home in Grasmere), 5–6 to Leigh Hunt, and 7–8 to Haydon. Hunt had proclaimed Haydon the successor to Michelangelo and Raphael so *Raphael* is probably the painter (but he might possibly be the archangel who in *PL* explains the works of God to Adam and Eve).

45 l. 13. 'Of mighty Workings in a distant Mart' (*L*); the mute half line seems to have been Haydon's suggestion (*L*. i. 118).

To G.A.W.

Composed, possibly (as Woodhouse notes) at George Keats's instigation, in Dec. 1816. Georgiana Augusta Wylie married George in May 1818. Published in *1817*.

To Kosciusko

Composed Dec. 1816; published in the *Examiner*, 16 Feb. 1817 and in *1817*. Tadeusz Kosciusko (1746–1817) was a Polish patriot. Having fought with Washington in the American Army in 1776 (and been made an honorary American citizen) he returned to Europe to find his native Poland's independence compromised and threatened. In 1794 he led a patriot insurrection against the Russians and defeated the Russian army at Raclawice. Later in the same year he was defeated and taken as a prisoner to Russia. In 1798 he settled in France where he resisted Napoleon's attempts to make Poland a pawn whilst enlisting his support for the cause of independence. Kosciusko was a hero to many English liberals. Leigh Hunt kept a bust of him in his cottage and Coleridge included a sonnet on him in his *Sonnets on Eminent Characters* (1774). In one of his lectures Astley Cooper described the division of the sciatic nerve in a wound of Kosciusko (A. Ward, *John Keats: The Making of a Poet* (New York, 1963), p. 52).

l. 7. *And change.* Are changed (*Examiner*); And changed (*1817*).

'I stood tip-toe upon a little hill'

46 Completed Dec. 1816; published in *1817* as the first poem in the volume. The poem was originally titled, and referred to in letters as, 'Endymion' (ll. 113–24 and 181–210 explain why).

Motto: Leigh Hunt, *The Story of Rimini* iii. 430.

47 ll. 47–52. Rhythmically and grammatically these lines recall Wordsworth's 'Ode: Intimations of Immortality' and anticipate Keats's later development of the ode form.

47–9 ll. 61–106. According to Cowden Clarke this passage recalls 'our having frequently loitered over the rail of a foot-bridge that spanned . . . a little brook in the last field upon entering Edmonton' (*Recollections*, p. 138).

48 l. 89. *sleek.* A verb.

49–51 ll. 125–204. This passage, which traces the origins of poetry to an animistic response to nature owes much to Wordsworth, *The Excursion* iv. 845–81 and to *A Midsummer Night's Dream* v. i. 11–22.

49 l. 134. *vases.* To rhyme with *faces*; see 'Hadst thou liv'd in days of old' 55–6 (above, p. 17).

50 l. 147. *lamp.* By which Psyche illicitly glimpsed, and thereby forfeited, her divine lover.

50–1 ll. 181–204. See *Endymion*; these lines show that Keats understood this myth as a key to the nature of poetry and go far to explain his very extensive treatment of it in *Endymion*.

51 l. 218. The Apollo Belvedere; the Venus de Medici is suggested in 220.

Written in Disgust of Vulgar Superstition

52 'Written in 15 minutes' (according to note on autograph draft) on 22 Dec. 1816; published in R. M. Milnes (ed.), *The Poetical Works of John Keats* (London, 1876); present text from Tom Keats's transcript. The first line seems to echo the opening of Gray's 'Elegy in a Country Churchyard'.

l. 7. *Lydian airs*. Milton, 'L'Allegro' 136.

ll. 12–13. These lines recall the mournful and complaining departure of the routed pagan gods in Milton's 'On the Morning of Christ's Nativity' 172–96.

On the Grasshopper and Cricket

53 Composed 30 Dec. 1816. 'The author and Leigh Hunt challenged each other to write a sonnet in a quarter of an hour [on this subject]. . . . Both performed the task within the time allotted' (Woodhouse in *1817*). 'Keats won as to time' (*Recollections*, p. 135). Published in *1817*. At least three other poems were composed in competition with Hunt. The previous poem may also have been composed under these circumstances—or the speed of its composition may have suggested the idea of the contest.

Title: perhaps suggested by La Fontaine's 'Le Cigale et la Fourni' (The Grasshopper and the *Ant*).

'*God of the golden bow*'

Composed late 1816 or early 1817; published in the *Western Messenger* 1 June 1836; present text from Keats's fair copy (Morgan Library). Woodhouse describes the occasion in a note in W2:

> As Keats & Leigh Hunt were taking their wine together after dinner, at the house of the latter, the whim seized them (probably at Hunt's instigation) to crown themselves with laurel after the fashion of the elder Bards.—While they were thus attired, two of Hunt's friends happened to call upon him—Just before their entrance H. removed the wreath from his own brows, and suggested to K. that he might as well do the same. K. however in his mad enthusiastic way, vowed that he would not take off his crown for any human being: and he accordingly wore it, without any explanation, as long as the visit lasted.—// He mentioned the circumstance afterwards to some of his friends, along with his sense of the folly (and I believe presumption) of his conduct—And he said he was determined to record it, by an apologetic Ode to Apollo on the occasion—He shortly after wrote this fragment.

See also Keats's letter to Bailey of 8 Oct. 1817 (above, p. 361) in which he quotes from a letter he had written to George in the spring: 'I hope Apollo is not angered at my having made a Mockery of him at Hunt's.' In his contrition on this occasion Keats wrote 3 poems (the other 2 not in the present volume).

54 l. 23. *germ*. Seed (Lat. *germen*).

'*After dark vapors have oppress'd our plains*'

Composed 31 Jan. 1817; published in the *Examiner*, 23 Feb. 1817 (the present text).

l. 14. *a poet's death*. 'I always somehow associate Chatterton with autumn' (Keats to Reynolds, 21 Sept. 1819; p. 493 above).

'*This pleasant Tale is like a little Copse*'

Composed Feb. 1817 'without the alteration of a single word . . . [it] was an extempore effusion' (*Recollections*, p. 139). Keats wrote the poem inside Cowden Clarke's copy of *The Poetical Works of Geoffrey Chaucer* (London,

1782), vol. xii while Cowden Clarke 'lay asleep on the sofa' (*KC* ii. 170). Published in the *Examiner*, 16 Mar. 1817 (the present text). The 'pleasant tale' is *The Floure and the Leaf*; see note to Motto to 'Sleep and Poetry' (above, note to p. 33).

ll. 1–8. The transposition of poetic narrative into spatial terms is characteristic of medieval and Renaissance poetry. Cf. Keats to Bailey (8 Oct. 1817; above, p. 361). 'Do not the Lovers of Poetry like to have a little Region to wander in where they may pick and choose'.

ll. 13–14. An allusion to the Babes in the Wood who are given a blanket of leaves by the robins.

To Leigh Hunt Esq.

Composed Feb. 1817 when the final proof sheets of *1817* had arrived from the printer. Keats 'withdrew to a side-table . . . [and] in the buzz of mixed conversation, he composed . . . the Dedication Sonnet' (*Recollections*, p. 138). Published, as 'Dedication', in the front of *1817*.

l. 1. Echoes Wordsworth, 'Ode: Intimations of Immortality' 18 'There hath passed away a glory from the earth'.

ll. 5–8. Jack suggests Poussin's *The Triumph of Flora* as a stimulus to these lines (p. 117, pl. x).

On seeing the Elgin Marbles

56 Composed 1 or 2 Mar. 1817 after visiting the British Museum with Haydon to see the Parthenon frieze which Lord Elgin had recently acquired for the nation. According to Severn Keats went 'again and again to see the Elgin Marbles, and would sit for an hour or more at a time beside them rapt in revery' (William Sharp, *Life and Letters of Joseph Severn* (London, 1892), p. 23). Published in both the *Champion* and the *Examiner* on 9 Mar. 1817, and in *Annals of the Fine Arts*, 3 (Apr. 1818; the present text).

l. 5. Haydon, to whom this and another sonnet on the subject were sent, comments 'I know not a finer image' (*L* i. 122); eagles were supposedly able to look unblinking at the sun.

l. 10. *Hamlet* v. ii. 206 'how ill all's here about my heart'.

On a Leander which Miss Reynolds my kind friend gave me

Composed probably during Mar. 1817; published in the *Gem*, 1829, repr. in [anon. ed.], *The Poetical Works of Coleridge, Shelley, and Keats* (Galignani, Paris, 1829); present text from Keats's draft (Harvard). A 'Leander' was one of James Tassie's paste reproductions of gems engraved with classical scenes. This one depicted Leander swimming across the Hellespont to Hero (see Jack, pl. ix*b*). Woodhouse records Keats's intention at one time to write 'a series of sonnets & short poems on some of these gems' (*W2*). At some point in 1819 Keats gave his sister a set of Tassie's cameo gems (see above, p. 447).

On the Sea

57 Composed probably on 17 Apr. 1817 while Keats was lodging at Carisbrooke on the Isle of Wight. A copy was included in Keats's letter to Reynolds of 17–18 Apr. 1817 (above, p. 350). Published in the *Champion*, 17 Aug. 1817 (the first poem of Keats's to be printed in a periodical other than the *Examiner*).

l. 3. Hecate's 'spell' is her influence over the tides.

l. 9. *vex'd.* A latinate use (from *vexare* to shake, agitate).

l. 14. *if.* Woodhouse's clerk's transcript in letter-book; omitted, probably in error, from the *Champion*.

'*Unfelt unheard unseen*'

Composed before 17 Aug. 1817 when l. 9 was quoted by Reynolds in the *Champion*; published in *1848*; present text from Keats's fair copy (Morgan Library). The quotation in l. 12 has not been identified.

'*You say you love; but with a voice*'

58 Date of composition unknown. There is a speculative link between this, the previous, and the following poems with Isabella Jones. Published *TLS* 16 Apr. 1914. Text from Charlotte Reynolds's transcript. Sidney Colvin suggests an echo of the Elizabethan 'A Proper Wooing Song', reprinted by Thomas Park (a Hampstead neighbour) in *Heliconia* (London, 1815), ii. 76 (S. Colvin, *John Keats* (London, 1917), pp. 157–8).

'*Hither hither Love*'

59 Date of composition unknown; published in the *Ladies' Home Companion* (New York), 1837; reprinted in Lowell. Present text from holograph (Yale). J. H. Payne, who was given the manuscript by George Keats and who was the first to print the poem, described his source as 'one of those unpremeditated effusions in the handwriting of John Keats, just scribbled as if playing with his pen'.

Endymion

Composed between late Apr. 1817 and 28 Nov. 1818; revised and recopied between Jan. and Mar. 1818; published, as separate volume, in 1818. Keats began work on the poem while at Carisbrooke with a clear, workmanlike programme for its completion: see i. 39–77 and the spring letter to George from which he quotes to Bailey (8 Oct. 1817; above, p. 361) in which the poem is described as 'a test, a trial of my Powers of Imagination and chiefly of my invention . . . by which I must make 4000 Lines of one bare circumstance and fill them with Poetry'. Shelley's cousin Medwin (whose evidence is unreliable) stated that Shelley and Keats had each agreed to write the long poems they were planning in 1817 in a spirit of friendly rivalry (S. Colvin, *John Keats* (London, 1917), p. 73); but Shelley's *Alastor* was a more important poetic provocation to *Endymion* than the projected *Laon and Cythna*. As both the published and rejected Prefaces show, Keats thought of *Endymion* rather as 'an endeavour than a thing accomplished' (above, p. 348), seeing the poem as a necessary but flawed apprentice piece. In a letter to Haydon (28 Sept. 1817) written soon after the completion of Book iii Keats admits 'I would write the subject thoroughly again, but I am tired of it' (*L* i. 168). See also his letter to Shelley of 16 Aug. 1820 (above, p. 535) in which he refers to the mind of *Endymion*'s author as 'like a pack of scattered cards'. The reviews in the Tory press (notably *Blackwood's*, Aug. 1818; the *British Critic*, June 1818, and *QR* Apr. 1818) were not favourable. It is they that are alluded to in the sentimental myth that hostile criticism was the cause of Keats's early death (see Keats's letter to Hessey of 8 Oct. 1818; above, p. 417–18). A favourable review—possibly by Haydon—in the *London Magazine* (Apr. 1820) initiated a widespread reaction against the Tory attacks on the poem, and Keats is identified on the title-page of *1820* as 'Author of Endymion'. The 'bare circumstance' of which Keats was to make 4,000 lines is the legend of Endymion (see Glossary) which he had adumbrated in 'I stood tip-toe' 181–94.

Motto to Preface: misquoted from Shakespeare, *Sonnets* xvii. 12.

61 *Books i and ii*: drafted between late Apr. and the end of Aug. 1817. During this time Keats stayed at Carisbrooke, Margate, Canterbury, Hastings, and London.

62 l. 47. *little boat*. See note to 'Calidore' 19.

64 l. 129. *goodly company*. Recalls the 'compaignye' of Chaucer's pilgrims; Keats left for Canterbury (l. 134) on 17 May.

l. 136. *burden*. Refrain (though 'bearing' makes a pun on the usual sense).

ll. 141–4. Apollo spent a period of exile living as a Thessalonian shepherd (Ovid, *Metamorphoses* ii. 677–82).

65 l. 158. *Leda's love*. Jupiter disguised as a swan; cf. Spenser, 'Prothalamion' 43–4.

66 l. 208. *scrip*. Satchel (cf. *FQ* I. vi. 35).

ll. 232–306. This stanzaic 'Hymn to Pan' is a working towards the form of the later odes. Shelley saw in it a 'promise of ultimate excellence' (G. Matthews (ed.), *Keats: The Critical Heritage* (London, 1971), p. 124); however when Keats, encouraged by Haydon, recited it to Wordsworth 'in his usual half-chant . . . walking up & down', Wordsworth's only comment was that it was 'a very pretty piece of paganism'. According to Haydon, not one to play down emotional drama, Keats 'felt it *deeply* . . . and though he dined with Wordsworth after at my table—he never forgave him' (*KC* ii. 143–4).

67 l. 248. *Passion*. A verb; John Croker singles this out for censure in *QR* (Apr. 1818).

68 l. 283. *huntsmen*. Fair copy; huntsman *Endymion*.

l. 285. 'All the strange, mysterious and unaccountable sounds which were heard in solitary places, were attributed to Pan' (Baldwin, *Pantheon*, p. 105).

69 l. 320. *genitors*. Progenitors.

ll. 327–31. Lemprière, but not Ovid, mentions Zephyr's role in the death of Narcissus.

l. 334. *raft*. An archaic form of 'reft' (torn); useful for the rhyme.

l. 347. Forman and De Selincourt suggest that Keats got the idea of Apollo appearing to the Argonauts at sea from Shelley since the incident—a favourite with Shelley—does not appear in any of Keats's usual sources of classical knowledge.

70 l. 392. *famish'd scrips*. Depleted lunch-boxes.

71 ll. 405–6. Probably 'The History of the Young King of the Black Isles' in the *Arabian Nights*, in which a young man appears whose lower half is marble.

l. 408. *Peona*. Keats's creation.

72 l. 460. *mazy world. A Midsummer Night's Dream* II. i. 113 ('mazed world').

ll. 466–7. 'He said, Dear Maid, may I this moment die,| If I feel not this thine endearing Love' (fair copy); see Keats's letter to George and Tom of 23–4 Jan. 1818 (above, p. 373) in which he relates Hunt's objection that this conversation 'is unnatural . . . for the Brother & Sister'.

73 l. 510. *Paphian*. From Venus' temple at Paphos.

74 ll. 512–14. Ovid, *Metamorphoses* iii. 155–252.

l. 517. *bland*. Mild.

75 l. 555. *ditamy, and poppies*. Plants sacred to Diana.

ll. 574–8. Cf. *A Midsummer Night's Dream* IV. i. 204–16 (Bottom's unutterable dream).

76 l. 614. *gordian'd*. Knotted—Keats's own coinage; cf. iii. 494.

77 ll. 648–50. Cf. *Measure for Measure* III. i. 123–5.

l. 657. *count*. A serious play on 'cunt'.

78 l. 683. *dew-dabbled. A Midsummer Night's Dream* III. ii. 443; *Venus and Adonis* 703. *ouzel*. Blackbird (also in *A Midsummer Night's Dream* III. i. 125).

79 l. 726. *bard*. A metonym for what the bards will sing?

80 ll. 776–81. 'To fret at myriads of earthly wrecks,| Wherein lies happiness? In that which becks | Our ready minds to blending pleasurable:| And that delight is the most treasurable | That makes the richest Alchymy. Behold | The clear Religion of Heaven! Fold' (fair copy). See Keats's letter to Taylor of 30 Jan. 1818 (above, p. 376) for his comments on this passage.

l. 786. *Eolian*. As in the music of an 'Aeolian' wind-harp.

l. 792. The war between the Titans and the Olympians—the starting point of *Hyperion*.

81 l. 815. *pelican*. The pelican was fabled to wound its breast in order to feed its young with its own blood. As such it is a type of Christ.

ll. 832–42. The churchman Bailey censured Keats's inclination 'to that abominable principle of *Shelley's*—that *Sensual Love* is the principle of *things*. Of this I believe him to be unconscious, & can see how by a process of imagination he might arrive at so false, delusive, & dangerous conclusion' (*KC* i. 35). According to T. Medwin, Byron's summary of what 'Keats somewhere says' was 'that "flowers would not blow, leaves bud" &c, if man and woman did not kiss. How sentimental!' (*Conversations of Lord Byron* (London, 1824), p. 239).

83 l. 907. *sloth*. Sloths are in fact herbivorous, but Keats puns on the sense of 'idleness' to make sloth a deadly predator.

84 l. 924. *amber studs*. Marlowe, 'The Passionate Shepherd to his Love' 17–18.

Book ii

86 Completed *c.*28 Aug. 1817 at Hampstead.

l. 13. *close*. Embrace; used by Shakespeare as a noun (e.g. *Twelfth Night* v. i. 158) and as a verb (*Troilus and Cressida* III. ii. 49).

l. 23. Plutarch, in his *Life of Themistocles*, tells of the appearance of an owl on the ship's mast when Themistocles was trying to persuade his reluctant officers to do battle, 'and that hereupon all the other Grecians dyd agree to his opinion and prepared to fight by sea' (*The Lives of the Noble Grecians and Romans*, trans. Thomas North (Oxford, 1928), i. 316–17).

ll. 24–5. Alexander crossed the Indus with his army in 326 BC.

ll. 31–2. *Hero . . . Imogen . . . Pastorella*. Thwarted lovers—from Marlowe's *Hero and Leander* (or possibly Shakespeare's *Much Ado About Nothing*), *Cymbeline*, and *FQ* VI. xi.

ll. 38–40. Cf. George Herbert, 'The Pulley' 16–17 'Yet let him keep the rest,| But keep them with repining restlessnesse'.

87 l. 60. *pight*. Archaic form of 'pitched'.

88 ll. 90–1. Cf. *Troilus and Cressida* III. iii. 78–9 'men, like butterflies,| Show not their mealy wings but to the summer'; see also *Endymion* ii. 996.

l. 118. *Meander*. See 'Sleep and Poetry' 74 n. (above, p. 564).

89 l. 138. *burr*. 'a nebulous or nimbous disk of light enfolding' the moon (*OED*).

90 ll. 186–95. In Drayton's *The Man in the Moone* 430–5 Endymion rides through the air in Phoebe's chariot.

91 ll. 211–14. These lines may reflect Keats's recent reading of Dante's *Divine Comedy* in Cary's translation.

l. 230. *antre*. Cave (*Othello* I. iii. 140).

ll. 230–1. Vulcan was blacksmith to the gods and as such capable of fashioning a cast-iron woven rainbow.

92 l. 251. *sphering time*. Time taken to circle the sun.

l. 277. *fog-born elf*. Wil-o'-the-wisp; cf. *King Lear* III. iv. 51–3.

93 l. 282. *raught*. Reached.

l. 287. Perhaps a description of wild fritillaries, whose petals are so chequered.

95 l. 362. *lyre*. Unrhymed because of revisions. In the draft l. 363 continues 'Dire | Was the lovelorn despair to which it wrought | Endymion'.

l. 373. *the Carian*. Endymion.

ll. 389–427. Some of the particulars of this bower of Adonis come from *FQ* III. vi.

96 l. 407. *Disparts*. Opens up (Spenser).

l. 417. *virgin's bower*. The plant old man's beard.

97 ll. 458–80. This version of the Venus and Adonis story is based largely on Shakespeare's poem (in which Keats delighted—see p. 368 above), *FQ* III. i. 34–8 and vi. 46–9, and Ovid's *Metamorphoses* x. 708–39. There are echoes of Shakespeare's poem in ll. 474 and 490.

99 l. 535. *Love's self*. Cupid. *superb*. Proud (a Latinism).

l. 537. *quell*. Means to quell—i.e. his bow.

100 l. 569. *zoned*. Clasped; Keats's own coinage from 'zone' (girdle or belt); used in the rejected sexually explicit lines of 'St. Agnes Eve' (see p. 553 l. 324 above).

ll. 579–84. Cowden Clarke recalls 'that Sunday afternoon, when [Keats] read to Mr. Severn and myself the description of the "Bower of Adonis"; and the conscious pleasure with which he looked up when he came to the passage that tells of the ascent of the car of Venus' (*KC* ii. 151).

102 l. 658. *eagle*. Emblem of Jupiter.

l. 674. *Hesperean*. Perhaps meaning 'westward' (where Hesperus is); but more likely 'Hesperidean' is meant, i.e. 'as in the garden of the Hesperides'.

103 ll. 689–90. The Pleiades.

105 l. 761. *dov'd Ida*. The mountain near Troy where Paris judged Venus to be the fairest; here a metonym for Venus.

106 ll. 823–4. The perception behind 'Ode on Melancholy' (above, p. 290).

ll. 830–9. This aetiology of myth is influenced by Wordsworth's account in *The Excursion*; see 'I stood tip-toe upon a little hill' 125–204 n. (above, p. 567).

107 l. 854. *former chroniclers*. Earlier historians of this tale (e.g. Drayton); this

technique (which Keats uses again in 'Isabella') of referring back to his 'authors' is Chaucerian.

l. 866. *Eolian tuned.* Tuned like an Eolian harp to be sensitive to the slightest stir or shift of air; see 'Ode to Apollo' 34 n. (above, p. 558).

108 ll. 885–912. A poetic adumbration of Keats's conception of 'The vale of Soul-making' (above, p. 473).

109 ll. 938–48. Keats's treatment of this Ovidian myth is itself Ovidian (cf. Ovid, *Amores* II. xv). This mode of imaginative sympathy was much used by 16th- and 17th-c. writers. Cf. Marlowe's 'Hero and Leander' ii. 183–92 and Lovelace's 'On Lucasta taking the Waters at Tunbridge'.

110 l. 961. *Oread-Queen.* Diana.

Book iii

111 Composed during Sept. 1817—mostly while staying with Bailey at Magdalen Hall, Oxford. In a letter to Haydon of 28 Sept. 1817 Keats records 'within these last three weeks I have written 1000 lines—which are the third Book of my Poem. My Ideas with respect to it I assure you are very low—and I would write the subject thoroughly again. but I am tired of it and think the time would be better spent in writing a new Romance which I have in my eye for next summer' (*L* i. 168).

ll. 1–21. Described as a 'jacobinical apostrophe' by the reviewer in the *British Critic* (Sept. 1818). Woodhouse notes in his copy of *Endymion* 'K said, with much simplicity, "It will easily be seen what I think of the present Ministers by the beginning of the 3rd book"'. Bailey later wrote:

> he had written the first few introductory lines . . . before he became my guest. I did not then, and I cannot now very much approve that introduction. The 'baaing vanities' have something of the character of what was called 'the cockney school'. Nor do I like the forced rhymes, & the apparent effort, by breaking up the lines, to get as far as possible in the opposite direction of the Pope school. (*KC* ii. 269)

ll. 7–8. *fire-branded foxes.* Judg. 15: 4–5.

l. 11. Hunt had recently pointed to the French clergy's acceptance of cardinals' hats as symbols of tyranny: 'The Roman purple . . . the garb of the Antonines,—and of the Neros!' (*Examiner*, 31 Aug. 1817).

112 ll. 16–18. Perhaps recalling the loud festivities after Napoleon's abdication (6 Apr. 1814).

114 ll. 97–9. Each braved a hostile element for love's sake (Leander water, Orpheus the underworld, Pluto the upper air).

ll. 119–36. Cf. *Richard III* I. IV. 22–6 (Clarence's fearful dream of drowning).

116 l. 192. *an old man.* Glaucus.

117 ll. 222–5. 'I remember his upward look when he read of the "magic

ploughs"' (Bailey, *KC* ii. 271). There are parallels between Keats's Glaucus and the leech-gatherer in Wordsworth's 'Resolution and Independence' (published 1807).

l. 234. *Thou art the man.* 2 Sam. 12: 7.

l. 243. *that giant.* Typhon.

119 ll. 318–638. The story of Glaucus and Scylla is based on Ovid's *Metamorphoses* (xiii. 890–968, xiv. 1–74).

121 l. 406. Mount Oetna, where Hercules burnt himself on a pyre.

122 l. 414. *Phoebus' daughter.* Circe.

123 ll. 449–72. In Ovid's account Glaucus resists Circe who consequently turns him into a monster; Scylla is also turned into a monster whereas here (621–35) she falls into a trance. Keats's version makes Glaucus more like Endymion, who is apparently unfaithful to his love.

124 l. 504. *penny pelf.* Keats has anglicized Charon's two-obol fee for ferrying souls. (Pennies were used to close the eyes of English dead.)

125 ll. 551–2. Cf. *Hamlet* I. ii. 129.

126 ll. 571–9. Perhaps recalls the tone of some of the melodramas at Drury Lane.

ll. 590–9. Cf. Lyly's *Endymion* II. iii. 29–36.

128 l. 685. *Atlas-line.* So called because of its great burden of meaning: cf. 'To Charles Cowden Clarke' 63 (above, p. 30).

129 l. 703. *tempest-tost. Macbeth* I. iii. 25.

l. 706. *Macbeth* V. v. 19–21.

130 ll. 728–44. Cf. the description of Satan's host in *PL* i. 544–67.

133 l. 845. *mere.* Pure.

ll. 882–7. The palace floor is only distinguishable from air by the reflections it gives.

137 l. 1000. *Egean seer.* Nereus.

Book iv

138 Composed between *c.*5 Oct. and 28 Nov. 1817. The book was begun at Hampstead and completed at Burford Bridge, near Boxhill in Surrey, where Keats went on 21 Nov.

ll. 1–29. Keats copied these lines into his letter to Bailey of 28–30 Oct. 1817 (*L.* i. 171–2) his recent reading of *PL* may have inspired the invocation to the Muse.

l. 10. *eastern voice.* The Old Testament (the letter draft has 'hebrew voice').

l. 15. *Ausonia.* Ancient name for Italy.

l. 26. *shrives.* Confesses (by metonymy from its proper meaning of 'grants absolution').

l. 27. *poets gone*. Burns and Chatterton?

141 l. 129. *gorgon*. Petrifying.

ll. 146–81. See Keats's comments on this 'Ode to Sorrow' in his letter to Bailey of 22 Nov. 1817 (above, p. 365).

142 l. 157. *spry*. Spray.

ll. 167–8. Echoes Sabrina's song in Milton's 'Comus' (ll. 897–9).

ll. 182–3. Cf. Ps. 137 :1: 'By the rivers of Babylon there we sat down, yea, we wept when we remembered Zion.'

ll. 186. Cf. Milton's 'Lycidas' 150.

143 ll. 193–272. The particulars of Keats's description of Bacchus' triumphal progress are from various sources, including Titian's *Bacchus and Ariadne* (see *Finney*, pp. 276–91). Milton's 'On the Morning of Christ's Nativity' is an informing presence (cf. ll. 265–7 with Milton's l. 191).

l. 203. Bacchus was also called Lyaeus, 'the deliverer from care'.

144 ll. 251–2. The couplet has a Popish ring and 252 echoes 'An Essay in Criticism' 372 'Swift Camilla scours the plains'.

147 ll. 356–8. Rom. 8: 39: 'Nor height, nor depth, nor any other creature, shall be able to separate us from the love of God.'

148 l. 394. *Skiddaw*. A mountain in the Lake District, known to Keats through his reading of Wordsworth.

149 l. 441–2. Icarus.

150 l. 459. *daedale*. One of Keats's mythological coinages, derived from Daedalus, father of Icarus and master artificer. Spenser uses it to mean 'skilful' (*FQ* iii, Prol. 13). Here the meaning is probably 'labyrinthine' (because of the maze that Daedalus built).

153 l. 581. *belt of heaven*. The zodiac.

ll. 582–3. Aquarius, the water-carrier, is sometimes identified with Ganymede, cup-bearer to Jupiter.

155 l. 685. Woodhouse notes in his *Endymion*: 'The dew-claw, is the small short claw in the back part of the animal's leg, above the foot'; Keats is more likely to have intended 'clawed by the dew'.

l. 686. *syrinx flag*. A reed like the one into which Syrinx grew.

156 l. 713. *Delphos*. Oracle (like the one at Delphi).

158 l. 774. *brother*. Apollo, who became Endymion's brother-in-law when his sister Diana marries the boy.

160 ll. 877–8. Matt. 10: 29–31.

162 l. 943. *Titan's foe*. Jupiter.

l. 950. *seemlihed*. Seemliness (Spenser).

ll. 956–7. Cf. 'Hyperion' i. 89–90, 92–4.

On Oxford

164 Composed during Keats's stay with Bailey at Magdalen Hall, Sept. 1817; ll. 10 and 11 may contain references to Magdalen College which then, as now, supported a boys' choir and had a deer park in its grounds; published Forman (1883) (from a now-lost transcript by Brown) in the form given here. Keats included the poem in a letter to Reynolds of Sept. 1817 with this introduction: 'Wordsworth sometimes, though in a fine way, gives us sentences in the Style of School exercises—for Instance

The lake doth glitter
Small birds twitter &c.

Now I think this is an excellent method of giving a very clear description of an interesting place such as Oxford is—' (*L* i. 151–2).

l. 9. A 'trencher' is a mortar-board; a 'common hat' might mean a commoner's hat (noblemen wore gold tassels) but, since this would be the same as the trencher, the meaning here is probably 'the hat of an ordinary citizen'.

'Think not of it, sweet one, so'

Composed *c.*11 Nov. 1817; published in *1848*. The original draft (now in the Morgan library) is written onto the last leaf of the original manuscript of *Endymion*. Present text from W3.

l. 14. *rill.* Keats's draft; 'hill' *W3*.

'In drear nighted December'

165 Composed Dec. 1817; published in the *Literary Gazette*, 19 Sept. 1829. Present text from holograph fair copy (University of Bristol library).

166 l. 21. The *Literary Gazette* has the more abstract, more rational, less Keatsian line 'To know the change and feel it'. Woodhouse is probably responsible for the revision. 'I plead guilty . . . of an utter abhorrance of the word "feel" for feeling. . . . But Keats seems fond of it, and will ingraft it "in aeternum" on our language' (in a letter to Taylor of 23 Nov. 1818; *KC* i. 64). John Jones discusses Keats's use of this word in *John Keats's Dream of Truth* (London, 1969), pp. 35–68.

l. 23. *steel.* Keats wrote 'steal'.

'Before he went to live with owls and bats'

Date of composition uncertain, but probably 1817 (see below); published in W. R. Nicoll and T. J. Wise (eds.), *Literary Anecdotes of the Nineteenth Century* (London, 1896) ii, pp. 277–8. Present text from Charles Brown's transcript. Aileen Ward ('Keats's Sonnet, "Nebuchadnezzar's Dream"', *Philological Quarterly*, 34 (1955), 177–88) identifies this poem as an act of solidarity with William Hone, the bookseller who had been tried for writing political parodies of the Creed, Catechism, and Litany: see Keats's comments on the trial in his letter to George and Tom of 21–7 Dec. 1817 (above, p. 369). Nebuchadnezzar's dream is in Dan. 2–4. Daniel (here probably a type of Hone) interprets the king's nightmares as foreseeing the overthrow of his kingdom (or Tory rule).

l. 4. *Naumachia.* Miniature mock sea-battle (an item in the peace celebrations of 1814).

l. 5. Quoted from *Romeo and Juliet* III. i. 77.

To Mrs. Reynoldse's Cat

Composed 16 Jan. 1818; published in Thomas Hood's *Comic Annual*, 1830 (Jane Reynolds married Hood in 1825); present text from holograph fair copy (James Fraser Gluck Collection in the Rare Book Room of Buffalo and Erie County Public Library).

l. 1. *grand climacteric.* 63 years for a man so, if one cat year is equal to seven human years, this cat must have reached nine.

167 l. 12. *have.* W2; holograph 'has'.

l. 14. In the 19th c. garden walls were often topped with broken glass bottles to deter intruders.

Lines on seeing a Lock of Milton's hair

Composed 21 Jan. 1818; published *PDWJ* 15 Nov. 1838; reprinted in *1848*; present text from holograph fair copy (Keats House). See Keats's letter to Bailey, 23 Jan. 1818 (above, p. 372) for context.

l. 18. *delian.* From Delos (Apollo's birthplace); cf. 'delphian' ('Hence Burgundy, Claret and port' (above, p. 170) 10).

l. 35. *the simplest vassal of thy power.* The lock of hair, subject ('vassal') to Milton's vital force.

On Sitting Down to Read King Lear *Once Again*

168 Composed 22 Jan. 1818; published *PDWJ* 8 Nov. 1838; reprinted in *1848*; present text from holograph fair copy (Keats House); the poem is written inside Keats's folio Shakespeare. See Keats's letter to his brothers of 23 Jan. 1818 for context and comments (above, p. 374).

'When I have fears that I may cease to be'

Composed Jan. 1818 (and described as 'my last Sonnet' in a letter to Reynolds of 31 Jan.; *L* i. 222); published in *1848*; present text from letter (as transcribed by Woodhouse's clerk). The form of this sonnet, and the way in which the form structures the argument, is Shakespearian. Keats was to favour this form from this point on. See *KC* i. 128–30 for Woodhouse's account of Keats's method of composition as illustrated by this poem.

'O blush not so, O blush not so'

169 Composed probably on 31 Jan. 1818 when it appears in the same letter to Reynolds as the previous poem and is introduced by the sentence 'an inward innocence is like a nested dove; or as the old song says. . .'.

Published in Forman (1883); present text from letter (as transcribed by Woodhouse's clerk). The poem is written in the manner of an Elizabethan song (such as *Much Ado about Nothing* II. iii. 62 ff. which is echoed in l. 9 here). Swinburne, in a letter to Rossetti of 23 May 1870, called it 'a short bawdy song which was unfit for publication'.

l. 5. *want*. Other transcripts of the poem give 'won't', which destroys the contrast; they also give 'nought' (not 'naught') in l. 7 and lose the play on 'naughty'.

ll. 10–13. See Keats's remarks about Eve's apple in his letter of 14 Feb.–3 May 1819 (above, p. 451).

'Hence Burgundy, Claret and port'

170 Composed, like the 2 previous poems, late Jan. 1818 and included in a letter to Reynolds of 31 Jan. with the introduction 'Now I purposed to write to you a serious poetical Letter—but I find that a maxim I met with the other day is a just one "on cause mieux quand on ne dit pas *causons*" . . . Yet I cannot write in prose, It is a sun-shiny day and I cannot so here goes' (*L* i. 220). Published in *1848*; present text from the letter.

l. 3. *couthly*. Well-known, familiar (Brown's transcript and *1848* give 'earthly').

l. 10. *delphian*. Probably meaning 'oracular' (as from Delphi, site of Apollo's oracle).

l. 11. *Caius*. Reynolds, who thus signed his articles in the *Yellow Dwarf*.

'God of the Meridian'

Details of composition as for previous 3 poems. This is often printed as a continuation of the previous poem but the metrical differences, plus the gap between the poems in the letter transcription, suggest two distinct poems. Published in *1848*; present text from the letter. The 'god of the Meridian' is Apollo ('meridian' meaning midday when the sun is at its height).

ll. 14–15. A literal image of 'rapture' (perhaps suggested by depictions of the rape of Ganymede such as Rembrandt's).

171 l. 25. *unalarm'd*. Cf. Wordsworth, *The Excursion*, Pref. 35.

Lines on the Mermaid Tavern

171 Composed towards the end of Jan. 1818 after an evening spent at the Mermaid Tavern, Cheapside, famous meeting-place of poets such as Shakespeare, Jonson, Beaumont, and Fletcher; Keats wrote of this evening in a now-lost letter to George (*L* i. 225). This poem was copied in Keats's letter to Reynolds of 3 Feb. 1818 (above, p. 377); published in *1820*.

l. 12. *sup and bowse*. Both mean 'drink' ('sip' and 'booze').

Robin Hood

172 Composed *c.*3 Feb. 1818 'in a Spirit of Outlawry' and included in a letter

to Reynolds of that date (see p. 377 above), headed 'To J.H.R. In answer to his Robin Hood Sonnets'. These latter were published in the *Yellow Dwarf*, 21 Feb. 1818. Robin Hood, in his defiance of the Norman barons, was a patriot hero to early 19th-c. liberals. To Joseph Ritson, who published a collection of ballads relating to Robin Hood in 1795, Hood was 'a man who, in a barbarous age, and under a complicated tyranny, displayed a spirit of freedom and independence, which has endeared him to the common people whose cause he maintained' (*Robin Hood: A Collection of all the Ancient Poems, Songs and Ballads* (1795; repr. 1846), p. 4). See J. Barnard, 'Keats's "Robin Hood", John Hamilton Reynolds, and the "Old Poets"' (*Proceedings of the British Academy*, 75, (1989)). Published in *1820*.

l. 10. *rent nor leases*. Probably a reference to Reynolds's employment as a lawyer.

l. 30. *pasture Trent*. Most editors suggest that 'pasture' is here used adjectivally to mean 'pastoral'. But it is simpler to understand 'Trent pasture'.

l. 34. *The Tale of Gamelyn* was a 14th-c. metrical romance about a band of forest outlaws.

173 l. 36. *'grenè shawe'*. Green wood; the phrase occurs in Chaucer's *Friar's Tale* (l. 88).

l. 55. *tight*. Clever, neat, skilful; cf. 'Where be ye going you Devon Maid' (above, p. 181) 3.

'Time's sea hath been five years at its slow ebb'

Composed 4 Feb. 1818 and addressed, according to Woodhouse, to a 'Lady whom [Keats] saw for some few moments at Vauxhall' who also, again according to Woodhouse, occasioned 'Fill for me a brimming bowl' and ll. 9–10 of 'When I have fears that I may cease to be'. Published in *Hood's Magazine*, 2 (1844); present text from *W2*. This is Keats's second Shakespearian sonnet.

To the Nile

174 Composed 4 Feb. 1818 during a sonnet-writing contest between Keats, Shelley, and Hunt. Published in *PDWJ* 19 July 1838; present text from Brown's transcript (ll. 6–8 in Keats's hand). Woodhouse records that 15 minutes was the time allowed for composition and that Hunt ran over. In a letter to his brothers of (?)14 Feb. Keats writes 'The Wednesday before last Shelley, Hunt & I each wrote a Sonnet on the River Nile' (*L* i. 227–8). Shelley's begins 'Month after month the gather'd rains descend'; Hunt's, 'It flows through old hush'd Ægypt and its sands'.

l. 1. *moon-mountains*. The mountains at the Nile's source were called the Mountains of the Moon.

ll. 10–11. Cf. *Twelfth Night* IV. ii. 43.

'Spenser, a jealous Honorer of thine'

Composed 5 Feb. 1818; published in *1848*; present text from holograph fair copy (Morgan Library). The 'jealous Honorer' of Spenser is Reynolds, referred to as a 'forester' on account of his Robin Hood sonnets.

l. 7. *quell*. For 'quell' meaning 'means to quell' see *Endymion* ii. 537; *1848* has 'quill' and Keats may have intended an acoustic pun.

'Blue!—'Tis the life of Heaven—the domain'

175 Composed 8 Feb. 1818, entitled 'An Answer', in response to Reynolds's sonnet 'Sweet poets of the gentle antique line' which ends 'Dark eyes are dearer far | Than those that mock the hyacinthine bell.' Published in *1848*; present text from *W2*.

'O thou whose face hath felt the Winter's wind'

Composed 19 Feb. 1818 and included in a letter to Reynolds of that date (above, p. 379); published in *1848*; present text from letter (holograph in Robert Taylor Collection, Princeton University library). These are the words that the thrush seemed to say. The thought in this sonnet is close to that in Luke 12: 22–31.

Extracts from an Opera

176 Composed sometime in 1818; published in *1848*; present text from Brown's transcript. No opera to accommodate these songs has been identified. Perhaps they represent part of one of Keats and Brown's collaborative theatrical schemes.

177 *IV*. Cf. Shakespeare, *Sonnets* cxxx. 1–8 (marked by Keats in his copy).

178 *V*. Keats's first attempt at a ballad.

The Human Seasons

179 Composed at Teignmouth between 7 and 13 Mar. 1818 (when Keats sent it in a letter to Bailey—see p. 383 above for comments); published in Leigh Hunt's *Literary Pocket Book* for 1819 where it is signed simply 'I' (for Iohannes?). The title was probably given by Hunt.

'For there's Bishop's Teign'

Composed *c.*21 Mar. 1818 at Teignmouth and enclosed in a letter of this date to Haydon (above, p. 385); published in Tom Taylor's *Life of Benjamin Robert Haydon from his Autobiography and Journals* (London, 1853), i. 362–3; present text from holograph letter (Harvard).

ll. 1–25. Bishopsteignton, Kingsteignton, and Combeinteignhead are all villages near Teignmouth; Arch Brook (l. 7) and Wildwood's Point (l. 13) are on the south side of the Teign estuary. The marshland around the estuary used to extend into Newton Abbot (l. 19); the 'Barton' of l. 25 is a farm.

180 l. 35. *plight*. Fold or (more generally, as used by Spenser) dress.

l. 38. *Soho*. An area of central London.

l. 39. *dack'd-haired*. Short-haired (i.e. docked).

l. 42. *Prickets*. Technically, buck deer in their second year; here a sexual pun.

'Where be ye going you Devon maid'

181 Composition as for previous poem; published (without stanza 2) in *1848*; present text from holograph letter.

l. 6. *junkets*. Milk dishes traditional to Devon, where they are served with clotted cream. 'Junkets' was Hunt's punning nickname for John Keats.

ll. 13–14. Keats's underlinings suggest sexual meanings.

'Over the hill and over the dale'

Either drafted or copied in a letter to Rice of 24 Mar. 1818 (Dawlish fair had taken place on 23 Mar., which was Easter Monday); see letter (above, p. 388) for context. Lines 1–4 were published in *1848*. The complete poem was first published in Lowell, i. 610–11; present text from holograph letter.

l. 2. Dawlish is about three miles from Teignmouth.

l. 5. *Rantipole*. Wild, rakish.

l. 16. *venus*. Prostitute.

To J. H. Reynolds Esq.

182 Composed 25 Mar. 1818 and sent from Teignmouth on that date as a verse letter to Reynolds who was unwell in London, 'In the hopes of cheering you through a Minute or two'. Keats asks Reynolds to excuse 'the unconnected subject and careless verse—You know, I am sure, Claude's Enchanted Castle and I wish you may be pleased with my remembrance of it' (*L* i. 263). Published in *1848* (without the final 4 lines); present text from *W2*.

ll. 7–10. Types of the world upside down: the contemplative Voltaire is seen as a man of action; the active soldier Alexander as a slugabed; the great unworldly philosopher as a mere social creature; Hazlitt disliked the novelist Maria Edgeworth (1767–1849) and would presumably have avoided her cat too.

l. 11. Junius Brutus Booth (1796–1852), the actor; 'so so' means drunk.

l. 16. *wild boar tushes*. Such as those that puncture Venus's joy in Adonis.

l. 18. *Aeolian harps*. See 'Ode to Apollo' (above, p. 7) 34.

183 ll. 20–2. The details recall Claude's *Landscape with the Father of Psyche Sacrificing at the Milesian Temple of Apollo* which had been exhibited at the British Institution in 1816.

l. 21. *Gloams*. Darkens (a Scots verb); *1848* and other editors amend to 'Gleams'; as Barnard points out (p. 595), 'the "o" sound echoes throughout ll. 20–2.'

ll. 26–66. The reference is to Claude's *Enchanted Castle* now in the

National Gallery; probably known to Keats through the 1782 engraving by François Vivares and William Woollett.

l. 29. *Urganda*. The enchantress Urganda the Unknown in the 16th-c. romance *Amadis of Gaul*.

l. 42. *santon*. A Mohammedan monk or hermit; the word reflects the fashionable orientalism of the time.

l. 44. *Cuthbert de Saint Aldebrim*. Probably a made-up name (like Coleridge's 'Roland de Vaux of Tryermaine').

l. 46. *Lapland Witch*. See *PL* ii. 665.

184 ll. 72–4. This image of self-declaration was, as Miriam Allott notes, suggested by North's *Life of Alcibiades*: 'Alcibiades set up straight his flagge on in the toppe of the galley of his admirall, to shewe what he was', *North's Plutarch* (1579; repr. Oxford, 1928), ii. 149.

l. 88. *lampit*. Limpit (a Scots spelling).

185 l. 106. Wordsworth had entitled a group of his poems 'Moods of my own Mind'.

l. 108. The Kamschatka peninsula is on the east coast of Russia; Keats had read about the Christian mission there in W. Robertson's *History of America* (Dublin, 1777), and the Count de Buffon's *Histoire Naturelle* (English trans. London, 1792); see A. D. Atkinson, 'Keats and Kamschatka', *N&Q* 196 (1951), 340–6.

l. 112. *Centaine*. 100 lines.

l. 113. 'Soft! here follows prose' (*Twelfth Night* II. v. 142).

Isabella

Composed between Feb. and Apr. 1818 (completed by 27 Apr.); published in *1820*; a fair copy of the poem in the British Museum is entitled 'The Pot of Basil'. The story of Isabella comprises the Fifth Novel of the Fourth Day in *The Novels and Tales of the Renowned John Boccaccio* (i.e. *The Decameron*) which Keats read in the 1620 translation (5th edn., London, 1684). The idea of retelling a Boccaccio tale may have been prompted by Hazlitt's lecture on Dryden and Pope on 3 Feb. 1818, in which he said that 'a translation of some of the . . . serious tales in Boccaccio . . . if executed with taste and spirit, could not fail to succeed in the present day' (*Works*, v. 82). Keats and Reynolds had originally intended to collaborate over a volume of tales from Boccaccio, but other preoccupations kept Reynolds from the work (though he subsequently published two tales in his *The Garden of Florence* (London, 1821), where he refers in the Advertisement to the earlier joint plan). Charles Lamb, in his review of *1820*, was to call 'Isabella' 'the finest thing in the volume' (*New Times*, 19 July 1820), but Keats came to be dissatisfied with the poem and uneasy about its tone. See his letter to Woodhouse of 21–2 Sept. 1819 (above, p. 496) where he describes the poem as 'too smokeable'. Keats had met Isabella Jones at

Hastings the previous summer. Perhaps the community of names furthered his attraction to this particular tale.

187 l. 44. *ruddy tide*. Blood engorging his vocal chords. The phrase has an 18th-c. ring (like 'finny droves') but is far from euphemistic.

l. 62. *fear*. Frighten.

188 l. 95. *Theseus' spouse*. Ariadne.

l. 99. *Aeneid* vi. 450–1.

189 ll. 105–20. Bernard Shaw wrote that these two stanzas 'contain all the Factory Commission Reports that Marx read, and that Keats did not read because they were not yet written in his time' (*The John Keats Memorial Volume* (London, 1921), p. 175).

l. 107. *swelt*. Swelter.

l. 113. *Ceylon diver*. Diving for pearls to enrich the brothers; ll. 113–18 echo Dryden's 'Annus Mirabilis' 9–12.

l. 123. *orange-mounts*. Plantations of orange trees (or, possibly, heaps of gold).

l. 124. *lazar stairs*. Stairs in a lazar house, occupied by the poor and the sick.

l. 125. *red-lin'd accounts*. Account books showing income, expenditure, and profit. George and Tom's experience as employees in Abbey's counting-house may inform these lines.

l. 126. *richer*. Cf. Keats's use of 'rich' in his journal-letter to George and Georgiana of 14 Feb.–3 May 1819 in referring to the pains he has taken over 'Ode to Psyche': 'I think it reads the more richly for it' (above, p. 475).

l. 131. *that land inspired*. Palestine; Keats is succumbing to an anti-Semitic cliché in this line. The 'ducats' of l. 134 suggest that Shylock may be behind this.

190 l. 140. *pest*. Exod. 10: 21.

l. 150. *ghittern*. A kind of guitar.

ll. 161–8. The brothers' motivation, which is not so explained in Boccaccio, may have been suggested by Webster's *The Duchess of Malfi*.

192 l. 209. *murder'd man*. Charles Lamb was the first to admire this prolepsis in print: 'The anticipation of the assassination is wonderfully conceived in one epithet' (*New Times*, 19 July 1820).

193 l. 262. '[Ahaz] burnt incense in the valley of the son of Hinnom, and burnt his children in the fire, after the abhominations of the heathen' (2 Chron. 28: 3).

195 l. 321. This, like many 'Adieux' in Keats's poetry, recalls the departure of the Ghost in *Hamlet*.

196 l. 344. *forest-hearse*. For Isabella the forest exists only as a bearer of Lorenzo.

197 l. 370. She had embroidered a design on it.

l. 374. Her breasts.

l. 381. *horrid.* The Latin *horridus* means 'bristling'. Keats uses the word etymologically.

198 l. 393. *Perséan sword.* Given to Perseus by Mercury for decapitating Medusa.

l. 396. *harps.* Bards, minstrels.

199 l. 451. *Baälites of pelf.* Worshippers of the false god money.

'Mother of Hermes! and still youthful Maia!'

201 Composed at Teignmouth on 1 May 1818 and enclosed in a letter to Reynolds of 3 May (see p. 395 above) in which Keats refers to it as an ode; published in *1848*; present text from Woodhouse's clerk's transcript of the letter.

l. 3. *Baiæ.* A reference to Tasso (see 'To Charles Cowden Clarke' (above, p. 29) 29.

l. 5. *Sicilian.* The dialect of Theocritus.

'Give me your patience Sister while I frame'

202 Composed at the 'Foot of Helvellyn' on 27 June 1818 and copied into the journal-letter to George and Georgiana of Sept. 1819 (*L* ii. 195). The initial letters of each line form an acrostic of GEORGIANA AUGUSTA KEATS; published in the *New York World*, 25 June 1877; present text from holograph letter.

l. 10. *Anthropopagi.* 'Anthropophagi', as in *Othello* I. iii. 144.

'Sweet sweet is the greeting of eyes'

Composed 28 June 1818 while at Keswick, and included in a letter of that date to George and Georgiana; published in Lowell, ii. 28; present text from the letter.

On Visiting the Tomb of Burns

203 Composed 1 July 1818 at Dumfries and included in a letter to Tom of 29 June–2 July. Keats writes:

> Burns' tomb is in the Churchyard corner, not very much to my taste, though on a scale, large enough to show they wanted to honour him . . . This Sonnet I have written in a strange mood, half asleep. I know not how it is, the Clouds, the sky, the Houses, all seem anti Grecian & anti Charlemagnish. (*L* i. 309)

Published in *1848*; present text from John Jeffrey's transcript of Keats's lost letter.

l. 9. *Minos-wise.* With the discrimination of Minos.

l. 11. *Fickly.* Jeffrey; *1848* and other editors give 'Sickly' but, as J. C. Maxwell points out, 'Fickly' is a bona-fide word meaning 'deceitful', and it avoids the repetition of 'sick' (*KSJ* 4 (1955), 77).

l. 12. *Cast* conjectured in *1848*; Jeffrey's transcript has a blank.

Old Meg she was a Gipsey

Written for Fanny Keats and enclosed in a letter to her of 3 July 1818; published *PDWJ* 22 Nov. 1838; present text from Keats's letter to Tom of 3–9 July 1818. During their walk through Kirkudbrightshire, the setting of Scott's novel *Guy Mannering* (1815)—which Keats had not read—Brown described to Keats the character of Meg Merrilies who appears in that novel. Brown records that Keats was 'much interested in the character. There was [a] little spot, close to our path-way,—"There", he said, in an instant positively realising a creation of the novellist, "in that very spot, without a shadow of doubt, has old Meg Merrilies often boiled her kettle!"' (*KC* ii. 61).

204 l. 25. *Margaret Queen*. Probably referring to the wife of Henry VI as portrayed by Shakespeare.

l. 28. *chip hat*. Hat made of thin strips of wood.

'There was a naughty Boy'

Composed 3 July 1818 at Kirkudbright and included in a letter to Fanny Keats of 2–5 July (see p. 406 above) where it is introduced as 'a song about myself'. Published in Forman (1883); present text from letter.

l. 20. *revetted*. Rivetted.

206 l. 76. *Miller's thumb*. A small freshwater fish, like the stickleback (childishly pronounced 'Tittle bat') of the following line.

'Ah! ken ye what I met the day'

207 Drafted or copied in a letter to Tom of 10 July 1818 (*L* i. 327) from Ballantrae, Ayrshire. Published in Forman (1883); present text from letter. The poem is written in pseudo-Scots with some genuine dialect words ('brig' for 'bridge' l. 9, 'daffed' for 'daunted' l. 35) and the archaic 'yeve' for 'give' (l. 5) amongst the phonetic spellings.

Sonnet to Ailsa Rock

208 Composed on 10 July 1818 at Girvan, Ayrshire, and copied in a letter to Tom in which Keats also describes his first sight of the rock (*L* i. 329); published in Leigh Hunt's *Literary Pocket Book* for 1819 (1818) where, like 'The Human Seasons', it is signed simply 'I'. The sonnet was summarized in *Blackwood's* (Dec. 1819) as 'Mr. John Keats standing on the sea-shore at Dunbar, without a neckcloth, according to the custom of Cockaigne, and cross-questioning the Crag of Ailsa.'

'This mortal body of a thousand days'

209 Keats composed this sonnet, of which he was not proud, whilst visiting Burns's birthplace at Ayr, 'for the mere sake of writing some lines under the roof': see letters to Reynolds and Tom of 11–13 July and 10–14 July

(above, pp. 411–12, *L* i. 332). Brown, who preserved a transcript of the poem, commented that the 'conversion [of the cottage] into a whiskey-shop, together with its drunken landlord, went far towards the annihilation of his poetic power' (*KC* ii. 62). Published in *1848* (the only source).

'There is a joy in footing slow across a silent plain'

Composed before 22 July 1818 when it was copied into a letter to Bailey and described as 'cousin-german to the Circumstance' of visiting Burns's cottage (*L* i. 344). Published in the *Examiner*, 14 July 1822. The earlier of two holograph texts is entitled 'Lines written in the highlands after a visit to Burns's Country'. Present text from the letter. Keats's use of rhyming fourteeners reflects his reading of Renaissance literature, such as Chapman's Homer. But George copied the poem into quatrains of alternating four and three stress lines.

'All gentle folks who owe a grudge'

211 Composed on 17 July 1818 at Cairndow, Argyllshire after bathing in Loch Fyne 'a saltwater Lake . . . quite pat and fresh but for the cursed Gad flies' (*L* i. 334); published in Forman (1883); present text from letter (in Keats House).

l. 21. *Lowther*. See Index of Correspondents.

212 l. 29. *Southey*. Robert Southey (1774–1843) became Poet Laureate in 1813; disliked by radicals because of his political apostasy.

ll. 30, 33. *Mr. D——*, *Mr. V——*. Not identified.

l. 40. *Mister Lovels*. Hero of Scott's *The Antiquary* (1816).

l. 47. *chouse*. Cheat (see Jonson, *The Alchemist* I. ii. 26).

Not Aladin magian

213 Composed between 24 July 1818 (when Keats visited Fingal's Cave on Staffa) and 26 July when it is inserted into the journal-letter to Tom of 23–6 July from the Isle of Mull. Published in the *Western Messenger*, 1 July 1836; present text from letter to Tom. (Keats copied a slightly different version in his letter to George and Georgiana of 17–27 Sept. 1819.) In his letter Keats tells Tom:

> I am puzzled how to give you an Idea of Staffa. It can only be represented by a first rate drawing—The finest thing is Fingal's Cave—it is entirely a hollowing out of Basalt Pillars. Suppose now the Giants who rebelled against Jove had taken a whole Mass of black Columns and bound them together like bunches of matches—and then with immense Axes had made a cavern in the body of these columns . . . For solemnity and grandeur it far surpasses the finest Cathedrall. (*L* i. 348–9)

l. 3. *wizard of the Dee*. Merlin.

l. 5. Cf. Rev. 1: 9–12 for St John's description of his vision on Patmos.

l. 24. *thrice*. Trice—Keats's spelling may reflect the incorrect belief that the word means 'a third of a second'.

l. 25. *Lycidas*. See Milton, 'Lycidas' 154–8.

l. 32. *Finny palmers*. Fish, who are palmers (pilgrims) because they swim to a shrine.

214 l. 39. *Pontif priest*. High priest; if Keats had by now made a start at learning Greek he might have intended a play on the Greek πόντος (sea).

l. 51. *cutters*. Small passenger boats.

After drafting or copying the poem in his letter to Tom, Keats writes 'I am sorry I am so indolent as to write such stuff as this.'

'Read me a Lesson muse, and speak it loud'

Composed 'on the top of Ben Nevis' (*L* i. 357) 2 Aug. 1818; published in *PDWJ* 6 Sept. 1838; present text from holograph letter. Brown recalled that Keats 'sat on the stones, a few feet from the edge of that fearful precipice, fifteen hundred feet perpendicular from the valley below, and wrote this sonnet' (*KC* ii. 63).

'Upon my Life Sir Nevis I am piqued'

215 Composed 3 Aug. 1818 at Letterfinlay, Inverness-shire in a letter to Tom. Published in Forman (1883); present text from the letter. Keats tells Tom about their effort in climbing Ben Nevis, put in perspective by:

> one Mrs. Cameron of 50 years of age and the fattest woman in all inverness shire who got up this Mountain some few years ago—true she had her servants but then she had her self—She ought to have hired Sysiphus. . . .'T is said a little conversation took place between the mountain and the Lady—After taking a glass of Wiskey as she was tolerably seated at ease she thus begun [the poem follows] (*L* i. 354)

l. 2. *reek'd*. Sweated.

l. 4. *bate*. Rest.

216 l. 30. *how the gemini*. A tacit pun on 'how the deuce' (gemini–twins–dual-deuce).

l. 32. As humans quake, so mountains earthquake.

l. 43. *Buss*. Kiss.

l. 52. *gust*. Both 'blast' and 'taste' ('gusto' was a fashionable concept; see Hazlitt, *Works*, iv. 77–80).

On Some Skulls in Beauley Abbey, near Inverness

217 Written collaboratively with Brown, probably during early Aug. 1818, before Keats's ill health forced him to curtail his tour. (They were at Inverness from 6 to 8 Aug.) Published in the *New Monthly Magazine*, 4 (January, 1822); present text from Brown's fair copy (Morgan library). S. Colvin prints the poem in an appendix (*John Keats* (London, 1917),

pp. 553–6) and identifies the lines which Keats told Woodhouse were his own. These are the lines printed here in roman type. Hamlet's contemplation of skulls (v. i) is behind this poem which may, in turn, have suggested Thomas Hardy's 'Voices from things growing in a Churchyard'.

Mottoes. The first is from Wordsworth's sonnet 'Beloved Vale' (1807), which Wordsworth later revised; the second is from Clarence's account of his dream in *Richard III* I. iv. 33.

l. 6. *creed's undoing*. The Reformation and dissolution of the monasteries.

219 l. 43–5. This monk was an illuminator of missals.

l. 61. *King Lear* III. ii. 52.

Nature withheld Cassandra in the Skies

220 A 'free translation of a Sonnet of Ronsard' done at some point in Sept. 1818 and copied into a letter to Reynolds of 22 Sept. (above, p. 417). Published in *1848*; present text from holograph (Keats House). Woodhouse had lent Ronsard's works to Keats in mid-Sept. The Ronsard original is the second sonnet of *Le Premier Livre des amours* (1587):

> Nature ornant Cassandre qui devoit
> De sa douceur forcer les plus rebelles,
> La composa de cent beautez nouvelles
> Que dés mille ans en espargne elle avoit.
> De tous les biens qu'Amour au ciel couvoit
> Comme un tresor cherement sous les ailes,
> Elle enrichit les graces immortelles
> De son bel oeil, qui les Dieux esmouvoit.
> Du Ciel à peine elle estoit descendue
> En devint folle, & d'un si poignant trait
> Amour coula ses beautez en mes veines,
> Qu'autres plaisirs je ne sens que mes peines,
> Ny autre bien qu'adorer son pourtrait.

' 'Tis "the witching time of night" '

221 Composed in a letter to George and Georgiana, 14 Oct. 1818 (above, see p. 425) where it is introduced as a prophecy 'that one of your Children should be the first American poet'. The American writer John Howard Payne (1781–1852) wrote to Milnes in 1847 that 'The writer does not seem to have known that we have had, and then possessed, many poets in America; though he possibly meant the word "first" to be understood as *greatest*' (*KC* ii. 224–5), published in *1848*; present text from the letter. The first line is from *Hamlet* (III. ii. 388).

l. 20. The copy text reads 'Though the linnen then that will be'. The British economy at this time rested on cotton.

'And what is Love?—It is a doll dress'd up'

222 Composed during 1818; published in *1848* (with the title 'Modern Love'); present text from *W2*.

l. 8. *Wellingtons*. Boots—not then made of rubber—named after the Duke of Wellington *c.*1817; modish at the time of writing.

223 ll. 15–16. Cleopatra was fabled to have dissolved and drunk a pearl in a toast to Antony.

l. 17. *beaver hats*. Hats made of beaver fur. On 17 Mar. 1819 Keats wrote to George and Georgiana that Abbey 'began again (he has done it frequently lately) about that hat-making concern . . . he wants to make me a Hat-maker' (*L* ii. 77). Abbey may have already broached the subject at the time this poem was written.

Fragment: 'Welcome joy, and welcome sorrow'

Composed during 1818; published in *1848*; present text from Brown's transcript. The motto misquotes *PL* ii. 898–90 'For Hot, Cold, Moist, and Dry, four champions fierce | Strive here for mastery, and to battle bring | Their embryon atoms; they around the flag | Of each his faction'. Keats underlined this passage in his copy of *PL* at Hampstead. Cf. Keats's description of the poetical character in his letter to Woodhouse of 27 Oct. 1818 (above, p. 418).

l. 2. The contrast is between dull obliviousness and mercurial sharpness.

ll. 12–15. These verbal emblems reflect Keats's reading in Renaissance literature.

224 l. 23. This line recurs in 'What can I do to drive away' (above, p. 329) 54.

l. 34. Echoes *Hamlet* IV. v. 31–2.

Fragment: 'Where's the Poet? Show him! show him!'

Composed during 1818; published in *1848*; present text from Brown's transcript. As for previous poem, cf. letter of 27 Oct. 1818 (above, pp. 418–19).

ll. 8–10. Cf. 'if a Sparrow come before my Window I take part in its existence and pick about in the Gravel' (letter to Bailey of 22 Nov. 1817; above, p. 366).

To Homer

Composed during 1818; published in *1848*; present text from Brown's transcript. See Keats's remarks about learning Greek and reading Homer in his letter to Reynolds of 27 Apr. 1818 (above, p. 393).

l. 1. Keats compares his own 'giant ignorance' to the traditional blindness of Homer, the inward seer.

225 l. 12. *triple sight*. Making three worlds visible: Jove's heaven, Neptune's ocean, and Pan's earth.

Hyperion

Composed between late 1818 (Keats possibly refers to this work in a letter to Woodhouse of 27 Oct. 1818 and definitely refers to it in a letter of 18 Dec. to George and Georgiana, see pp. 419 and 435–6 above) and Apr.

1819 when it was abandoned unfinished. Published in *1820*. The subject of the poem—the overthrow of the Titans by the Olympian gods and the establishment of Apollo as god of the sun, music and healing, in the place of the Titan sun-god Hyperion—had been on Keats's mind for some time and was certainly part of the plan for future work at the time he was completing *Endymion* (see iv. 774, and notes to 'Not Aladin Magian' (above, pp. 158, 588). See also Keats's letter to Haydon of 23 Jan. 1818 (above, p. 371) where he writes of the differences between *Endymion* and the projected 'Hyperion'. Keats's sources for the history of the overthrow of the Titans were principally Hesiod's *Works and Days* and *Theogony* (in Cooke's 1728 translation) and Hyginus' *Fabulae* (included in *Auctores Mythographici Latini*, a copy of which Keats acquired in 1819). His more modern reference works included Andrew Tooke's *Pantheon* (London, 1722), J. Lemprière's *Bibliotheca Classica* (Reading, 1788), and Baldwin's *Pantheon*. Keats's reading of Dante and Milton strongly inform the poem while Shakespeare's *King Lear* influences the portrayal of Saturn. The writing of the poem was so bound up with the period of Tom's illness and death that Keats may have lacked heart to complete it—though he was to revive the project in 'The Fall of Hyperion' later. It was published as 'A Fragment' (by then a recognized genre). The following Advertisement appears in *1820*:

> If any apology be thought necessary for the appearance of the unfinished poem of *Hyperion*, the publishers beg to state that they alone are responsible, as it was printed at their particular request, and contrary to the wish of the author. The poem was intended to have been of equal length with *Endymion*, but the reception given to that work discouraged the author from proceeding.
>
> Fleet Street, 26 June 1820

Keats wrote against this in one copy 'This is none of my doing—I was ill at the time. This is a lie' (referring to the final sentence). The poem was greatly admired by contemporaries. Even Byron, who on the whole disliked Keats's work, wrote that 'His fragment on *Hyperion* seems actually inspired by the Titans and is as sublime as Aeschylus' (MS note dated 12 Nov. 1821, quoted in G. Matthews (ed.), *Keats: the Critical Heritage* (London, 1971), p. 128).

ll. 1–7. Bailey gives these lines as an instance of Keats's notions on the 'principle of melody in Verse . . . particularly in the management of open & close vowels . . . Keats's theory was, that the vowels should be so managed as not to clash one with another so as to mar the melody,—& yet that they should be interchanged, like differing notes of music to prevent monotony' (*KC* ii. 277).

ll. 8–9. See note to *fleecy Crowns* on p. 335 (p. 613 below) for the genesis of these lines.

l. 1. See Keats's comment on the sound of the word 'vale' in his note to *PL* i. 321 (above, p. 338).

226 l. 31. *Memphian sphinx*. Memphis was a major city in ancient Egypt. The British Museum had acquired some Egyptian sculptures in 1818; see Keats's comments in his letter of 14 Feb.–3 May 1819 (above, p. 456).

l. 61. *reluctant*. See K's comment on *PL* vi. 58–9 (above, p. 343).

227 l. 86. Cf. Keats's description of Fingal's Cave in headnote to 'Not Aladin magian' (above, p. 588).

228 l. 129. *gold clouds metropolitan*. Miltonic word order; the clouds form a metropolis for the gods.

l. 137. *Druid locks*. Edward Davies's *Celtic Researches* (London, 1804) connected the Celts with the Titans; but the vowel play achieved in this line may have influenced the choice of epithet.

229 l. 147. *The rebel three*. Saturn's sons Jupiter, Neptune, and Pluto.

l. 167. Cf. *PL* x. 272–3.

230 ll. 196–200. Cf. *PL* i. 767–71.

231 ll. 252–4. Cf. *PL* i. 142–5.

ll. 259–63. Cf. *PL* ix. 180–90.

232 l. 274. *colure*: an astrological term for 'each of two great circles which intersect each other at right angles at the poles, and divide the equinoctial and the ecliptic into four equal parts' (*OED*); Keats found the word in *PL* ix. 66.

ll. 281–3. Cf. Shelley's 'Ozymandias'.

233 l. 323. *first born*. Saturn.

234 ll. 354–7. The manner of this book's ending recalls Dante's conclusions of many cantos in the *Divine Comedy*.

Book ii

ll. 7–12. This description is informed by Keats's experience of the Lake District through his own travels and through his reading of Wordsworth.

235 ll. 33–8. See Keats's description of a Druid temple in his letter to Tom of 29 June 1818 (above, p. 403); see also note to i. 137.

ll. 39 ff. The description of the Titans' suffering reflects Keats's recent reading of Dante's *Inferno*.

238 l. 161. *engine*. Mobilize.

ll. 168–9. Oceanus is self-taught.

l. 170. *not oozy*. Unlike the 'oozy Locks' of Milton, 'Lycidas' 175.

240 l. 232. *young God of the Seas*. Neptune.

241 l. 280. According to Severn, 'a beautiful air of Glucks . . . furnished the groundwork of the coming of Apollo in Hyperion' (*KC* ii. 133).

242 l. 341. *winged thing*. Victory was traditionally represented with wings.

243 ll. 374–6. Both Baldwin and Lemprière record that the statue erected to Memnon near Thebes 'had the peculiar property of uttering a melodious

sound every morning when touched by the first beams of the day as if to salute his mother [Aurora]; and every night at sunset, it gave another sound, low and mournful as lamenting the departure of the day' (Baldwin, *Pantheon*, p. 269).

Book iii

244 l. 12. *Dorian flute*. Milton writes of 'the *Dorian* mood' in *PL* i. 550 (marked by Keats); the Dorian mode is one of the ancient musical modes. Here the Delphic harp and Dorian flute usher in another era than that of shell music (i. 131, ii. 270).

l. 13. Apollo.

l. 29. Hyperion.

ll. 31–2. *mother . . . twin-sister*. Latona and Diana.

245 l. 46. *awful Goddess*. Mnemosyne.

ll. 77–9. Mnemosyne has staked her future on Apollo and abandoned the Titans, once her peers.

246 ll. 82–120. Keats told Woodhouse that Apollo's speech 'seemed to come by chance or magic—as if it were something given to him' (*KC* i. 129).

ll. 86–7. Cf. Milton, 'Samson Agonistes' 80–2.

ll. 113–20. See Keats's letter to Reynolds of 3 May 1818 (above, p. 396) for thoughts on wisdom and sorrow.

247 l. 136. Woodhouse noted (in his copy of *Endymion*) that 'The poem, if completed, would have treated of the dethronement of Hyperion, the former God of the Sun, by Apollo—and incidentally of Oceanus by Neptune, of Saturn by Jupiter &c and of the war of the Giants for Saturn's reestablishment—with other events, of which we have but very dark hints in the Mythological poets of Greece & Rome. In fact, the incidents would have been pure creations of the Poet's brain.'

Fancy

Composed late 1818; published in *1820*. Keats copied this (in a slightly longer form) and the following poem in his letter to George and Georgiana on 2 Jan. 1819 (above, p. 441) as 'specimens of a sort of rondeau which I think I shall become partial to'.

248 l. 21. *shoon*. Shoes.

249 l. 81. *Ceres' daughter*. Proserpine, or Persephone, captured by Pluto, the 'God of Torment'.

Ode: 'Bards of Passion and of Mirth'

250 See note for previous poem; composed late 1818; published in *1820*. Keats describes this as 'on the double immortality of Poets' (see p. 442 above).

ll. 23–4. See Keats's letter to Bailey of 22 Nov. 1817 (above, p. 365) where he speculates 'that we shall enjoy ourselves here after by having what we called happiness on Earth repeated in a finer tone and so repeated'.

'I had a dove and the sweet dove died'

251 Composed during Dec. 1818 or the very beginning of Jan. 1819 (it was copied in a letter to George and Georgiana on 2 Jan.); published in *1848*; present text from the letter.

Faery Song: 'Ah! woe is me! poor Silver-wing!'

Composed 1818 or 1819; published in *PDWJ* 18 Oct. 1838; this text from Brown's transcript.

l. 11. *favonian*. Favonius is the west wind; so 'mild, balmy'.

Song: 'Hush, hush, tread softly, hush, hush my dear'

252 Composed during 1818; published in *Hood's Magazine*, Apr. 1845; present text from Fanny Brawne's transcript (Keats House). 'Textually this is the strangest poem in the Keats canon' (J. Stillinger, *The Texts of Keats's Poems*, (Cambridge, Mass., 1974), p. 211); no two versions are substantively identical. Charlotte Reynolds told H. B. Forman that Keats 'was passionately fond of music, and would sit for hours while she played the piano to him. It was to a Spanish air which she used to play that the song "Hush, hush! tread softly!" was composed' (Forman (1883), i, pp. xxix–xxx). An anonymous reviewer of the Paris opera season of autumn 1859 refers to the tune by Daniel Steibelt (1765–1823) to which Keats wrote this song (*Athenaeum*, 15 Oct. 1859, p. 505). Gittings (*The Living Year* (London, 1954), pp. 57–60) believes that the poem echoes Keats's flirtation with Isabella Jones. If so it is strange that he should have given the poem to Fanny Brawne to transcribe.

l. 20. *dream*. Perhaps a copying error; every other version has 'sleep'.

The Eve of St. Agnes

Drafted during the last two weeks of Jan.—and perhaps the first few days of Feb.—1819 while Keats was at Chichester and Bedhampton; significantly revised at Winchester during early Sept. 1819 (Keats read the poem to Woodhouse on 12 Sept.). Published in *1820*. On 19 Sept. Woodhouse wrote to John Taylor that Keats had:

> had the Eve of St A. copied fair: He has made trifling alterations, inserted an additional stanza early in the poem to make the *legend* more intelligible, and correspondent with what afterwards takes place, particularly with respect to the supper & the playing on the Lute.—he retains the name of Porphyro [Lionel in the earliest drafts]—has altered the last 3 lines to leave on the reader a sense of pettish disgust, by bringing Old Angela in (only) dead stiff & ugly.—He says he likes that the poem should leave off with this Change of Sentiment—it is what he aimed at, & was glad to find from my objections to it that he had succeeded.—I apprehend he had a fancy for trying his hand at an attempt to

> play with his reader, & fling him off at last . . . There was another alteration which I abused for 'a full hour by the *Temple* clock.' You know if a thing has a decent side, I generally look no further—As the Poem was originally written, *we* innocent ones (ladies & myself) might very well have supposed that Porphyro, when acquainted with Madeline's love for him, & when 'he arose, Etherial flushd &c &c' . . . set himself at once to persuade her to go off with him . . . to be married, in right honest chaste & sober wise. But, as it is now altered, as soon as M. has confessed her love, P. (instead) winds by degrees his arm round her, presses breast to breast, and acts all the acts of a bonâ fide husband, while she fancies she is only playing the part of a Wife in a dream. This alteration is of about 3 stanzas; and tho' there are no improper expressions but all is left to inference, and tho' profanely speaking, the Interest on the reader's imagination is greatly heightened, yet I do apprehend it will render the poem unfit for ladies, & indeed scarcely to be mentioned to them among the 'things that are.'—He says he does not want ladies to read his poetry:—that he writes for men—& that if in the former poem there was an opening for doubt what took place, it was his fault for not writing clearly & comprehensibly—that he shd despise a man who would be such an eunuch in sentiment as to leave a maid, with that Character about her, in such a situation: & shd despise himself to write about it &c &c &c—and all this sort of Keats-like rhodomontade. (*L* ii. 162–3)

Taylor in his reply of 25 Sept. 1819 entirely concurred with Woodhouse:

> This Folly of Keats is the most stupid piece of Folly I can conceive. . . . I don't know how the Meaning of the new Stanzas is wrapped up, but I will not be accessary (I can answer also for H[essey] I think) towards publishing any thing which can only be read by Men. . . . So far as he is unconsciously silly in this Proceeding I am sorry for him, but for the rest I cannot but confess to you that it excites in me the Strongest Sentiments of Disapprobation—Therefore . . . if he will not so far concede to my Wishes as to leave the passage as it originally stood, I must be content to admire his Poems with some other Imprint. (*L* ii. 182–3)

Woodhouse notes (in *W2*) that 'K. left it to his Publishers to adopt which [versions] they pleased, & to revise the Whole.'

In a letter to Taylor of (?)11 June 1820 Keats objects to an editorial emendation to stanza 7 (l. 58) which misunderstood his very concrete intention. 'I do not use *train* for *concourse of passers by* but for *Skirts* sweeping across the floor' (*L* ii. 295), so we can assume that the form the poem took in *1820* had gained his assent though it may not have represented his preference. For a complete text of 'St. Agnes Eve' in a version which includes the contested stanzas see Appendix I below.

Woodhouse records that the poem was written at the suggestion of Isabella Jones; but the extent of her contribution is unclear. In his letter to Bailey of 14 Aug. 1819 (above, p. 487) Keats describes the poem as 'on a popular superstition'. This he would have found in John Brand's *Observa-*

tions on Popular Antiquities (2 vols., London, 1813) and in Burton's *The Anatomy of Melancholy*, III. 2. iv. i. Brand writes that on St Agnes' Eve (20 Jan.) 'many kinds of divination are practised by virgins to discover their future husbands . . . This is called fasting St. Agnes' Fast' (i. 32). Keats always referred to the poem as 'St. Agnes Eve'. For his remarks on the poem see his letters to George and Georgiana of 14 Feb. 1819 and to Woodhouse of 21–2 Sept. 1819 (above, p. 450, 496). L. Hunt printed the whole poem, with his comments, in his *Imagination and Fancy* (London, 1844).

253 ll. 14–15. According to Woodhouse 'the stone figures of the Temple Church probably suggested these lines' (*W2*); but Keats may be remembering the arrangement of the sculpted tombs at Chichester Cathedral.

255 l. 70. *amort*. Dead.

l. 71. *lambs unshorn*. St Agnes appeared to her parents in a vision after her death in the company of a white lamb—thereafter her emblem. On her day lambs'-wool was offered on the altar to be spun and woven by nuns (see l. 117).

258 l. 171. Possibly a reference to *FQ* III. iii. 7–11; see however Karen J. Harvey ('The Trouble about Merlin: the Theme of Enchantment in "The Eve of St. Agnes" *KSJ* 34 (1985), 83–94) who suggests that Merlin's debt is to Vivien who, as a fay, is also a 'demon'. Amongst Brown's inventory of Keats's books was a 'History of Arthur'.

l. 173. *cates*. Edible delicacies (Elizabethan); there is a semantic link with 'catering' (l. 177).

259 l. 206. *tongueless nightingale*. Evokes the myth of Procne and Philomel in Ovid, *Metamorphoses* vi; see J. C. Stillinger, 'The Hoodwinking of Madeline', *SP* 58 (1961), 533–55 (reprinted in W. J. Bate (ed.), *Keats: A Collection of Critical Essays* (Englewood Cliffs, NJ, 1964) and in J. Stillinger, *The Hoodwinking of Madeline* (Urbana, Ill., 1971)) for an illumination of this poem's dark side.

ll. 208–16. Gittings (*The Living Year* (London, 1954), pp. 73–82) suggests that Keats has incorporated architectural details from the Gothic chapel at Stansted Park.

260 l. 241. The homophony of 'pray' and 'prey' clinches an ambiguity of tone.

261 l. 257. A sleep-inducing charm.

l. 261. Keats told Cowden Clarke that this line 'came into my head when I remembered how I used to listen in bed to your music at school' (*Recollections*, p. 143).

l. 266. *soother*. Keats's compound of 'smoother' and 'more soothing'; cf. Ben Jonson, *A Celebration of Charis* iv. 15–16 'Doe but marke, her forehead's smoother | Than words that sooth her!'

l. 267. 'Here is delicate modulation . . . [making] us read the line delicately, and at the tip-end, as it were, of one's tongue' (L. Hunt, *Imagination and Fancy* (London, 1844), p. 337).

l. 280. *unnerved*. Weak, slack.

262 l. 292. *'La belle dame sans mercy'*. Title of a poem written by Alain Chartier in 1424; Keats's own poem of this title had not yet been composed.

263 l. 325. *flaw-blown sleet*. Cf. Cary's Dante, *Inferno* vi. 9–10 'Large hail, discolour'd water, sleety flaw | Through the dun midnight air stream'd down amain'.

l. 344. *haggard*. A haggard is a hawk that has refused training; so, 'wild', 'intractable'.

264 l. 355. *darkling*. An Elizabethanism for 'dark', 'obscure', which Keats made his own.

The Eve of St. Mark

265 Composed between 13 and 17 Feb. 1819, shortly after Keats's return from Chichester and Bedhampton; published in *1848*; present text from holograph draft in the British Library. Keats transcribed ll. 1–114 of the poem in his Sept. 1819 letter to George and Georgiana (above, p. 509) and clearly thought the poem apropos the description of Winchester he had just given and 'quite in the spirit of Town quietude'. St Mark's eve is on 24 Apr. The practice attached to it is described in John Brand's *Observations on Popular Antiquities* (London, 1813), i. 166:

> It is customary in Yorkshire . . . for the common people to sit and watch in the church porch on St. Mark's Eve, from eleven o'clock at night till one in the morning. The third year (for this must be done thrice), they are supposed to see the ghosts of all those who are to die the next year, pass by into the church. When any one sickens that is thought to have been seen in this manner, it is whispered about that he will not recover, for that such, or such an one, who watched St. Mark's Eve, says so.

The poem as it stands contains no reference to this custom, but the following poem—a fragment almost certainly intended to be part of the present poem—does. R. Gittings (*The Living Year* (London, 1954), pp. 86–92) links details in the poem with Keats's memories of Stansted Chapel and Isabella Jones's room. The poem's Middle English pastiche recalls Chatterton, though Keats's manner is more parodic and light-hearted. See also the prose fragment 'Whenne Alexandre the Conqueroure' (above, p. 333) and 'He is to weet a melancholy Carle' (above, p. 271).

l. 28. *golden broideries*. The gold illuminations of the volume; the patterns of these are often intricately intertwined.

l. 33. *Aaron's breastplate*. Exod. 28: 15–30, Lev. 8: 8.

ll. 33–4. *seven Candlesticks*. Rev. 1: 13–20. The seven-branched candelabra, representing the seven Churches of Asia, is, along with the Ark of the Covenant (ll. 36–8) depicted in Lewis Way's window at Stansted Chapel. Both are Jewish symbols, redefined by Christians.

266 l. 38. *golden Mice*. 1 Sam. 6: 4, 11, 18.

l. 39. *Bertha*. Keats may have got the name from the heroine of T. Chatterton's *Aella, a Tragic Interlude* (1777).

267 l. 79. *Lima Mice*. Phonetic spelling of 'lemur mice'.

l. 80. *legless*. Traditionally, birds of Paradise are always in the air and never alight.

l. 81. *av'davat*. The amadavat, an Indian song-bird.

l. 98. In copying the fragment in his letter to George and Georgiana Keats breaks off at the end of this line to introduce the remainder: 'What follows is an imitation of the Authors in Chaucer's time—'t is more ancient than Chaucer himself and perhaps between him and Gower.'

268 l. 117. *holy shrine*. St Mark's basilica at Venice.

'Gif ye wol stonden hardie wight'

Date of composition uncertain but the only copy is written on the back of ll. 99–114 of the Morgan Library draft of the previous poem. Published in E. de Selincourt (ed.), *The Poems of John Keats* (5th edn., London, 1926), p. 584; present text from Morgan library draft.

'Why did I laugh tonight? No voice will tell'

Composed Mar. 1819 (before 19 Mar. when Keats copied it into his letter to George and Georgiana (above, p. 465); published in *1848*; present text from letter (which see for comments).

269 ll. 9–12. The figure of Marlowe's Doctor Faustus, who dies at midnight having made 'a deed of gift of body and soul' to Mephostophilis, is behind these lines and the imagery of the whole poem; l. 12 recalls Faustus's 'See, see where Christ's blood streams in the firmament' (xix. 146).

'When they were come unto the Faery's Court'

Composed as 'a little extempore' in a letter to George and Georgiana on 15 Apr. 1819 (above, p. 468); ll. 1–17 published in *Macmillan's Magazine*, Aug. 1888; complete text first published in H. B. Forman (ed.), *Poetry and Prose of John Keats* (London, 1890); present text from the letter (the only source). The dwarf and the ape in this poem are possibly the same as those in Brown's 'The Fairies' Triumph' (see note to 'Shed no tear—O shed no tear' (below, p. 603).

l. 12. *Persian feathers*. Turbanned head-dresses pinned with feathers were part of the vogue of orientalism.

l. 14. *Otaheitan*. From Tahiti (Keats got the name from Cook's *Voyages*).

270 l. 66. *cup biddy*. 'Come up Biddy' (i.e. 'gee up').

ll. 67–8. Allott refers to *Hamlet* IV. ii. 16–18 'such officers do the King best service in the end: he keeps them, like an ape, in the corner of his jaw'.

l. 71. *wards*. 'the ridges projecting from the inside plate of a lock, serving to prevent the passage of any key the bit of which is not provided with incisions of corresponding form and size' (*OED*).

271 l. 80. *King Lear* IV. vi. 107; *Romeo and Juliet* III. i. 136.

'He is to weet a melancholy Carle'

Composed 16 Apr. 1819 in Keats's journal-letter to George and Georgiana as an extempore portrait of Charles Brown 'in the manner of Spenser' (see p. 468 above); published in *1848*; present text from the letter. Wordsworth had used Spenserian stanzas, and occasionally diction, in 'Within our happy Castle there dwelt one' (published 1815) which includes a self-portrait and a portrait of Coleridge.

272 l. 10. *half and half.* 'a mixture of ale and beer, or beer and porter, in equal quantities' (Partridge).

l. 14. *lewd ribbalds. FQ* II. i. 10.

l. 15. *lemans.* Lovers; see *FQ* VI. viii. 21 'Thus I triumphed long in lovers paine,| And sitting carelesse on the scorners stolle,| Did laugh at those that did lament and plaine'; Keats is here making light of Brown's flirtatiousness which, when Fanny Brawne was the object, caused him such pain (see his letter to Fanny Brawne, 5 July 1820 (above, p. 532)).

ll. 16–17. Ps. 13: 1.

l. 21. *olden Tom or ruin blue.* Two terms for gin—Blue Ruin was cheap gin.

l. 22. *nantz.* Cherry brandy (originally from Nantes).

ll. 26–7. 'The daughters of Zion are haughty . . . walking and mincing as they go, and making a tinkling with their feet' (Isa. 3: 16). Burton quotes this in a passage marked by Keats in the *Anatomy of Melancholy* (London, 1806 edn., ii. 245).

A dream, after reading Dante's Episode of Paolo and Francesca

Composed Apr. 1819 (on or before the 16th when Keats copied the poem in his journal-letter to George and Georgiana (above, p. 469)); published in the *Indicator*, 28 June 1820, p. 304. See the letter for introductory remarks. One of the holograph manuscripts of the poem is on a blank leaf at the end of vol. i of Cary's Dante.

273 l. 10. *world-wind. Indicator*; extant MSS give 'whirlwind' but, as Jerome McGann argues, 'world-wind' is unlikely to be a copyist's or printer's error for the more expected word. Keats was staying with Hunt (editor of the *Indicator*) at the time of going to press. (J. McGann, 'Keats and the Historical Method in Literary Criticism', *MLN* 94 (1979), 1003–5). The *Indicator* text is signed 'Caviare'—a reference to *Hamlet* II. ii. 457 possibly suggested by the review of *Endymion* in the *Chester Guardian* which deemed it 'caviar to the general' (D. Hewlett, *A Life of John Keats* (2nd edn., London, 1949), p. 180).

La belle dame sans merci

Composed 21 Apr. 1819 in Keats's journal-letter to George and

Georgiana (see p. 472 above); published in the *Indicator*, 10 May 1820 where it, like the previous poem, is signed 'Caviare'. The present volume's policy of printing those poems published during Keats's lifetime in their latest published form has not been followed in this case, where the letter text has been used. The alterations made in the *Indicator* version may have been made by Hunt. If they are Keats's own revisions they were made at a time of serious ill health which may well have led to a failure of nerve and loss of confidence in earlier judgements. The letter version given here is, justly, the better known of the two (but see McGann, art. cit. (above, n. 10 to previous poem), pp. 1005–8); the *Indicator* text is given as Appendix II below. The title of the poem is that of a medieval ballad written by Alain Chartier in 1424 which, according to Hunt in the *Indicator*, appears in a translation at the end of Chaucer's *Works* (London, 1782 edn.), attributed to Chaucer.

Song of four Fairies: Fire, Air, Earth, and Water

274 Composed 21 Apr. 1819 in Keats's journal-letter to George and Georgiana (see p. 472 above); published in *1848*; present text from holograph fair copy (Harvard) which Woodhouse identifies as Keats's 'copy for the press'—an indication that the poem may have been intended at one time for *1820*. The names of the fairies reveal their natures, like the names of the fairies in *A Midsummer Night's Dream*: a salamander is fabled to be a fire-dweller; a zephyr is a wind (hence 'air'); Dusketha stands for dusk-dark earth; and Breama takes her name from the water-dwelling bream.

278 l. 100. *twilight*. Used as a verb (cf. 'earthquake' in 'Upon my Life Sir Nevis I am piqued' (above, p. 216) 32).

Sonnet to Sleep

Composed late Apr. 1819 ('lately written' when transcribed in a letter on 30 April; see p. 474 above); published in *PDWJ* 11 Oct. 1838; present text from holograph album transcript (Berg collection). Keats drafted the first 12 lines on the flyleaf of vol. ii of *PL*. The precise order in which this and the 4 following poems were composed is uncertain. They represent a period of concentrated technical endeavour and experiment with patterns of rhyme and consonance to bind a stanza.

l. 11. *hoards*. The letter holograph and all transcripts save *W1* give 'lords'. Woodhouse has pencilled 'hoards' over 'lords' in *W2* and Keats adopted this when he transcribed the poem into the album.

Ode to Psyche

Composed late Apr. 1819 (it is referred to as 'the last I have written' on 30 Apr. when Keats transcribed it into his journal-letter to George and Georgiana (above, p. 475)); published in *1820*. See the letter for introductory comments. Keats had referred to the legend of Cupid and Psyche in 'I stood tip-toe' (above, pp. 49–50) 141–50; his treatment is based on the episode in Apuleius' *Golden Ass* in William Adlington's translation (1566).

279 l. 14. *Tyrian*. Purple (a dye that comes from Tyre).

l. 21. *winged boy*. Cupid (not named *because* the speaker knew him).

ll. 28–35. The thought, diction, and rhythm of these lines recall Milton's description of the noisy rout of pagan gods in 'On the Morning of Christ's Nativity' (esp. ll. 173–80).

l. 41. *lucent fans*. Gleaming wings.

280 ll. 54–5. Cf. Keats's description of Derwent Water which 'ooses out from a cleft in perpendicular Rocks, all fledged with Ash & other beautiful trees' in a letter to Tom of 29 June 1818 (*L* i. 306).

l. 65. *shadowy thought*. Cf. Keats's ideas about the advantages of 'being in uncertainties' and making 'up ones mind about nothing' (letters of 21, (?)27 Dec. 1817 and of 24 Sept. 1819; above, p. 370, 515).

On Fame ('Fame like a wayward girl will still be coy')

Composed 30 Apr. 1819 when it is described as 'just written' in a letter to George and Georgiana (14 Feb.–3 May 1819) (above, p. 474); published in the *Ladies' Companion* (New York), 1837; present text from the letter.

l. 10. *Potiphar*. Gen. 39.

On Fame ('How fever'd is that Man who cannot look')

281 Composed 30 Apr. 1819 as an extempore in Keats's journal-letter to George and Georgiana (of 14 Feb.–3 May 1819) while Brown was transcribing the previous sonnet (above, p. 474); published in *1848*; present text from the letter.

l. 13. *teasing*. The copy text reads 'leasing' while later transcripts give 'teasing'—a much more characteristic word which also makes more sense. I have assumed that Keats failed to cross his *t*.

'If by dull rhymes our English must be chain'd'

Composed late Apr. or early May 1819 and enclosed in Keats's journal-letter to George and Georgiana of 14 Feb.–3 May (above, p. 475); published in *1848*; present text from Charles Brown's transcript. See the letter for Keats's introductory remarks about the sonnet form.

ll. 3, 5. *Fetter'd . . . Sandals*. Plays on the idea of metrical feet.

Two or three Posies

282 Composed probably on 1 May 1819 in a letter to Fanny Keats (above, p. 449) for which this date is conjectured. Published in Forman (1883); present text from the letter.

l. 20. *Mrs.— mum!* The self-censored name is, as the rhyme suggests, 'Abbeys'. Fanny was still living under the restrictive care of her guardian Richard Abbey and his wife at the date of this letter. In a letter to George and Georgiana of 17–27 Sept. 1819 (above, p. 510) Keats writes 'Does not the word mum! go for ones finger beside the nose—I hope it does'.

Ode on Indolence

283 Composed during spring 1819; published in *1848*; present text from Brown's transcript. A prose description of the mood which this poem treats occurs in the 19-Mar. passage of Keats's journal-letter to George and Georgiana (above, p. 463): 'Neither Poetry, nor Ambition, nor Love have any alertness of countenance as they pass by me: they seem rather like three figures on a greek vase'. On 9 June 1819 Keats refers to this poem in a letter to Mary-Ann Jeffery (above, p. 478) as 'the thing I have most enjoyed this year'. The mood of the poem is linked, reactively, to the pressure which Keats was under to make himself financially eligible to be Mrs Brawne's son-in-law.

Motto. Matt. 6: 28.

l. 10. *Phidian lore*. The mysteries of plastic art; Phidias (5th c. BC) was the creator of the Elgin Marbles.

284 l. 43. *lawn*. Both fabric (picking up on 'embroider'd', l. 42) and grass.

l. 54. *pet-lamb*. In K's letter to Mary-Ann Jeffery of 9 June (above, p. 478) he writes 'I hope I am a little more of a Philosopher than I was, consequently a little less of a versifying Pet-lamb.'

'Shed no tear—O shed no tear'

285 Date of composition uncertain, but probably some time in 1819; published in *PDWJ* 18 Oct. 1838; present text from holograph (Harvard). The song was written to be included in Brown's unfinished 'The Fairies' Triumph' (manuscript at Keats House). The princes Elury and Azameth have heedlessly plucked a flower and seen it shrivel to dust. In their grief they 'heard a most enchanting melody breathed forth; and looking up they saw, perched on a slender bough, a bird of lovely form, and brilliant plumage, and it gazed down on them with its mild dove-like eyes, and warbled its song to cheer them'. That song is the poem. See J. Stillinger, 'The Context of Keats's "Fairy Song"', *KSJ* 10 (1961), 6–8.

Ode to a Nightingale

Composed May 1819; published in *Annals of the Fine Arts*, July 1819 and in *1820*. Brown described the circumstances of the poem's composition in his 'Life' of Keats (written 17 years after the event):

> In the spring of 1819 a nightingale had built her nest near my house. Keats felt a tranquil and continual joy in her song; and one morning he took his chair from the breakfast-table to the grass-plot under the plum-tree, where he sat for two or three hours. When he came into the house, I perceived he had some scraps of paper in his hand, and these he was quietly thrusting behind the books. On inquiry, I found those scraps, four or five in number, contained his poetic feeling on the song of our nightingale. The writing was not well legible; and it was difficult to arrange the stanzas on so many scraps. With his assistance I succeeded, and this was his 'Ode to a Nightingale', a poem which has been the delight of every one. (*KC* ii. 65)

The only known holograph of the poem comprises only two sheets of paper (in the Fitzwilliam Museum, Cambridge) but Brown's memory may have been false. Keats would have known Coleridge's 'The Nightingale: A Conversation Poem' (1798) (though it is Coleridge's 'This Lime Tree Bower my Prison' which most influences the movement of thought in the present poem). Nightingales also head the list of topics broached by Coleridge during the two miles on which Keats accompanied him in Apr. 1819 (see above, p. 468).

ll. 1–4: echo Horace, *Epode*, xiv. 1–4 'Mollis inertia cur tantam diffuderit imis| Oblivionem sensibus,| pocula Lethaeos ut si ducentia somnos| arente fauce traxerim' ('Why spineless langour on all my innermost self | Has spread such deep forgetfulness,| As if with gullet all dry I had drained to the dregs | Such cups as bring Lethean sleep'; trans. Lord Dunsany and Michael Oakley, *The Collected Works of Horace* (London, 1961), p. 129).

l. 2. *hemlock*. A sedative or poison; cf. Marlowe's rendering of Ovid, *Elegies* III. vi. 14 'like as if cold hemlock I had drunk'.

286 l. 15. *warm South*. Southern wine.

l. 26. Tom Keats had died of tuberculosis on 1 Dec. 1818. Cf. also Wordsworth, *The Excursion* iv. 760 'While man grows old, and dwindles, and decays'. This stanza has many echoes in the letters; see e.g. 10 Jun. 1818 to Bailey (above, p. 399): 'were it my choice I would reject a petrarchal coronation—on account of my dying day, and because women have Cancers'.

287 ll. 46–9. Cf. *A Midsummer Night's Dream* II. i. 249–52 and Keats's remarks about the 'delightful forwardness' of the flowers (in his journal-letter of spring 1819; above, p. 475).

l. 51. *PL* iii. 38–40, marked by Keats: 'the wakeful Bird | Sings darkling, and in shadiest Covert hid | Tunes her nocturnal Note'.

l. 60. *requiem*. Prolepsis, anticipating the speaker's death.

ll. 65–7. Ruth was driven from her native Moab by famine and worked in Bethlehem as a gleaner (Ruth 2: 1–3). Hazlitt, in his lecture 'On Poetry in General', writes that 'The story of Ruth . . . is as if all the depth of natural affection in the human race was involved in her breast' (*Works*, v. 16); Wordsworth's 'The Solitary Reaper' (published in 1807) also informs the image.

Ode on a Grecian Urn

288 Composed during 1819; published in *Annals of the Fine Arts*, 15 (January, 1820), and in *1820*. The Keats–Shelley Memorial House in Rome has a drawing, or tracing, made by Keats of the Sosibios Vase taken from the Musée Napoléon. See Jack, pp. 214–24 for illustrations of vases and other Greek works of art of the kind which inform the poem. The 'heifer lowing at the skies' (l. 33) is almost certainly inspired by the heifer being led to sacrifice in the south frieze of the Elgin Marbles (Jack, pl. xxxiii).

289 l. 28. 'Greek statues . . . are marble to the touch and to the heart . . . In

their faultless excellence they appear sufficient to themselves. By their beauty they are raised above the frailties of passion or suffering. By their beauty are they deified' (Hazlitt, *Works*, v. 11). In the *Examiner* (26 May, 1816), defining the singular kind of gusto possessed by Greek statues, Hazlitt writes 'It seems enough for them *to be*, without acting or suffering . . . their beauty is power' (*Works*, iv. 79).

l. 41. *brede . . . overwrought*. The principal meanings are concerned with decoration and embroidery, but the biological and emotional senses of the homophones are played on.

l. 48. *friend to man*. Cf. Keats's remark about Milton being 'an active friend to Man all his Life and . . . since his death' (in his letter to Rice of 24 March 1818; above, p. 387).

ll. 49–50. The inverted commas are not present in any surviving transcript of the poem and do not appear in the *Annals* text. *1820*, which Keats saw through the press, is the sole authority for this unowned quotation.

Ode on Melancholy

290 Composed during 1819; published in *1820*. Brown's transcript of the poem contains a cancelled preliminary stanza in which Gothic self-dramatizing is presented in all its extravagance:

> Though you should build a bark of dead men's bones,
> And rear a phantom gibbet for a mast,
> Stitch creeds together for a sail, with groans
> To fill it out, bloodstained and aghast;
> Although your rudder be a Dragon's tail,
> Long sever'd, yet still hard with agony,
> Your cordage large uprootings from the skull
> Of bald Medusa; certes you would fail
> To find the Melancholy, whether she
> Dreameth in any isle of Lethe dull.

ll. 6–7. Psyche (the soul) was often represented as a butterfly; the death's head moth, whose markings resemble the human skull, is presented as a deathly antitype.

The Fall of Hyperion

291 Begun as a revision of 'Hyperion' probably during July 1819 while Keats was at Shanklin, Isle of Wight; 'given up' *c.*21 Sept. 1819 (see Keats's letter to Reynolds of that date (above, p. 493)) though Keats probably continued to toy wth the poem after that date. (Brown records that Keats was 'deeply engaged in remodelling his poem of "Hyperion" into a "vision" while also writing *The Cap and Bells*' (*KC* ii. 72). Published (without i. 187–210) by R. M. Milnes ('Another Version of Keats's "Hyperion"', *Miscellanies of the Philobiblion Society*, 3 (1856–7)), and at the same time issued privately by him as a separate pamphlet. The omitted lines were first published in E. de Selincourt's *Hyperion: a Facsimile*

(Oxford, 1905); present text from *W2*. The poem attempts a reconstruction of the abandoned 'Hyperion' and retains many passages from the earlier poem. Here Keats casts the narrative in the form of a dream vision (a genre with medieval precedents, such as Chaucer's *The Book of the Duchess*). Keats's professed reason for abandoning the poem was his dislike of its 'Miltonic inversions' (letter to Reynolds, 21 Sept. 1819; above, p. 493). Keats felt Milton's influence to be creatively suffocating (see p. 515 above) and it was clearly so pervasive in this poem as to be ineradicable. (This is equally, if not more, true of 'Hyperion'.) Keats's rapidly deteriorating health may also have contributed to his abandonment of the poem which had reached a position of impasse. His friends tended to prefer the earlier 'Hyperion' and Keats decided against including the new poem in *1820*. Lines i. 1–11, 61–86, and ii. 1–4, 6 are quoted in a letter to Woodhouse of 21 Sept. 1819 (above, p. 495).

Canto i

ll. 31–4. Cf. *PL* v. 303–7, 326–8 (both marked in his copy by Keats).

292 l. 48. *Caliphat*. Keats probably meant an individual caliph (Muslim ruler) rather than his abstract office (caliphate); the line smacks of Gothic melodrama.

l. 56. *Silenus*. See Jack, pl. xxxvii for illustration of the Borghese vase with such a detail.

l. 70. *faulture*. Weakness (the letter draft has 'failing'); the word carries a sense of geological fault.

l. 74. *asbestus*. 'the unquenchable stone' (*OED*).

l. 75. Matt. 6: 20.

293 l. 97. *mid-May*. 'midway' *W2*; 'midday' *Miscellanies of the Philobiblion Society*, 3 (1856–7); A. E. Housman suggested the emendation (*TLS* 8 May 1924, p. 286). The emended phrase links with the 'Maian incense' of 103.

294 l. 137. 'The most ancient alters were adorned with horns' (John Potter, *Antiquities of Greece* (Edinburgh, 1813), i. 229.

l. 144–5. *Dated on | Thy doom*. Postponed your death.

295 l. 157. Cf. 'Sleep and Poetry' 122–5.

ll. 187–210: omitted in the *Miscellanies* text; Woodhouse records 'Keats seems to have intended to erase this.' Bate (p. 599) cautions against our forgetting 'that these lines are departing from the course of the poem—to some extent conflicting with it (as Keats himself saw)—and then attempt-[ing] to use them as a means of explaining the context'.

296 l. 205. *Iliad* i. 9–12 (Apollo spreads a plague among the Greeks).

ll. 207–8: perhaps Hunt (or Thomas Moore), Wordsworth, and Byron.

l. 222. *war*. The war of the Titans against the rebel Olympians.

l. 226. *Moneta*. Another name for Mnemosyne (the name used in 'Hyperion'); see Finney, p. 500.

297 ll. 249–50. Cf. Dante, *Purgatorio* xxx. 79–80 in which Beatrice appears an awesome mother to Dante; the fearful figure in Coleridge's 'The Ancient Mariner' is also recalled.

298 l. 294. 'Hyperion' began here.

300 ll. 382–3. Cf. *Purgatorio* xii. 16–18; also Shelley's 'Ozymandias'.

301 l. 411. Pan is solitary because the Golden Age has passed.

302 l. 465. *Antichamber*. Cf. Keats's account of the 'chambers' of thought in his letter to Reynolds of 3 May 1818 (above, p. 397). The vocabulary recalls that of Renaissance 'faculty psychology' and may reflect Keats's reading of Burton, as well as his direct knowledge of the 'cerebral apartments'.

Canto ii

ll. 1–3. Cf. *PL* v. 571–4.

ll. 1–6. See Keats's comment in his letter to Woodhouse of 21 Sept. 1819 (above, p. 495).

303 l. 50. *Mnemosyne*. Moneta; her seat and posture may recall Dürer's *Melancholia*.

Lamia

305 Begun late June or early July 1819 while at Shanklin, and completed at Winchester during late Aug. or early Sept. Revised Mar. 1820; published in *1820*. There was a 6-week interval between the composition of the two parts, during which Keats was occupied with 'Otho the Great' and 'The Fall of Hyperion'. Keats wrote the poem with an eye to its reception, wishing to create a poem less 'weak-sided' and 'smokeable' than his other two narrative poems (see letters to Reynolds of 11 July 1819 and to Woodhouse of 21, 22 Sept. 1819 (above, pp. 481 and 496). Woodhouse, in a letter to Taylor of 19 and 20 Sept. 1819, reports that Keats had read him 'Lamia' 'which he has half fair Copied'. He gives a summary of the plot and comments:

> The metre is Drydenian heroic—with many triplets, & many alexandrines. But this K. observ'd, & I agreed, was required, or rather quite in character with the language & sentiment in those particular parts.—K. has a fine feeling when & where he may use poetical licenses with effect. (*L* ii. 164–5)

Brown, in his *Life of John Keats*, says 'He wrote it with great care, after much study of Dryden's versification' (*KC* ii. 67). Keats placed the poem first in *1820*. The prime source for the story is the passage from Burton's *Anatomy of Melancholy* (III. 2. i. 1), which was printed at the end of the poem in *1820*, as it is here. The character of Lamia owes something to that of Geraldine in Coleridge's 'Christabel' and also to the enchantress in Peacock's *Rhododaphne* (1818).

Part i

ll. 1–145. The episode of Hermes and the nymph derives from Ovid, *Metamorphoses* ii. 708 f.

l. 11. *his great summoner* i.e. Jove, whose cup he bears.

306 ll. 36–7. Cf. Chaucer, 'The Knightes Tale' 903 'For pitee renneth soon in gentil herte'; also *Inferno* v. 100 (in Cary's translation): 'Love, that in gentle heart is quickly learned'.

l. 46. *cirque-couchant*. Lying coiled—a neologism disguised as an archaic heraldic term.

l. 55. *penanced*. Transformed from her former shape as penance or punishment.

307 l. 81. *star of Lethe*. One of Hermes' roles is that of *psychopomp* (conductor of the souls of the dead).

l. 114. *psalterian*. Like the music of the psaltery.

l. 115. *Circean*. Dangerously enchanting (like Circe); or, perhaps, 'transformed' (like one of Circe's victims).

308 ll. 126–7. Cf. Keats's comparison of the imagination to Adam's dream (letter to Bailey, 22 Nov. 1817; above, p. 365).

ll. 146–70. Woodhouse to Taylor: 'The change is quite Ovidian, but better' (*L* ii. 164).

309 l. 163. *rubious-argent*. Reddish silver—another pseudo-heraldic coinage (Shakespeare uses 'rubious').

l. 168. Keats's comment on an earlier draft of this line was:

Shore Shore Shore Shore
Jane Jane (*KC* i. 112)

(Nicholas Rowe's *Jane Shore* (1714) had been performed at Drury Lane during Apr. 1819.)

l. 174. *Cenchreas*. Cenchrea was a Corinthian port.

l. 179. *Cleone*. A village south of Corinth.

311 l. 256. He is her captive.

312 l. 279. 'Thou art a scholar. Speak to it Horatio' (*Hamlet* I. i. 42); Marlowe's Doctor Faustus also used his scholarship to speak with spirits—scholars were reputedly versed in spirit lore.

l. 284. Perhaps suggested by Montaigne's *Apologie de Raimond Sebond* (1575–80) in which the reliability of sensory evidence is queried on the grounds that there may be other senses of which we have no experience.

313 l. 320. In Marlowe's *Hero and Leander* (i. 91–3, 131–4) the lovers first sight one another at the feast of Adonis (usually held in Venus' temple).

l. 329. *Peris*. Persian genii—familiar through pantomime.

ll. 334–5. A Byronic (*Don Juan*-style) and also a Marlovian couplet (see also ii. 24–5).

l. 347. *comprized*. Absorbed.

314 l. 375. *Apollonius*. Apollonius of Tyana, Pythagorean philosopher b. *c.* 1 AD credited with occult knowledge.

ll. 388–93. Woodhouse, in a letter to Taylor of 20 Sept. 1819, compares the secret palace to 'the Cavern Prince Ahmed . . . found in the Arabian Nights, when searching for his lost arrow' (*L* ii. 164).

Part ii

317 l. 80. *The serpent.* Python.

l. 81. Woodhouse wrote to Taylor on 19, 20 Sept. 1819 that Keats commented 'Women love to be forced to do a thing, by a fine fellow—*such as this*' (*L* ii. 164).

318 ll. 122–62. Keats copied a draft of these lines and of 199–220 in a letter to Taylor of 5 Sept. 1819 (*L* ii. 157–8). The draft text differs considerably from *1820*.

l. 136. *viewless.* Invisible (cf. 'Ode to a Nightingale' (above, p. 286) 33).

319 l. 160. *daft.* Bemused, foxed.

ll. 181–5. John Potter (*Antiquities of Greece* (Oxford, 1697), ii. 376–7) gives information about the design of classical Greek tables which Keats may have used here.

320 ll. 199–220. The letter draft of these lines makes the contrast between the gross guests and the ethereal banquet the source of social satire:

> 'Where is that Music?' cries a Lady fair,
> 'Aye, where is it my dear? Up in the air'?
> Another whispers 'Poo!' saith Glutton 'Mum!'
> Then makes his shiny mouth a napkin for his thumb.

ll. 226. Bacchus' wand ('thyrsus') is wound with vine and ivy.

ll. 229–37. Hazlitt, in his 1818 lecture 'On Poetry in General' had said that 'the progress of knowledge and refinement has a tendency to circumscribe the limits of the imagination, and to clip the wings of poetry' (Works, v. 9). Haydon recorded in his diary that during a dinner-party at his house on 28 Dec. 1817 Keats had agreed with Lamb that Newton had 'destroyed all the Poetry of the rainbow by reducing it to a prism' (*The Diary of Benjamin Robert Haydon*, ed. W. B. Pope (Cambridge, Mass., 1960), ii. 72).

321 l. 236. *gnomed mined.* Cf. Wordsworth's 'Peter Bell' (published 1819), ll. 976–80 where the sound of miners is interpreted as diabolic spirits.

'Pensive they sit, and roll their languid eyes'

323 Composed at Winchester in a letter to George and Georgiana on 17 Sept. 1819 (above, p. 503); published in the *World* (New York), 25 June 1877; present text from the letter. In spite of having by this time declared his love for Fanny Brawne Keats, wilfully preserving his integrity and ill able to provide for himself, let alone a wife, was resistant to romantic love during his stay in Winchester. The present poem is apropos Haslam's courtship of his future wife (see the letter for comments).

l. 9. *humane society.* The Royal Humane Society was founded in 1774.

l. 10. *Mr. Werter.* Sensitive hero of Goethe's cult novel *Die Leiden des jungen Werther* (1773), trans. Richard Graves as *The Sorrows of Young Werter: a German story* (London, 1783). Werter is conscious that 'Every moment I am myself a destroyer. The most innocent walk deprives of life thousands of poor insects' (i. 144–5).

l. 14. *Romeo*. Cf. 'And what is Love?—It is a doll dress'd up' (above p. 222) where the incongruity of modern love with fine sensibility is also contemplated.

l. 15. *Cauliflower*. The shape of the untrimmed wick.

l. 16. *winding sheet*. 'A mass of solidified drippings or grease clinging to the side of a candle, resembling a sheet folded in creases, and regarded as an omen of death or calamity' (*OED*).

324 l. 23. *Wapping*. A district in East London, near the Thames. East London has long been associated with garment manufacture.

To Autumn

Composed at Winchester on 29 Sept. 1819; published in *1820*.

See Keats's letter to Reynolds of 21 Sept. 1819 (above, p. 493) for a description of the poem's occasion.

l. 25. *bloom*. A transitive verb.

l. 26. *stubble-plains*. The personification implied by 'stubble' recalls Shakespeare, *Sonnets* xii. 7–8.

'Bright Star, would I were stedfast as thou art'

325 Composed during 1819; published *PDWJ* 27 Sept. 1838. The present text is from a copy which Keats made inside his 1806 edition of the *Poetical Works of Shakespeare* (now at Keats House). According to Severn this copy was made on board the *Maria Crowther* off the British coast *en route* for Italy in late Sept. or early Oct. 1820. Until the discovery of an earlier transcript by Brown, dated 1819, this was believed to be Keats's last poem.

ll. 1–4. Cf. Keats's description of Lake Windermere in his letter to Tom of 25–7 June 1818 (above, p. 401).

On Coaches (from *The Jealousies*)

Composed probably during Nov. and Dec. 1819 at Hampstead. This extract, with this title, was published in the *Indicator*, 23 Aug. 1820 (the present text). The (unfinished) whole was published in *1848* with the title 'The Cap and Bells; Or, the Jealousies. A Faery Tale. Unfinished'. Brown describes the circumstances under which the work was composed:

> By chance our conversation turned to the idea of a comic faery poem in the Spenser stanza, and I was glad to encourage it. He had not composed many stanzas before he proceeded to write with spirit. It was to be published under the feigned authorship of Lucy Van Lloyd and to bear the title of 'The Cap and Bells', or, which he preferred, 'The Jealousies'. This occupied the mornings pleasantly. He wrote it with the greatest facility; in one instance I remember having copied (for I copied as he wrote) as many as twelve stanzas before dinner. In the evenings, at

as he wrote) as many as twelve stanzas before dinner. In the evenings, at his own desire, he was alone in a sitting-room, deeply engaged in remodelling his poem of 'Hyperion' into a 'Vision'. (*KC* ii. 71–2)

The poem remained unfinished because of Keats's ill health. In a letter to Reynolds of 28 Feb. 1820 Keats hopes he 'shall be soon well enough to proceed with my fairies' (*L* ii. 268). He also refers to the poem in letters to Brown of May and June 1820 (*L* ii. 289–90, 299) and, in August (?)1820 he writes to Brown, apropos 'the offence the ladies take at [him]' that 'If ever I come to publish "Lucy Vaughan Lloyd", there will be some delicate picking for squeamish stomachs' (*L* ii. 327–8). The longer poem deals—lightly, for once—with the Keatsian topic of miscegenation between different orders of being; in this case between fairy and mortal. The fairy emperor Elfinan, preferring the substantial embrace of the mortal Bertha to that of his betrothed, the fairy Bellanaine, attempts to evade his arranged marriage by capturing and eloping with his beloved. (The situation has parallels both with the Prince Regent's marriage to the Princess Caroline and with Byron's unhappy marriage. Byron is actually quoted at one point.) Elfinan summons his page Eban, 'A fay of colour, slave from top to toe', on an errand to Hum the soothsayer whose assistance he requires in planning his elopement. The extract describes Eban's mode of conveyance.

326 l. 10. *the string.* The check-string which passengers pulled to signal to the driver to stop.

l. 11. *Jarvey.* The word used for a driver of a hackney coach (like the modern 'cabbie'); here the speaker addresses the coach and driver as one. *hack.* A hackney cab and also the horse.

l. 13. *linsey-wolsey.* A mixture of wool and flax; a cheap cloth.

l. 19. *crop.* Stomach.

ll. 35–6. *Tilburies . . . Phaetons . . . Curricles . . . Mail-coaches.* All forms of light carriage.

327 l. 40. *Louted.* Bowed (Spenserian, like the stanza).

'The day is gone, and all its sweets are gone'

Composed during 1819—probably towards the end of the year after Keats's return to Hampstead and the propinquity of Fanny Brawne to whom this and the following poems are addressed. Published *PDWJ* 4 Oct. 1838; present text from Brown's transcript. Cf. 'I eternally see her figure eternally vanishing' (in letter to Brown of 30 Sept. 1820; above, p. 539).

'What can I do to drive away'

Composed probably late in 1819; published in *1848* (the sole source). For a contemporary poetic comment on this poem see Tony Harrison, *A Kumquat for John Keats* (Newcastle upon Tyne, 1981).

328 l. 33. *wrecked.* Often emended to 'wretched'; Keats was probably playing on a medieval pronunciation of 'wrecched' whose etymology he may have misunderstood.

l. 42. *flowers*. Ed.; 'bad flowers' *1848*. Forman conjectured that Keats first wrote 'bud' (misread as 'bad') and forgot to cancel the word before emending to 'flowers'.

'I cry your mercy—pity—love!—aye, love'

Composed during 1819 (so dated in Brown's transcript); probably late in the year. Published in *1848*; present text from Brown's transcript.

To Fanny

329 Composed late 1819 or early 1820—most probably after 3 Feb. when Keats was confined indoors after a second haemorrage and suffered agonies of jealousy at the neighbouring Fanny's relative freedom (see for example the letter of ?5 July 1820 (above, p. 532)). Published in *1848*. Present text from R. M. Milnes's transcript (the only complete source text).

l. 1. *let my spirit blood!* Blood-letting was then, as for the previous 3,000 years, a common therapeutic practice and one which Keats was regularly enduring at the time he wrote this poem. ('Whenever Keats lost blood, they opened a vein in his arms and removed still more blood' (Bate, p. 637)). The principle was that pressure would thereby be eased.

l. 8. *wintry air*. Allott cites this line as evidence for assigning the poem to Feb. 1820. But the winter may be entirely metaphorical (as in 'Isabella' (above, p. 187) 65–6).

330 l. 27. *wreath*. Perhaps a conscious or unconscious play on 'reef'.

l. 40. *blow-ball*. Dandelion clock.

'This living hand, now warm and capable'

331 Composed probably late 1819; published in H. B. Forman (ed.), *The Poetical Works of John Keats* (London, 1898); Keats drafted or copied the lines on to the outside of a sheet on whose other side he drafted ll. 45–51 of 'The Jealousies'. Bate (p. 626) suggests that the lines may have been written for inclusion in some future play. It is equally possible to read them as expressions of the jealousy afflicting the previous poem and in the manner of Donne's 'The Visitation'.

l. 1. Cf. 'The Fall of Hyperion' (above, p. 291) i. 18.

In after time a Sage of mickle lore

Keats wrote these lines at the end of the *FQ* in the now lost copy of Spenser which he gave to Fanny Brawne. Brown describes this as 'the last stanza of any kind that [Keats] . . . wrote before his lamented death.' Published in *PDWJ* 4 July 1839; present text from Brown's transcript (in his copy of Spenser at Keats House).

l. 2. *the Giant*. See *FQ* v. ii. 30–50 where the Giant travesties justice in his revolutionary attempt to make all things equal. In *FQ* Sir Artegall (the knight of Justice) and Talus, his iron squire, oppose and destroy the giant.

Here Keats envisages the giant's re-education into a triumphant champion of real freedom. (See Introduction (above, p. xxii) for Brown's comments on the poem's politics.)

PROSE

'Whenne Alexandre the Conqueroure'

333 Composed probably late 1815 while Keats was a student at Guy's. The only source for this fragment is Walter Cooper Dendy, *The Philosophy of Mystery* (London, 1841), pp. 99–100 where it is quoted at the end of a chapter on the pathology of 'Poetic Phantasy or Frenzy'. Keats is introduced as an example of one self-blighted, eschewing medicine, 'that science, that might have nursed and fortified [his] mind', in favour of poetry. Dendy describes this piece as 'a quaint fragment which he one evening scribbled in our presence, while the precepts of Sir Astley Cooper fell unheeded on his ears' (p. 98). Dendy (who entered Guy's and Thomas's as a student in 1811) was not an exact contemporary of Keats, and his evidence may not be reliable. Nevertheless, the tone and diction of the passage and Keats's known pleasure in medieval pastiche suggest that the fragment is genuine.

Keats's marginalia to the Shakespeare Folio

Keats possessed a copy of the 1808 reprint of the Shakespeare First Folio (1623). The title-page of this volume (now in Keats House) is inscribed in bold strokes 'John Keats | 1817'. Below, in smaller letters 'to F.B. | 1820'. The dates of the marginalia cannot be ascertained. This was not the edition of Shakespeare that Keats took with him to the Isle of Wight in 1817 (Whittingham's 7-vol. edition of 1814; a full account of these volumes is available in Spurgeon, *Keats's Shakespeare* (2nd edn., Oxford, 1929)). The tenor of some of Keats's marginalia evinces an interest in the kind of textual criticism which engaged Dilke.

334 *a storm*. The Folio and Quarto editions of this play give 'A-scorn[e]'; Rowe and later editors emended to 'a storm'; Hazlitt conjectured 'ascaunt'.

335 *fleecy Crowns*. Cf. 'Hyperion' (above, p. 225) i. 8–9; Finney (pp. 513–14) describes the process by which Shakespeare's image and Keats's annotation on it enabled Keats to arrive at these lines in 'Hyperion'.

ayrie ayre. The Folio texts of this play give 'ayrie ayre', but the Quarto and later texts omit 'ayrie' and some editors give 'verie air'.

Keats's Marginalia to Paradise Lost

336 These marginalia are written in Keats's 2-vol. edition of *Paradise Lost* (Edinburgh, 1808), now at Keats House.

But his face . . . &c. PL i. 600–1

337 *pervade*. Forman gives 'here aid', assuming that Keats wrote 'here ade'. The handwriting could be interpreted either way, but 'pervade' with its

Latin sense of 'going through' conveys the idea of the sensors of the imagination feeling out its limits.

338 *Others more mild . . . &c.* ii. 546–7.

341 '*homely fac'd*'. From Chapman's 'Hymn to Pan' 73–4 'Yet the most useful Mercury embraced,| And took into his arms, his homely-faced'; the previous quotation is from *The Tempest* III. ii. 137.

other instances. These are *PL* ii. 817, ii. 812, ii. 777–8 ('Pensive here I sat | Alone'), and i. 620.

342 *Apollonian*. Cf. Keats's remarks in his letter to Haydon of 23 Jan. 1818 (p. 371 above) about treating 'Hyperion' in 'a more naked and grecian manner' than 'Endymion'.

343 *Reluctant*. From *reluctari* 'to struggle against'; thus 'struggling', 'opposing', 'writhing', as well as 'unwilling' (see *PL* x. 515).

344 '*So . . . stalk*'. *PL* v. 479–80 (there is no 'the' before 'springs').

345 *ache*. Cf. 'The Eve of St. Agnes' (above, p. 253) 17–18.

prosiable. 'possible' veering to 'probable'? Cf. 'purplue' (above, p. 524).

Mr. Kean

Composed 19 or 20 Dec. 1817; published in the *Champion*, 21 Dec. 1817. Keats refers to this review in his letter to George and Tom of 21, ?27 Dec. (above, p. 369). Kean had been ill and away from the stage for some time. His 'reappearance' as Richard was on 15 Dec.

Habeas Corpus'd. *Habeas Corpus* had been suspended in Mar. 1817 and was restored in Jan. 1818.

'*which . . . cover*'. *Timon of Athens* v. i. 217–18.

Riches. Sir James Bland Burges' play (see notes to the letter).

346 '*A thing . . . tell*'. Coleridge, 'Christabel' 253 'A *sight* to dream of . . .'.

hieroglyphics. '[Shakespeare's] language is hieroglyphical. It translates thoughts into visible images' (Hazlitt, *Works*, v. 54–5). The notion of 'gusto' is also one which Hazlitt developed.

'*hybla . . . honeyless*'. *Julius Caesar* v. i. 34–5.

'*put up . . . rust them*'. *Othello* I. ii. 59 ('*Keep* up . . .').

'*gorging . . . limb*'. Byron, 'The Siege of Corinth' 411.

'*Be stirring . . . Norfolk*'. *Richard III* v. iii. 56 ('Stir with the lark . . .').

'*like . . . tediously away*'. *Henry V* IV, Chor. 22.

'*does . . . gently*'. *The Tempest* I. ii. 298.

'*And sole . . . fierce*'. From Cary's translation of *Inferno* iv. 126.

'*Great . . . conquered us*'. Edward Young, *The Revenge* I. i ('Great let me call him; for he conquered me').

347 *mystery*. See note to 'Lamia' (above, p. 609) ii. 229 f.

Rejected Preface to Endymion

See Keats's letter to Reynolds of 9 Apr. 1818 (above, p. 390) in which he responds to criticisms that Reynolds and others have clearly made of this preface. In a letter sent the following day (10 Apr. 1818; *L* i. 269) Keats enclosed the preface that was printed with *Endymion*, saying 'I am anxious you sho[d] find this Preface tolerable. if there is an affectation in it 'tis natural to me.'

348 '*let it be . . . malice me*'. From Marston's address 'To My Equal Reader' prefaced to *The Fawn*.

Letter to C. C. Clarke, 9 October 1816

Text from holograph.

the Author . . . the Sun. Cowden Clark.

Darwin. Erasmus Darwin (1731–1802); physician, botanist, and poet.

349 '*never . . . God*'. Horace, *The Art of Poetry*, as translated by the Earl of Roscommon, ll. 227–8 ('But for a *business* worthy of a God').

Meeting. A Baptist chapel which stood between 28 and 29 Dean St.

Letter to J. H. Reynolds, 17, 18 April 1817

Text from transcript in Woodhouse's letter-book.

350 *the sea, Jack, the sea*. Severn recalled Keats's pleasure in the 'inland sea' created by the wind over grass. '"The tide! the tide!" he would cry delightedly, and spring on to some stile or upon the low bough of a wayside tree, and watch the passage of the wind upon the meadow grasses of young corn' (William Sharp, *The Life and Letters of Joseph Severn* (London, 1892), p. 20. Cf. also Xenophon, *Anabasis*, iv. vii. 24 where the Greeks, returning wearily home from their campaign in the interior, cry out Θάλαττα, Θάλαττα' ('the sea, the sea'), rejoicing at the sight.

quick freshes. *The Tempest* III. ii. 67.

Confinement. In 1647–8.

'*Do you not hear the Sea?*'. *King Lear* IV. vi. 4 ('Do you hear the sea?').

351 *In . . . abysm of time*. *The Tempest* I. ii. 326, I. ii. 50.

'*The noble Heart . . . Glory excellent*'. *FQ* I. v. i.

Letter to Leigh Hunt, 10 May 1817

352 Text from holograph.

coasted crab. *A Midsummer Night's Dream* II. i. 48 (where the crab is *roasted*).

Rimini. Hunt's *The Story of Rimini*, a long poem based on the story of Paolo and Francesca. The 2nd edn. came out in 1817.

last Sunday. Hunt had written on 'a most deplorable instance of religious fanaticism' in the previous Sunday's *Examiner* (4 May). Another part of the paper reported on the Petzelian sect of Upper Austria who preached '*the purification by blood*, and enforced the *horrible doctrine of sacrificing men* for the purpose of *purifying others from their sins.*'

that sentence. In his review of 'A Letter to William Smith Esq. from Robert Southey Esq.' (*Examiner*, 4 May 1817) Hazlitt writes 'Why should one not make a sentence of a page long, out of the feelings of ones whole life?'

353 *the Nymphs*. A poem by Hunt.

my Poem. 'Endymion'.

Shelley. Hunt's host at Great Marlow, Bucks. (to where this letter is addressed). On one occasion when Hunt and Shelley were travelling in a stage-coach Shelley suddenly cried to Hunt 'Hunt, I pray thee, let us sit upon the ground and tell strange stories of the deaths of kings' (G. Gilfillan, *A Gallery of Literary Portraits* (3rd edn., Edinburgh, 1851), pp. 70–1).

354 *Miss Kent*. Hunt's sister-in-law.

Letter to B. R. Haydon, 10, 11 May 1817

Text from holograph.

Let Fame . . . eternity. *Love's Labour's Lost* I. i. 1–7.

'*one that . . . trade*'. *King Lear* IV. vi. 15.

355 *a Shakespeare*. The engraving of Shakespeare's head referred to in the letter to Reynolds of 17, 18 Apr. (above, p. 349).

Alfred . . . the highest. Cf. 'To Charles Cowden Clarke' (above, p. 30) 70.

356 *to have mine in*. Haydon's grand painting of 'Christ's Entry into Jerusalem' includes portraits of Wordsworth, Keats, and Hazlitt.

Manuscript. Haydon had discussed Napoleon's 'Manuscrit venu de St Helene' in the *Examiner* (4 May 1817).

357 *Bertram*. Napoleon's friend General Bertrand.

'*Yet . . . the story*'. *Antony and Cleopatra* III. xiii. 46.

the North. Wordsworth.

John Hunt. Leigh Hunt's brother.

Letter to Jane and Mariane Reynolds, 14 September 1817

Text from holograph.

The Mountains . . . Lambs. Ps. 114: 4.

358 *Timotheus*. Dryden, 'Alexander's Feast, or the Power of Musique' 131.

though inland far I be. Wordsworth, 'Ode: Intimations of Immortality' 166. ('. . . *we* be')

'*opening . . . streams*'. *A Midsummer Night's Dream* III. ii. 393 ('salt *green* streams').

The sun . . . shall be. A contemporary jingle.

'*to . . . western foam*'. Spenser, 'Epithalamion' 282.

359 *a mighty . . . Lord*. Gen. 10: 9.

Mr. W.D. William Dilke (as opposed to Mr W.H., 'onlie begetter' of Shakespeare's sonnets).

me. Conjectured, as there is a tear in the paper at this point.

'honest a Chronicler'. Henry VIII IV. ii. 72.

Letter to Benjamin Bailey, 8 October 1817

360 Text from holograph.

Lambs Conduit Street. The Reynoldses lived at No. 19.

neighbours. In Lisson Grove North.

The web . . . mingled Yarn. All's Well That Ends Well IV. iii. 71.

361 *Mockery . . . at Hunt's.* See notes to 'God of the golden bow' (above, p. 568).

362 *Ax Will.* ask Will (a college servant?).

Mercury. Probably as a cure for gonorrhoea; see Gittings, *John Keats*, App. 3.

Letter to Benjamin Bailey, 3 November 1817

Text from holograph.

villainy. Bailey had been disappointed of a curacy in the diocese of Lincoln because of a delay in his ordination.

363 *subscription.* To pay for Haydon's protégé Cripps's expenses.

Magazine. 'On the Cockney School of Poetry, No. 1', *Blackwood's*, Oct. 1817, pp. 38–41. This, and the subsequent article in this series, was prefaced thus:

> Our talk shall be (a theme we never tire on)
> Of Chaucer, Spenser, Shakespeare, Milton, Byron,
> (Our England's Dante)—Wordsworth—HUNT, and KEATS,
> The Muses' son of promise; and of what feats
> He may yet do.

The lines were signed 'Cornelius Webb'.

Letter to Benjamin Bailey, 22 November 1817

364 Text from holograph. See Stillinger, *The Hoodwinking of Madeline* (Urbana, Ill., 1971), App. 1 for the philosophical background to this letter.

365 *Men of Power.* Keats's interest in the nature of power and genius was shared and stimulated by Hazlitt. See e.g. Hazlitt's discussion of *Coriolanus* which Keats quotes in his letter to George and Georgiana of 14 Feb.–3 May 1819 (above, pp. 460–2).

my first Book. Keats is referring to *Endymion* and the 'Ode to Sorrow' (from *Endymion* iv).

he awoke . . . truth. PL viii. 478–84.

366 *philosophic Mind.* Wordsworth, 'Ode: Intimations of Immortality' 189.

367 *Christie*. J. H. Christie (d. 1876), later famous for mortally wounding John Scott in a duel occasioned by Lockhart's *Blackwood's* reviews. He was defended in court by Reynolds and Rice.

Letter to J. H. Reynolds, 22 November 1817

Text from transcript in Woodhouse's letter-book.

368 *this place*. Dorking.

When . . . beard. Shakespeare, *Sonnets* xvii. 5–8.

As the snail . . . head. Shakespeare, *Venus and Adonis* 1033–8 ('Shrinks *backward in*' l. 1034).

a poets . . . song. Shakespeare, *Sonnets* xvii. 11–12.

369 *rounce*. Someone who rounces (i.e. flounces around)? The copyist left a blank here and the word (conjectured as *rounce* by Gittings) was pencilled in later.

late Princess. Princess Charlotte Augusta died 6 Nov. 1817.

hinc atque illinc. 'from here and from there' (Virgil, *Georgian* iii. 257).

Letter to George and Tom Keats, 21, ?27 December 1817

Text from transcript by Jeffrey.

&&. Here the copyist (who wrongly dated the letter 1818) omitted a word.

Riches. Sir James Bland Burges, *Riches: or the Wife and Brother*, a play based on Massinger's *The City Madam*, in which Kean played Luke Traffic. See Keats's review, 'Mr. Kean' (above, p. 345).

essential service. William Hone (1780–1845), an antiquarian bookseller, was tried in Dec. for blasphemous libel on account of his anti-ministerial liturgical parodies (an example of which can be found in S. Macoby, *English Radicalism 1786–1832* (London, 1955), p. 336). Nearly 100,000 copies of his parodies were sold. Hone successfully defended himself against Lord Chief Justice Ellenborough (1750–1818). Thomas Wooler ((?)1786–1853), editor of the radical *Black Dwarf*, had been tried and acquitted for libels on the Ministry in June.

Death on the Pale Horse. Benjamin West's painting *Death on a Pale Horse* caused a considerable stir when first exhibited at Pall Mall. Hazlitt discussed it in the Dec. 1817 issue of the *Edinburgh Magazine* (*Works*, xviii. 135–40).

370 *Christ Rejected*. Another painting by West (1738–1820), an American and president of the RA.

Shelley's poem. 'Laon and Cythna'.

Letter to B. R. Haydon, 23 January 1817

371 Text from holograph.

business. Taylor had suggested that Haydon illustrate an episode from 'Endymion' ('the written tale') for use as a frontispiece. Haydon had counterproposed an engraving of Keats's head (which he never performed).

Letter to Benjamin Bailey, 23 January 1817

Text from holograph.

unfortunate Family. Bailey noted in the margin of this letter: 'This letter opens the excellent feelings of an excellent heart. "The unfortunate family" mentioned was most kindly treated by poor Keats.'

By heavens . . . Drachmas. *Julius Caesar* IV. iii. 73.

372 *binding*. Of apprenticing him to Haydon.

Letter to George and Tom Keats, 23, 24 January 1818

373 Text from transcript by Jeffrey.

'*thus . . . Land*'. *Richard III* V. ii. 3.

374 *Rox of the Burrough*. Probably a copyist's mistake for 'Cox', a medical bookseller who lived in Borough. John Hunt was Leigh Hunt's brother and editor of the *Yellow Dwarf*; Bob Harris was probably connected with the management of Covent Garden (see *L* i. 199 n.).

Furioso. The punctuation obscures the fact that 3 plays were offered: George Colman's *John Bull* (1803), a comedy; *The Review*, a musical farce; and William Barres Rhodes' *Bombastes Furioso* (1810), 'a burlesque tragic opera' (not in fact performed that night).

Scott. The articles in *Blackwood's* on 'The Cockney School of Poetry' (see letter to Bailey, 3 Nov. 1817 (above, p. 363)), signed by 'Z' were in fact written by J. G. Lockhart and not by John Scott (who had opposed Hunt's views on Byron in the *Champion*).

Letter to John Taylor, 30 January 1818

375 Text from holograph.

'*chime a mending*'. *Troilus and Cressida* I. iii. 159.

376 *next Work*. This was Taylor's *The Identity of Junius with a Distinguished Living Character Established*, 2nd edn. (1818). In this work Taylor identifies the author of the *Letters of Junius* as Sir Philip Francis.

Letter to J. H. Reynolds, 3 February 1818

Text from transcript by Woodhouse in his letter-book. Reynolds had sent Keats 2 of his 'Robin Hood' sonnets by the second post. These sonnets, plus a third on the same subject (addressed to Reynolds's future wife), were published in the *Yellow Dwarf*, 21 Feb. 1818. Keats quotes from l. 5 of the first sonnet ('[Is] no arrow found,—foil'd of its antler'd food'); Reynolds emended the phrase Keats finds fault with, making Marian 'young as the dew' (and not 'tender and true') in the published version.

Mast. Collective name for fruit of the oak, beech, etc.—esp. to denote pig fodder.

Sancho. Cervantes' Sancho Panza, Don Quixote's squire.

377 *primrose*. Cf. Wordsworth, 'Peter Bell' 218–20.

Manasseh . . . Esau. The tribe of Manasseh occupies much of the Book of Joshua where it is often described as 'the half tribe'; Esau, as John Barnard points out was, like Robin Hood, a hunter. 'Keats's "Robin Hood", John Hamilton Reynolds, and the "Old Poets"', *Proceedings of the British Academy*, 75 (1989).

'*nice Eyed wagtails*'. Leigh Hunt, 'The Nymphs' ii. 170.

'*Cherub Contemplation*'. Milton, 'Il Penseroso' 54.

'*Matthew with a bough*' . . . '*under an oak*'. Wordsworth's 'The Two April Mornings' 59 is set against *As You Like It* II. i. 31.

Catkins. Long thin poems instead of plump sonnets.

Harpsicols. Probably a musical evening was in store.

Letter to J. H. Reynolds, 19 February 1818

378 Text from holograph.

Pallaces. The Buddhist 'places of delight'.

'*an odd angle of the Isle*'. *The Tempest* I. ii. 223.

'*girdle round the earth*'. *A Midsummer Night's Dream* II. i. 175.

'*Spirit . . . good*'. Wordsworth, 'The Old Cumberland Beggar' 77.

Letter to John Taylor, 27 February 1818

379 Text from holograph.

380 *quiet*. These remarks refer to *Endymion* i. 149, 247; the later remarks refer to i. 334, 495.

Go-cart. At this time the name for a bottomless wooden frame with rollers in which a child could learn to walk.

O for a Muse . . . ascend. Henry V, I. i. 1.

Percy Street. Home of the artists De Wint and Hilton.

Letter to Benjamin Bailey, 13 March 1818

381 Text from holograph.

Pulvis Ipecac. Simplex. An emetic.

382 *Acrasian*. Acrasia is creator of the deceptively alluring Bowre of Blisse in *FQ* II.

383 '*consecrate . . . look upon*'. Cf. Shelley, 'Hymn to Intellectual Beauty' 13–15.

Pecten. plectrum.

Greig. For 'Gleig'.

Letter to J. H. Reynolds, 14 March 1818

Text from transcript in Woodhouse's letter-book.

parapet. A parapet stone had at one time fallen and injured Brown in the leg.

herculaneum. Herculaneum, near Naples, was overwhelmed with lava at the same time as Pompeii.

384 *dash*. Sudden heavy rainfall.

Lydia Languish. Heroine of Sheridan's *The Rivals*.

Radcliffe. Mrs Anne Radcliffe (1764–1823), author of Gothic novels.

Letter to B. R. Haydon, 21 March 1818

385 Text from holograph.

seal. Haydon had written to Keats (4 Mar.) that a gold ring and a seal, thought to have belonged to Shakespeare, had been found in a field at Stratford-upon-Avon.

B-hrell. Presumably, by analogy, 'Bitcherell'.

386 *Dentatus*. *The Assassination of Dentatus*, Haydon's second major work, had won the British Institution prize for the best historical picture in May 1810.

dart. The River Dart.

Millman. Henry Hart Milman (1791–1868), whose poetic drama *Fazio* (later called *The Italian Wife*) he described as 'an attempt at reviving the old national drama with greater simplicity of plot'.

Letter to James Rice, 24 March 1818

Text from holograph.

Answer to Salmasius. Milton's *Pro Populo Anglicano Defensio* (1651), a response to Claudius Salmasius's attack on the Commonwealth. So pointed was it, it was thought to have undermined Salmasius's health.

387 *can't*. Keats wrote 'and'.

Bucks . . . Castelreaghs. Charles Bucke (1741–1846), dramatist; *Hengist* is a play, its author unknown; Robert Castelreagh (1769–1822), Foreign Secretary from 1812 to 1822 but associated also with the government's repressive measures at home.

388 *kit o' the german*. 'Kit' is early 19th-c. slang for 'penis'. The phrase still remains obscure but Gittings notes that 'the whole passage is . . . [a] piece of sexual joking, indicating that . . . Rice had left some bastards behind in that part of Devon' (Gittings, *Letters*, p. 402).

Letter to B. R. Haydon, 8 April 1818

Text from holograph.

nonsense. The Devon poems.

galligaskins. Loose breeches.

389 *compositions and decompositions*. Cf. Hazlitt: 'In Shakespeare there is a con-

tinual composition and decomposition of [the elements of character]' ('On Shakespeare and Milton', *Works*, v. 51); the idea of a snail-horn's sensitivity comes from Shakespeare's 'Venus and Adonis' 1033–8 (see Keats's remarks to Reynolds (letter of 22 Nov. 1817, above, p. 368)).

That. For 'What'.

'See . . . Wall'. *3 Henry VI* v. i. 17.

bit of Italian. Haydon had described a Mrs Scott as 'con occhi neri' (with black eyes).

fit . . . Kingston's. Plays on *The Merry Wives of Windsor* I. iv. 27; E. Blunden suggests that this is 'a hit at the set of John Warren, who published some volumes of verse by Reynolds and others' (*Reprinted Papers* (Tokyo, 1950), p. 246).

Letter to J. H. Reynolds, 9 April 1818

390 Text from transcript in Woodhouse's letter-book.

the thing. The original preface to 'Endymion' (above, see p. 347).

like lime-twigs . . . Book. Cf. *2 Henry VI* III. iii. 16 ('my winged *soul*').

from Jove. Herrick, 'Evensong' 1.

391 *Babbicun*. Babbacombe; Kent's Cavern is famous for flints and bones.

Letter to John Taylor, 24 April 1818

Text from holograph.

392 *'Ah! art awake'*. The lines referred to are iii. 429, 449; a list of *errata* was added to the end of the letter.

Letter to J. H. Reynolds, 27 April 1818

393 Text from transcript in Woodhouse's letter-book.

'that watery labyrinth'. Lethe; *PL* ii. 584.

publish or no. See headnote to 'Isabella' (above, p. 584); Keats and Reynolds had planned a joint volume of poems based on Boccaccio's tales.

Letter to J. H. Reynolds, 3 May 1818

394 Text from transcript in Woodhouse's letter-book.

Common. At Kennington, where Reynolds was then staying (south of the river).

'I have . . . head'. The words are said by Slender in *The Merry Wives of Windsor* (I. i. 123) and not, as Keats seems to recall, by Sir Andrew Aguecheek in *Twelfth Night*.

'Notus . . . Sierra-leona'. *PL* x. 702–3.

stockings . . . for. The smelly stockings belonged to the Bentleys' children.

395 *Spencerian*. Either Reynolds's poem 'The Romance of Youth' (later published in *The Garden of Florence* (London, 1821) or, as Gittings suggests, his lodgings at Spencer Place. His 'office' is that of solicitor.

Pepins. Pippins (see *The Merry Wives of Windsor* I. ii. 12); Keats may be punning on 'Pip-civilian' ('an amateur lawyer' (Gittings, *Letters*).

Burden . . . Mystery. Wordsworth, 'Lines Written a few Miles above Tintern Abbey' 38.

running one's rigs. Running riot.

396 *region of his song*. 'The Recluse' I. i. 793–4.

'Knowledge . . . is Sorrow'. 'Manfred' I. i. 10 ('Sorrow is knowledge'—Keats has misremembered Byron and attributed the thought to himself).

draughts. This letter survives only in transcript but, presumably, the original was 'crossed' and resembled a chequer board.

Patmore. Peter George Patmore (1786–1855), journalist friend of Hazlitt (who addressed the *Liber Amoris* to him).

Coleman. George Coleman (1762–1836), dramatist.

Little. pseudonym for Thomas Moore; this sentence recalls Pope's 'Essay on Man' iv. 380.

398 *Nom^e: Musa*. 'Nominative: Musa' (the words of a student beginning to recite the declension of the Latin noun *musa*).

Letter to Benjamin Bailey, 10 June 1818

399 Text from holograph.

Oxford Paper. Bailey had written two articles on 'Endymion' in the *Oxford University and City Herald and Midland City Chronicle*, 30 May and 6 June 1818 (reprinted in M. B. Forman (ed.), *The Poetry and Prose of John Keats* (London, 1938–9), ii. 237–43); in the first he calls 'upon this age to countenance and encourage this rising genius, and not to let him pine away in neglect'.

400 *carey*. H. F. Cary's translation of the *Divine Comedy* (in 3 32mo volumes) had just been reprinted by Taylor and Hessey.

'amiable Mister Keats'. In 'A Letter from Z to Leigh Hunt King of the Cockneys' Lockhart writes of the 'amiable but infatuated bardling Mister John Keats' (*Blackwood's*, May 1818).

'Foliage'. The title of Hunt's volume, reviewed in *QR* Jan. 1818, pp. 324–35.

Letter to Tom Keats, 25–27 June 1818

Text from *Western Messenger* (Louisville, Kentucky), June 1836, i. 772–7. Charles Brown kept a journal during the walking tour on which he and Keats embarked in June 1818 and which Keats begins to describe in this letter. Brown later published part of this journal in *PDWJ* 1, 8, 15, 22 Oct. 1840 (reprinted in *L* i. 421–42).

Bowne's. Bowness, on Lake Windermere.

401 *Wordsworth versus Brougham*. William Lowther (1787–1872) stood as Tory candidate against the Whig Henry Brougham (1778–1868) in the 1818

General Election. Lowther (Lord Lonsdale) had considerable parliamentary influence. Wordsworth's support represents a rejection of his early radicalism.

402 *'mazy . . . shades'*. *PL* iv. 239.

Letter to Tom Keats, 29 June, 1, 2 July 1818

403 Text from transcript by Jeffrey.

404 *'the ancient . . . Crag'*. Wordsworth, *Poems on the Naming of Places* ii ('To Joanna'), l. 56.

'no new . . . France'. Burns, 'Tam o' Shanter' 116 ('Nae cotillion brent new frae *France*').

405 *by Burns*. This is as far as Jeffrey, the copyist, has transcribed.

Letter to F. Keats, 2, 3, 5 July 1818

Text from holograph.

406 *a ℥*. This is the sign for a fluid ounce.

you have heard. See headnote to 'Old Meg she was a Gipsey' (above, p. 587).

parliament. A thin flat gingerbread.

Lady's fingers. Pink ringed peppermints.

Letter to Tom Keats, 3, 5, 7, 9 July 1818

407 Text from holograph.

408 *little Susannas*. They are persecuted by Elders. The story of Susanna and the Elders is in the Septuagint and Theoditian versions of the book of *Daniel* and in the Authorized Version Apocrypha.

409 *'Before . . . go'*. From the ballad 'Robin Hood and the Bishop of Hereford'.

Hummums. An hotel in Great Russell Street, London.

the sound of the Shuttle. A 'disgusting' sound because of the conditions under which hand-loom weavers were obliged to work.

Letter to J. H. Reynolds, 11, 13 July 1818

410 Text from transcript in Woodhouse's letter-book.

411 *cutty-stool*. 'The stool of repentance' (*OED*).

Caliph Vatheck. From William Beckford's novel *Vathek* (1786).

412 *as if . . . spies*. *King Lear* v. iii. 17.

Vingt-un. The card game *vingt-et-un*.

413 *Bailey . . . Cumberland*. Bailey's delayed ordination took place at Carlisle in Aug. 1818.

Letter to Mrs. James Wylie, 6 August 1818

Text from transcript by Jeffrey.

Mahomet . . . spilt. Cf. Addison, *Spectator*, 18 June 1711.

414 *Jessy of Dumblane*. Cf. the Scots song 'Jessie the Flow'r o' Dumblane'.

Letter to C. W. Dilke, 20, 21 September 1818

415 Text from holograph.

three pages, and a half. The sheet is folded in half and the letter inscribed on the right recto and both verso sides; approximately half of the left recto remains after the address has been inscribed.

'*with retractile claws*'. Cary's Dante, *Inferno* xvii. 101.

Eustace. J. C. Eustace, author of *A Classical Tour Through Italy* (London, 1817).

Thornton. Thomas Thornton (d. 1814) author of *The Present State of Turkey* (London, 1807).

Bath. A kind of letter-paper.

against Blackwood. An article ('Hazlitt Cross-Questioned') in the Aug. 1818 issue of *Blackwood's* had called Hazlitt 'a mere quack . . . and a mere bookmaker'. Hazlitt took out a suit against them which he dropped on *Blackwood's* agreeing to pay costs and damages.

416 *Mrs. D*. Dilke was staying with the Snooks at Bedhampton (to where this letter is addressed). His wife remained at Wentworth Place in Hampstead.

hateful . . . contraries. *PL* ix. 121–2.

Charles. Dilke's son.

'*Love . . . veins*'. 'Nature withheld Cassandra in the Skies' (above, p. 220) 12.

Letter to J. H. Reynolds, ?22 September 1818

Text from transcript in Woodhouse's letter-book.

Gather the rose &c. 'Gather the rose of love while yet thou mayest' (*Godfrey of Buloigne* (i.e. *Gerusalemme Liberata*, trans. Fairfax, London, 1600) xvi. 157).

417 *that woman*. Probably Jane Cox, a cousin of Reynolds. See letter to George and Georgiana Keats of 14–31 Oct. 1818 (above, p. 422).

Letter to J. A. Hessey, 8 October 1818

Text from transcript in Woodhouse's letter-book.

the Chronicle. The *Morning Chronicle* of 3 and 8 Oct. 1818 contained letters from 'J.S.' and R.B.' defending Keats against Croker's vindictive review of Endymion' in *QR*.

Letter to Richard Woodhouse, 27 October 1818

418 Text from holograph. Keats had attended Hazlitt's lecture 'On Shakespeare and Milton' (*Works*, v. 44–68) earlier in the year. Its stimulus on his thought is evident in this letter.

419 *saturn and Ops*. Who appear in 'Hyperion'.

Letter to George and Georgiana Keats, 14, 16, 21, 24, 31 October 1818

420 Text from holograph.

421 *your Mother*. Mrs Wylie.

Crutched Friars. Site of the convent of Crouched or Crutched Friars in London; Keats is perhaps suggesting that Mrs Millar told a Gothic tale as if she wrote Gothic script; or that she tells stories lamely.

Settlement. Morris Birkbeck's settlement in Illinois which George hoped to join.

422 *Lightermen*. Men employed on lighters (flat-bottomed boats for unloading cargo vessels).

Alfred Exeter paper. Reynolds' article in the *Alfred West of England Journal* (6 Oct. 1818) was reprinted in the *Examiner* (12 Oct. 1818, p. 648); 'J.S.' (John Scott) had reviewed *Endymion* kindly, though not uncritically, in the *Morning Chronicle* (3 Oct. 1818).

coming the Richardson. Taking on the epistolary style of Samuel Richardson.

Cousin of theirs. Jane Cox.

423 *particular*. Displaying partiality; flirtatious.

John . . . cradle. John Howard (?1726–1790) was a philanthropist, and Richard Hooker (?1554–1600) a theologian but no bishop; Keats is creating a kind of Renaissance emblem here.

This is Lord Byron. In fact a misquotation from Leigh Hunt's *The Story of Rimini* (iii. 121–2).

424 *like Hunt*. Probably Henry ('Orator') Hunt. He had contested the Westminster seat in the 1818 General Election but won only 84 votes.

Sir F. Burdett. Burdett (1770–1844) entered parliament in 1796 where he was active (and twice imprisoned) in the cause of reform.

divine right Gentlemen. Champions of the monarchy.

Godwin perfectibily man. William Godwin, in the second edition of *Political Justice* (1796), introduced 5 propositions which express a position of extreme optimistic rationalism (Bk. I, ch. v, sect. 5: 1976 Penguin edn., pp. 140–6).

425 *what day of the Month*. 16 Oct.

426 *dowling*. A schoolboy slang term (used at Westminster etc.) for 'fagging' or 'labouring' (from the Greek δοῦλος 'slave'). Keats has just mentioned 'the horrid System . . . of the fagging at great Schools' and so uses the word to express his own efforts.

Capper and Hazlewood. Stockbrokers and forwarding agents for letters to George Keats.

Rackets. A bat-and-ball game played against a wall by two people; similar to squash.

Lady. Mrs Isabella Jones whom Keats first met at Hastings in May or June 1817.

428 *'like . . . waftage'*. *Troilus and Cressida* III. ii. 9–10 (*'strange* soul').

to make the angels weep. *Measure for Measure* II. ii. 122.

Letter to B. R. Haydon, 22 December 1818

429 Text from holograph. Haydon, whose persistent eye-problems had slowed down the completion of his huge *Christ's Entry into Jerusalem* and worsened his finances, had asked Keats for a loan. For his alacritous response to Keats's offer see *L* i. 415.

430 *with*. For 'without'.

Salmon. Haydon's servant and model.

Letter to George and Georgiana Keats, 16–18, 22, ?29, 31 December 1818, 2–4 January 1819

Text from holograph.

B. The previous letter to George and Georgiana had been called 'A' (see p. 429 above).

431 *Miss Keasle*. Miss Millar's lodger (as is the Miss Waldegrave mentioned later).

432 *Bethnal green*. Where Haslam lived with his parents.

Mr and Mrs . . . A tear in the paper has made the name unreadable.

this thin paper. Colvin notes that this is 'paper of the largest folio size, used by Keats in this letter only, and containing some 800 words a page of his writing'. (S. Colvin ed., *The Letters of John Keats* (London, 1891).)

Hopner . . . Poles. H. P. Henry Parkyns Hoppner (1795–1833) sailed in the *Alexander* which, along with Sir John Ross's *Isabella*, rediscovered Baffin's Bay in an attempt to make the North-West Passage. Ross (1777–1851) published *A Voyage of Discovery* in 1819.

stuff you can imagine. Leigh Hunt's *The Literary Pocket-Book; or Companion for the Lover of Nature and Art* (1819) was something between a diary and an anthology; its purpose both instructive (it contained a chronological list of eminent persons) and useful (it listed the addresses of print and plaster-cast shops in London, and also hackney-coach fares). It also included two poems by Keats—'The Human Seasons' and 'Sonnet to Ailsa Rock'.

edinburgh Reviewer. For the *Edinburgh Magazine*.

Archer. Archibald Archer, a painter.

433 *daughter senior*. Frances (Fanny) Brawne.

This morning. 17 Dec.

Allegany ridge. The Allegheny Mountains run through West Virginia and North Carolina.

Savoy. The Haute Savoie in the French Alps.

434 *Gifford's*. In fact it was Croker's.

Misses Porter. Jane and Anna Maria Porter, writers of romances such as Jane's *The Pastor's Fire-Side* (London, 1817) and Anna Maria's *The Hungarian Brothers* (London, 1807).

435 *Lions*. At the Tower of London.

436 *Redall*. G. S. Reddall, sword-cutler, 236 Piccadilly (Gittings, *Letters*).

Bartolozzi. Francesco Bartolozzi (1727–1815), Florentine engraver resident in England 1764–1802.

not seventeen. In fact she was eighteen.

437 *Gattie*. John Byng Gattie (1788–1828); worked at the Treasury.

Altam and his Wife. A play by Charles Ollier (London, 1818).

La Biondina. First words of a popular Italian song.

What... absorbs me quite. From Pope, 'The Dying Christian to his Soul' 9.

438 *Drewe family*. Reynolds was engaged to Eliza Powell Drewe of Exeter.

Twisse's. Horace Twiss (1789–1849), barrister, author (of *Posthumous Parodies of the Poets* (1812), etc.), and politician.

Liston. John Liston (1776–1846), comic actor with an exceptionally expressive face.

Charles Kemble. An actor (1775–1854) of great range who worked mainly at Covent Garden.

scotch novels. Scott's Waverly novels.

Book of Dubois's. Edward Dubois, *My Pocket Book* (2nd edn., London, 1808), satirizing the travels of Sir John Carr.

439 *my large poem*. 'Hyperion'.

What. Want.

440 *Bactrians*. Not camels but inhabitants of Bactria, site of an ancient civilization in South-East Asia.

Mrs. Tighe. Mary Tighe (1772–1810), author of 'Psyche, or the Legend of Love', the story of Cupid and Psyche told in Spenserian stanzas, published with other poems after her death in 1811.

Beattie. James Beattie (1735–1803), poet, essayist, and moral philosopher; 'The Minstrel', in Spenserian stanzas, is his most famous poem.

father Nicholas. Henry Mackenzie's *The Story of Father Nicholas* appeared in the Edinburgh *Lounger*, 82–4 (1786) and was reprinted in *Works* (London, 1816), pp. 154–69.

Book of Prints. Carlo Lasinio's *Pitture a fresco del Campo Santo di Pisa*

(Florence, 1812); it includes drawings based on Orcagna's *Triumph of Death*.

441 *democratic paper*. The *Democratic Press, Philadelphia*, 24 Oct. 1818 reprinted an article on 'Dinner to Thomas Moore Esq.,' from the *Dublin Evening Post*.

442 *'He is . . . of woe'*. Keats is quoting from Hazlitt's lecture 'On the English Novelists' (printed in the *Examiner*, 28 Dec. 1818, pp. 825–6: *Works*, vi. 130–1).

443 *Sunday*. 3 Jan. 1819.

Manker. Moncur, a friend of Brown.

Kirkman. George Buchanan Kirkman was a relative of George Felton Matthew.

letter. 'To the Electors of Westminster'; Hobhouse, along with Burdett and Hunt, was contesting the Westminster seat.

a Book . . . about it. Thomas Edward Bowdich, *Mission from Cape Coast Castle to Ashantee, with a statistical Account of that Kingdom* (London, 1819).

444 *Sir Lucius*. Sir Lucius O'Trigger in Sheridan's *The Rivals*.

Letter to B. R. Haydon, 8 March 1819

445 Text from Forman, *Letters*.

Examiner. 'Attacks on Mr Haydon' (*Examiner*, 7 Mar. 1819, pp. 157–8), in which Haydon replies to critics of an exhibition of his pupils' work.

Letter to Joseph Severn, 29 March 1819

446 Text from holograph.

to be hurt. Severn did exhibit his miniature of Keats in the Royal Academy May exhibition, as well as his painting *Hermia and Helena*.

Letter to Fanny Keats, 12 April 1819

Text from holograph.

answers. In his previous letter to Fanny (not in this volume) Keats had sent her a Catechism and answered various scriptural and doctrinal questions for her. She was preparing for Confirmation and had asked her brother to help her with her homework; see Marie Adami, *Fanny Keats* (London, 1937), pp. 71–3.

447 *rout*. Large evening party.

Seals. These are Tassie's 'gems'; Keats had promised to get some for Fanny in a letter of 13 Mar. (see note to 'On a Leander that Miss Reynolds my kind friend gave me' (above, p. 569).

Letter to B. R. Haydon, 13 April 1819

Text from holograph.

448 *jot*. The signature has been cut off the manuscript.

Letter to Fanny Keats, May 1819

Text from holograph.

Letter to George and Georgiana Keats, 14, 19 February, ?3, 12, 13, 17, 19 March, 15, 16, 21, 30 April, 3 May 1819

449 Text from holograph.

C. See pp. 420 and 430 above for letters A and B.

450 *dead born.* Samuel Rogers, 'Human Life'.

stupe. Medical term for compress.

Cobbett . . . Settlement. 'To Morris Bikbeck' in *Cobbett's Weekly Register*, 6 and 13 Feb. 1819.

451 *Apostate Man.* Richard Lalor Sheil (1791–1851), referred to as 'Shield' later in this letter, author of *The Apostate* (1817) and of *Evadne* (produced at Covent Garden 10 Feb. 1819).

must be Ollier's. The author of the 'fuzz' was Charles Lamb.

452 *Davenport's.* Burridge Davenport, merchant and Hampstead neighbour.

Carlisle. Richard Carlile (1790–1843) 'suffered and achieved more for the Liberty of the press than any other Englishman of the nineteenth century' (G. M. Trevelyan, *British History in the Nineteenth Century* (London, 1922), p. 162); for a full account of Carlile's career see W. H. Wickwar, *The Struggle for the Freedom of the Press 1819–1832* (London, 1928), pp. 67–79.

mother Radcliff. As in the Gothic novels of Ann Radcliff.

Mr Way. The Revd Lewis Way (1772–1840); see Gittings, *The Living Year* (London, 1954), pp. 76–7.

453 *18.* In fact 19.

Speck. Speculation.

454 *my poem. Endymion.*

455 *Jeremy Taylors.* Jeremy Taylor, *The Golden Grove* (London, 1655).

456 *On Monday.* The next two paragraphs are from a transcript by Jeffrey of a leaf which Keats accidentally omitted from this letter and forwarded in his letter of 17 Sept. 1819.

15.2. 'fifteen two' (a score in cribbage).

457 *bam.* As in 'bamboozle'; the Latin means (as he says) 'I do not wish him to have her, for'.

St. Lukes. A lunatic asylum.

de Cockaigne. A play on his 'cockney' label.

Velocepede. The velocipede or 'swift walker' was patented in England in 1818.

458 *Milnes.* Joseph Milner's *History of the Church of Christ*, 4 vols. (York, 1794–1809).

Letter to Gifford: *A Letter to William Gifford, Esq. from William Hazlitt Esq.* (*Works*, ix. 13–59). Extracts appeared in the *Examiner* on 7 and 14 Mar. 1819 (pp. 156, 171–3); Gifford was the editor of the *QR* from which he had attacked not only Hazlitt, but a generation of young writers.

dedicate . . . leaves. Romeo and Juliet I. i. 152–3 'Ere he can spread his sweet leaves to the air or dedicate his beauty to the sun'.

459 *They keep . . . swallow'd. Hamlet* IV. ii. 17–19 'he keeps them, like an ape an apple, in the corner of his jaws, first mouth'd to be last swallow'd'.

'*have . . . saint Peter*'. *Othello* IV. ii. 91.

'*Keep a corner . . . and gender in*. A conflation of *Othello* III. iii. 272 and IV. ii. 62.

flattering unction . . . souls. Hamlet III. iv. 145.

I cry . . . for. Othello IV. ii. 88.

memorial to Congress. Extracts from Thomas Moore's satirical poem 'Tom Crib's Memorial to Congress' appeared anonymously (with a glossary of slang) in the *Examiner* (11, 18 Apr. 1819, pp. 237–8, 253–4).

460 *consistent*. Hazlitt wrote 'inconsistent'.

461 *effect*. Keats copied this line twice.

'*pride . . . virtue*'. *Othello* III. iii. 350.

462 *fictitious*. Hazlitt wrote 'factitious'.

463 *Segar*. Cigar.

the False one. By Beaumont and Fletcher.

ettrick shepherd. James Hogg (1770–1835) was a contributor to *Blackwood's*.

Neither Poetry . . . disguisment. Cf. 'Ode on Indolence' (above, p. 283), stanza 3.

464 *In wild Nature . . . Worms*. Cf. 'To J. H. Reynolds Esq.' (above, p. 185), 102–5.

'*We have . . . one human heart*'. 'The Old Cumberland Beggar' 153.

465 '*How . . . lute*'. Milton, 'Comus' 475–7.

466 '*whole . . . casing air*'. *Macbeth* III. iv. 22 ('as *broad* and general').

The Framptons. Haslam's father's employers.

Amena. This was the name with which Charles Wells—a school friend of the Keats brothers—signed the spoof letters to Tom Keats which Tom believed to be genuine. One of these letters is reproduced in D. Hewett, *A Life of John Keats* (2nd edn., London, 1949), pp. 377–81.

the Boys. Brown's nephews.

467 *Sir John Leicester's gallery*. His private collection at Hill Street, Berkeley Square. It was open to the public once a week from mid-March to mid-May 1819.

Northcote. James Northcote, RA (1746–1831).

published this morning. 16 April, by Taylor and Hessey.

Simon Pure. The play is by Susanna Centlivre (1667–1723); in it Simon Pure, 'a Quaking Preacher', is impersonated by Colonel Fainwell.

Tragedy. *The Italians; or the Fatal Accusation*, performed at Drury Lane 3 and 12 Apr. 1819.

468 *Germany!* The murder of the German playwright August von Kotzebue (1761–1819, author of *Lover's Vows*, the shocking play in *Mansfield Park*) was reported and discussed at length in the *Examiner* (4 and 11 Apr. 1819, pp. 219, 225, 233). The murderer was Karl Ludwig Sand.

playing young gooseberry. Wrecking.

469 *all wound with Browns*. *The Tempest* II. ii. 13 (where Caliban is 'all wound with *adders*').

Skinner. Brown's solicitor.

John Brown. Brown's brother.

470 *bever and wet and snack*. All names for between-meal intakes of food (in the first and last case) and drink ('wet').

give . . . a cold Pig. 'to awaken by sluicing with cold water or pulling off bedclothes' (Partridge).

471 *real florimels*. As opposed to false Florimells such as the soulless and deceitful Florimell look-alike created by Archimago in *FQ* III. viii.

Barbara Lewthwaite. Child heroine of Wordsworth's 'The Pet Lamb'.

Fell . . . Gale . . . Foy. From Wordsworth's 'The Idiot Boy': the words 'and who can wonder at it?' follow with a line crossed through them.

muse. The words 'who sits aloof in a cheerful sadness, and' follow with a line crossed through them.

heart of Mid Lothian. By Daniel Terry, with a score by Sir Henry Rowley Bishop.

north Pole. Henry Aston Barker's panorama 'representing the North Coast of Spitzbergen' opened at Leicester Square on 12 Apr. 1819.

472 *America*. William Robertson, *The History of America* (Dublin, 1777).

'*a poor forked creature*'. *King Lear* III. iv. 107–8 ('forked *animal*').

Letter to Mary-Ann Jeffery, 31 May 1819

475 Text from holograph.

476 *India for a few years*. Dilke later recalled that 'he *wrote* about surgeon of an Indiaman, but *talked* about a South sea Whaler, and, as if to bid defiance to fortune, would have fixed on something more hateful, could his imagination have helped him to it' (*KC* ii. 223).

Fanny. Mary-Ann's sister Francis.

Nothing . . . flower. Wordsworth, 'Ode: Intimations of Immortality' 181–2.

Letter to Mary-Ann Jeffery, 9 June 1819

477 Text from A. F. Sieveking, 'Some Unedited Letters of John Keats', *Fortnightly Review*, 54 (1893), 734–5.

a friend . . . attached. James Rice.

Boyardo. Matteo Maria Boiardo (1434–94), author of the *Orlando Inammorato*; Keats may have acquired the anecdote from Leigh Hunt who (later) produced *Stories from the Italian Poets* (London, 1846), which includes a section on Boiardo who was, according to Marilyn Butler, in the 19th-c. liberals' pantheon of past heroes (*Romantics, Rebels and Revolutionaries* (Oxford, 1981), p. 122).

478 *chequer-work.* A crossed letter in which the lines of script run vertically as well as horizontally so that twice as much can be written on a single sheet.

Letter to Fanny Brawne, 1 July 1819

Text from Forman (1883).

479 *Should.* Forman supplies 'think me'.

Pam. Knave of Clubs; highest trump card in five-card Loo.

To see . . . expression. Massinger, *The Duke of Milan* I. iii. 203–8 ('To see those eyes I prize above my own | Dart favours, though compelled, upon another;| Or those sweet lips, yielding immortal nectar, | Be gently touch'd by any but myself. | Think, think Marcelia, what a cursed thing | I were, beyond expression!')

480 *Margaret.* Fanny's younger sister, also known as 'Toots'.

Letter to Fanny Brawne, 8 July 1819

Text from holograph.

481 *your Comet.* A comet appeared on 3 July and had presumably been seen by Fanny Brawne.

Letter to J. H. Reynolds, 11 July 1819

Text from (incomplete) transcript in Woodhouse's letter-book.

sauntering Jack & Idle Joe. Cf. Matthew Prior, 'An Epitaph'.

Act. of *Otho the Great.*

Letter to Fanny Brawne, ?15 July 1819

482 Text from Forman (1883).

oriental . . . color. 'The History of the Basket' from Henry Weber, *Tales of the East* (Edinburgh, 1812), ii. 666–82; 'the remembrance of [the] garden' which the men visit 'renders all the pleasures of the world insipid to them' (p. 677); the circumstances of the tale recall *Endymion* and 'La Belle Dame sans Merci'.

483 *stories.* Narrative poems.

Letter to Fanny Brawne, 25 July 1819

484 Text from Forman (1883).

abstract Poem. 'The Fall of Hyperion'.

Sam. Fanny Brawne's brother; the identity of 'the Bishop' is unknown—perhaps he was a persistent rat at Wentworth Place.

Letter to Fanny Brawne, 5, 6 August 1819

485 Text from holograph.

Tragedy. *Otho the Great*; 'the lover' is Leudolph.

Maleager. Meleager, renowned for courage and strength, his life coextensive with a firebrand guarded by his mother Althaea; Keats may have seen an engraving of the statue in the Vatican Museum.

Letter to Benjamin Bailey, 14 August 1819

487 Text from holograph: only a fragment of this letter survives.

Letter to Fanny Brawne, 16 August 1819

Text from holograph.

August 17th. The postmark is for 16 Aug.

488 *Idalia*. Venus.

489 *I think they spell it*. So they did in 1819.

uncrystallize and dissolve. Cf. Stendhal's vocabulary of 'cristallisations' in *De L'Amour* (1822).

Letter to John Taylor, 23 August 1819

Text from holograph.

490 *chancery suit*. The suit was threatened by his aunt Margaret Midgley Jennings; Abbey persuaded Keats that this would involve a freeze of his grandmother's estate. This was not true; nor was the suit ever begun.

louting. bowing low.

491 *on the other side*. Brown's letter to Taylor, in which he offers his name as guarantor of whatever debts Keats should incur, is printed in *L* ii. 145.

Letter to J. H. Reynolds, 24 August 1819

Text from holograph.

August 25th. Keats dates all his Winchester letters one day ahead of the correct date.

able. Woodhouse adds 'to bear'.

Letter to J. H. Reynolds, 21 September 1819

492 Text from transcript by Woodhouse in his letter-book.

'*kepen in solitarinesse*'. 'The Eve of St. Mark' 106.

ferril. Ferrule; a metal cap at the end of the cane.

493 *Dian skies*. Diana being goddess of chastity.

composed upon it. Woodhouse, who transcribed this letter, notes that this is an allusion to the 'Ode to Autumn' which Keats sent him in the letter which follows.

leave the country. Reynolds was staying at the Woodhouse home in Bath.

494 *stage*. Portion of a stage-coach journey.

'You'll . . . jocular'. As the 'naturally jocose' Master Vellum says in Addison's *The Drummer* (1716).

Letter to Richard Woodhouse, 21, 22 September 1819

Text from holograph.

as wordsworth says. 'The Idiot Boy' 289.

495 *Sed . . . Heigh ho la*. Plays on Virgil, *Eclogues* iii. 79 'et longum "formose, vale, vale" inquit "Lolla"' (and long 'Farewell, farewell, beautiful Lollas' he called).

double postage. Postage costs were paid by the recipient, not the sender, of a letter. The charges were high and based on letter weight. The technique of crossing letters saved paper and hence postage (see note to letter of 9 June 1819, p. 633 above).

circumvallation. These lines begin at right angles to the passage from 'The Fall of Hyperion'. They are squeezed into the remaining space on the paper so that the final lines curve round.

Undine. George Soane, *Undine: or the Spirit of the Waters . . . translated from F. H. C. de la Motte Fouqué's German* (London, 1818).

American . . . so much of. Charles Brockden Brown (1771–1810); see Hazlitt, *Works*, xvi. 318–20.

Armenian. Schiller, *The Armenian; or the Ghost Seer*, trans. William Render, 4 vols. (Dublin, 1800).

496 *unpoeted I write*. F. E. L. Priestley (*TLS* 4 Feb. 1939, p. 73) suggests an analogy with *King Lear* III. i. 14 ('unbonneted he runs'); R. Rogers, in reply (*TLS* 25 Feb. 1939, pp. 121–2) thinks Keats is more likely to have written 'unposted I write'.

agrestrural. One of Keats's verbal portmanteaux, from *agrestis* (having to do with the country) and *rus* (country).

smokeable. Open to mockery.

otiosus-peroccupatus. Indolent-preoccupied.

Bramble. Matthew Bramble, from Smollett's *Humphrey Clinker* (1771); Keats proceeds to furnish a continuation of this novel, set, like Woodhouse, in Bath. Chowder is Tabitha Bramble's dog.

497 *piminy*. 'You have only, when before your glass, to keep pronouncing to

yourself nimini-pimini—the lips cannot fail of taking their plie' (John Burgoyne, *The Heiress* (1786), III. ii). Mrs Humphrey's face is becoming similarly pinched.

Cupid and Veney in the Spectator. The *Spectator*, 11 Mar. 1712 contained the diary of 'a Maiden Lady of good Fortune'. Her Friday entry reads: '*Eight in the Morning. A-bed*, read over all Mr. *Froth's* Letters, *Cupid* and *Veny*.'

Letter to Charles Brown, 22 September 1819

The text of this letter is from Brown's *Life of John Keats* (*KC* ii. 68–9), where it is not given in full. The omitted names are (in order): Reynolds, (?) Miss Brawne, impossible to guess, (?) Chichester, (?) Bedhampton, (?) Well Walk.

Letter to C. W. Dilke, 22 September 1819

499 Text from holograph.

on the common. Prostitute myself.

500 *certainly breed*. His greed being like that of a pregnant woman (cf. Webster, *The Duchess of Malfi* II. ii. 1–3).

'*as good as the times allow, Sir*'. Massinger, *A Very Woman* III. i. 104 ('*Strong*' as the time allows sir').

the present public proceedings. In the wake of the Peterloo Massacre (16 Aug.) the pressure for justice and reform increased. The Common Council of London had demanded an inquiry into the events of 16 Aug. and the punishment of the guilty. This demand was taken to the Throne on 17 Sept. and turned down.

suspicious looking letters. Sheridan, *The Rivals* IV. i. 28, 'a designing and malicious-looking letter'.

Westmonisteranian. Dilke's adored son Charles attended Westminster School.

Letter to George and Georgiana Keats, 17, 18, 20, 21, 24, 25, 27 September 1819

501 Text from holograph.

Mrs. Jennings. Margaret Midgley Jennings; widow of their uncle Captain Jennings.

Gliddon the partner. Walton & Gliddon was a legal firm.

Fry. Probably the other trustee of the Keats property.

Audubon. John James Audubon (1785–1851), the famous naturalist. He had persuaded George to invest in a commercial river-boat: the boat sank and George lost all his money.

502 *weaver boy*. Henry Hunt's colleague in St Peter's Fields Samuel Bamford (1788–1872) published poems under the name 'Weaver Boy'. For other weaver poetry see E. P. Thompson, *The Making of the English Working Class*

(Harmondsworth, 1968), pp. 322–4. The *Blackwood's* reviews had been condescending about Keats's social status.

503 *Spectator*. (6 May 1712) tells of a 'wit' who put together specialist dinner parties, including one of 'oglers' and one of 'stammerers'.

504 *O . . . elf*. Ben Jonson, *The Sad Shepherd* II. viii. 53.

'Still . . . daughter'. *Hamlet* II. ii. 188.

505 *coppy for you*. *The Anatomy of Melancholy*, III. 2. iii. 1; for a detailed demonstration of the influence of this work on Keats see Gittings, *The Living Year* (London, 1954), pp. 215–23.

nare simo patuloque. 'with a flat and spreading nose'.

pendulis mammis. 'with drooping breasts'.

bloody-falln. Chilblained.

aufe. Changeling, half-wit (an early form of 'oaf').

si qua patent meliora puta. Burton wrote 'latent' (not 'patent'); 'think what appears is better' ('what is not seen' in Burton).

506 *remedium amoris*. 'love cure'.

Irus' daughter, Thersite's sister. Irus is a beggar in the *Odyssey*, Thersites a deformed Greek officer: both are here emblems of personal repulsiveness.

Grobian's Scholler. A Grobian is a sloven.

Mathews. Charles Mathews (1776–1835) the comedian.

one against literary ambition. *Don Juan*, Canto i, stanza 218.

coffee-german. A relative of Abbey in the coffee trade?

508 *Hunt's triumphal entry into London*. Henry Hunt had been arrested at the meeting at St Peter's Fields in Manchester on 16 Aug. Having been released on bail he returned to London to a triumphant reception on 13 Sept. *The Times* on 14 Sept. estimated that he was met by 300,000, excluding those who watched from windows.

Crown and anchor. In the Strand.

Colnaghi's. Paul and Dominic Colnaghi were print dealers at 23 Cockspur Street.

any but yourselves—. . . At this point Keats copied the acrostic 'Give me your patience Sister while I frame' and a long portion of his 23, 26 July 1818 letter to Tom (not in this volume).

a large Q. Perhaps indicating a general question or enquiry as to the welfare of George and Georgiana.

my Poem. *Endymion*.

Monday. Keats copied the description of Winchester which follows in his letter to Reynolds of 21 Sept. (above, p. 492).

509 *Lady Bellaston*. From Fielding's *Tom Jones*.

Ut tibi placent. 'may they be pleasing to you'.

Fusili. Henry Fuseli (1741–1825); Keats puns on 'fusil' (a light musket).

510 *Pinxit . . . Sculpsit*. Latin terms used with the signatures on engravings to distinguish the creative artist (*pictor*) from the engraver (*sculptor*) who copies his work on to the plate.

studied in the Life-Academy. Been involved with naked women.

goodman Delver. *Hamlet* V. i. 14.

511 *Wise woman of Brentford*. *Merry Wives of Windsor* IV. v. 27. The wise woman of *Brainford* was notoriously fat.

fag. From 'fatigue' (i.e. lesson, session).

512 *Hook the farce writer*. Theodore Edward Hook (1788–1841), also novelist and editor of *John Bull*. The joke is on 'hook and eye'.

emblems. *Divine Emblems* (as they were called after 1724; before then *A Book for Boys and Girls*, 1686).

Hammond. Thomas Hammond, the surgeon to whom Keats had been apprenticed.

515 *'Political Justice'*. By William Godwin (London, 1793); see Keats's earlier remarks about Dilke being a 'Godwin perfectibility Man' (above, p. 424).

516 *like the manchester weavers*. In Sept. 1818.

Pun mote. A play on *bon mot*.

518 *make me sneeze*. Because 'Rapee' was also a brand of snuff.

519 *'Not as . . . can have'*. *Otho the Great* I. III. 24–9.

Letter to Fanny Brawne, 13 October 1819

Text from holograph.

520 *'to reason . . . my Love'*. Ford, *'Tis Pity She's a Whore* I. III. 78.

Letter to John Taylor, 17 November 1819

Text from holograph.

gradus ad Parnassum altissimum. 'step to the highest Parnassus'.

Holingshed's Elisabeth. In Raphael Holinshed, *Chronicles of England, Scotland and Ireland* (1577).

Letter to Fanny Keats, 8 February 1820

521 Text from holograph.

tongues and the Bones. Play on 'tongs and the bones' (*A Midsummer Night's Dream IV*. i. 29); the fable is by Aesop.

Letter to James Rice, 14, 16 February 1820

522 Text from holograph.

babble . . . green fields. *Henry V* II. iii. 17.

Hogarth's heads. Brown's copies of heads from Hogarth's *The Rake's Progress* are in Keats House.

Letter to Fanny Brawne, ?February 1820

523 Text from holograph.

Letter to Fanny Brawne, ?February 1820

Text from holograph.

524 *purple*. As Keats explains, the word was first written 'purplue'.

Letter to Fanny Brawne, ?27 February 1820

Text from holograph.

Rosseau and two Ladies. *Correspondance originale et inédite de J. J. Rousseau avec Mme Latour de Franqueville et M. du Peyrou*, 2 vols. (Paris, 1803).

his famous novel. *La Nouvelle Heloïse* (1761); Keats owned a copy of this.

Letter to C. W. Dilke, 4 March 1820

525 Text from holograph.

he sent me that also. Barry Cornwall's *Dramatic Scenes* were published in 1819; *Marcian Colonna, an Italian Tale, with Three Dramatic Scenes and other Poems* in 1820.

526 *to his Son*. M. B. Forman notes that Lord Chesterfield on 28 Jan. 1751 reproved his son for signing a draft 'in the worst and smallest hand I ever saw in my life' (B. Dobrée (ed.), *The Letters of Philip Dormer Stanhope, Fourth Earl of Chesterfield* (London, 1932), iv. 1668).

plumpers. A plumper is 'a vote given solely to one candidate at an election (when one has the right to vote for two or more)' (*OED*). Cobbett had returned from America the previous Nov. and was standing for election to parliament as a member for Coventry. (He did not get into parliament until the year of its reform, 1832.)

Letter to Fanny Brawne, ?March 1820

Text from Forman (1883).

Letter to Fanny Brawne, ?March 1820

527 Text from holograph.

as Shallow says. *2 Henry IV*, III. ii. 37 ('*all shall* die').

Letter to Fanny Brawne, ?March 1820

528 Text from Forman (1883).

preface. *1820* appeared with no motto and a preface written by Taylor (see headnote to 'Hyperion' (above, p. 592).

Letter to Fanny Brawne, ?March 1820

Text from holograph.

Letter to Fanny Brawne, ?May 1820

529 Text from holograph. This letter and the 3 following were sent from Kentish Town.

How my senses have ached at it. Othello IV. ii. 69 ('the sense aches at thee').

mames. For 'names'.

Letter to Fanny Brawne, ?June 1820

530 Text from holograph.

531 *reading*. Keats first wrote 'writing' and then crossed it out.

'*true . . . truth*'. *Troilus and Cressida* III. ii. 176.

Letter to Fanny Brawne, ?5 July 1820

532 Text from Forman (1883).

Letter to Fanny Brawne, ?August 1820

533 Text from holograph.

this. 'My dearest Girl'.

Elm Cottage. Home of the Brawnes during the end of 1818 and the start of 1819.

534 *Go to a Nunnery. Hamlet* III. i. 120.

his indecencies. Brown's housekeeper Abigail O'Donaghue was also his lover and bore him a son, Carlino.

Letter to John Taylor, 13 August 1820

Text from holograph.

Letter to Percy Bysshe Shelley, 16 August 1820

535 Text from holograph. Shelley had written to Keats from Italy on 27 July 1820 to invite him to spend the winter with himself and Mary at Pisa and so avoid the ill effects of another English winter. Shelley's letter is printed in *L* ii. 310–11. It includes some remarks on *Endymion* which he had reread 'ever with a new sense of the treasures of poetry it contains, though treasures poured forth with indistinct profusion'. He also hoped that Keats had received a copy of *The Cenci*.

load . . . with ore. Cf. *FQ* II. vii. 285.

536 *metap*[cs]. Metaphysics.

Letter to Charles Brown, ?August 1820

Text from Brown's *Life of John Keats* (*KC* ii. 78–9). Brown explains that the omitted words are 'a continuation of the secret of his former letter, ending with a request that I would accompany him to Italy'.

Dr. Clarke. Later Sir James Clarke (1788–1870); in later life he became physician to Queen Victoria. Clarke obtained rooms for Keats and Severn in Piazza di Spagna 26, opposite his own house, and took great pains to look after Keats on their arrival.

Lucy Vaughan Lloyd. 'The Jealousies'.

Letter to Fanny Keats, 23 August 1820

537 Text from holograph.

Seal-breaking business. In his previous letter to his sister (13 Aug.; *L* ii. 313–14) Keats had mentioned that a servant at Hunt's house had opened and failed to deliver a note that had arrived for him. This caused Keats such distress that he left Mortimer Terrace and returned to Wentworth Place.

'*turning a Neuk*'. From Burns, 'To Miss Ferrier' 15; a 'neuk' is a corner.

Letter to Charles Brown, 30 September 1820

538 Text from holograph.

28. Keats mistook the date.

'*while . . . liking*'. *1 Henry IV* III. iii. 6.

Letter to Mrs Samuel Brawne, ?24 October 1820

539 Text from holograph.

at the Health Office. The Naples authorities kept the ship in quarantine on account of a minor outbreak of typhus in London; this quarantine was protracted by the thoughtless arrival on board of 6 British sailors in search of compatriot company who then had to stay in the already cramped quarters of the *Maria Crowther* for another 7 days.

Tootts. Fanny's sister Margaret.

Letter to Charles Brown, 1 November 1820

541 Text from Brown's *Life of John Keats* (*KC* ii. 83–4).

q——. The word is probably 'quiff' (fuck).

Letter to Charles Brown, 30 November 1820

542 Text from Brown's *Life of John Keats* (*KC* ii. 85–6).

the little horse. Dr Clarke recommended horseback riding for Keats's health and a horse was hired at great cost. Keats often rode in the company of Lt. Elton, a young English army officer also suffering from TB.

we shall all die young. Only Fanny Keats survived into old age.

FURTHER READING

MAJOR EDITIONS

John Keats, *Poems* (London, 1817).
——*Endymion* (London, 1818).
——*Lamia, Isabella, The Eve of St. Agnes, and Other Poems* (London, 1820).
R. M. Milnes (ed.), *Life, Letters, and Literary Remains of John Keats*, 2 vols. (London, 1848).
R. M. Milnes, 'Another Version of Keats's "Hyperion"', *Miscellanies of the Philobiblion Society*, 3 (1856–7).
H. B. Forman (ed.), *The Poetical Works and Other Writings of John Keats*, 4 vols. (London, 1883).
—— (ed.), *The Letters of John Keats to Fanny Brawne* (London, 1878).
John Gilmer Speed (ed.), *The Letters and Poems of John Keats*, 5 vols. (New York, 1883).
H. B. Forman (ed.), *The Poetry and Prose of John Keats* (London, 1890; rev. edn. by M. B. Forman, 8 vols. (London, 1938–9: the 'Hampstead Keats')).
Sidney Colvin (ed.), *The Letters of John Keats to his Family and Friends* (London and New York, 1891).
H. B. Forman, *The Letters of John Keats* (London, 1895).
M. B. Forman (ed.), *The Letters of John Keats*, 2 vols. (Oxford, 1931, 4th edn. 1952).
E. de Selincourt (ed.), *The Poems of John Keats* (London, 1905, rev. edn. 1926).
H. W. Garrod (ed.), *The Poetical Works of John Keats* (Oxford, 1939, rev. edn. 1958).
H. E. Rollins (ed.), *The Letters of John Keats*, 2 vols. (Cambridge, Mass., 1958).
Robert Gittings (ed.), *Letters of John Keats* [selected] (Oxford, 1970, repr. 1975).
Miriam Allott (ed.), *The Poems of John Keats* (London, 1970).
John Barnard (ed.), *John Keats: the Complete Poems* (Harmondsworth, 1973).
Jack Stillinger (ed.), *John Keats: Complete Poems* (Boston, Mass., 1978).

WORKS OF REFERENCE

Michael G. Becker *et al.*, *A Concordance to the Poems of John Keats* (New York, 1981).
J. R. MacGillivray, *Keats: A Bibliography and Reference Guide with an Essay on Keats's Reputation* (Toronto, 1949, repr. 1986).
Sister Pio Maria Rice, 'John Keats: A Classified Bibliography of Critical Writings . . . 1947–1961', *Bulletin of Bibliography*, 24 (1965), 167–8, 187–92.
Jack Stillinger, *The Texts of Keats's Poems* (Cambridge, Mass., 1974).

An annual bibliography has appeared in the *Keats–Shelley Journal* from 1952 and covers works dating back to July 1950. These bibliographies from July 1950 to June 1962 are reprinted in D. B. Green and E. G. Wilson (eds.), *Keats,*

Shelley, Byron, Hunt and their Circles: A Bibliography (Lincoln, Nebr., 1964) and those covering July 1962 to December 1974 in R. A. Harley (ed.), *Keats, Shelley, Byron, Hunt and their Circles: A Bibliography* (Lincoln, Nebr., 1978).

CRITICISM AND BIOGRAPHY

John Bailey, 'Keats and Reality', *Proceedings of the British Academy* (London, 1962).
W. J. Bate, *The Stylistic Development of John Keats* (London, 1958).
—— (ed.), *Keats: a Collection of Critical Essays* (Englewood Cliffs, NJ, 1964).
—— *John Keats* (London, 1979).
Bernard Blackstone, *The Consecrated Urn* (London, 1959).
Harold Bloom, *The Visionary Company* (Garden City, 1961).
Cleanth Brooks, 'History without Footnotes: An Account of Keats's Urn', in Cleanth Brooks, *The Well Wrought Urn* (New York, 1947).
Sidney Colvin, *John Keats: His Life and Poetry, His Friends, Critics and After-Fame* (London, 1917).
Morris Dickstein, *Keats and his Poetry: a Study in Development* (Chicago, 1971).
Walter Evert, *Aesthetic and Myth in the Poetry of Keats* (Princeton, 1965).
Claude Lee Finney, *The Evolution of Keats's Poetry* (New York, 1936).
Robert Gittings, *The Living Year* (London, 1954).
—— *The Mask of Keats* (London, 1956).
—— *John Keats* (London, 1968).
Donald C. Goellnicht, *The Poet-Physician: Keats and Medical Science* (Pittsburgh, 1984).
Ian Jack, *Keats and the Mirror of Art* (Oxford, 1967).
John Jones, *John Keats's Dream of Truth* (Oxford, 1969).
Judy Little, *Keats as a Narrative Poet* (Lincoln, Nebr., 1975).
Jerome McGann, 'Keats and the Historical Method in Literary Criticism,' *MLN* 94 (1979), 988–1032.
G. Matthews (ed.), *Keats: the Critical Heritage* (London, 1971).
John Middleton Murray, *Keats and Shakespeare* (Oxford, 1926).
E. C. Pettet, *On the Poetry of Keats* (Cambridge, 1957).
Christopher Ricks, *Keats and Embarrassment* (Oxford, 1984).
M. R. Ridley, *Keats's Craftsmanship* (Oxford, 1933, repr. London, 1963).
Caroline Spurgeon, *Keats's Shakespeare*, 2nd edn. (Oxford, 1929).
Bernice Slote, *Keats and the Dramatic Principle* (Lincoln, Nebr., 1958).
Stuart M. Sperry, *Keats the Poet* (Princeton, NJ, 1973).
Jack Stillinger, *The Hoodwinking of Madeline* (Urbana, Ill., 1971).
C. D. Thorpe, *The Mind of John Keats* (New York, 1926).
Lionel Trilling, 'The Poet as Hero: Keats in his Letters', in Lionel Trilling, *The Opposing Self* (New York, 1955), pp. 3–49.
Helen Vendler, *The Odes of John Keats* (Cambridge, Mass., 1983).
Aileen Ward, *John Keats: The Making of a Poet* (New York, 1963).
R. S. White, *Keats as a Reader of Shakespeare* (London, 1987).
Susan Wolfson, Introduction to 'Keats and Politics: A Forum', *Studies in Romanticism*, 25 (1986), 171–81. This includes Morris Dickstein's 1983

Modern Language Association paper 'Keats and Politics' (pp. 175–81) and essays by William Keach (pp. 182–96), David Bromwich (pp. 197–210), Paul H. Fry (pp. 211–19), and Alan J. Bewell (pp. 220–8).

HISTORICAL CONTEXT

Marilyn Butler, *Romantics, Rebels and Reactionaries* (Oxford, 1981).

S. Macoby, *English Radicalism 1786–1832* (London, 1955).

E. P. Thompson, *The Making of the English Working Class* (London, 1963, rev. edn. Harmondsworth, 1968).

W. H. Wickwar, *The Struggle for the Freedom of the Press 1819–1832* (London, 1928).

GLOSSARY OF CLASSICAL NAMES

Actaeon: a huntsman who came across Diana bathing naked with her attendants. The goddess transformed him into a stag and he was pursued and devoured by his own hounds.

Adonis: a beautiful huntsman loved by Venus, against whose advice he hunted a boar who impaled and killed him. Venus, in sorrow at his death, changed him into an anemone. Proserpine restored him to life so that he might spend half the year in the underworld with her, the other half above ground with Venus. The myth of Adonis is consequently involved with regeneration and renewal (as in Spenser's 'Garden of Adonis', *FQ* III. vi). The 'Adonian feast' referred to in *Lamia* (i. 320) was an annual fertility rite, held in Venus' temple. Shakespeare's *Venus and Adonis* tells the first part of the story.

Aeolus: king of storms and winds.

Aeothon: one of the horses that drew the chariot of the sun.

Alecto: one of the Furies; her head is covered with snakes.

Alpheus: a river in Arcadia who fell in love with the nymph Arethusa when she bathed in him. Diana changed her into a fountain which the river Alpheus pursued under the sea as far as the land of Ortygia where the fountain Arethusa rose.

Amalthea: Cretan princess who fed Jupiter with goat's milk.

Amphion: son of Jupiter; lyrist of such power that the walls of Thebes were constructed by the sound of his lyre. Keats conflates him with Arion whose music drew helpful dolphins.

Amphitrite: wife of Neptune; mother of Triton.

Andromeda: tied to a rock jutting over the sea to be devoured by a monster in sacrifice to Neptune, she was rescued by Perseus who petrified the monster with Medusa's head.

Apollo: son of Jupiter and Latona, brother of Diana. He took over the role of sun-god from the Titan Hyperion and, as such, drives his chariot daily through the heavens. He is also the god of music and poetry (and often represented with a lyre), a prophet, and the god of medicine.

Arcady: Arcadia; home of poet shepherds and of Pan, their god.

Arethusa *see* Alpheus.

Argonauts: heroes who sailed with Jason in search of the Golden Fleece.

Argus: a king of Argos. He had one hundred eyes, only two of which slept at any time. Mercury lulled him to sleep with his lyre and slew him (on Jupiter's orders) while he was supposed to be watching over Io on Juno's behalf.

Ariadne: lover of Theseus (who left her) and of Bacchus who crowned her with seven stars which became a constellation at her death.

Arion: ancient lyrist whose music was so sweet that dolphins were drawn by it and, on one occasion, carried him across the sea to safety.

Atlas: one of the Titans who fought against the Apollonian gods; later changed by Perseus into a mountain in north Africa so high it was said to bear the heavens.

Aurora: goddess of dawn; dawn itself.

Bacchus: son of Jupiter and Semele; god of the vine. He is represented as a young man, crowned with ivy and vine leaves and carrying a thyrsus (a pine-staff twined round with ivy and vine leaves). The panther is sacred to him. He is associated with (and paralleled to) the Egyptian god Osiris. He is represented at the head of his rout, drawn in a chariot by a lion and a tiger, accompanied by Pan, Silenus (his foster-father), satyrs, and Bacchantes (followers inspired by divine fury).

Boreas: god of the North Wind.

Briareus: son of Coelus and Terra; one of the Titans who fought the Apollonians. He had a hundred hands and fifty heads.

Caf: Keats makes him a Titan, father to Asia. Keats found the name of Kaf, a mountain which surrounds the world, in H. Weber's *Tales of the East* (1812) and S. Beckford's *Vathek* (1816).

Caria: home of Endymion.

Castor and Pollux: twin sons of Jupiter and Leda; constellated after death as Gemini.

Ceres: daughter of Saturn and Vesta; goddess of corn and harvest; mother (by Jupiter) of Proserpine.

Circe: daughter of Sol and Perseis. A powerful witch, she lived in Aeaea where, after feasting men as guests, she transformed them into beasts. She did this to all Odysseus' men, but not to Odysseus who became her lover and stayed away from Penelope for another year.

Clio: daughter of Jupiter and Mnemosyne; Muse of history.

Coelus: son (and later husband) of Tellus. His sons were Titans. He is also called Uranus.

Cupid: son of Venus and Jupiter; Eros, the god of love. He is usually represented as a child with a quiverful of arrows with which to wound mortals with love. But as Psyche's lover he should be thought of as a young man.

Cybele: sister and wife of Saturn, daughter of Coelus and Tellus. She is often taken for her mother and invoked as mother earth and mother of the gods.

Cynthia: another name for Diana.

Cytherea: Cyprus, Venus' birthplace.

Daedalus: a great artificer and inventor. He created the Cretan labyrinth and, in order that he and his son Icarus might escape King Minos, wings of wax and feathers.

Danaë's son: Perseus; Jupiter entered the imprisoned Danaë as a shower of gold whence Perseus was conceived.

Daphne: a daughter of the river Peneus. When she was pursued by an amorous Apollo she invoked Diana who changed her into a laurel.

Deucalion: a son of Prometheus. He was king of Thessaly at the time when Jupiter punished mankind for its wickedness by flooding the world. Advised by his father, Deucalion built a boat and saved himself and his wife Pyrrha.

Diana: daughter of Jupiter and Latona, sister of Apollo (and Moon to his Sun). She has several offices, all associated with her role of moon-goddess, and is goddess of childbirth, chastity, and hunting. As Diana she is represented as a hunter attended by nymphs. In the heavens she is known as Luna, and in hell as Hecate where she is associated with magic and enchantment.

Dido: Queen and founder of Carthage; lover of Aeneas who deserted her under divine compulsion. Dido burned herself to death.

Dis: Pluto.

Dolor: a Titan to Keats (though not to Hesiod or Hyginus). In Hyginus is the phrase 'ex Aethere et Terra, Dolor' ('from Air and Earth, Grief').

Doris: daughter of Oceanus and Tethys; married to her brother Nereus. Their fifty daughters are the Nereides.

Dryad: wood-nymph.

Dryope: there are several nymphs of this name, two of them invoked in *Endymion*. One is mother of Pan by Mercury; the other a nymph raped by Apollo and, along with her son Amphisus, changed into a lotus.

Echo: daughter of Air and Tellus, formerly attendant to Juno who punished her indiscretion by limiting her speech to answers. She fell in love wth Narcissus and turned to stone in her grief at his lack of response. The stone retained her voice.

Elysium: paradisal home of the virtuous dead.

Enceladus: son of Titan and Tellus; the most powerful of the Titans who fought the Apollonians. He was eventually overwhelmed under Mount Aetna whose flames are thought to be his breath.

Endymion: a Carian shepherd with whom Diana fell in love when she saw him naked on the top of Mount Latmos. Diana visited him nightly while he slept. Baldwin explains: 'the meaning of the fable is that Endymion was a great astronomer; he passed whole nights upon mount Latmos contemplating the heavenly bodies' (*Pantheon*, p. 206).

Erebus: son of Chaos and Darkness; an infernal deity used by Keats and others to signify hell itself.

Eurydice: wife of Orpheus, who entered the underworld to recover her at her death. She was allowed to follow him on the condition that he did not turn back during the ascent to earth. Orpheus failed and lost.

Favonius: the West Wind.

Flora: the Roman goddess of flowers and gardens.

Ganymede: a beautiful boy, snatched to heaven by Jupiter who made him cupbearer to the gods.

Glaucus: sea deity who fell in love with the Nereid Scylla and applied to Circe for help (*see* Scylla).

Hamadryads: female deities of particular trees. Their lives are coextensive with those of their trees.

Hebe: daughter of Jupiter and Juno; goddess of youth; cupbearer to the gods until she slipped at an important function and was replaced by Ganymede.

Hecate: Diana's infernal name.

Helicon: a mountian in Boeotia sacred to the Muses.

Hermes: Mercury.

Hesperides: the three daughters of Hesperus; guardians of the golden apples given by Juno to Jupiter at their wedding.

Hesperus: son of Iapetus; the name given to the planet Venus when it appears after sunset.

Hippocrene: fountain that flows from Mount Helicon. According to Baldwin the waters are 'violet-coloured, and are represented as endowed with voice and articulate sound' (*Pantheon*, p. 49).

Hyacinthus: a boy loved by Apollo and Zephyrus. Jealous of Apollo, whom Hyacinthus loved, Zephyrus killed the boy with a quoit. Apollo changed the boy's blood into a flower and inscribed it with his own lament.

Hybla: mountain in Sicily; site of odoriferous flowers and a good place for honey.

Hyperion: son of Coelus and Tellus; a Titan. God of the sun until supplanted by Apollo.

Iapetus: a Titan; son of Coelus and Tellus.

Icarus: son of Daedalus. He flew too near the sun during his flight from Crete, the wax binding his wings melted, and he fell into the sea.

Ida: mountain near Troy where Paris gave Venus the prize as the most beautiful. Hence a synonym for Venus.

Iris: goddess of the rainbow.

Ixion: banished from heaven by Jupiter and eternally tied to a revolving wheel in hell.

Jove: Jupiter.

Juno: daughter of Saturn and Ops; sister and jealous wife of Jupiter; queen of the gods. The peacock is sacred to her.

Jupiter: son of Saturn and Ops (who saved him when Saturn devoured all his sons); educated on Mount Ida. After defeating Saturn he divided the world between himself and his brothers and became ruler of heaven.

Latmos: a mountain in Caria where Endymion encountered the moon.

Latona: mother of Diana and Apollo.

Leander: lover of Hero, he swam the Hellespont to spend the night with her and was drowned. Marlowe tells their story in *Hero and Leander* (completed by Chapman).

Leda: wife of Tyndarus, King of Sparta. She was impregnated by Jupiter (who came to her as a swan) and bore two eggs, one of which contained Helen of Troy.

Lethe: a river in hell; the drinking of its waters induces forgetfulness of all that went before.

Lucifer: the name of the planet Venus when it appears before dawn.

Lycaeus: a mountain in Arcadia sacred to Jupiter and Pan.

Maia: goddess of the month of May; mother of Mercury.

Mars: god of war; lover of Venus (with whom he was caught *in flagrante* by Vulcan).

Melpomene: muse of tragedy.

Mercury: the Roman name for Hermes, son of Maia and Jupiter. He is the messenger of the gods and wears a winged hat (the *petasus*) and has wings (*talaria*) on his heels; he also carries a wand (the *caduceus*) with wings at the top and two intertwined serpents on the stem. He is noted for his nimbleness and wit (he is the god of eloquence) and also for his magical powers. As 'psychopomp' he conducts the souls of the dead.

Midas: King of Phrygia. In recompense for the hospitality he had shown towards Silenus, Bacchus undertook to grant whatever he wished. Midas asked that all he touched should turn to gold. This he regretted when he was unable to eat.

Minerva: goddess of wisdom, war, and the liberal arts; daughter of Jupiter from whose brain she sprang fully armed.

Minos: one of the three judges of hell.

Mnemosyne: orginally a Titan; daughter of Coelus and Terra. By Jupiter she is mother of the nine Muses. Her name means 'memory'.

Momus: son of Nox; god of satire.

Moneta: although this is a name given to Juno, Keats's Moneta is identified with Mnemosyne on the authority of Hyginus.

Morpheus: son and minister of Somnus, god of sleep.

Mulciber: Vulcan.

Naiad: female deity of woods and streams.

Nais: sea nymph; mother of Glaucus.

Narcissus: beautiful boy who fell in love with his own reflected image and pined to death.

Nemesis: daughter of Nox; infernal goddess of vengeance.

Neptune: son of Saturn after conquering whom he took over the rule of the sea. He wields a trident.

Nereids: sea nymphs; daughters of Nereus and Doris.

Nereus: son of Oceanus and Tellus; a sea deity resident in the Aegean sea. He is a prophet and is usually represented as an old man.

Niobe: mother by Amphion of ten sons and ten daughters. All but one of these were killed by Diana and Apollo on behalf of their mother Latona whom Niobe had insulted. In her grief she turned to stone and is almost an emblem of sorrow ('like Niobe, all tears' (*Hamlet* I. ii. 149).

Nox: daughter of Chaos; one of the most ancient deities.

Oceanus: son of Coelus and Tellus; he presided over every part of the sea. According to Homer he was the father of all gods.

Olympians: members of Jupiter's court on the top of Mount Olympus.

Ops: daughter of Coelus and Tellus: often identified with Cybele.

Oread: a mountain nymph, such as one of those who attend Diana.

Orion: a giant who was blinded by King Oenopion. He recovered his sight by turning to face the full beam of the rising sun.

Orpheus: son of Calliope and Apollo; a Thracian shepherd whose lyric music had the power to move animals and stones as well as Pluto, who allowed him to attempt the recovery of Eurydice.

Osiris: son of Jupiter and Niobe; husband of Isis. He is a great Egyptian deity and often identified with Bacchus.

Pallas: Pallas Athene; another name for Minerva.

Pan: the Greek πᾶν means 'all' or 'everything'. Pan is the son of Mercury and Dryope. He is the god of shepherds; a satyr—goat from the waist down and with horns on his head. His chief home is Arcadia. He attempted to rape the nymph Syrinx, and when she eluded him by changing into a reed, he made the reed into a 'pan-pipe' and played it.

Paphos: Venus' birthplace in Cyprus.

Pegasus: a winged horse born of Medusa's blood. He lives on Mount Helicon and is an emblem of the soaring imagination.

Philomel: the name often refers to any nightingale. The original Philomel was raped by her brother-in-law Tereus who cut out her tongue to keep her quiet. She conveyed her history to her sister Procne by weaving a tapestry. Procne punished Tereus by serving their son, Itylus, to him as meat. Before Tereus could stab them Procne was turned into a swallow and Philomel into a nightingale.

Phoebe: another name for Diana (though, properly, her Titan mother).

Phoebus: Apollo; the sun.

Phorcus: son of Pontus and Tellus; father (with his sister Ceto) of the Gorgons and the dragon who guarded the apples of the Hesperides. Keats thought of him as a Titan.

Pleiad: one of the Pleiades, Atlas' seven daughters who were constellated at their deaths.

Pluto: son of Saturn and Ops; ruler of the infernal kingdom after the division of Saturn's realm. His wife is Proserpine.

Pollux *see* Castor.

Polyphemus: King of the Cyclops (one-eyed shepherd giants); he was outsmarted by Odysseus who blinded him and escaped with those of his men whom Polyphemus had not already eaten.

Pomona: a nymph presiding over fruit trees and gardens.

Porphyrion: son of Coelus and Tellus: a Titan.

Prometheus: son of Iapetus and Clymene. He stole fire from heaven and with it animated the clay which made the first man and woman. For this he was punished by being tied to a rock while a vulture preyed upon his liver.

Proserpine: daughter of Ceres and Jupiter; resident of Sicily until Pluto carried her down to the underworld and made her his queen. Ceres sought her everywhere in vain. To comfort her Jupiter arranged for Proserpine to spend half the year above ground with her mother.

Proteus: a sea deity with prophetic gifts. He seldom uses these because of his ability to change shape and elude those who would consult him.

Psyche: a nymph with whom Cupid fell in love and whom he visited unseen at night. Psyche, incited by the suggestion that her lover was a serpent, hid a lamp in order to see him. A drop of hot oil fell from this lamp and woke the god who, once seen, fled. After a period of expiatory suffering Psyche was reunited with Cupid and granted immortality by Jupiter. Consequently she is a recent addition to the Olympian hierarchy. The Greek ψυχή means 'soul' and also 'butterfly' (the latter an emblem of the former).

Pyrrha *see* Deucalion.

Pythia: priestess of Apollo at Delphi and a prophetic medium. When divinely inspired she would often seem possessed by furies and was fearful to witness.

Python: a huge serpent which grew from the silt after Deucalion's flood. Apollo shot him dead with arrows and instituted the Pythian games to mark his victory.

Rhadamanthus: son of Jupiter and Europa. He was created one of the judges of hell on account of the justice with which he had ruled on earth.

Saturn: son of Coelus (or Uranus) and Tellus; leader of the Titans, with whom he was overthrown by his sons under Jupiter's leadership. According to traditional accounts the banished Saturn fled to Italy where his civilizing and beneficent rule came to be known as the Golden Age. Keats antedates this Golden Age to before the overthrow of the Titans.

Scylla: a Nereid with whom Glaucus fell in love. Circe, to whom Glaucus had applied for help, wanted Glaucus for herself and transformed Scylla into a monster. In her terror Scylla threw herself into the sea and was changed into the rock which bears her name.

Semele: mother of Bacchus.

Silenus: tutor and companion to Bacchus. He is usually represented as fat, jolly, and drunk, riding on an ass.

Syrinx: a nymph who was transformed into a reed in her flight from Pan.

Tartarus: 'the abode of woe' (Baldwin, *Pantheon*, p. 149); the part of the underworld that is not Elysium.

Tellus: the earth; mother of the Titans. Only Chaos is older.

Tempe: a valley in Thessaly celebrated for its cool shades and pleasant landscape.

Tethys: daughter of Uranus and Tellus; wife of Oceanus and mother of the great rivers and the Oceanides.

Thalia: the Muse of festivals and of pastoral and comic poetry; she is one of the three Graces.

Themis: for Keats, following Hesiod, one of the Titans.

Thetis: a sea goddess; daughter of Nereus and Doris; mother of Achilles.

Titans: the forty-five sons of Coelus and Tellus. They include Saturn, Hyperion, Oceanus, Iapetus, Briareus, and Cottus. They were giants of enormous strength.

Triton: son of Neptune and Amphitrite; a sea deity. He is usually represented blowing a conch shell.

Typhon: for Keats, a Titan.

Urania: daughter of Jupiter and Mnemosyne; Muse of astronomy. She is usually represented in an azure robe and crowned with stars.

Uranus: another name for Coelus.

Venus: sea-born goddess of beauty; mother of Cupid (by Mars). Her girdle (or 'zone') has the power to impart beauty and excite love. Among the many she has loved is Adonis. Venus is also the name of the evening star.

Vertumnus: a Roman deity presiding over spring and orchards. He wooed Pomona—at first without success—and eventually married her.

Vesper: another name for Venus, the evening star.

Vesta: goddess of the hearth.

Vulcan: divine smith and artificer; fashioner of Jupiter's thunderbolts. He was married to Venus.

Zephyrus: the West Wind, son of Atreus and Aurora. He is able to generate fruit and flowers with his sweet breath and is represented as a delicate winged youth.

INDEX OF KEATS'S CORRESPONDENTS AND OTHERS TO WHOM HE FREQUENTLY REFERS

Literary figures already well known, such as Hazlitt, Lamb, and Shelley, have not been included.

Richard Abbey. A north-country acquaintance of Keats's grandmother Alice Jennings, and appointed by her trustee of her estate and guardian of the Keats children until they came of age. By profession a tea-merchant, he employed George and Tom Keats for a short time in his counting-house. He was certainly shifty and probably dishonest in his handling of the Keats affairs and he and his wife were restrictive guardians to Fanny Keats.

Benjamin Bailey (1791–1853). Bailey, a friend of Rice and Reynolds, met Keats in the spring of 1817. He was then an undergraduate at Magdalene Hall, Oxford where he invited Keats later that year. He and Keats visited Stratford-upon-Avon together. He defended Keats in the *Oxford Herald* (30 May, 8 June 1818) and in the same summer was ordained and obtained a curacy at Carlisle. In 1819 he married the daughter of Bishop Gleig. This led to Keats's disenchantment with Bailey who had previously seemed to be courting Mariane Reynolds.

The **Bentleys**. Mr Bentley was a postman and he and his wife rented rooms to all three Keats brothers at 1 Well Walk, Hampstead. Keats complained of the smell of their children's stockings.

William Bewick (1795–1866). A portrait artist and historical painter; at one time a student of Haydon (whose studio he attended almost daily between 1817 and 1820). He was an excellent copyist.

Morris Birkbeck (1764–1825). Founded the town of Albion in Illinois where he had bought 16,000 acres of land. George and Georgiana Keats hoped to join his settlement when they went over to America but, on arrival, found only heavily forested land left.

Fanny Brawne (1800–65). She, her widowed mother Mrs Samuel Brawne, and her younger siblings Samuel and Margaret ('Toots') probably met Keats some time in the summer of 1818 when they rented Charles Brown's half of Wentworth Place. From about April 1819 until 1829 they rented Dilke's part of the house and so were next-door neighbours to Keats when he moved in with

Brown in October. Keats guarded the secret of his intense love for her very closely. It was clearly reciprocated and Fanny Brawne was later to refer to 25 December 1818, when Keats almost certainly proposed to her, as the happiest day of her life. Though Keats lacked the financial stability then thought desirable in a son-in-law, Mrs Brawne accepted Keats's engagement to her daughter (in October 1819) and the two Brawne women nursed Keats lovingly during his last month in England. After Keats's death Fanny Brawne opened a correspondence with Keats's sister. Her letters reveal tact, sympathy, quick-wittedness, and a capacity for self-mockery (see F. Edgcumbe (ed.), *Letters of Fanny Brawne to Fanny Keats* (Oxford, 1936)). In 1833 she married Louis Lindo (later Lindon).

Charles Armitage Brown (1787–1842). In 1805 he went to St Petersburg and speculated in bristles with little or no capital. In 1810 he returned, bankrupt, to England. In 1814 his comic opera *Narensky, or the Road to Yaroslaf* was performed at Drury Lane, securing him £300 and a life ticket to that theatre. He met Keats before September 1817 and the following summer they took a walking holiday in the north together. After Tom's death Keats became Brown's tenant at Wentworth Place (the house which he had co-built with his school friend Dilke). In September 1819 he was married in an illegal service to his housekeeper Abigail O'Donaghue. She bore him a son, Carlino. Brown was a practical, kind, and jealous friend to Keats.

Charles Cowden Clarke ('C.C.C.') (1787–1877). His father John Clarke ran the school at Enfield which the Keats boys attended. He taught at the school himself and, as teacher and friend, was a greatly formative influence on Keats, introducing him to the *Examiner* and liberal politics, feeding and encouraging his love of literature, and introducing him to practising writers (including Hunt and Lamb). In 1817, when his father gave up the school, he moved to Ramsgate and was less frequently in touch with Keats. He married Vincent Novello's daughter Mary.

Barry Cornwall, pseudonym for **Bryan Procter** (1787–1874). Solicitor and writer; friend of Leigh Hunt, Lamb, and Hazlitt. His writings included plays and narrative poems (including one on 'The Fall of Saturn: a Vision' published in 1823).

Charles Cripps. A young artist whom Haydon noticed copying an altar-piece at Oxford. Haydon thought he might have talent and proposed to train him if a subscription could be raised to support the penniless artist during his apprenticeship. Keats and Bailey were caught up in this scheme and had much of the responsibility devolved upon them. Nothing came of it.

Burridge Davenport. A merchant and near neighbour in Hampstead (living in Church Row), with whom Keats was on visiting terms.

Charles Wentworth Dilke (1789–1864). Dilke worked in the Navy Pay Office until 1836 but was a keen amateur scholar. Between 1814 and 1816 he brought out his continuation of Dodsley's *Old Plays*. Charles Brown was an old schoolfellow and the two built and shared, as two residences, Wentworth Place (now Keats House) in Hampstead. He and his wife Maria met Keats at some point before the autumn of 1817 from when Keats was a frequent visitor. Keats particularly enjoyed Maria's sense of humour. In April 1819 the Dilkes moved to Westminster in order to be near their son Charles at Westminster School. After Keats's death Dilke supervised Fanny Keats's finances and initiated the action against Abbey.

Edward Dubois (1774–1850). Barrister, amateur writer, and contributor to various periodicals (the *Morning Chronicle* in particular). He satirized the travels of Sir John Carr in the *Monthly Mirror*. He also translated the *Decameron*, and wrote art notices and dramatic criticism.

Mary Frogley. A cousin of Richard Woodhouse and a member of the Mathews' social circle. She saw a lot of Keats during 1814–15. In 1820 she became Mary Neville.

William Gifford (1756–1826). Gifford came from a poor background and was bought out of his apprenticeship (to a shoemaker) by a benefactor who thus enabled him to study. In 1797–8 he wrote for the *Anti-Jacobin* and in 1809 became the founder editor of the *Quarterly Review*, supported by eminent Tories. He also edited Jacobean plays (including Jonson and Massinger) and encouraged Dilke's work in this line. Byron approved of Gifford's classicism but most of the Keats circle disliked him for his politics.

George Robert Gleig ('Gleg[g]' to Keats) (1796–1888). Gleig was formerly a soldier but returned to Magdalene Hall, Oxford, after the Battle of Waterloo. He was ordained in 1820. His father was George Gleig, Bishop of Brechin, and his sister Hamilton Gleig married his fellow student Benjamin Bailey. He went on to write novels and military history.

William Haslam (1795–1851). A schoolfellow of the Keats boys at Enfield, Haslam continued 'a most kind and obliging and constant friend' (letter of 14–31 Oct. 1818). He worked as a solicitor and lived with his family in Bethnal Green until his father's death. He then assumed his father's post with Frampton and Sons, wholesale grocers in the City. He was very much a family friend, accompanying Keats on visits to Fanny Keats and conveying to her the news of her brother's death. Keats, defensively perhaps, was amused by Haslam's courtship of his future wife Mary. She was close to childbirth when the time came for Keats to go to Italy and it was Haslam, unable to accompany Keats himself, who arranged for Severn to make the journey and accompanied Keats and Severn to Gravesend. He wrote to Severn that 'If I know what it is to love, I truly love John Keats.'

Benjamin Robert Haydon (1786–1846). Born in the West Country, he came to London to study at the RA in 1804. He produced huge paintings on epic themes but constantly struggled against poor eyesight and poverty (he was thrice imprisoned for debt). He was among the first to appreciate the beauty and importance of the Elgin Marbles and campaigned for the nation to buy them.

He met Keats at Leigh Hunt's during October 1816, took him up, encouraged him, and included his portrait in his painting *Christ's Entry into Jerusalem*. Their enthusiastic friendship cooled when Haydon failed to repay a loan that Keats had ill been able to afford. It was Haydon who originated the myth that Keats sprinkled his throat with cayenne the better to enjoy the coolness of claret.

Henry. Henry Wylie, Georgiana's brother.

James Augustus Hessey ('Mistessey') (1785–1870). Partner to John Taylor in the publishing firm Taylor and Hessey (established in 1806 at 93 Fleet Street). The firm took over *1817* from Charles and James Ollier and published Keats's two subsequent volumes. Hessey, who saw to the retail side of the business, figures less prominently than his more literary partner, but both men were generous personal friends to Keats as well as solicitous and far-seeing publishers.

Hodgkinson. Richard Abbey's junior partner.

William Hone (1780–1843). In 1817 Hone, an antiquarian bookseller off Ludgate Hill, produced the second *Reformist's Register*. In the same year he wrote and published his anti-ministerial liturgical parodies of which nearly 100,000 copies were sold. He was brought to trial for blasphemous libel by Lord Ellenborough. Hone was acquitted after conducting his own defence—'a thing, which not to have been, would have dulled still more Liberty's Emblazoning' (letter of 21, 27 Dec. 1817).

Henry ('Orator') **Hunt** (1773–1835). Imprisoned in 1810 for assaulting a gamekeeper, he shared prison rooms with William Cobbett. He took part in the Spa Fields meetings of 1816 and established himself as a charismatic popular leader and orator. In 1818 he contested the Westminster seat in Parliament (but was not elected until 1830). After the Peterloo Massacre on 16 August 1819 he was arrested and sentenced to two years' imprisonment—but not before his triumphant entry into London. Samuel Bamford, the 'weaver poet' and a political colleague of Hunt, later recalled the egoism that vitiated Hunt's politics.

Leigh Hunt (1784–1859). Became editor of the *Examiner* (a weekly newspaper on the side of reform). In 1813 he and his brother John were sentenced to two years' imprisonment for intending 'to traduce and vilify his Royal Highness [the Prince Regent]'. He was to Keats at this time a hero of liberty and he is referred to in some of Keats's poems as 'Libertas'. Clarke introduced Keats to Hunt in 1816; Hunt showed Keats's poetry to Hazlitt and began to publish it in the *Examiner*. He was a generous, if sometimes over-assiduous, patron. Keats's public declaration of their association (in the dedicatory sonnet to *1817*) presented the Tory press with a handle with which to identify his politics with Hunt's. Keats's enthusiasm for Hunt and his poetry diminished as his own voice grew more assured, but Hunt remained a kind and solicitous friend both before and after Keats's death, and is one of his most perceptive critics.

Mary-Ann Jeffery (b. 1798). Daughter of the family with whom Tom, accompanied first by George and then by Keats, stayed in Teignmouth in the first part of 1818. Her sister Sarah Francis was known both as Sarah and Fanny (hence the belief of editors before Gittings that there were three sisters). Local tradition favours the idea that she was in love with Keats and that her poem 'Si deseris pereo' (published in a volume of 1830) was addressed to Keats in 1818.

Edmund Kean (1787–1833). Kean revolutionized Shakespearean acting. Eschewing the stylized incantatory delivery of his predecessors he established a style that was shockingly immediate. In 1814 he played Shylock at Drury Lane without adopting the usual anti-Semitic caricature. Shylock, Richard III, and Sir Giles Overreach (in Massinger's *A New Way to Pay Old Debts*) were his greatest roles. Stories of his licentious life provided vicarious pleasure to the respectable. He founded his own club—the Wolf Club—in 1815. In November 1820 he went to New York for the first time. It is doubtful whether Keats ever met him but certain that his admiration for Kean was intense and probably self-identifying.

Fanny Keats (1803–89). As the youngest of her family, and a female, she was kept on a tight rein by Abbey and suffered more than her siblings under his guardianship. In 1816 she was sent to a boarding school at Walthamstow kept by the Misses Tuckey. She left two years later. Abbey disapproved of Keats's visits but Keats nevertheless came when he could and contrived as much pleasantness as possible for her, sending her plants and amusing letters. When she came of age in 1824 Dilke forced Abbey to hand over her share of her grandmother's estate. She was the only Keats child to survive into old age (in 1826 she married the Spanish novelist Valentin Llanos).

George Keats (1797–1841). After leaving the Clarke school at Enfield George worked for a short time at Abbey's counting-house. He left in 1816 after a quarrel with Hodgkinson. From this time until his departure for America he lived with Keats and Tom—in the City and in Well Walk, Hampstead. He also accompanied Tom to Paris (in 1817?) and to Teignmouth in 1818. In the same year he married Georgiana Wylie and in June the couple sailed for America where they hoped to join the Birkbeck settlement. Finding only heavily forested land in Illinois they went on to Hendersonville, Kentucky where George lost money on a business venture proposed by the naturalist Audubon. Thence they moved on to Louisville where George built a lumber mill, and later a flour mill and came in time to be a respected and thriving citizen; but not during Keats's life. In 1820 he returned to England in the hope of securing his share of Tom's estate and whatever else Keats could spare. He did not know of Keats's engagement to Fanny Brawne nor of the injury he did him in taking as much as was offered. Nevertheless Keats felt some resentment towards George for this and his marvellous journal-letters to his brother and sister-in-law end after this visit.

Tom Keats (1799–1818). For a short time after leaving the Enfield school Tom worked with George in Abbey's counting-house. But the tuberculosis which was to kill him made him unable to work. He visited Paris with George and Lyons on his own at some time in 1817. In 1818 he stayed in Teignmouth, accompanied and nursed by George and Keats in turn. His nature was gentle, generous, and humorous. Keats loved him dearly and, from his return to Scotland until Tom's death, was his constant and devoted nurse.

John Kingston. A civil servant (Deputy Comptroller of Stamps) who cultivated artistic society and whom Keats greatly disliked.

Charles Landseer (1799–1879), **Edwin Landseer** (1802–73), and **Thomas Landseer** (1795–1880). Sons of **John Landseer** (1769–1852), the engraver. They were acquaintances of Haydon (Charles was his pupil) and artists of historical scenes, animals, and engravings respectively.

Mr Lewis. A Hampstead neighbour living in Well Walk.

William Lowther (1787–1872), Lord Lonsdale. Represented the county of Westmorland as an opponent of reform in 1813, 1818, 1820, and 1826. No fewer than nine members of his family held parliamentary seats (S. Macoby, *English Radicalism 1786–1832* (London, 1955), 332). Wordsworth had dedicated *The Excursion* to his father, the previous Lord Lonsdale.

John Martin. A partner in the publishing firm Rodwell and Martin. He was a friend of Reynolds and Rice and a member of their 'Saturday Club'.

George Felton Mathew (b. 1795). Met Keats in 1815 while the latter was a student at Guy's. He was introduced by his cousins the Misses Mathew, who were friends of George. Mathew wrote sentimental verse and disliked Keats's politics, but Keats briefly ignored their differences and, knowing no or few others at the time, embraced Mathew as a brother poet. Mathew reviewed *1817* in the *European magazine* (May 1817) and expressed a preference for those (in fact weaker) poems written at the time of his brief intimacy with Keats.

Mrs Millar. Georgiana Keats's aunt.

Thomas Moore (1779–1852). Dublin-born poet and writer of disreputable verse. His *Irish Melodies*, with music by Sir John Stevenson, appeared in 1808 and he was effectively Ireland's national lyrist. He lampooned the Regent because of his failure to uphold the Irish Catholic cause. He was a friend of Leigh Hunt (who had also upheld the Irish Catholics in the *Examiner*) and of Byron. His eastern poem *Lalla Rookh* (1817) was hugely successful.

Vincent Novello (1781–1861). Organist, composer and editor of music, until 1820 he lived in Oxford Street. Lamb describes an evening at his house in the 'Chapter on Ears' in the *Essays of Elia*. His daughter Mary married Charles Cowden Clarke (who was by then his business partner).

Charles Ollier (1788–1859). He and his brother James were publishers in Vere Street from *c.*1816. Charles was the more literary of the two. They published Hunt's *Foliage* and the second edition of *The Story of Rimini* as well as Keats's *1817*. Keats's friends and brothers felt that the Olliers had done too little to publicize and sell the volume while they, in turn, felt they had been unwise to take on so unmarketable an author.

James ?Peachey. Schoolfellow of Keats.

John Hamilton Reynolds (1794–1852). Reynolds began his working life in the Amicable Insurance office though his real passion was literature. In 1817 he entered into legal partnership with James Rice. (Rice and Reynolds were to defend Patmore, Scott's second, in the aftermath of the Scott–Christie duel.) He met Keats through Hunt and introduced him to many of his future friends, including Rice, Bailey, Brown, and Dilke. He published several volumes of verse, amongst them *Safie, an Eastern Tale* (1817) dedicated to Byron, *The Naiad* (1816), and what Shelley called 'the ante-natal Peter', a parody of Wordsworth's *Peter Bell*. He was an enthusiast of boxing (as was Hazlitt) and published *The Fancy* in 1820, a miscellany of 'poetical remains' from a fictitious boxer, Peter Corcoran. Keats's 'Robin Hood' sonnets were written in response to poems of Reynolds and they planned a collaborative volume of poems based on Boccaccio's tales (*The Garden of Florence* (1821) represents Reynolds's side of this.) He defended Keats against the *Quarterly*'s attack in the *Alfred West of England Journal* (6 Oct. 1818). He was one of Keats's dearest and most intimate friends though there seems to have been some cooling latterly due to his family's disapproval of Keats's engagement to Fanny Brawne. Reynolds described himself as 'one of those who pant for distinction but have not within them that immortal power to command it'.

The **Reynolds** Family. For some time Keats was on good terms with the whole of this (J. H. Reynolds's) family and in particular with the mother, to whose cat he addressed a poem, and with two of the daughters, Jane (1791–1846, who later married Thomas Hood), and Mariane (1797–1874). However, Keats grew to dislike the two sisters, finding them superficial and ungenerous towards women whom he admired (their cousin Jane Cox and Fanny Brawne).

James Rice (1792–before 1833). Rice was introduced to Keats by Reynolds with whom he went into legal partnership in 1817. Along with Reynolds and Bailey he was a member of the Zetosophian ('I seek wisdom') Society which asked its members to write and read one essay a month. He was also a member of the Saturday Club (founded 25 Oct. 1817) which Keats and Reynolds attended. Keats had a great respect and fondness for Rice whom he described as 'the most sensible, and even wise Man I know' (letter of 17–27 Sept. 1819). He suffered persistent ill health but 'his good heart and good spirits kept him up' (Dilke to George Keats in 1833). Some of his whimsical and comic writings were published by Dilke in the *Athenaeum* (after Keats's death).

Joseph Ritchie (?1788–1819). A surgeon turned explorer, Ritchie had met Tom Keats in Paris and met Keats through Haydon. He promised to carry *Endymion* to Africa and fling it into the mountains of the Sahara (appropriately, the 'Mountains of the Moon'). His expedition was mismanaged by the authorities at home and he died at Murzuk, south of Fezzan. His companion George Francis Lyon described the expedition in *A Narrative of Travels in Northern Africa* (London, 1821). Ritchie wrote of Keats that 'if I am not mistaken he is to be the great poetical luminary of the age to come.'

Sawrey. Keats's doctor.

Joseph Severn (1793–1879). Severn met Keats and his brother George (of whom he became a close friend) around 1816 when he was a struggling painter. They may have been introduced by Haslam who lived in Bethnal Green, near to Severn's home in Hoxton. In December 1818 he won an RA competition with his Spenserian *Cave of Despair*. His *Hermia and Helena* was also exhibited at the RA but he had little success in early life with paintings other than miniatures. He was a friend of Hunt, Reynolds, and Brown. When Haslam suggested that he accompany Keats to Rome (a place advantageous to his career as a painter) he accepted with alacrity in spite of his father's violent opposition to the journey. He travelled with Keats on the *Maria Crowther* to Naples and was his attentive nurse to the end. After Keats's death he was taken up by the British in Rome and subsequently (1861) became British consul. He is buried, at his own request, beside Keats.

Horace Smith (1779–1849). Author of *Rejected Addresses* (with his brother James, London, 1812) and *Horace in London* (London, 1813), novelist, and wit. He was a close friend of Leigh Hunt and Shelley and an acquaintance of Keats.

John Snooke. Dilke's brother-in-law (married to Dilke's sister Letitia). He lived at Bedhampton where he was visited by Keats.

John Taylor (1781–1864). Partner in the publishing firm Taylor and Hessey. He was the author of works on the identity of Junius (the pseudonym used by the author of letters of political invective published between 1769 and 1771). From 1821 he edited the *London Magazine* with Thomas Hood. He and his partner Hessey were generous publishers to Keats and personal friends. It was they who raised the funds to make the journey to Rome possible.

Charles Wells (1800–79). Schoolfellow of Tom Keats and at one time a friend of Keats. A compulsive practical joker, he addressed a series of letters to Tom in 1816 purporting to be from a young woman, 'Amena Bellafila', who was in love with him. Tom's reaction is unknown for Keats only discovered the letters after Tom's death and thought Wells guilty of the blackest villainy for attempting this deceit.

Richard Woodhouse (1788–1834). Scholar, linguist (he produced a *Grammar of the Spanish, Portuguese and Italian Languages* in 1815) and lawyer. He was legal and literary adviser to Taylor and Hessey. The date of his meeting with Keats is uncertain but he greatly admired *1817* and *Endymion* and devoted much time and assiduity to his scholarly collection of Keatsiana (for which all readers and students of Keats owe him a debt of gratitude). He was an immensely kind and generous friend to Keats, and in August 1819 gave him £50 to be presented in the form of an advance from his publishers (see *L* ii. 1503). He wrote to Taylor that 'Whatever people [say they] regret that they could not do for Shakespeare or Chatterton, because he did not live in their time, that I would embody into a Rational principle, and . . . do for [Keats].' Taylor described him as 'abstemious to a remarkable degree, of great industry, averse to pleasure (in the London acceptation of the word) . . . reads much, and with the strictest attention . . . extremely attentive to religious duties . . . and possesses more real humanity than . . . any one I know.'

Georgiana Augusta Wylie (1801/2–1879). Daughter of James Wylie, adjutant of the Fifeshire Regiment of Fencible Infantry. Married George Keats around May 1818 and soon after accompanied him to America. They had eight children, two (Georgiana and Rosalind) during Keats's lifetime. She had an unusual and independent mind and Keats was very fond of her and felt unusually at ease, both in her company and in correspondence with her. After George and Georgiana had left Keats maintained regular contact with the family. The mother, **Mrs Wylie**, of whom he was particularly fond, lived in Romney Street with her son Charles. Her other son Henry lived with his aunt Mrs Millar.

Index of Poem Titles and First Lines